PRINCIPLES OF WATER QUALITY MANAGEMENT

PRINCIPLES OF WATER QUALITY MANAGEMENT

W. Wesley Eckenfelder, Jr.

CBI Publishing Company, Inc.
51 Sleeper Street
Boston, Massachusetts 02210

CBI

To my wife Kathy, whose love and patience for five months in Belgium, and countless nights and weekends, made this book possible.

Cover Photo: 45-mgd water treatment facility in New Bedford, Massachusetts, designed by Camp Dresser & McKee Inc. Photo reproduced courtesy of Camp Dresser & McKee Inc.

Text/Cover Design: Jack Schwartz

Composition: Woodland Graphics, a division of W. E. Andrews Company

Text Printing: Alpine Press

Library of Congress Cataloging in Publication Data
Eckenfelder, William Wesley, 1926–
  Water quality engineering for practicing engineers.

  Bibliography.
  Includes index.
  1. Water quality management.   I.   Title.
TD365.E25   1979   628'.3   79–20509
  ISBN 0–8436–0338–0

Printed in the United States

Printing (last digit): 9 8 7 6 5 4 3 2 1

CONTENTS

v

PREFACE

Recent legislation and more stringent effluent criteria are placing increased emphasis on effective water quality management. The past few years have seen increased emphasis on research and development into both conventional wastewater treatment processes as well as new and advanced technology. The purpose of this book is to present a concise summary of present theory, with emphasis on the application of that theory to both municipal and industrial wastewater treatment problems. This book, in large measure, is an updating of the author's previous books, *Biological Waste Treatment,* published by Pergaman Press, *Industrial Water Pollution Control,* published by McGraw-Hill, *Water Quality Engineering for Practicing Engineers,* published by Cahners Books.

This book should be useful as a text for advanced undergraduate and graduate courses in Water Pollution Control, Industrial Waste Treatment, and Water Quality Management. The content of this book has been used as a text in Continuing Education Courses sponsored by Manhattan College, Vanderbilt University, and the American Institute of Chemical Engineers. A series of problems applying the theories as developed in the text are presented where appropriate. For in-depth study of specific topical areas the reader is referred to the cited book references at the end of the text.

General Concepts of
Water Quality Management

Over the past decade, water pollution control has progressed from an art to a science. Increased emphasis has been placed on the removal of secondary pollutants, such as nutrients and refractory organics, and on water reuse for industrial and agricultural purposes. This, in turn, has generated both fundamental and applied research, which has improved both the design and operation of wastewater treatment facilities.

Solving water pollution problems today involves a multidisciplinary approach in which the required water quality is related to agricultural, municipal, recreational, and industrial requirements. In many cases a cost-benefit ratio must be established between the benefit derived from a specified water quality and the cost of achieving that quality.

Wastewaters emanate from four primary sources:

1. Municipal sewage
2. Industrial wastewaters
3. Agricultural runoff
4. Storm-water and urban runoff

Estimating municipal wastewater flows and loadings can be done in one of several ways, based on knowledge of past and future growth plans for the community, sociological patterns, and land-use planning. Two possible ways are:

1. Population-prediction techniques. Several mathematical techniques are available for estimating population growth. Caution should be employed in the use of these procedures, particularly in areas subject to rapid industrial expansion, rapid suburban development, and changing land-use patterns.

2. Saturation population from zoning practice. Percentages of a satura-

tion population can be estimated for fully developed areas based on zoning restrictions (single dwelling residential, multiple dwelling residential, commercial, etc.).

Provisions should be included for infiltration in the case of separate sewers, and storm flows in the case of combined sewers.

The average characteristics of domestic sewage are listed in table 4.1. The estimated sewage contributions from commercial and industrial establishments are summarized in table 4.5.

As municipal and industrial wastewaters receive treatment, increasing emphasis is being placed on the pollutional effects of urban and agricultural runoff. The range of concentration of pertinent characteristics in these wastewaters is given in table 1.1. Present research on storm-water treatment considers large holding basins in which the storm waters are treated in the municipal facility after the storm (an *in situ* treatment by screening, sedimentation, chlorination, and so on). In the future, water quality management in highly urbanized areas will have to consider storm water as a major pollutant.

TABLE 1.1

Pollution from Urban and Agricultural Runoff

Constituent	Urban Runoff[a] (Storm Water)	Agricultural Runoff[b]
Suspended solids, mg/l	5–1200	–
Chemical oxygen demand (COD), mg/l	20–610	–
Biological oxygen demand (BOD), mg/l	1–173	–
Total phosphorus, mg/l	0.02–7.3	0.10–0.65
Nitrate nitrogen, mg/l	–	0.03–5.00
Total nitrogen, mg/l	0.3–7.5	0.50–6.50
Chlorides, mg/l	3–35	–

[a] From Weibel *et al.* [1]
[b] From Sylvester [2]

Agricultural runoff is a major contributor to eutrophication in lakes and other natural bodies of water. Effective control measures have yet to be developed for this problem. Runoff of pesticides is also receiving increasing attention.

A procedure for the development of an effective water quality management program is shown in figure 1.1. Such a program is directed toward establishing the most economic long-range solution for specified water quality needs.

After establishing the water quality criteria, design considerations should include combined or separate treatment, ocean outfalls in coastal areas, and flexibility for expansion of the facilities to upgrade the effluent quality as the loading to the facilities is increased.

WATER QUALITY STANDARDS

Water quality standards are usually based on one of two primary criteria, stream standards or effluent standards. Stream standards are based on dilution requirements or the receiving-water quality based on a threshold value of specific pollutants or a beneficial use of the water. Effluent standards are based on the concentration of pollutants that can be discharged or on the degree of treatment required.

Stream standards are usually based on a system of classifying the water quality based on the intended use of the water. Table 1.2 shows typical classifications, the primary quality criteria, and the usual degree of treatment needed to meet these criteria.

Although stream standards are the most realistic in light of the use of the assimilative capacity of the receiving water, they are difficult to administer and control in an expanding industrial and urban area. The equitable allocation of pollutional loads for many industrial and municipal complexes also poses political and economic difficulties. A stream standard based on minimum dissolved oxygen at low stream flow intuitively implies a minimum degree of treatment. One variation of stream standards is the specification of a maximum concentration of a pollutant (that is, the BOD) in the stream after mixing at a specified low flow condition. Procedures for determining the assimilative capacity of natural water bodies are discussed in chapter 3.

Note that the maintenance of water quality and hence stream standards are not static but subject to change with the municipal and industrial environment. For example, as the carbonaceous organic load is removed by treatment, the detrimental effect of nitrification in the receiving water increases. Eutrophication may also become a serious problem in some cases. These considerations require an upgrading of the required degree of treatment.

Effluent standards are based on the maximum concentration of a pollutant (mg/l) or the maximum load (lb/day) discharged to a receiving water. These standards can be related to a stream classification.

In 1972 the U.S. Legislature passed Public Law 92-500 which requires certain levels of treatment for industrial wastewater discharges by specific time periods. Effluent guideline criteria (expressed as kilograms of pollutant per unit of production) have been developed for each industrial category to be met by specified time periods. By July 1, 1977 each industry was required to obtain a level of treatment defined as Best Practicable Control Technology Currently Available (BPCTCA). By 1983 each industry should achieve a level of treatment defined as Best Available Treatment Economically Achievable (BATEA).

The BPCTCA is defined as the level of treatment that has been proven to

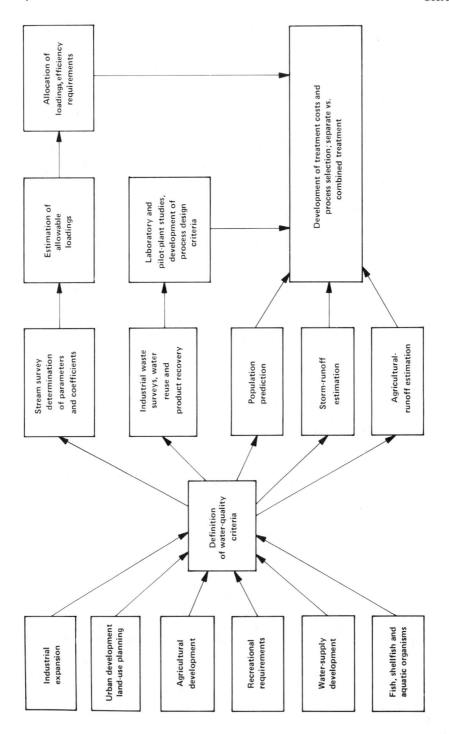

FIGURE 1.1

Development of a Water Quality Management Program

TABLE 1.2

Stream Classification for Water Quality Criteria[a]

Class	Use	Quality Criteria	Required Treatment
A[b]	Water supply, recreation	Coliform bacteria, color, turbidity, pH, dissolved oxygen, toxic materials, taste- and odor-producing chemicals, temperature	Secondary (tertiary in some cases to meet criteria) plus disinfection
B[b]	Bathing, fish life, recreation	Coliform bacteria, pH, dissolved oxygen, toxic materials, color and turbidity (at high levels), temperature	Secondary plus disinfection
C	Industrial, agricultural, navigation, fish life	Dissolved oxygen, pH, floating and settleable solids, temperature	Primary and, in some cases, secondary
D	Navigation, cooling, water	Nuisance-free conditions, floating material, pH	Primary

[a]From [3,4]
[b]May require nutrient (nitrogen and phosphorus) removal

be successful for a specific industrial category and that is currently in full-scale operation. Sufficient data exists for this level of treatment so that it can be designed and operated to achieve a level of treatment consistently and with reliability. For example, in the pulp and paper industry BPCTCA has been defined as biological treatment using the aerated lagoon or the activated sludge process with appropriate pretreatment.

The BATEA is defined as the level of treatment beyond BPCTCA that has been proven feasible in laboratory and pilot studies and that is, in some cases, in full-scale operation. BATEA in the pulp and paper industry may include such processes as filtration, coagulation for color removal, and improved in-plant contol to reduce the waste load constituents.

In general, effluent guidelines are developed by considering an exemplary plant in a specific industrial category and multiplying the wastewater flow per unit production by the effluent quality attainable from the specified BPCTCA process to obtain the effluent limitation in pounds or kilograms per unit of production. The effluent limitations consider both a maximum thirty-day average and a one-day maximum level. In general the daily maximum is two to three times the thirty-day average. For example the average wastewater flow from an exemplary plant is 30,000 gallon/ton of production and the average effluent BOD is 30 mg/l.

The effluent limitation can then be computed:

$$30000 \text{ gal/ton} \cdot 8.34 \cdot 10^{-6} \cdot 30 \text{ mg/l} = 7.5 \text{ lbs/ton}$$

It is recognized that the wastewater volume and characteristics from a specific industrial category will depend on such factors as plant age, size, raw materials used and in-plant processing sequences. An example of the variability in raw waste load from plants in the sulfite pulp industry is shown in figure 1.2. In some cases this may require modification to the effluent limitations.

Some effluent limitations promulgated by state environmental control agencies are in terms of concentrations (mg/l). These are usually based on the effluent level attainable by the application of prevalent technology. Some effluent limitations employed by the state of Illinois are shown in table 1.3.

The U.S. Environmental Protection Agency (EPA) has also developed pretreatment guidelines for those industrial plants which discharge into municipal sewer systems. In general, compatible pollutants such as BOD, suspended solids, and coliform organisms can be discharged providing the municipal plant has the capability of treating these wastewaters to a satisfactory level. Noncompatible pollutants, such as grease and oil, heavy metals, etc., must be pretreated to specified levels. Rigid limitations have been developed for the discharge of toxic substances to the nation's waterways.

In several cases, such as shellfish areas and aquatic reserves, the usual water quality parameters do not apply because they are nonspecific as to det-

TABLE 1.3

Illinois State Effluent Quality Limitations

Parameter	Concentration mg/l
Arsenic (total)	0.25
Barium (total)	2.00
Cadmium (total)	0.15
Chromium (total hexavalent)	0.30
Chromium (total trivalent)	1.00
Chromium (total)	
Copper (total)	1.00
Cyanide	0.025
Cyanide (total)	
Fluoride (total)	15.00
Iron (total)	2.00
Iron (dissolved)	0.50
Lead (total)	0.10
Manganese (total)	1.00
Mercury (total)	0.0005
Nickel (total)	1.00
Oil (hexane solubles on equivalent)	15.00
pH (range)	5–10
Phenols	0.30
Selenium (total)	1.00
Silver	0.10
Zinc (total)	1.00
Total Suspended Solids (other than deoxygenating wastes as covered in Rule 404)	15.00
Total Dissolved Solids	750/3500

rimental effects on aquatic life. For example, COD is an overall measure of organic content but it does not differentiate between toxic and nontoxic organics. In these cases, a species diversity index has been employed as related to either free-floating or benthic organisms. The index indicates the overall condition of the aquatic environment. It is related to the number of species in the sample. The higher the species diversity index, the more productive the aquatic system. The species diversity index, K_D, is computed by the equation $K_D = (S-1)/\log_{10} I$, where S is the number of species and I the total number of individual organisms counted.

The water quality criteria for various industrial uses are summarized in table 1.4. The surface water quality criteria for public water supplies have been summarized in reference 5. Color should not exceed 75 units and odors should

FIGURE 1.2

Variability in Raw Waste Load from a Sulfite Pulp Plant

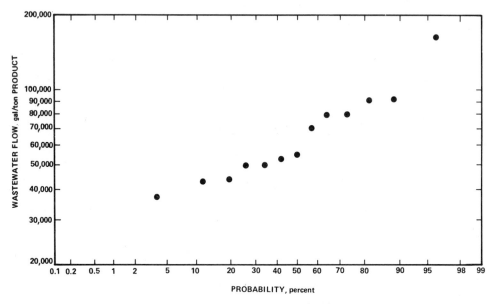

(a) Log Probability Plot of Flow

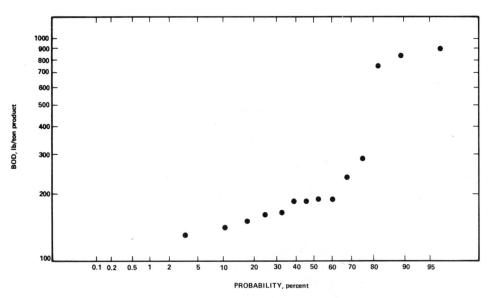

(b) Log Probability Plot of BOD

Unit Operations	Total RWL	In-Plant Process Alternatives		
		A	B	C
Wood preparation		Wet Debarking	Dry Debarking	
	Flow[a]	5.6	2.0	
	BOD[b]	9.0	4.0	
Pulping + washing		No chemical recovery	SSL[d] treatment with barometric condenser spill collection	SSL treatment with surface condenser, spill collection
	Flow	25.8	33.5	10.9
	BOD	850.0	185.0	185.0
Bleaching		CEH[c] (brightness level = 84-86)		
	Flow	26.3		
	BOD	34.0		
Papermaking		Conventional with moderate white water reuse	Conventional with extensive white water reuse and fiber recovery	
	Flow	20.0	10.0	
	BOD	12.0	8.0	

[a]Flow units are expressed as gallons/ton.

[b]BOD units are expressed as pounds/ton.

[c]CEH = Chlorine, Caustic Extraction and Hypochlorite.

[d]SSL = Spent Sulfite Liquor

(c) Factors Affecting Variability in Raw Waste Load Shown in (a) and (b)

TABLE 1.4
Industrial Water Quality Limits[a]

Industry	Turbidity (units)	Color (units)	Hardness	Temp. (°F)	pH	TDS	SS	S_iO_2	F_e	M_n	Cl	SO_4	Alkalinity
Textiles (SIC 22)	0.3–5	0–5	0–50	–	–	100–200	0–5	25	0–0.3	0.01–0.05	100	100	50–200
Pulp and paper (SIC 26)													
Fine paper	10[b]	5	100	–	–	200	–	20	0.1	0.03	–	–	75
Kraft paper													
Bleached	40[b]	25	100	–	–	300	–	50	0.2	0.10	200	–	75
Unbleached	100[b]	100	200	–	–	500	–	100	1.0	0.50	200	–	150
Groundwood papers	50[b]	30	200	–	–	500	–	50	0.3	0.10	75	–	150
Soda and Sulfite pulp	25[b]	5	100	–[c]	–	250	–	20	0.1	0.05	75	–	75
Chemicals (SIC 28)	–	500	1,000	–	5.5–9.0	2,500	10,000	[c]	10	2.00	500	850	500
Petroleum (SIC 2911)	–	25	900	–	6.0–9.0	3,500	5,000	85	15	–	1,600	900	500
Iron and steel (SIC 33)	–	–	–	100	5.0–9.0	–	100[d]	–	–	–	–	–	–
Food canning (SIC 2032 & 2033)	–	5	250	–	6.5–8.5	500	10	50	0.2	0.20	250	250	250
Tanning (SIC 3111)													
Tanning processes	[e]	5	150	–	6.0–8.0	–	–	–	50	–	250	250	[c]
Finishing processes	[e]	5	[f]	–	6.0–8.0	–	–	–	0.3	0.20	250	250	[c]
Coloring	[e]	5	[e]	–	6.0–8.0	–	–	–	0.1	0.01	–	–	[c]
Soft drinks (SIC 2086)	–	5	[g]	–	[g]	[g]	–	–	0.3	0.05	500	500	85

[a] From [5]; units are mg/l unless otherwise specified.
[b] Units in mg/l as S_iO_2.
[c] Not considered a problem of concentrations encountered.
[d] Settleable solids.
[e] Not detectable by test.
[f] Lime softened.
[g] Controlled by treatment for other constituents.

be virtually absent. Ammonia nitrogen, nitrite nitrogen, and nitrate nitrogen should not exceed 0.5 mg/l, 1 mg/l, and 10 mg/l, respectively. Chloride should not exceed 250 mg/l, and pH value should be between 5.0 and 9.0. Sulfate should not exceed 250 mg/l, and the geometric means of fecal coliform and total coliform densities should not exceed 2,000/100 ml and 20,000/100 ml, respectively.

For freshwater aquatic life the pH should be between 6 and 9 and the alkalinity should not be decreased more than 25 percent below the natural level. For most fish life, the dissolved oxygen should be in excess of 5.0 mg/l. For wildlife the pH should be between 7.0 and 9.2, and the alkalinity should be between 30 and 130 mg/l.

For irrigation the pH should be between 4.5 and 9.0, and the sodium adsorption ratio should be within the tolerance limits determined by the U.S. Soil Salinity Laboratory Staff. For continuous use on all soils the metal contents for aluminum, cadmium, chromium, cobalt, copper, iron, lead, and zinc should be no more than 5.0 mg/l, 0.01 mg/l, 0.1 mg/l, 0.05 mg/l, 0.2 mg/l, 5.0 mg/l, 5.0 mg/l and 2.0 mg/l, respectively.

REFERENCES

1. Weibel, S. R. *et al.* 1964. Urban land runoff as a factor in stream pollution. *Journal Water Pollution Control Federation* 36:914.
2. Sylvester, R. O. Nutrient content of drainage water from forested, urban and agricultural areas. *Transactions Seminar Algae Metropolitan Wastes.* Cincinnati, Ohio: U.S. Public Health Service, R. A. Taft Center.
3. McKee, J. E. and Wolf, H. W. 1963. *Water quality criteria.* Report to the California Water Quality Control Board, SWPCB Publ. 3A, 2d ed. Sacramento, CA.
4. McGauhey, P. H. *Engineering Management of Water Quality,* New York: McGraw-Hill.
5. U.S. EPA. 1972. *Report of the Committee on Water Quality Criteria. Water Quality Criteria 1972.* Washington, D.C.

Sewage and Industrial Waste Characteristics

The increased emphasis on water quality for multipurpose use has defined a number of parameters of special significance in municipal sewage and industrial wastewaters. These are:

1. BOD (biochemical oxygen demand)—defines the biodegradable organic content of the waste
2. COD (chemical oxygen demand)—measures the total organic content, both degradable and refractory
3. TOC (total organic carbon) and TOD (total oxygen demand)—measure the total organic content
4. suspended and volatile suspended solids
5. total solids
6. pH—alkalinity and acidity
7. nitrogen and phosphorus
8. heavy metals and inorganic solids

Figure 2.1 schematically compares domestic wastewater and an industrial wastewater from the organic chemicals industry. Though all wastewaters differ in composition and in concentration, certain general conclusions can be drawn. Figure 2.1 shows the distinct and major differences between an industrial wastewater and domestic sewage. A major portion of the BOD in domestic sewage is present in colloidal or suspended form, while industrial wastewaters are usually soluble in character. The nondegradable COD in domestic sewage is low (usually less than 60 mg/l) while industrial wastewaters may have a nondegradable COD level in excess of 500 mg/l. While domestic sewage has a surplus of nitrogen and phosphorous relative to the BOD present, many industrial wastewaters are deficient in nitrogen and phosphorous. This can be an asset, since the surplus in the domestic sewage serves as a supplementary nutrient for the industrial wastewa-

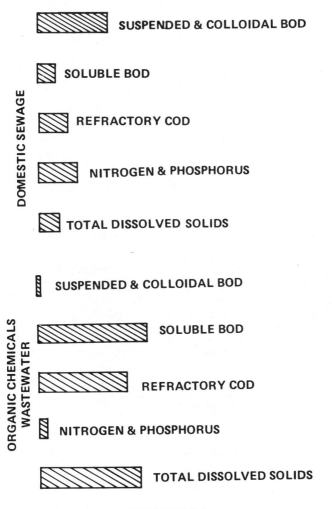

FIGURE 2.1

A comparison of the Characteristics of Domestic
Sewage and an Organic Chemicals Wastewater

ters. Total dissolved solids (inorganic salts) in domestic sewage primarily reflect
the concentration in the carrier water, while many industries substantially increase the TDS through their process wastewaters.

Certain industrial wastes will contain parameters of special significance,
such as phenol or cyanide. The significance of these parameters with respect to
water quality management is listed in table 2.1.

TABLE 2.1
Undesirable Characteristics of Industrial Wastes

1. Soluble organics: dissolved oxygen depletion in streams and estuaries; discharge relative to assimilative capacity of water body or by effluent standard; result in tastes and odors in water supplies; e.g., phenol.
2. Toxic materials and heavy metal ions: usually rigid standards as to discharge of such materials; e.g., cyanide, Cu, and Zn.
3. Color and turbidity: esthetically undesirable; imposes increased loads on water-treatment plants; e.g., color from pulp and paper mills.
4. Nutrients (nitrogen and phosphorus): enhance eutrophication of lakes and ponded areas; critical in recreational areas.
5. Refractory materials: results in foaming in streams; e.g., ABS.
6. Oil and floating material: regulations usually require complete removal; esthetically undesirable.
7. Acids and alkalis: neutralization required in most regulatory codes.
8. Substances resulting in atmospheric odors: e.g., sulfides.
9. Suspended solids: results in sludge bank in streams.
10. Temperature: thermal pollution resulting in depletion of dissolved oxygen (lowering of saturation value).

ESTIMATING THE ORGANIC CONTENT OF WASTEWATERS

The organic content of a waste can be estimated by each of four tests, although considerable caution should be exercised in interpreting the results.

1. The BOD test measures the biodegradable organic carbon and, under certain conditions, the oxidizable nitrogen present in the waste.
2. The COD test measures the total organic carbon with the exception of certain aromatics, such as benzene, which are not completely oxidized in the reaction. The COD test is an oxidation reduction reaction, so other reduced substances, such as sulfides, sulfites, and ferrous iron, will also be oxidized and reported as COD.
3. The TOC test measures all carbon as CO_2, and hence the inorganic carbon (CO_2, HCO_3^-, and so on) present in the wastewater must be removed prior to the analysis or corrected for in the calculation.
4. The TOD test measures organic carbon and unoxidized nitrogen and sulfur. Remember to exercise considerable caution in interpreting the test results and in correlating the results of one test with another.

Biochemical Oxygen Demand

The BOD is conventionally reported as the five-day value and is defined as the amount of oxygen required by living organisms engaged in the utilization and stabilization of the organic matter present in the wastewater. The standard test involves seeding with sewage, river water, or effluent, and incubating at 20° C.

Reaction in the BOD Bottle

The reaction in the BOD bottle is the same as in all aerobic reactions and occurs in two separate and distinct phases, as shown in Figure 2.2. Initially, the organic matter present in the wastewater is utilized by the seed microorganisms for energy and growth. This results in a utilization of oxygen and the growth of new microorganisms. When the organics originally present in the wastewater are removed, the organisms present continue to use oxygen for autooxidation or endogenous metabolism of their cellular mass. When the cell mass is completely oxidized, only a nonbiodegradable cellular residue remains and the reaction is complete. The oxygen consumed in the two phases is defined as the ultimate BOD. The oxidation of BOD is therefore a two-phase reaction.

The removal and oxidation of the organics present in the wastewater is usually complete in 18 to 36 hours (phase 1). The total oxidation of the cell mass will take more than 20 days (phase 2). The rate of reaction during the first or assimilation phase is 10 to 20 times the rate of endogenous oxidation. These relative rates are also shown in Figure 2.2.

Formulation of the BOD

The BOD has been classically formulated as a continuous first-order reaction of the form:

$$\frac{dL}{dt} = -kL \tag{2-1}$$

which integrates to:

$$L_t = L_o e^{-kt} \tag{2-2}$$

Defining:

$$y = L_o - L_t \tag{2-3}$$

equation (2–2) can also be expressed as:

$$y = L_o(1 - e^{-kt}) \tag{2-4}$$

or classically formulated as:

$$y = L_o(1 - 10^{-kt}) \tag{2-5}$$

The relation between k and k_{10} is:

$$k = 2.303k_{10}$$

where y = amount of oxygen consumed or BOD exerted after any
                time t.
        L_o = ultimate BOD (carbonaceous demand only) or the total
                amount of oxygen consumed in the reaction.
        L_t = BOD remaining at time t.

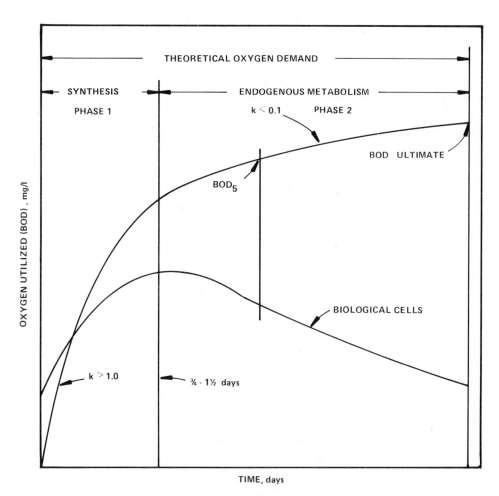

FIGURE 2.2
Reactions Occurring in the BOD Bottle

k, k_{10} = average reaction-rate constant, base e and base 10, respectively.

t = time of incubation.

It should be emphasized that because the reaction in the BOD bottle consists of two separate and distinct phases with an order-of-magnitude difference in reaction rate, the mean k in equation (2–4) will vary markedly, depending on the quantity and nature of organics present in the wastewater. For example, consider a waste that is treated through an activated sludge process. In the raw waste a rapid consumption of oxygen occurs (phase 1, Figure 2.2) followed by the slower endogenous rate, yielding a high mean k in equation (2–4).

In the treated effluent most of the organics originally present in the wastewater have been removed, so most of the oxygen is consumed at the slower endogenous rate. This, in turn, yields a lower mean k as compared to the untreated wastewater.

Typical values of the mean rate constant k_{10} are summarized in table 2.2. Note from table 2.2 that when comparing wastes or treatment-process efficiencies or when estimating the ultimate BOD, the reaction rate k must be considered. Typical BOD curves for a raw waste and a treated effluent are shown in figure 2.3.

TABLE 2.2
Average BOD Rate Constants at 20° C

Substance	k_{10}, day^{-1}
Untreated wastewater	0.15–0.28
High-rate filters and anaerobic contact	0.12–0.22
High-degree biotreatment effluent	0.06–0.10
Rivers with low pollution	0.04–0.08

Both k and L_o are unknown in the BOD reaction [equation (2–4)], so they must be calculated indirectly. Several procedures have been developed for this, three of which are summarized below.

1. **Method of Moments** (Moore et al.) [1]. From a smoothed curve of the data, tabulation of t, y, and ty is made for a given numerical sequence of days (that is, 1, 2, 3). From prepared charts, k or k_{10} is determined from the ratio $\Sigma y/\Sigma ty$ and L_o from $\Sigma y/L_o$. To minimize error, the data should be plotted, a smooth curve drawn through the points, and k or k_{10} and L_o computed from figure 2.4.

2. **Log-difference Method.** The BOD equation (2–4) expressed in differential form is:

$$\frac{dy}{dt} = r = L_o k e^{-kt} \tag{2–6}$$

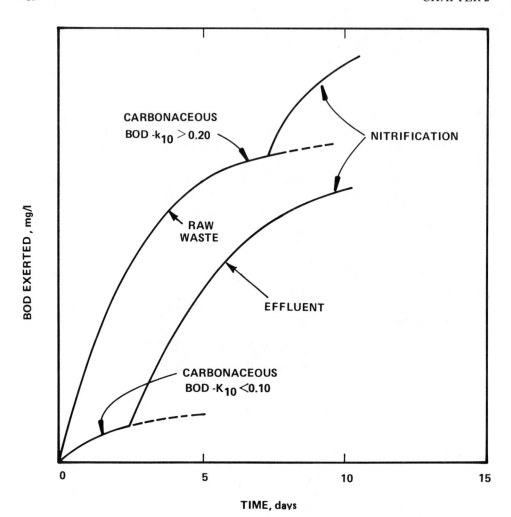

FIGURE 2.3

Comparison of BOD Curves for a Raw Waste and a
Treated Effluent

in which r is the rate of oxygen utilization with time. Applying logarithmic form:

$$\ln r = \ln(L_o k) - kt \qquad (2\text{–}7)$$

or

$$\log r = \log(L_o k) - k_{10} t \qquad (2\text{–}8)$$

Both equations represent straight lines where the slope and intercept are defined by b and a, respectively. From equation (2–7):

$$b = -k$$

$$a = \ln L_o k$$

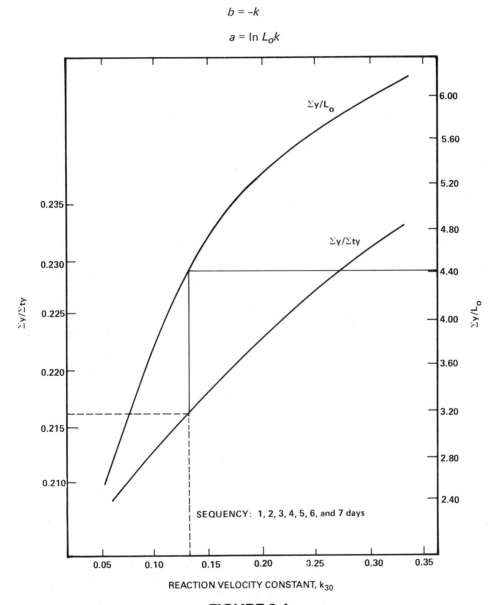

REACTION VELOCITY CONSTANT, k_{30}

FIGURE 2.4

Calculation of BOD Constants from the Method
of Moments. (After Moore et al. [1].)

From equation (2–8):

$$b = -k_{10}$$

$$a = \log L_o k$$

From the corrected values of y, differences are calculated and tabulated. The differences are plotted against time on semilog paper. From this plot k or k_{10} and L_o are computed. This method can take into account nonconformance with first-order kinetics.

3. **Graphical Method** (Thomas) [2]. $(t/y)^{1/3}$ is plotted as the ordinate against t as the abscissa. From the plot:

$$k_{10} = 2.61\frac{b}{a}, \ L_o = \frac{1}{2.3k_{10}a^3}$$

where b = the slope of the line
       a = the intercept

PROBLEM 2–1

Plot y against t on cartesian-coordinate paper. Draw a curve of best fit through the observed points, including lags and plateaus, if any, to get a picture of the deoxygenation curve. If a majority of observed values describe a smooth progression, those that fail to fit may be considered erratic and curve values may be used for subsequent calculations. Lags should be eliminated by curve fitting and taking the observed points after the lag termination. The results are shown in table 2.3. Calculate k and L_o.

TABLE 2.3
Daily Differences

t	y	Daily difference from smoothed curve
0	0	–
1	7.3	7.3
2	12.8	5.5
3	16.0	3.3
4	20.1	4.3
5	22.5	2.4
6	23.8	1.3
7	25.3	1.5

Plot the daily difference, corrected if necessary, on semilog paper with time on the linear scale and the daily difference on the log scale (see figure 2.5). The differences are conventionally plotted as .5, 1.5, 2.5 days, and so on, to illustrate intervals rather than points (table 2.3).

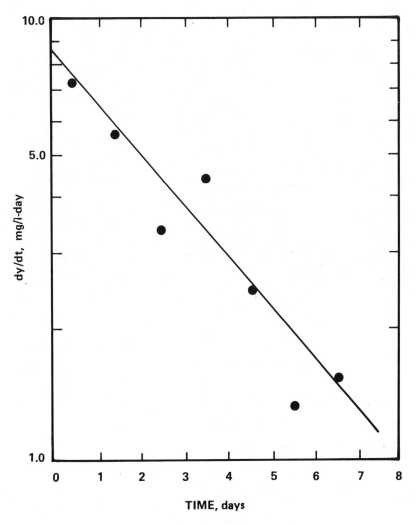

FIGURE 2.5

Calculation of k and L_o by the Difference Method

From figure 2.5 the value of the intercept of the line with the daily difference ordinate at zero time is represented by $L_o k = 8.3$, and taking another point (7, 1.3) from the line, the slope is as follows:

$$-k_{10} = \frac{\log{(8.3)} - \log{(1.3)}}{0 - 7}$$

$$k_{10} = 0.115/\text{day}$$

and:

$$k = 2.3 \cdot 0.115$$

$$= 0.265/\text{day}$$

Therefore:

$$L_o = \frac{8.3}{0.265}$$

$$= 31.3 \text{ mg/l}$$

The ratio between the five-day and the ultimate BOD as related to the reaction rate k_{10} is shown in figure 2.6.

Temperature Effects on the BOD

The BOD reaction rate, k, is directly affected by temperature. L_o is slightly affected because oxidizability increases with temperature. The relationship, derived from van't Hoff's Law, is:

$$k_{(T)} = k_{(20)} \theta^{(T-20)}$$

$$\theta = 1.047 \text{ (Phelps) (inaccurate at low temperatures)}$$

$$= 1.056 \ (20 - 30^\circ \text{ C) (Schroepfer [3])}$$

$$= 1.135 \ (4 - 20^\circ \text{ C) (Schroepfer [3])}$$

Factors Affecting the BOD

Several factors affect the BOD test and should be considered, particularly when dealing with industrial wastes.

Seed. The acclimated organisms present below industrial outfalls provide an excellent source of seed for the BOD determination. With few exceptions, carefully selected seed from the receiving stream will yield the highest BOD values. When stream water is used for seed, nitrification difficulties may increase.

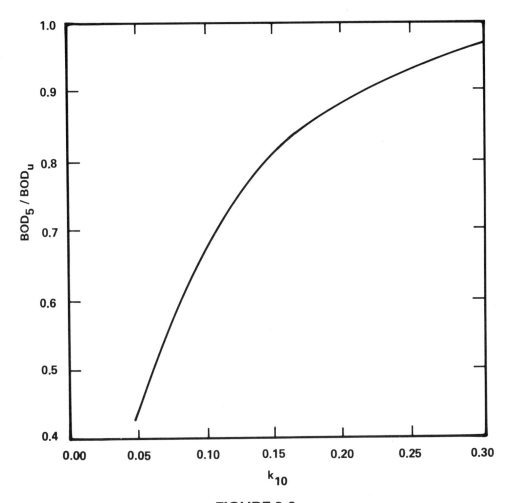

FIGURE 2.6

Relationship between k_{10} and BOD_5/BOD_u

Occasionally it is necessary to artificially develop a microbial culture that will oxidize the industrial waste. To develop an acclimatized heterogenous microbial culture, start with settled domestic wastewater containing a large variety of organisms and add a small amount of the industrial effluent. The amount of waste added is increased until a culture develops that is adapted to the waste. The mixture of domestic wastewater and the industrial waste is aerated by bubbling air continuously through the liquid. A noticeable increase in the cloudiness or turbidity of the aerating mixture generally indicates an acclimated culture. If a DO (dissolved oxygen) probe is available, the oxygen uptake can be evaluated daily to determine when an acclimated culture has developed. The effect of acclimation is shown in figure 2.7 and figure 2.8.

The amount of seed required to produce a normal rate of oxidation must be determined experimentally. The most frequent error is the use of insufficient seed. The effects of seed concentration on the BOD are illustrated in figure 2.9.

Large numbers of algae in stream water that is used for dilution water may produce significant changes in the oxygen content. When stream samples containing algae are incubated in the dark, the algae survive for a time. Short-term BOD determinations may show the influence of oxygen production by the algae. After prolonged lack of light, the algae die and the algal cells contribute to the total organic content of the sample and increase the BOD. Therefore, samples incubated in the dark may not be representative of the deoxygenation process in the stream, because the benefits of photosynthesis are lacking. On the other hand, samples incubated in the light, under conditions of continual photosynthesis, yield low BOD values. The influence of algae in the BOD test is difficult to evaluate, and extreme care should be taken when stream water that contains large numbers of algae is used for dilution water.

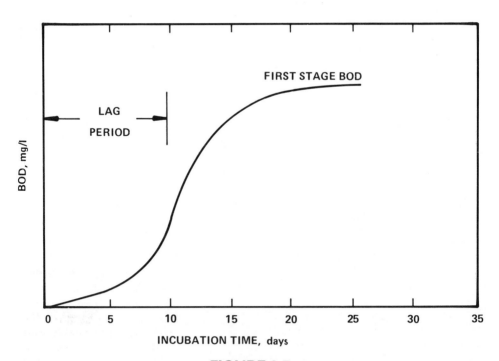

FIGURE 2.7
Effect of Unacclimated Seed on BOD Exertion

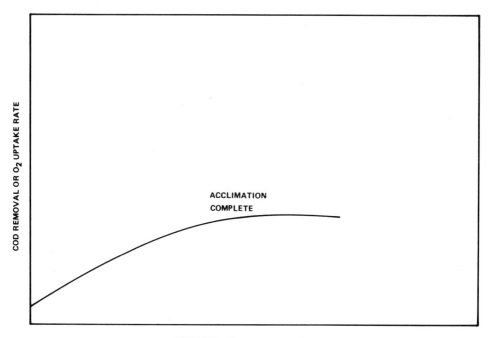

FIGURE 2.8
Acclimation Results

Time of Incubation. The importance of the time variable is indicated in the basic BOD equation. The standard BOD test is performed using a five-day period of incubation and results are usually reported on this basis as BOD_5. However, complete stabilization of the organic matter in a wastewater depends on nature of the waste, the rate of its utilization, the microbial population, temperature, lack of toxic materials and availability of nutrients.

With most industrial wastes, biological degradation is not complete with the five-day incubation period. Biodegradation of industrial wastes may vary from 25 percent to 95 percent in five days, whereas biodegradation of domestic waste is relatively constant in five days and generally represents about 70 percent of the ultimate BOD.

Toxicity. Various chemical compounds are toxic to microorganisms. At high concentrations the substances kill the microbes, and at sublethal concentrations the activity of microbes is reduced. The effects of heavy metals on the BOD are illustrated in figure 2.10. Toxicity is usually evidenced by an increase in BOD with increasing dilution. The effect of heavy metals can be minimized by the addition of a chelating agent such as EDTA.

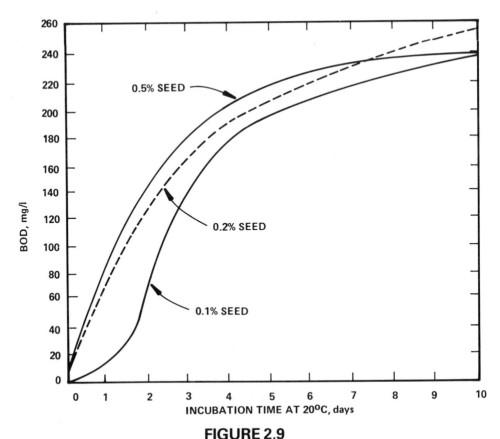

FIGURE 2.9

Effect of Seed Concentration on BOD Exertion

Nitrification. The oxidation process described by the BOD equation $y = L_o (1 - e^{-kt})$ represents the oxidation of carbonaceous matter:

$$C_xH_yO_z \longrightarrow CO_2 + H_2O$$

The oxidation of nitrogenous material may be shown as:

$$NH_3 \longrightarrow NO_2^- \longrightarrow NO_3^-$$

The rate constant is usually less than in the case of the carbonaceous matter.

Under some circumstances these two oxidations can proceed simultaneously, and the resultant BOD curve is a composite of the two reactions. Normally, however, the nitrification will not begin until the carbonaceous demand has been partially satisfied, yielding a curve similar to figure 2.11. Mathematically the reactions can be described by:

$$y = L_o (1 - e^{-k_1 t}) + L_N (1 - e^{-k_2 t})$$

where L_o = ultimate carbonaceous demand
L_N = ultimate nitrogenous demand
k_1 = rate constant for carbonaceous demand
k_2 = velocity constant for nitrogenous demand

Nitrification occurs most often in effluents that have undergone partial oxidation of the waste components. Nitrification represents a demand on the oxygen resources of the receiving stream; therefore, it should be recognized as part of the total demand of the waste.

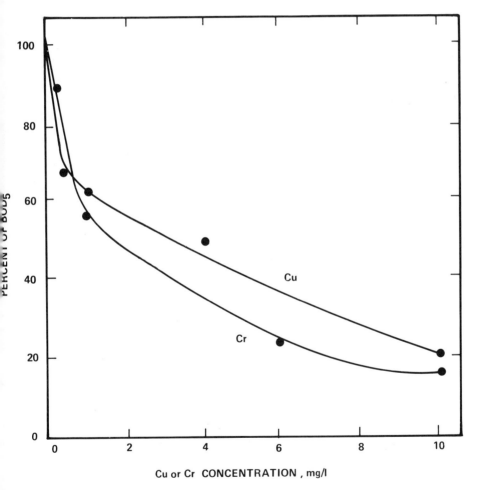

FIGURE 2.10

Effect of Metal-ion Concentration of BOD Exertion
(After Morgan and Lackey [4])

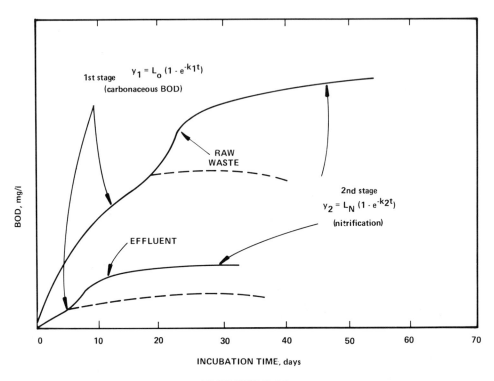

FIGURE 2.11
First and Second Stage BOD in Raw Waste and
Treated Effluent

Nitrification can be eliminated by pasteurizing and reseeding or by the addition of methylene blue or thiourea. The rate of nitrification can be determined by a parallel set of BOD samples, one with and one without nitrification suppression.

Other Procedures for the Evaluation of the BOD

Rapid BOD techniques have been suggested:
1. Arbitrary selection of a shorter time with incubation at 20° C or higher.
2. Correlation of BOD with chemical oxygen demand (COD) (permanganate or dichromate reflux), total organic carbon (TOC), total oxygen demand (TOD), acidimetry, and color.
3. Use of a massive inocula of mixed or pure culture.
4. Use of the concept of a plateau value for complete stabilization or organic matter [5]
5. Respirometric methods.

A common respirometric method uses the Hach respirometer shown in figure 2.12. As oxygen is consumed in the test the partial pressure of oxygen in the gas phase is reduced resulting in a pressure differential on a manometer. This differential is directly proportional to the BOD. Because of its simplicity and the generation of a BOD exertion curve with time, this unit is very applicable to wastewater treatment plant control.

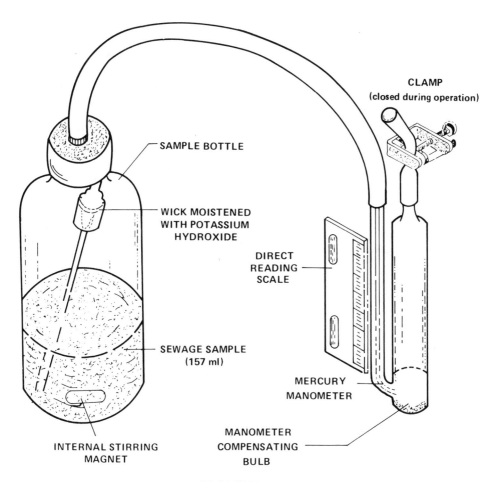

FIGURE 2.12

Diagram of Hach Manometric BOD Apparatus,
Showing One Cell

(Courtesy Hach Co.)

Chemical Oxygen Demand

The procedure for chemical oxygen demand as given in *Standard Methods for the Examination of Water and Wastewater* [10] employs potassium dichromate with reflux. Silver sulfate ($Ag_2 SO_4$) is added as a catalyst, and when chlorides are present, $HgSO_4$ can be added to complex the chlorides and eliminate the need for a chloride correction. All organic compounds are not oxidizable chemically by the dichromate procedure. Sugars, branched-chain aliphatics, and substituted benzene rings are completely oxidized with little or no difficulty. However, benzene, pyridine, and toluene are not oxidized by this method. Other compounds (straight-chain acids, alcohols, and amino acids) can be completely oxidized in the presence of the silver sulfate catalyst.

Chlorides interfere with the COD analysis; the interference is eliminated by using mercuric sulfate, which complexes the chlorides. The theoretical COD of organic compounds may be calculated if the reaction is known. The oxidation of 1000 mg of phenol is used as an example:

$$C_6H_5OH + 7O_2 \longrightarrow 6CO_2 + 3H_2O$$

$$\text{theoretical COD} = \frac{(1000)(224)}{94} = 2383 \text{ mg}$$

When the constituents of a waste are known, it is at times helpful to estimate the theoretical COD to determine the yield of the dichromate refluxing procedure. In this manner the ability of the COD test to provide a representative estimate of the strength of the waste is established. The preparation of the required reagents, the required apparatus, and the procedure are discussed in detail in *Standard Methods* [10].

A short-term rapid chemical oxygen demand test [6] can be used for routine plant control or for comprehensive waste surveys. This test employs a short oxidation period, so some of the more complex organics may be incompletely oxidized in this test. It is, therefore, desirable to compare the results of the short-term test with the *Standard Methods* test for a particular wastewater. Some comparative results reported by Jeris [6] are summarized in table 2.4.

A COD test using an automatic process has recently developed, giving continuous on-line reading of demand; it uses higher digester temperatures and sulfuric acid concentrations; and the dual cell potentiometric method reduces problems with color and turbidity seen in the photometric system.

PROBLEM 2-2

A wastewater shows the following analysis:

$$COD = 2140 \text{ mg/l}$$

$$BOD_5 = 835 \text{ mg/l}$$

$$k_{10} = 0.17 \text{ day}^{-1}$$

The treated effluent from an activated sludge plant with a sludge age of 12 days showed the following:

$$BOD_5 = 17 \text{ mg/l}$$

$$k_{10} = 0.08 \text{ day}^{-1}$$

Assume the COD generated by biooxidation is 2 percent of influent COD. Estimate the effluent COD.

Experimentally the ultimate BOD would be accounted for only 90½ of the measured COD. The effluent COD can be calculated by:

$$\text{Effluent COD} = \text{Influent COD} - \frac{\text{Removed COD}}{\text{by Biooxidation}} + \frac{\text{Generated COD}}{\text{by Biooxidation}}$$

TABLE 2.4
Rapid COD Test Results[a]

Compound	Theoretical[b]	COD, mg/l Rapid Method	Standard Method
Glucose	1299	1264	1258
Pyridine	2185	36	0
Ethanol	821	715	777
Acetic acid	772	729	727
Glycine	1243	250	505
Phenylalanine	1390	1330	1347
Nutrient broth		492	536
Activated sludge		2150	2266
Digested sludge		1409	1441
Raw sewage		183	202
Settled sewage		172	183

[a]After Jeris [6].
[b]Computed using the weight of compound.

The removed COD by biooxidation can further be expressed
by:

$$\text{Removed COD by Biooxidation} = \frac{1}{0.90} \text{ (Removed Ultimate BOD)}$$

Ultimate BOD for influent:

$$= \frac{835}{1 - 10^{-0.17 \cdot 5}} \text{ mg/l}$$

$$= 972 \text{ mg/l}$$

Ultimate BOD for effluent:

$$= \frac{17}{1 - 10^{-0.08 \cdot 5}} \text{ mg/l}$$

$$= 28 \text{ mg/l}$$

Therefore:

$$\text{Effluent COD} = 2140 - \frac{1}{0.90} \text{ (972 - 28)} + 0.02 \text{ (2140) mg/l}$$

$$= 1134 \text{ mg/l}$$

Total Organic Carbon

Total organic carbon (TOC) can be measured either manually or automatically. The manual or wet oxidation method involves oxidation of the sample in a
solution of chromic acid, potassium iodate phosphoric acid, and fuming sulfuric
acid. The combustion products are swept through a Pregl-type combustion tube
and the resulting carbon dioxide collected and weighed in an absorption train.

Carbon Analyzer

The recently developed carbon analyzer utilizes the concept of complete
combustion of all organic matter to carbon dioxide and water; then the gas stream
is allowed to pass through an infrared analyzer sensitized for carbon dioxide and
the response is recorded on a strip chart.

The carbon analyzer is shown in figure 2.13. A microsample of the wastewater to be analyzed is injected into the catalytic combustion tube, which is
maintained at a temperature of 900 to 1000° C. The sample is vaporized and the
carbonaceous material is completely oxidized in the presence of a cobalt catalyst
and the pure oxygen carrier gas. The oxygen flow carries the carbon dioxide and
steam out of the furnace; the steam is condensed; and the remaining carbon

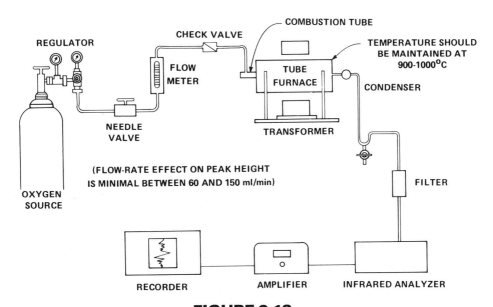

FIGURE 2.13

Flow Diagram of Carbon Analyzer (After Ford [8])

dioxide, oxygen, and water vapor mixture enters the infrared analyzer. As the amount of carbon dioxide is directly proportional to the initial sample carbon concentration, the peak-height response can be compared to a calibration curve and the value determined. This value includes carbon in the form of carbonates or bicarbonates which may have been present. To measure only organic carbon, all inorganic carbon must first be removed from the sample. This is accomplished by acidifying the sample, and purging with nitrogen gas prior to injection. The sample is measured for carbon before and after the acidification-purging step, any reduction being attributed to the loss of inorganic carbon and volatile organic compounds. The volatile organic fraction, if present, must be measured independently, and this is accomplished by using a closed-system diffusion-cell method [7].

A modified carbon analyzer has facilitated the measurement of total organic carbon, as shown in figure 2.14. A dual-combustion tube system allows direct differentiation between organic and inorganic carbon, making the preliminary acidification and scrubbing step unnecessary. A micro-sample is first injected into the high-temperature combustion tube and the total carbon concentration determined by the method shown in figure 2.13. An identical volume of sample is then injected into the low-temperature combustion tube, which contains quartz chips wetted with phosphoric acid. The operating temperature of 150° C is below the value at which organic matter is oxidized, but in the acid environment all carbonates are removed in the form of carbon dioxide. This gas then flows through the infrared analyzer and is measured as inorganic carbon. The total

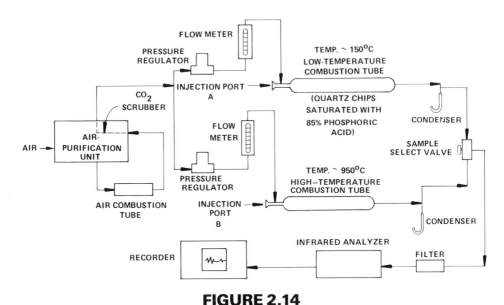

FIGURE 2.14

Flow Diagram of Modified Carbon Analyzer

organic carbon concentration is taken as the difference between the total carbon and total inorganic carbon values. An optional air-purification unit is available to provide a suitable carrier gas should a bottled source not be desirable.

Some interference in the infrared-energy-absorption pattern is possible if anions such as NO_3^-, Cl^-, SO_4^{--}, and PO_4^{---} are present in excess of 10,000 mg/l. Industrial wastewaters containing such concentrations should therefore be diluted with carbon-dioxide-free water prior to analysis.

Relationships Among BOD, COD, and TOC

When considering routine plant control or investigational programs, the BOD is not a useful test because of the long incubation time required to obtain a meaningful result. It is therefore important to develop correlations among BOD, COD, and TOC.

Let us first consider a completely biodegradable substance such as glucose. The ultimate BOD will measure about 90 percent of the theoretical oxygen demand. (Approximately 10 percent of the original organics end up as nonbiodegradable cellular residue and hence is not measured in the BOD.) The COD will measure the theoretical oxygen demand. Therefore, for these substrates:

$$COD = \frac{BOD_u}{0.9} = TOD$$

Organic Carbon-Oxygen Demand Relationship

In attempting to correlate BOD or COD of a wastewater with TOC, one should recognize those factors which may discredit the correlation. These include the following:

1. A portion of the COD of many industrial wastes is attributed to the dichromate oxidation of ferrous iron, nitrogen, sulfites, sulfides, and other oxygen-consuming inorganics.
2. The BOD and COD tests do not include many organic compounds that are partially or totally resistant to biochemical or dichromate oxidation. However, all the organic carbon in these compounds is recovered in the TOC analysis.
3. The BOD test is susceptible to variables that include seed acclimation, dilution, temperature, pH, and toxic substances. The COD and TOC tests are independent of these variables.

One would expect the stoichiometric COD/TOC ratio of a wastewater to approximate the molecular ratio of oxygen to carbon (32/12 = 2.66). Theoretically the ratio limits would range from zero, when the organic material is resistant to dichromate oxidation, to 5.33 for methane or slightly higher when inorganic reducing agents are present.

Reported BOD, COD, and TOC values for several organic compounds are given in table 2.5 and industrial wastewaters are listed in table 2.6. The COD/TOC ratio varies from 1.75 to 6.65 [8].

TABLE 2.5
Relationship Between COD and TOC for Organic Compounds[a]

Substance	COD/TOC (Calculated)	COD/TOC (Measured)
Acetone	3.56	2.44
Ethanol	4.00	3.35
Phenol	3.12	2.96
Benzene	3.34	0.84
Pyridine	3.33	Nil
Salicylic acid	2.86	2.83
Methanol	4.00	3.89
Benzoic acid	2.86	2.90
Sucrose	2.67	2.44

[a]After Ford [8].

TABLE 2.6
Oxygen Demand and Organic Carbon
of Selected Industrial Wastewater[a]

Waste	BOD$_5$ (mg/l)	COD (mg/l)	TOC (mg/l)	BOD/TOC	COD/TOC
Chemical[b]	–	4,260	640	–	6.65
Chemical[b]	–	2,410	370	–	6.60
Chemical[b]	–	2,690	420	–	6.40
Chemical	–	576	122	–	4.72
Chemical	24,000	41,300	9,500	2.53	4.35
Chemical-refinery	–	580	160	–	3.62
Petrochemical	–	3,340	900	–	3.32
Chemical	850	1,900	580	1.47	3.28
Chemical	700	1,400	450	1.55	3.12
Chemical	8,000	17,500	5,800	1.38	3.02
Chemical	60,700	78,000	26,000	2.34	3.00
Chemical	62,000	143,000	48,140	1.28	2.96
Chemical	–	165,000	58,000	–	2.84
Chemical	9,700	15,000	5,500	1.76	2.72
Nylon polymer	–	23,400	8,800	–	2.70
Petrochemical	–	–	–	–	2.70
Nylon polymer	–	112,600	44,000	–	2.50
Olefin processing	–	321	133	–	2.40
Butadiene processing	–	359	156	–	2.30
Chemical	–	350,000	160,000	–	2.19
Synthetic rubber	–	192	110	–	1.75

[a] After Ford [8].
[b] High concentration of sulfides and thiosulfates.

Although it has been difficult to correlate BOD with TOC for industrial wastes, relatively good correlation has been obtained for domestic wastewaters. This is reasonable when one considers the type and consistency of the waste constituents. A five-day BOD-TOC correlation for sewage has been reported by several investigators. A ratio of 1.87 has been reported by Wuhrmann [9]. Mohlman and Edwards [11] have reported a range of 1.35–2.62 for raw domestic waste. The calculated relationship between five-day BOD and TOC is:

$$\frac{BOD_5}{TOC} = \frac{O_2}{C} = \frac{32}{12}(0.90)(0.77) = 1.85$$

where:
1. The ultimate BOD exerts approximately 90 percent of the theoretical oxygen demand.
2. The five-day BOD is 77 percent of the ultimate BOD for domestic wastes.

A decrease in the COD/TOC and BOD_5/TOC ratios has been observed during the biological oxidation of both municipal and industrial wastewaters, as shown in table 2.7 and in figure 2.15. This can be attributed to:
1. The presence of inorganic reducing substances that would be oxidized in the biological process, thereby reducing the COD/TOC ratio.
2. Intermediate compounds may be formed during the biological process without significant conversion of organic matter to carbon dioxide. A reduction in COD may not be accompanied by a reduction in TOC.
3. The BOD reaction-rate constant k will be greater than 0.15/day in the raw waste and less than 0.1/day in the treated effluent. The BOD_5/BOD_u and hence the BOD_5/COD ratio depends on this rate. This is at least in part responsible for the reduction in the BOD_5/COD or BOD_5/TOC during biological oxidation.
4. The concentration of nonremovable refractory materials will account for a larger portion of the COD in the effluent than in the raw waste, thereby lowering the BOD_5/COD or the BOD_5/TOC ratio.

TABLE 2.7
Variation of COD/TOC and BOD_5/TOC Through Biological Treatment

Waste	COD/TOC		BOD₅/TOC	
	Raw	**Effluent**	**Raw**	**Effluent**
Domestic	4.15	2.20	1.62	0.47
Chemical	3.54	2.29	—	—
Refinery-chemical	5.40	2.15	2.75	0.43
Petrochemical	2.70	1.85	—	—

BOD, COD, and TOC measure the gross concentration of organics present in the wastewater but give no indication of their composition. Recently gel-chromatography has been employed for size separation of the soluble organics in wastewaters. Elution with water results in a series of samples in which organics with a molecular weight greater than 1,500 are in the first elution fraction, followed by a graduation from large to small in subsequent fractions. Those organics with an affinity for the gel are eluted with NH_4OH. The elutions have been divided

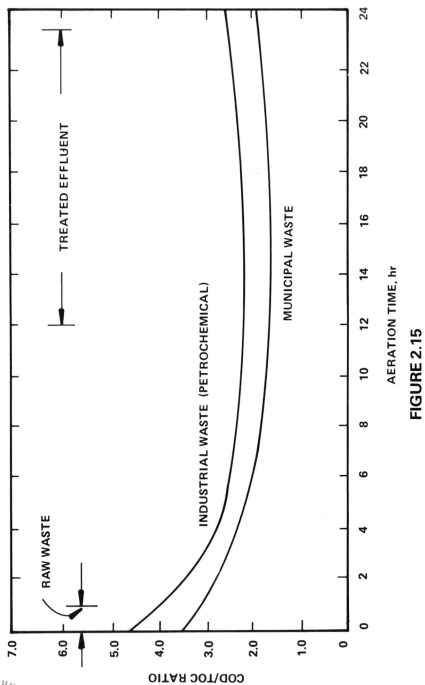

FIGURE 2.15

COD/TOC Ratio at Various Stages of Biological Oxidation

into six groups and the TOC and UV absorbance at 260 nm and 220 nm measured. A considerable number of organic substances can be characterized by the six groups, the behavior of the separated impurities corresponding to conversion in wastewater treatment processes and natural waterways.

Figure 2.16 shows the changes in wastewater composition through activated sludge, coagulation, and carbon adsorption. Figure 2.17 shows the changes through coagulation and adsorption alone. Figure 2.16 shows that biological treatment only removes a portion of the TOC insensitive to UV absorbance at 260 nm. High molecular weight impurities as in group 1 are removed by chemical coagulation. Carbon adsorption is effective for the removal of organics smaller than group 2 with TOC to E 260 (elution at 260 nm) ratios less than 50. Impurities in group 3 that are insensitive to UV absorbance at 260 nm cannot be effectively treated by physical chemical processes and require some form of biological treatment. Impurities in group 6 that exhibit UV absorbance at 220 nm are inorganic substances and therefore are not measured by TOC. These inorganic substances could be removed by such processes of demineralization as electrodialysis, ion exchange, and reverse osmosis.

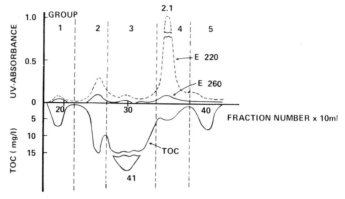

(a) GEL-CHROMATOGRAM OF RAW SEWAGE (x 10)

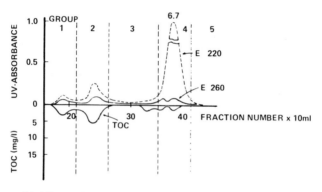

(b) GEL-CHROMATOGRAM OF SECONDARY EFFLUENT (x 10)

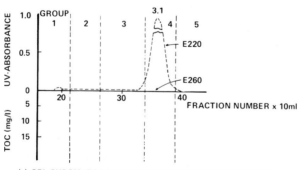

(c) GEL-CHROMATOGRAM OF COAGULATED AND ABSORBED
     SECONDARY EFFLUENT (x 10)

FIGURE 2.16

Gel-Chromatogram of Effluent from Biological and
Physical-Chemical Treatment of Sewage [12]

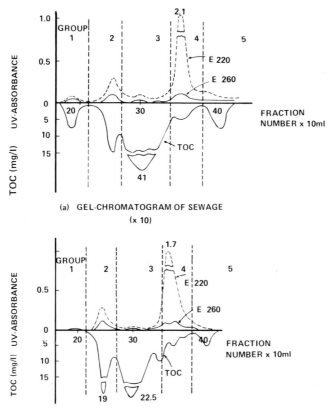

(a) GEL-CHROMATOGRAM OF SEWAGE
(x 10)

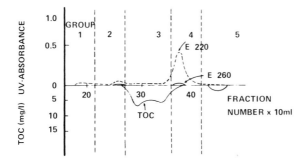

(b) GEL-CHROMATOGRAM OF COAGULATED RAW SEWAGE
(x 10)

(c) GEL-CHROMATOGRAM OF COAGULATED AND ABSORBED RAW SEWAGE (x 10)

FIGURE 2.17
Gel-Chromatogram of Effluent from Physical-Chemical
Treatment of Sewage [12]

REFERENCES

1. Moore, E. W. *et al.* 1950. Simplified method for analysis of BOD data. *Sewage Ind. Wastes* 22:1343.
2. Thomas, H. A. 1950. Graphical determination of BOD curve constants. *Water Sewage Works* 97:123.
3. Schroepfer, G. J. *et al.* 1964. The research program on the Mississippi River in the vicinity of Minneapolis and St. Paul. *Advances in Water Pollution Research* vol. 1. Oxford: Pergamon Press.
4. Morgan, G. B. and Lackey, J. B. 1958. BOD determinations in wastes containing cheleated copper or chromium. *Sewage Ind. Wastes* 30:283.
5. Busch, A. W. and Myrick, N. 1960. BOD progression in soluble substrates *Proc. 15th Ind. Waste Conf.* Lafayette, Ind.: Purdue University.
6. Jeris, J. S. 1967. A rapid COD test. *Water Wastes Eng.* 5:89.
7. Schaffer, R. B. *et al.* 1965. Application of a carbon analyzer in waste treatment. *Journal Water Pollution Control Federation* 37:11545.
8. Ford, D. L. 1968. Application of the total carbon analyzer for industrial wastewater evaluation. *Proc. 23rd Ind. Waste Conf.* Lafayette, Ind.: Purdue University.
9. Wuhrmann, K. 1964. *Hauptwirkungen and Wechsel wirk Kungen einiger Betriebsparameter im Belebschlammsystem Ergebnissemehrjahriger.* Zurich: Grossversuche Verlag.
10. APHA, AWWA, and WPCF. 1976. *Standard Methods for the Examination of Water and Wastewater.* 14th ed. Amer. Public Health Assoc.
11. Mohlman, F. W. and Edwards, G. P. 1931. Determination of carbon in sewage and industrial wastes. *Ind. Eng. Chem.* anal. ed. 3:1119.
12. Tambo, N. 1976. Application of gel-chromatography to the evaluation of removal efficiency of organic impurities in wastewater treatment processes. *Advanced Wastewater Treatment JRGWP Seminar.* Tokyo, Japan: The Japan Research Group of Water Pollution.

Analysis of Pollutional Effects in Natural Waters

The movement and reactions of waste materials through streams, lakes, and estuaries is a resultant of hydrodynamic transport and biological and chemical reactions by the biota, suspended materials, plant growths, and bottom sediments. These relationships can be expressed by a mathematical model that reflects the various inputs and outputs in the aquatic system. Considering the oxygen balance, the general relationships for the oxygen-sag curve are:

$$\frac{\partial C}{\partial t} = \epsilon \frac{\partial^2 C}{\partial X^2} - U \frac{\partial C}{\partial X} \pm \Sigma S \qquad (3\text{--}1)$$

where C = concentration of dissolved oxygen
   t = time at a stationary point
   U = velocity of flow in the X direction
   ε = turbulent diffusion coefficient
   S = sources and sinks of oxygen
   X = distance downstream

Equation (3–1) assumes that the concentration of any characteristic is uniform over the stream cross-section and that the area is uniform with distance. If this is not the case, equation (3–1) must be suitably modified.

The sources of oxygen are:

1. In incoming or tributary flow
2. Photosynthesis
3. Reaeration

The sinks of oxygen are:

1. Biological oxidation of carbonaceous organic matter.
2. Biological oxidation of nitrogenous organic matter.
3. Benthal decomposition of bottom deposits.
4. Respiration of aquatic plants.
5. Immediate chemical oxygen demand.

SOURCES OF OXYGEN

In Incoming or Tributory Flow

The quantity of oxygen in the incoming or tributary flow is considered as an initial condition in equation (3–1). The dissolved oxygen present in waste discharges should also be considered if the waste flow is large relative to the stream flow.

Photosynthesis

Photosynthesis is the production of oxygen from the growing of green plant life in water courses from sunlight, carbon dioxide, and other stream nutrients. The degree of photosynthesis depends upon sunlight, temperature, mass of algae and rooted plants, and available nutrients. It will exhibit a diurnal variation, as shown in figure 3.1.

Reaeration

By the process of natural reaeration oxygen is added to the water body. Reaeration is primarily related to the degree of turbulence and natural mixing in the water body (high in sections of rapids; low in impounded areas). The oxygen transfer from air to water can be defined as:

$$N = K_L A (C_s - C_L) \qquad (3-2)$$

where N = lb of O_2/hr
K_L = O_2 transfer coefficient
A = surface area
C_s = O_2 saturation concentration
C_L = O_2 concentration

and, in concentration units:

$$\frac{dC}{dt} = \frac{K_L A}{V} (C_s - C_L) \qquad (3-3)$$

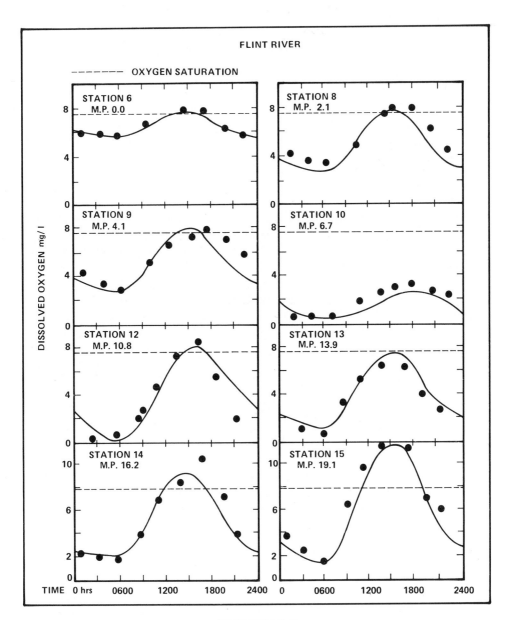

FIGURE 3.1

Observed and Calculated Temporal Profiles
of Dissolved Oxygen [31]

for a stream $A/V = 1/H$ and $(C_s - C_L) =$ oxygen deficit D and equation (3–3) reduces to:

$$\frac{dC}{dt} = \left(\frac{K_L}{H}\right) D = K_2 D \tag{3-4}$$

in which K_2 is the reaeration coefficient. Note that $A/V = 1/H$ applies only to a quiescent surface and would increase for a turbulent surface. Dobbins [30] estimates $A/V = 1.5/H$ for a highly turbulent stream surface.

The reaeration coefficient K_2 has usually been defined by a relationship of the type:

$$K_2 = \frac{CU^n}{H^m} \tag{3-5}$$

in which U is the average stream velocity and H the average stream depth. The coefficient C is a function of stream characteristics. Organics and surface-active agents present in the stream water affect K_2 and this effect is reflected in the constant C.

The exponents m and n also relate to stream conditions. Reported values of the coefficients are summarized in table 3.1. K_2 is affected by temperature and can be corrected by the relationship:

$$K_{2T} = K_{2 20°C} \, 1.028^{(T-20)} \tag{3-6}$$

TABLE 3.1

Summary of Coefficients
for the Reaeration Equation[a] at 20° C

$$K_2 = C \frac{U^n}{H^m} \text{ day}^{-1} \text{ (base 10)}$$

C	n	m	Reference
5.6	0.5	1.5	O'Connor & Dobbins [1]
5.0	1.0	1.67	Churchill *et al.* [2]
3.3	1.0	1.33	Langbein & Durum [3]
9.41	0.67	1.85	Owens *et al.* [4]
4.67	0.6	1.4	Bansal [5]
2.83	1.0	1.5	Isaacs *et al.* [6]
2.53	0.92	1.71	Kothandaraman & Ewing [7]
4.74	0.85	0.85	Negulescu & Rojanski [8]

[a] U is in fps and H in ft.

Very little consistency in K_2 values is observed using this formulation, because of the many factors that affect reaeration. Kothandaraman [10] has shown that the contaminants in river water may alter the reaeration rate coefficient by as much as ±15 percent as compared to the rate for distilled water.

Wilson and MacLeod [11] have shown that those equations derived from river results generally overpredict the reaeration coefficient while those derived from channel studies generally underpredict the reaeration coefficient.

Thackston and Krenkel [9] have introduced an energy surface renewal equation which has been reported to predict reaeration coefficients for most streams except white water rapids and very slow moving, deep reservoirs.

$$K_2, \text{day}^{-1} \text{ (base 10)} = 10.8 \left(1 + \left(\frac{U}{(gH)^{1/2}}\right)^{1/2}\right) \cdot \left(\frac{sg}{H}\right)^{1/2} \qquad (3\text{–}6a)$$

where U = velocity (ft/sec)
   g = gravity constant (32.2 ft/sec²)
   H = depth (ft)
   s = slope (ft/ft)

Tsivoglou and Neal [12] have conducted field tracer measurements of stream reaeration capacity on approximately 250 miles of 24 different streams over the past ten years. The studies have included a wide range of stream flows, water temperature and BOD, as well as a wide variety of hydraulic features such as waterfalls, shoals, rapids, pools, and relatively uniform reaches. The reaeration rate coefficient was found to be directly proportional to the rate of energy expenditure. Their relationship is:

$$K_2 = C\left(\frac{\Delta h}{t_f}\right) \qquad (3\text{–}7)$$

where C = constant (1/ft)
   Δh = water surface elevation change (ft)
   t_f = time of flow (hrs)
   K_2 = reaeration rate coefficient (base e) (1/hr)

Values of C have been experimentally determined and are shown in table 3.2. The

TABLE 3.2
Values of C for Equation (3–7)

C	Flow	Maximum C	Minimum C
0.054	$25 \leq Q \leq 3{,}000$ cfs	0.08	0.025
0.110	$1 \leq Q \leq 10$ cfs	0.16	0.06

value of C should be adjusted downward to the limiting minimum value in table 3.2 as the stream flow increases and adjusted upward to the limiting maximum value as the stream flow decreases. It should be emphasized that no predictive model for the reaeration coefficient is a satisfactory substitute for direct measurement of K_2.

SINKS OF OXYGEN

Biological Oxidation of Carbonaceous Organic Matter

In the biological oxidation of carbonaceous organic matter, the rate of removal, K_r, is related to the amount of unstabilized organics present:

$$L = L_0 e^{-K_r X/U} = L_0 e^{-K_r t} \tag{3-8}$$

where L is concentration of organics present at time t, L_0 is concentration of organics present at time zero, and K_r relates to the removal of organics by all mechanisms: sedimentation, oxidation, and volatilization. For oxidation alone, as might result from a soluble organic waste, equation (3–8) is expressed

$$L = L_0 e^{-K_1 X/U} = L_0 e^{-K_1 t} \tag{3-9}$$

It is recognized that the BOD exertion is not a first-order reaction, but rather a two-stage reaction (synthesis and endogenous respiration). As a result, Edeline and Lambert [13] have shown that the observed sag curve deviates from the model predicted by equation (3–9) in two ways; it starts with a greater oxygen consumption through adaption and growth of the biomass and recovers more slowly than predicted due to the continuing endogenous respiration. This fact should be recognized in interpreting oxygen sag data. K_r will be considerably greater than K_1 when suspended or volatile organics are present (see figure 3.2). The rate of removal of organics by mechanisms other than oxidation has been defined as K_3. The rate of oxidation of the organics in the BOD bottle is defined as k. And, k is usually less than K_1 because longitudinal mixing, the presence of bottom growths, and suspended biological solids will increase the reaction rate. (The concentration of seed organisms will also usually be higher in a stream.) Values of k depend upon the characteristics of the waste, decreasing with treatment or removal of readily oxidizable organics. Figure 3.3 shows results of k summarized by Novotny and Krenkel [14]. The reported range of values to be expected is given in table 3.3. By contrast, K_1 may have values in excess of 20 per day. Bosko [15] has related k to K_1 through the characteristics of the stream:

$$K_1 = k + \frac{U}{H} \eta \tag{3-10}$$

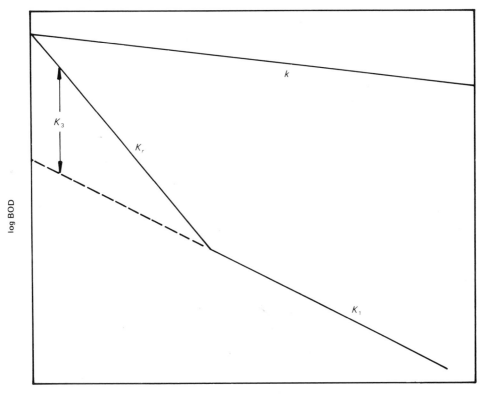

FIGURE 3.2

Deoxygenation Relationship in a Stream

or

$$K_1 = k + B$$

in which U is the velocity of stream flow, H the depth of the stream, and η a coefficient of bed activity that may vary from 0.1 for stagnant or deep waters to 0.6 or higher for rapidly flowing streams. Novotny [16] proposed the formula:

$$B = v \frac{S_e^{1/6}}{H^{3/4}} \tag{3–11}$$

where

H = depth
S_e = slope, ft/1,000 ft

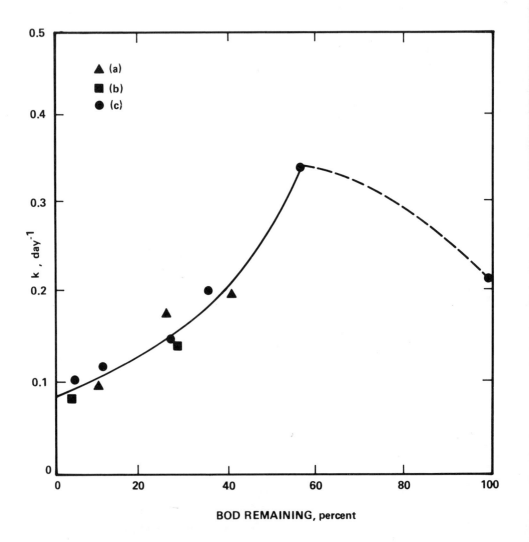

FIGURE 3.3

Effect of BOD Removal on the Magnitude of the
Laboratory Deoxygenation Coefficient k (Base 10). (a)
EPA Survey, 1969, (b) Waste Water Treatability Data,
1973, (c) Kittrell Kochtizky's Data, 1946, [14]

TABLE 3.3
Values of k

Substance	$k_{(e)}$, day^{-1}
Raw sewage, high-rate treatment effluent	0.35–0.60
High-degree biological effluent	0.10–0.15
Rivers with low pollution	0.10–0.12

and v, a coefficient to range from 0.1 for streams with moving bottoms (sand, mud) to 3.0 (B base e) for rocky bottoms in zones of higher pollution. These various reaction rates are compared in figure 3.2.

Biological Oxidation
of Nitrogenous Organic Matter

When unoxidized nitrogen is present in the wastewater, nitrification will result with time of passage or distance downstream. The rate of nitrification is less when untreated wastes are discharged to the stream and the concentration of nitrifying organisms is low, and the rate increases in the presence of well-oxidized effluents with a high seed of nitrifying organisms. The reactions are:

Oxidation of ammonia to nitrite by *Nitrosomonas:*

$$2NH_4^+ + 3O_2 \longrightarrow 2NO_2^- + 2H_2O + 4H^+$$

Oxidation of nitrite to nitrate by *Nitrobacter:*

$$2NO_2^- + O_2 \longrightarrow 2NO_3^-$$

Nitrifying organisms are sensitive to pH and function best over a pH range 7.5 to 8.0. The rate of nitrification decreases rapidly at dissolved oxygen levels below 2.0 to 2.5 mg/1, so that at low oxygen levels in the water body little or no nitrification will occur. Denitrification has been observed to occur in stretches of zero or near-zero oxygen concentration. The kinetics of nitrification can be expressed as an autocatalytic reaction.

$$\frac{dC}{dt} = -KC(N - C) \tag{3–12}$$

where C = concentration of ammonia or nitrate
        N = concentration initially present
        K = reaction rate coefficient

which integrates to:

$$C = \frac{N}{1 + e^{KN(t-\alpha)}} \tag{3-12a}$$

in which α = time for half completion of reaction. The equation can be linearized:

$$\frac{N - C}{C} = e^{KN(t-\alpha)} \tag{3-12b}$$

and plotted as log $[(N - C)/C]$ against t to evaluate the reaction rate K. Data can also be correlated as two forms of a first-order reaction:

$$\frac{dC}{dt} = -K(N - C) \tag{3-13}$$

and:

$$\frac{dC}{dt} = -KC \tag{3-13a}$$

at $t = \alpha$, $C = N/2$ for both curves, and integration yields:

$$C = N \left(1 - \frac{e^{K(t-\alpha)}}{2}\right) \tag{3-13b}$$

and:

$$C = \frac{N}{2} e^{-K(t-\alpha)} \tag{3-13c}$$

The temperature effect on the rate of nitrification in natural waters has been expressed [17] as:

Nitrosomonas (river water)

$$K_T = 0.47 \cdot 1.106^{(T-15)}$$

Nitrobacter (river water)

$$K_T = 0.79 \cdot 1.072^{(T-15)}$$

The usual operational equation for a stream may be considered:

$$N = N_0 e^{-K_n X/U} = N_0 e^{-K_n t} \tag{3-14}$$

It should be recognized that because the growth rate of the nitrifying organisms is considerably lower than the carbonaceous organisms, oxygen depletion through nitrification lags the deoxygenation from carbonaceous organics. When secondary sewage treatment plants are installed, the quantity of carbonaceous organics to be removed is greatly reduced, but much larger numbers of nitrifying organisms are present in the stream. Under these conditions nitrification is more rapid and may exert a significant oxygen demand.

Benthal Decomposition of Bottom Deposits

In polluted streams the river bottom may be covered by active biological materials, such as sludges and slimes. The growth and accumulation of these materials results from deposition of suspended organics and/or the transfer of soluble organics to the flowing slimes. The BOD is metabolized to new cell growth and aerobically or anaerobically decomposed. Deposition occurs at low velocities. At high velocities (above 0.3–0.45m/sec) deposited materials may be resuspended and cause a secondary increase in BOD. In areas where bottom deposits occur, oxygen will be used by diffusion into the upper layers of the deposit (the rate will increase in the presence of worms, which increases the porosity of the deposit) and from the diffusion of organic products of anaerobic degradation into the flowing stream water, which will increase the soluble organic oxygen demand of the water as shown in figure 3.4 (see also table 3.4). The oxygen consumption has been observed to increase somewhat with dissolved oxygen concentration. Invertebrates living in muds, such as midge larvae, increase the interchange between the mud and the overlying water. Novotny and Krenkel [14] developed a relationship to estimate the oxygen consumption of benthal deposits.

$$O_2 \text{ Uptake, } g/m^2/day = 0.5 \, (K_1 - k) LH \qquad (3\text{–}15)$$

in which L is the carbonaceous BOD of benthal deposits and H is the depth of benthal deposits. Figure 3.5 compares observed and computed results.

Ballinger and McKee [18] have examined more than 200 bottom sediment samples from rivers, lakes, and estuaries and have used analytical procedures to divide them into four general classes (see table 3.5). Edberg and Hofsten [19] have reported that there is no simple correlation between oxygen uptake and the content of organic matter in sediments.

Dissolved oxygen reduction over time is a measure of the uptake of the bottom muds. Results are usually reported in grams of O_2 uptake per square meter per day. Many of the *in situ* measurements that have been made up to this point are summarized in table 3.6.

The depth of the deposit is also of some interest since it appears that the uptake rate is dependent on this parameter. Oldaker *et al.* [20] found a linear relationship between estimates of the ultimate amount of oxygen needed for complete stabilization and depth of sediment (from domestic sewage) over the range of 1.5–2.0 cm. However, Edwards and Owens [21] working with relatively stable sediments, have indicated that uptake was independent of depth for depths greater than about 2.0 cm.

Respiration of Aquatic Plants

Oxygen will be removed from the water body by the respiration of aquatic plants. (Oxygen may also be contributed by photosynthesis.) The oxygen loss is

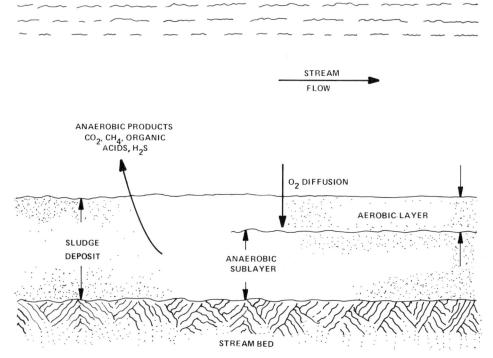

FIGURE 3.4

Benthal Decomposition in a Stream Bottom

TABLE 3.4

Characteristics of River Deposits

Deposit	Mean	Range
Dry solids, % of sediment	21	10–40
Volatile solids,		
% of dry	20	11–27
g/m^2	460	150–2,000
Organic nitrogen, % of volatile	3.5	1.3–5.1

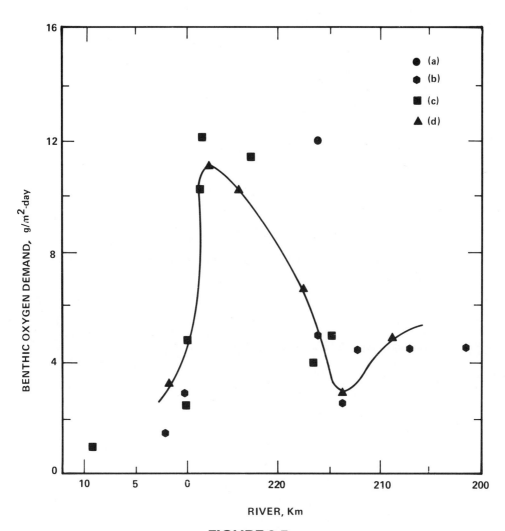

FIGURE 3.5

Comparison of Measured and Computed Benthic Oxygen Demand.
(a) June 1969—TVA. Band and Eddy Areas. (b) June 1969—EPA
Mid-Channel. (c) June 1969—EPA. Bank and Eddy Areas.
(d) Computed [14].

TABLE 3.5
Bottom Sediment Composition and Type

Typical Sediments	Organic Carbon (%)	Organic Nitrogen (%)	Type
Sand, silt, clay, loam	0.4– 2.1	0.02–0.10	I
Old stable sludge, peat, organic debris	2.0– 5.0	0.10–0.20	I
Paper mill wastes	6.0–15.0	0.10–0.30	II
Packinghouse wastes	2.8– 4.3	0.30–0.50	III
Fresh sludge, decaying algae, wastewater solids	5.0–40.0	0.70–5.00	IV

TABLE 3.6
Average Values of Oxygen Uptake of River Bottoms

Bottom Type and Location	Uptake (g of O_2/m^2/day) @ 20°C.	
	Range	Approximate Average
Sphaerotilus—(10 g of dry wt/m^2)	—	7.00
Municipal sewage sludge—outfall vicinity	2.00–10.0	4.00
Municipal sewage sludge—"aged" downstream of outfall	1.00– 2.0	1.50
Cellulosic fiber sludge[a]	4.00–10.0	7.00
Estuarine mud	1.00– 2.0	1.50
Sandy bottom	0.20– 1.0	0.50
Mineral soils	0.05– 0.1	0.07

[a]Calculated from reported values of 2.5 and 3.5 at 11°C.

usually expressed on a weight basis or on an area basis (g of O_2/m^2). Camp [22] states that just as photosynthesis can provide as much as two-thirds of the dissolved oxygen present in the water during the day, at night respiration of the algae and aquatic plants may deplete the oxygen resources of the river. Table 3.7 gives some values of the average gross production of oxygen due to photosynthetic activity.

TABLE 3.7

Average Values of Gross Photosynthetic Production of Dissolved Oxygen[a]

Water Type and Location	Photosynthetic Production g O_2/m²/day
Truckee River—Bottom attached algae	9
Tidal Creek—Diatom Bloom ($62 - 109 \cdot 10^6$ diatoms/l)	6
Delaware Estuary—summer	3.0–7.0
Duwamish River estuary— Seattle, Washington	0.5–2.0
Neuse River system— North Carolina	0.3–2.4

[a] After Thomann [23].

Nemerow [24] has tabulated several different organisms along with their rates of respiration and these are shown in table 3.8.

Oxygen is also consumed by the development of sewage fungus (*Sphaerotilus*) when available carbohydrates are present in the water.

TABLE 3.8

Oxygen Consumption of Some Fish and Representative Animals[a]

Water Living	cm³O_2/gram/hour of wt. of animal	Nonwater Living	cm³O_2/gram/hour of wt. of animal
Jelly fish	.0034–.005	Sparrow	6.7
Starfish	.03	Horse	.25
Leech	.023	Sheep	.34
Mussel	.0549	Dog	.83
Crayfish	.04		
Butterfly (at rest)	.6		
Eel	.04		
Goldfish	.07		
Trout	.22		

[a] After Nemerow [24].

The Water Pollution Research Laboratory [25] showed *Sphaerotilus* growths of 18 g/m^2/day at 20° C. They showed that maximum growth occurred at a stream velocity of 0.45 to 0.9 fps (0.14 to 0.27 m/sec). Higher velocities promote scour, and lower velocities have insufficient turbulence for distribution of the nutrient supply.

Immediate Chemical Oxygen Demand

Many industrial wastes contain chemicals that will exert an immediate oxygen demand, such as sulfites. This will usually be immediately apparent in the oxygen sag curve and the BOD curve in the stream.

Salinity

Salinity in a river or estuary causes a depression in the oxygen saturation level. As an example, at 25° C during summer critical conditions, if the salt concentration is increased to 10,000 mg/l the oxygen saturation value is reduced to 7.56 mg/l from 8.18 mg/l, which is about 8 percent. This phenomenon frequently occurs in tidal streams and waters that receive industrial wastewater such as oil field brines or cucumber brine from a pickling process. The salinity problem is also observed in many southwestern rivers and streams through irrigation practices in the United States.

MATHEMATICAL DEVELOPMENT OF THE BOD-OXYGEN SAG MODEL FOR STREAMS

When considering streams, the turbulent diffusion (i.e., longitudinal mixing) is generally insignificant and equation (3–1) reduces to:

$$\frac{\partial C}{\partial t} = -U \frac{\partial C}{\partial x} \pm \Sigma S \tag{3–1a}$$

in which $\partial C/\partial t$ is the change in concentration with time at a point source. Under steady-state conditions (i.e., there is no change in loading with time at a point source) equation (3–1a) becomes:

$$U \frac{\partial C}{\partial X} = \pm \Sigma S = \frac{dC}{dt} \tag{3–1b}$$

in which dC/dt is the change in oxygen concentration with time of flow downstream, U the velocity of flow, and X the distance downstream such that $t = X/U$. Assuming only deoxygenation by organic-matter oxidation and natural reaeration, equation (3–1a) becomes:

$$\frac{\partial C}{\partial t} = -\frac{U \partial C}{\partial X} - K_1 L + K_2 (C_s - C) \tag{3-1c}$$

Under steady-state conditions:

$$0 = -U \frac{dC}{dX} - K_1 L + K_2 (C_s - C) \tag{3-1d}$$

Dividing throughout by velocity yields:

$$-\frac{dC}{dX} = \frac{K_1 L}{U} - \frac{K_2}{U} (C_s - C)$$

and integrating yields:

$$C = C_s - \frac{j_1 L_0}{j_2 - j_r} (e^{-j_r X} - e^{-j_2 X}) - (C_s - C_0) e^{-j_2 X} \tag{3-1e}$$

where $j_1 = K_1/U$
$j_2 = K_2/U$
$j_r = K_r/U$

The critical point in the oxygen sag can be evaluated:

$$K_2 D_c = K_1 L_c = K_1 L_0 e^{-K_r t_c} \tag{3-1f}$$

and:

$$D_c = \frac{K_1}{K_2} L_0 e^{-K_r t_c} \tag{3-1g}$$

by differentiating the oxygen-sag equation and equating to zero:

$$t_c = \frac{1}{K_2 - K_r} \ln \frac{K_2}{K_1} \left[1 - \frac{D_0 (K_2 - K_r)}{K_1 L_0} \right] \tag{3-1h}$$

The effect of the various sources and sinks on a hypothetical river are shown in figure 3.6.

Equation (3-1d) can be modified to include other sources and sinks of oxygen, as previously discussed. The coefficients that are employed in the oxygen-sag relationships are summarized in table 3.9. An oxygen sag curve for the Holston River in Tennessee using Kittrell-Kochtitzky's [27] data is shown in figure 3.7. Edeline and Lambert [13] showed that when D_0 at or near zero (the stream at or near saturation above the wastewater discharge) equation (3-1h) reduces to:

$$t_c = \frac{1}{K_2 - K_r} \ln \frac{K_2}{K_r} = A$$

TABLE 3.9

Coefficients for the Evaluation of the
Assimilative Capacity of a Stream

Coefficient	Definition	Dependent on	Temperature Coefficient (θ)
K_1	Oxidation-rate coefficient for soluble organics	Concentration and nature of organics remaining	1.065–1.075
K_r	Removal-rate coefficient for all organics; includes oxidation, sedimentation, and immediate demands	Concentration of total organics remaining	1.00–1.075
K_3	Removal rate by sedimentation	Concentration of settleable organics	—
K_2	Reaeration coefficient	Stream velocity and depth	1.028
K_n	Rate coefficient for nitrification	Nitrogen concentration present; concentration of organics and presence of secondary sewage effluent	1.106
k	BOD-bottle-rate coefficient	Nature of organics, i.e. raw waste or treated effluent	1.065–1.075

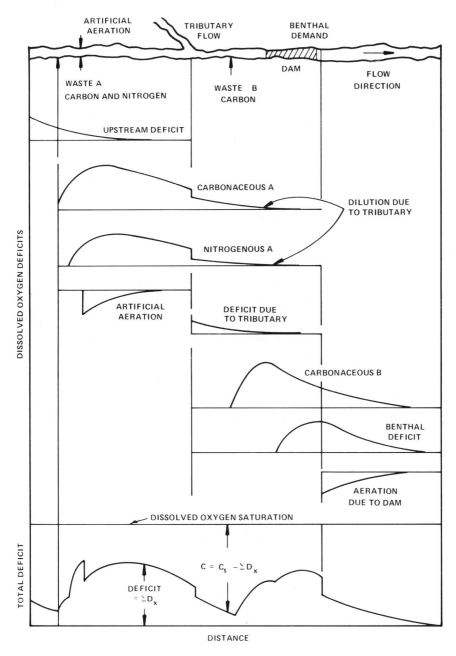

FIGURE 3.6

Sources and Sinks of Oxygen in a Natural Watercourse
(After O'Connor and DiToro [26].)

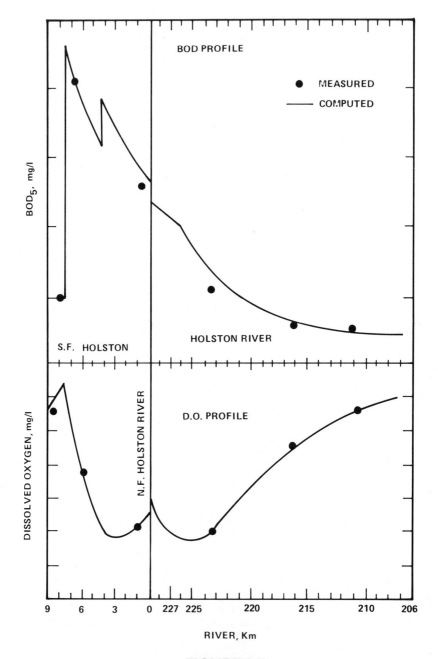

FIGURE 3.7
Verification of DO Sag Model for BOD$_5$ and DO
Using Kittrell-Kochtitzky's Data [14].

and:

$$D_c = \frac{K_1}{K_2} L_0 e^{-K_r A} = B L_0 e^{-K_r A} = C L_0$$

where *A, B,* and *C* are constants for the case in question. This yields a linear relationship between L_0 and D_c (as shown in figure 3.8) and is useful for predicting the effect of varying wastewater loads on the critical deficit. This relationship only applies for a single discharge under the imposed conditions.

STREAM SURVEY AND DATA ANALYSIS

The stream survey should be conducted during a period when dissolved oxygen is available at all locations over the survey course. A low-flow high-temperature period is desirable. (This may not be possible in cases of a highly polluted stream.) Flood-flow conditions should be avoided.

Sampling stations should be selected for convenience of location but should encompass all pertinent stretches of the stream and must be below all new wastewater discharges, tributary confluences, and impoundments. It is important that complete mixing of all discharges with the stream water be effected before sampling.

Physical Data to Be Collected

1. Stream cross section
2. Stream flow (from nearest gaging station)
3. Sections of impoundment, marshes, rapids, etc.
4. Location, identification, and measurement of pollutional sources, including waste flow variation and storm runoff

Sampling Data to Be Collected

1. Temperature
2. Dissolved oxygen
3. BOD (soluble and total)
4. Suspended and volatile suspended solids
5. Nitrogen (ammonia and nitrates)
6. Chlorophyl (in areas where algae are present)
7. Bottom deposits (where applicable)
8. Velocity of stream flow

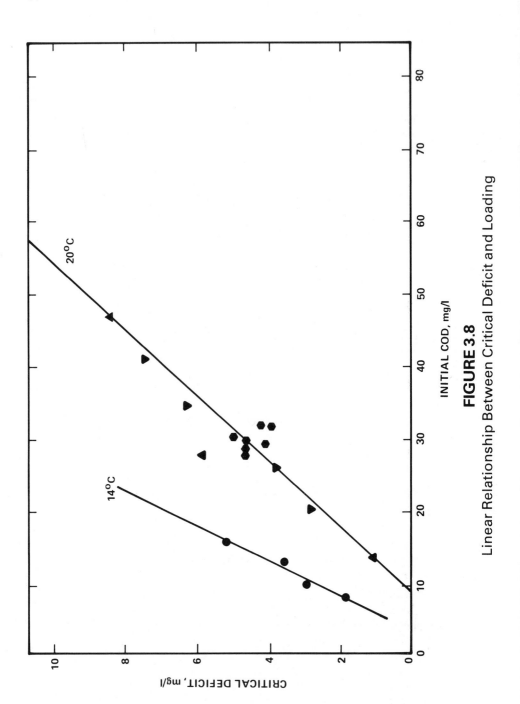

FIGURE 3.8

Linear Relationship Between Critical Deficit and Loading

Development of the Coefficients from the Survey Data

Conversion to the Ultimate BOD

The carbonaceous BOD data should be converted to ultimate values for evaluation of deoxygenation rates. Long-term BOD's should be run at pertinent stream stations (that is, below each major waste discharge) and the k and L_0 computed by one of the standard procedures (graphical, method of moments, etc.). In general, untreated wastes will have a k rate >0.15/day and treated wastewater a k rate <0.1/day (see figure 3.9).

Deoxygenation Rate, K_1

The deoxygenation rate, K_1, is determined from the slope of a plot of the log BOD_u against time of passage or distance. (This plot can be developed either as lb of BOD_u or concentration of BOD_u after mixing in the stream flow.) K_1 is

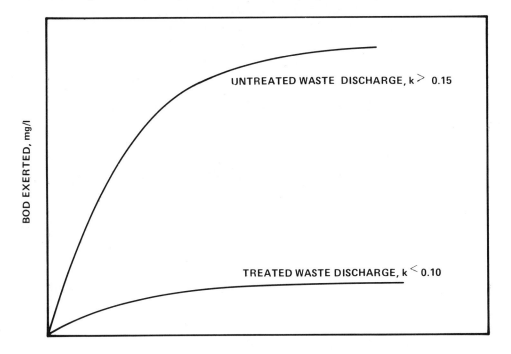

FIGURE 3.9

Variation in BOD Exertion for Treated and Untreated Wastes in a Stream determined over a stream stretch where only soluble BOD is exerted (see figure 3.10). As shown in figure 3.10, K_1 should be readjusted below each waste outfall.

Total Removal Rate, K_r

When volatile suspended solids are present, removal of BOD in the stream results both from oxidation and sedimentation of suspended organics on the stream bottom. The total removal rate is shown in figure 3.10. The coefficient K_3, the sedimentation rate, is computed as $K_3 = (K_r L - K_1 L_1)/L_3$ in which L_1 and L_3 are the BOD in soluble and suspended form respectively. When no solids are present in the wastewater discharge, $K_r = K_1$.

Nitrification, K_n

Evaluation of the nitrification coefficient K_n can be made in one of several ways. One method consists of calculating the ammonia nitrogen at each of several stations along a river. Assuming a first-order reaction, if a plot is made of the log of the ammonia nitrogen remaining versus time of passage, the slope of the straight line of best fit will define the nitrification coefficient. Possible error can be introduced if ammonia gas is lost to the atmosphere or if there is a high concentration of ammonia in the runoff that contributes to the stream flow. The calculations of K_n is shown in figure 3.11. The nitrification rate should be converted to terms of oxygen.

Upstream and Tributary Oxygen Contribution

Oxygen will be supplied from that present in upstream water and in tributary flows. At each tributary point or location of stream aeration, the oxygen content in the stream should be recalculated after mixing with the main stream flow using the following equation:

$$\frac{Q_1 C_1 + Q_2 C_2}{Q_3} = C_3 \tag{3-16}$$

where Q_1 = flow one
       Q_2 = flow two
       $Q_3 = Q_1 + Q_2$
       C_1 = concentration one
       C_2 = concentration two
       C_3 = concentration after mixing

Reaeration

The reaeration coefficient K_2 should be computed from equation (3-5), (3-6), or (3-7) and it should be corrected for stream temperature. Any coefficients

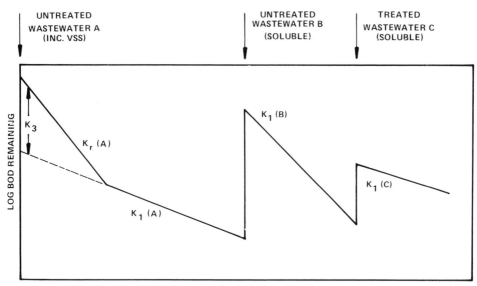

FIGURE 3.10
Deoxygenation Relationships in a Stream

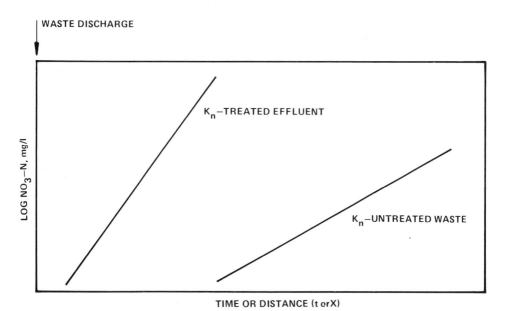

FIGURE 3.11
Nitrification Rates in a Stream

used in these equations should be selected from data that has been developed on a similar stream.

Photosynthesis

In areas where algae are present, photosynthesis results in an increase in dissolved oxygen during the daylight hours and a decrease during the night hours. The mathematical modeling of photosynthesis and respiration has generally been considered in one of three ways:

1. Determine the daily mean (gross or net) oxygen production and assume it to be a constant.
2. Model the oxygen production as a function of time and fit equations to the experimental data.
3. Assume the P-R (photosynthesis and respiration) is related to the amount of phytoplankton present in the water.

Biguria *et al.* [28] have proposed an equation of the form:

$$P\text{-}R = C \left\{ \exp\left[\frac{-E_a}{T + 273} \right] \right\} \left\{ \sin\frac{\pi t}{12} \right\}^n \tag{3-17}$$

and fitted data by regression analysis to determine the constants C and n and complex activation energy E_a. The authors reported good correlation could be obtained within the temperature range of the data (15–22° C).

Another method, first presented by Odum and used by O'Connor and DiToro [26] is to assume that the photosynthetic rate is directly proportional to the sunlight intensity that varies as a half sine wave. The rate of photosynthesis can then be expressed as a Fourier series:

$$P(t) = P_m \left\{ \frac{2t_p}{\pi} + 2 \sum_{n=1}^{\infty} a_n \cos\left[2\pi n(t - t_p/2)\right] \right\} \tag{3-18}$$

where $P(t)$ = photosynthetic rate
   P_m = maximum photosynthetic rate
   t_p = time period under consideration
   t = time
   a_n =

$$a_n = \frac{2\pi/t_p}{(\pi/t_p)^2 - (2n\pi)^2} \cos\left(n\pi t_p\right)$$

The respiration rate is considered to be a constant. This model was incorporated into an equation for dissolved oxygen and fitted to diurnal dissolved oxygen data from the Grand River in Michigan.

Benthal Demand

Benthal demand is estimated from an undisturbed core sample of bottom deposit or computed from equation (3–15). Benthic demands are usually expressed as grams of O_2/m^2 of stream bottom/day.

CORRELATION OF THE SURVEY DATA

The sources and sinks of oxygen are additive to develop the net oxygen balance in the stream equation (3–1). These can be individually plotted and the net oxygen-sag curve computed by the addition and subtraction of the sources and sinks of oxygen as shown in figure 3.6. The computed curve should then be compared to the observed field-dissolved oxygen measurements. If the agreement is poor, the computed coefficients should be readjusted to close the difference between the observed and calculated results.

CALCULATION OF ASSIMILATIVE CAPACITY OF A STREAM

The stream survey is used to develop the coefficients defining the sources and sinks of oxygen. These data are then used to define:

1. The oxygen-sag relationship under other conditions of stream flow and temperature
2. The effect on the oxygen sag of additional sources of pollution
3. The degree of treatment necessary for existing or future sources of pollution to meet specified dissolved oxygen levels under defined conditions of temperature and stream flow

To meet these requirements, the physical factors involved, such as stream flow and depth and the reaeration-rate coefficients, must be readjusted to meet the revised conditions.

Variation in Physical Parameters

When the stream flow is changed to meet critical conditions, it is reasonable to assume that the velocity and depth will also be affected, which in turn influence the reaeration coefficient. General relationships have been developed for the hydraulic characteristics of a stream [28]:

$$H \sim Q^b$$
$$U \sim Q^f$$
$$W \sim Q^m$$

The sum of the exponents should equal 1.0. Correlations for a given river can be determined as shown in figure 3.12.

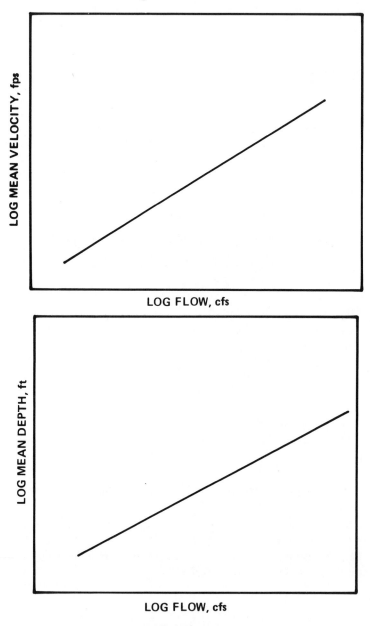

FIGURE 3.12
Hydraulic Characteristics of Natural Streams

The effect of depth variation on assimilative capacity is shown in figure 3.13. The effect of temperature and flow variation is shown in figure 3.14.

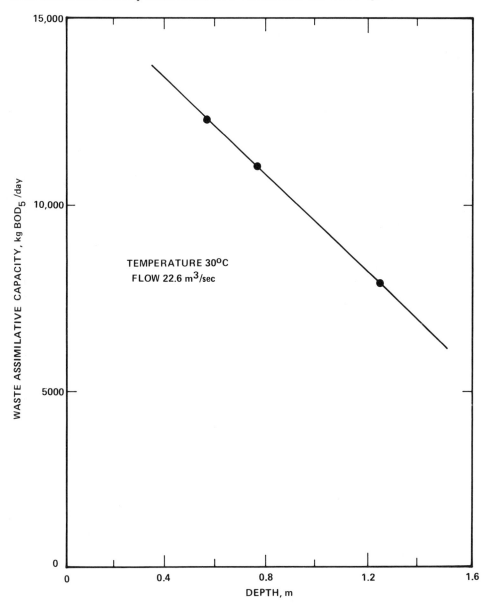

FIGURE 3.13

Effect of Depth Variation on Waste Assimilative
Capacity of Holston River [14]

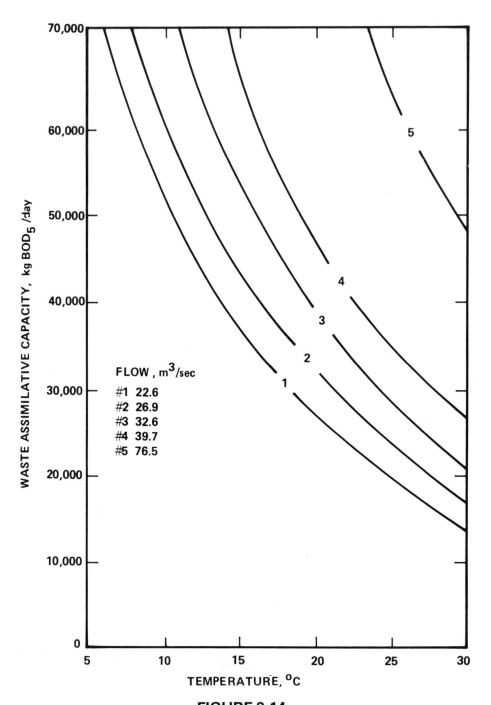

FIGURE 3.14

Effect of Temperature and Flow Variations on Waste
Assimilative Capacity of Holston River [14]

Variation in Rate Coefficients

The rate coefficients for sources and sinks of oxygen developed from the survey will be affected by changes in temperature, characteristics of waste discharges, and degree of waste treatment. The modifications of the coefficients needed for the prediction of other stream conditions are summarized in table 3.10.

TABLE 3.10

Changes in Rate Coefficients as a Result of
Environmental Effects

Coefficient	Temperature Coefficient	General Effects
K_r	1.0–1.075	Reduced by removal of suspended solids by treatment; temperature coefficient increases as $K_r \rightarrow K_1$
K_1	1.075	For complete removal of suspended solids $K_r = K_1$; increases by addition of readily assimilable waste; decreases with degree of treatment of wastewater; increases with increasing stream fertilization
K_n	1.106	Increases with degree of treatment (domestic sewage); low for nitrified effluent
K_2	1.028	Decreases with the addition of untreated wastes or surface-active agents; increases with degree of treatment of wastewaters; increases with velocity; decreases with increasing depth

Calculation of Assimilative Capacity of a Stream

1. Estimate the low-flow-stream condition. This will usually be selected from a statistical analysis of drought flow conditions. Estimate the velocity and depth from a plot such as figure 3.12. (In the absence of data on the stream in question, H may be assumed to vary as $Q^{0.6}$ and U as $Q^{0.4}$.)

2. Revise the coefficient as follows:
    K_r –   correct for temperature and removal of suspended solids by treatment

K_1 –   correct for temperature and *estimated* degree of treatment

K_n –   correct for temperature and *estimated* degree of treatment

K_2 –   correct for temperature, velocity, and depth at low-stream-flow condition; added wastes; or degree of treatment

3.   Modify benthic demand and algae for projected conditions.

4.   For the maximum allowable oxygen deficit, compute the maximum BOD loadings that can be discharged at each outfall from equation (3–1h) or modified for other sources and sinks of oxygen. The present and projected oxygen-sag curve and rate coefficients are schematically illustrated in figure 3.15. For simplicity, a stream with a single wastewater discharge is shown.

An assumed value of D_0/L_0 must be used (D_0 will be known). At t_c, the critical deficit, D_c, is computed as $(C_s – C_A)$, in which C_A is the minimum allowable dissolved oxygen at the sag point. D_c can be computed in pounds per day from the calculated flow.

At the critical point:

$$K_1 L_c = K_2 D_c$$

and:

$$L_0 = L_c e^{K_r t_c}$$

The computed L_0 should be checked against the original assumption (and recalculated if necessary). From this value the required treatment can be computed.

DEVELOPMENT OF THE BOD-OXYGEN SAG MODEL FOR ESTUARIES

An estuary is defined as that portion of a river that is under the influence of tidal action in which the dispersion factor in equation (3–1) is always significant. The advective term may or may not be significant, depending upon the rate of freshwater flow and the cross-sectional area.

Estimation of the Dispersion Coefficient, ε

For conservative substances such as chlorides or sulfates, equation (3–1) at a point source under steady-state condition becomes:

$$0 = \epsilon \frac{\partial^2 C}{\partial X^2} - U \frac{\partial C}{\partial X} \tag{3–19}$$

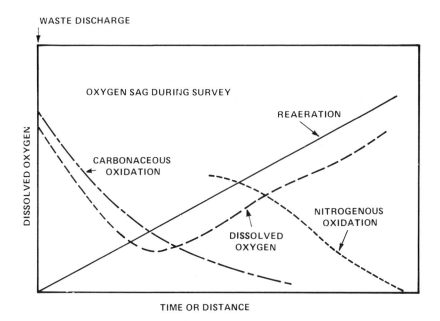

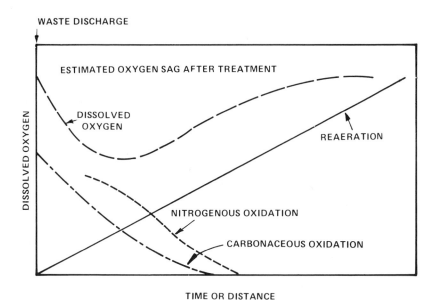

FIGURE 3.15
Oxygen Sag Before and After Treatment

which integrates to:

$$C = C_0 e^{UX/\epsilon} \tag{3-19a}$$

when $C = C_o$ at $X = O$.

Equation (3–19a) may be employed to estimate the turbulent diffusion coefficient ϵ from salinity or similar data. The velocity, U, is determined from the freshwater flow, Q, and the area, A, and the concentration, C, may be measured at various locations, X. The dispersion coefficient, ϵ, is calculated from the slope of a plot of the logarithm of the concentration against distance, X.

Estimation of the BOD Profile

For a nonconservative substance with a first-order decay, under steady-state conditions, equation (3–1) becomes:

$$0 = \epsilon \frac{\partial^2 L}{\partial X^2} - U \frac{\partial L}{\partial X} - K_r L \tag{3-20}$$

Equation (3–20) integrates to:

$$L = L_0 \exp\left[\frac{UX}{2\epsilon} (1 - m) \right] \tag{3-20a}$$

in which:

$$m = \sqrt{1 + \frac{4 K_r \epsilon}{U^2}} \tag{3-20b}$$

for the condition $L = L_0$ at $X = 0$ and $L = 0$ at $X = \pm \infty$. The concentration at $X = 0$ is obtained by taking a mass balance at this point. Equation (3–20a) may be expressed:

$$L = L_0 e^{j_1 X}$$

Estimation of the Dissolved Oxygen Profile

The steady-state distribution of dissolved oxygen is defined from equation (3–1) as:

$$0 = \epsilon \frac{\partial^2 C}{\partial X^2} - U \frac{\partial C}{\partial X} + K_2 (C_s - C) - K_1 L \tag{3-21}$$

Equation (3–21) integrates to:

$$D = F L_0 (e^{j_1 X} - e^{j_2 X}) + D_0 e^{j_2 X} \tag{3-21a}$$

where D = dissolved oxygen deficit

$$F = K_1(K_2 - K_r)$$

$$j_1 = \frac{U}{2\epsilon}\left(1 - \sqrt{1 + \frac{4K_r\epsilon}{U^2}}\right)$$

$$j_2 = \frac{U}{2\epsilon}\left(1 - \sqrt{1 + \frac{4K_2\epsilon}{U^2}}\right)$$

Equation (3–21a) is identical in form to equation (3–1e) except that it contains the dispersion coefficient ϵ. In estuaries where the freshwater flow, U, is very small and can be neglected, equation (3–20) becomes

$$0 = \epsilon\frac{\partial^2 L}{\partial X^2} - K_r L \tag{3–22}$$

which integrates to

$$L = L_0 \exp\left(-X\sqrt{K_r/\epsilon}\right) \tag{3–22a}$$

The dissolved oxygen profile is computed by modifying equation (3–21):

$$0 = \epsilon\frac{\partial^2 C}{\partial X^2} + K_2(C_s - C) - K_1 L \tag{3–23}$$

which integrates to:

$$D = FL_0\left[\exp\left(-X\sqrt{K_r/\epsilon}\right) - \exp\left(-X\sqrt{K_2/\epsilon}\right)\right] + D_0 \exp\left(-X\sqrt{K_2/\epsilon}\right) \tag{3–23a}$$

PROBLEM 3–1

Calculations of oxygen sag curve for a stream survey data were developed for a river (see figure I) and are summarized in table I. The long-term BOD values y for a sample from the stream are given in table II. The data from table II are plotted in figure II and the BOD rate constant k is found using the graphical method, where:

$$k = \frac{6b}{a} = 6\,\frac{.009}{.404} = 0.134/\text{day (base e)}$$

The ratio:

$$\frac{BOD_u}{BOD_5} = \frac{1}{(1 - e^{-(.134)5})} = 2.05$$

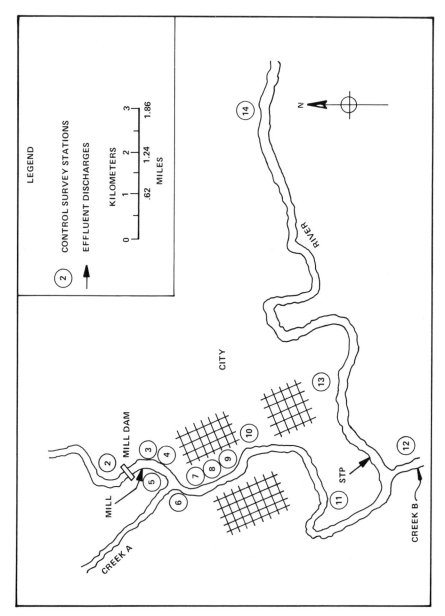

FIGURE I
Location Map

TABLE I

River Survey Data[a]

| Station | Flow Distance | | Drainage Area | | Flow Rate | | Temp. | DO | BOD$_5$ |
	Mile	(km)	Miles²	(km²)	cfs	(m³/sec)	(°C)	(mg/l)	(mg/l)
2	0.1	0.16	440	1140	111	3.11	26	6.5	2.5
Plant					40.5	1.13			
3	.2	0.32	440	1140	151	4.25	29	6.0	40.0
4	.5	0.805	440	1140	151	4.25	29	3.5	37.1
Creek A	0.6	0.96	166	430	30.4	0.85	22	8.5	0.5
6	.7	1.12	606	1570	182	5.10	28	3.7	28.9
7	1.0	1.61	606	1570	182	5.10	28	2.4	9.0
8	1.30	2.41	606	1570	182	5.10	28	1.9	28.9
9	1.8	2.9	606	1570	182	5.10	28	1.7	26.3
10	2.12	3.4	606	1570	182	5.10	28	1.4	23.9
11	3.58	5.75	606	1570	182	5.10	28	2.0	21.1
Creek B	5.1	8.2	159	410	25.4	0.71	22	8.0	0.5
13	5.8	9.3	765	1980	207	5.80	27	4.1	12.0
14	10.8	17.4	772	2000	207	5.80	27	5.7	6.3

[a] Mean U = 1 fps (30.48 cm/sec); mean H = 1 ft (30.48 cm).

TABLE II
Long-Term BOD

Days	y (mg/l)	t/y	(t/y)^{1/3}
1	14	.0714	0.414
2	26	.0770	0.424
3	38	.7900	0.430
4	46	.8700	0.443
5	56	.0893	0.446
6	62	.0967	0.459
7	65	.1076	0.475

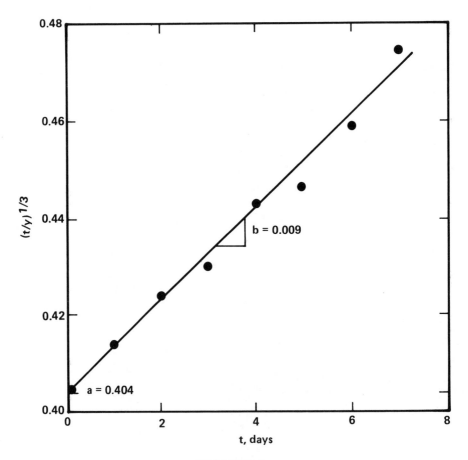

FIGURE II
Evaluation of k

Estimation of K_1

The BOD in lb/day at each station is calculated as in table III and plotted against distance in figure III. From figure III:

$$J_1 = 0.15/\text{mile}$$

but:

$$J_1 = K_1/U$$

The mean velocity $U = 1$ fps therefore $K_1 = 2.44/\text{day}$

Estimation of K_2

Assuming a functional relationship such that $K_2 = CU^n/H^m$ and using O'Connor's values from table 3.1:

$$C = 12.9 \text{ (base e)}$$

$$n = 1/2$$

$$m = 3/2$$

$$U = 1.0 \text{ ft/sec}$$

$$H = 1.0 \text{ ft.}$$

$$K_2 = \frac{12.9 \, (1.0)^{1/2}}{(1.0)^{3/2}}$$

$$K_2 = 12.9/\text{day @ 20°C}$$

at 28°C:

$$K_2 = 12.9 \, (1.028)^{(28-20)}$$

$$K_2 = 16.1/\text{day}$$

$$J_2 = K_2/U = 0.99/\text{mile}$$

Calculation of DO Profile

Assume $K_r = K_1$ and then $j_r = j_1$.
Analysis begins at Station 3.

$$J_1 = 0.15/\text{mile}$$

$$J_2 = 0.99/\text{mile}$$

$$F = \frac{J_1}{J_2 - J_1} = 0.178$$

TABLE III
Flow Rate

Station	Flow Distance Mile	(km)	Flow Rate cfs	(m³/sec)	BOD₅ (mg/l)	BOD₅ (1000 lbs/day)	BOD₅ (1000 kg/day)
2	0	0	111	3.12	2.5	1.48	0.672
3	.2	0.32	151	4.25	40.0	32.5	14.75
4	.5	0.805	151	4.25	37.1	30.0	13.6
6	.7	1.125	182	5.10	28.9	28.0	12.7
7	1.0	1.609	182	5.1	29.0	28.1	12.75
8	1.50	2.41	182	5.1	28.9	28.0	12.7
9	1.8	2.9	182	5.1	26.3	25.6	11.6
10	2.12	3.38	182	5.1	23.9	23.2	10.5
11	3.58	5.75	182	5.1	21.1	20.6	9.3
13	5.8	9.32	207	5.8	12.0	13.3	6.05
14	10.8	17.4	207	5.8	6.3	6.97	3.18

where F is the ratio of the rate of oxygen utilization to the net rate of oxygen availability. The initial BOD_5 at Station 3 = 40 mg/l; Mean T = 28°C; C_s = 7.9 mg/l; Initial DO concentration (C_0) = 6.0 mg/l; Ratio of BOD_u to BOD_5 = 2.05.

Station 3 to Creek A

L_0 = 2.05 (40) = 82 mg/l; FL_0 = 0.178 (82) = 14.6 mg/l; Initial deficit (D_0) = 7.9 − 6.0 = 1.9 mg/l.

Table IV is developed using:

$$\Sigma D = FL_0 \, \Delta + D_0 e^{-J_2 X}$$

where:

$$\Delta = e^{-J_1 X} - e^{-J_2 X}$$

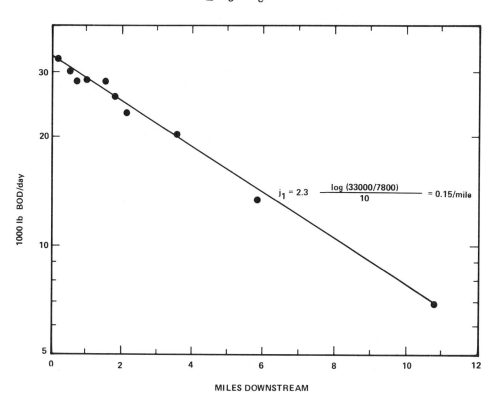

FIGURE III
Evaluation of K_1

TABLE IV

$$\Sigma D = FL_0\Delta + D_0 e^{-J_2 x}$$

Station	X (mile)	$e^{-J_1 x}$	$e^{-J_2 x}$	Δ	$FL_0\Delta$	$D_0 e^{-J_2 x}$	ΣD	D.O.
3	0							6.0
4	0.3	0.955	0.743	0.212	3.09	1.41	4.50	3.4
Creek A	0.4	0.941	0.673	0.268	3.91	1.28	5.19	2.7

Material balance at Creek A – stream confluence

From Creek A

$$Q = 30.4 \text{ cfs}$$

$$DO = 8.5 \text{ mg/l}$$

$$BOD_5 = 0.5 \text{ mg/l}$$

$$DO = \frac{151 \, (2.7) + 30.4 \, (8.5)}{182} = 3.7 \text{ mg/l}$$

BOD_5 remaining from Station 3 $= \leqslant_0 ^{-J_1 x} = 40 \, (0.941) = 37.6$ mg/l.

$$BOD_5 = \frac{151 \, (37.6) + 30.4 \, (0.5)}{182} = 31.2 \text{ mg/l}$$

Creek A to Creek B

$$L_0 = 2.05 \, (31.2) = 64 \text{ mg/l}$$

$$FL_0 = 0.178 \, (64) = 11.4 \text{ mg/l}$$

$$D_0 = 7.9 - 3.7 = 4.2 \text{ mg/l}$$

The sag-curve data are developed in table V.

Material balance at Creek B – stream confluence

From Creek B

$$Q = 25.4 \text{ cfs}$$

$$DO = 8.0 \text{ mg/l}$$

TABLE V

$$\Sigma D = FL_0 \Delta + D_0 e^{-J_2 X}$$

Station	X (mile)	$e^{-J_1 X}$	$e^{-J_2 X}$	Δ	$FL_0 \Delta$	$D_0 e^{-J_2 X}$	ΣD	DO
Creek A	0							3.7
6	0.1	0.985	0.904	0.081	0.92	3.80	4.72	3.2
7	0.4	0.941	0.673	0.268	3.06	2.82	5.88	2.0
8	0.7	0.901	0.502	0.399	4.55	2.11	6.66	1.2
9	1.2	0.835	0.304	0.531	6.05	1.28	7.33	0.6
10	1.5	0.798	0.225	0.573	6.54	0.95	7.49	0.4
11	3.0	0.638	0.051	0.587	6.70	0.21	6.91	1.0
Creek B	4.5	0.509	0.012	0.497	5.66	0.05	5.71	2.2

$$BOD_5 = 0.5 \text{ mg/l}$$

$$DO = \frac{182\,(2.2) + 25.4\,(8.0)}{207} = 2.9 \text{ mg/l}$$

$$BOD_5 \text{ from Creek A} = 31.2\,(0.509) = 15.9 \text{ mg/l}$$

$$BOD_5 = \frac{182\,(15.9) + 25.4\,(0.5)}{207} = 14.1 \text{ mg/l}$$

Creek B to Station 14

$$L_0 = 2.05\,(14.1) = 28.9 \text{ mg/l}$$

$$FL_0 = 0.178\,(28.9) = 5.14 \text{ mg/l}$$

$$D_0 = 7.9 - 2.9 = 5.0 \text{ mg/l}$$

The sag-curve data are developed in table VI and the calculated values of DO are plotted in figure IV.

TABLE VI

$$\Sigma D = FL_0 \Delta + D_0 e^{-J_2 X}$$

Station	X (mile)	$e^{-J_1 X}$	$e^{-J_2 X}$	Δ	$FL_0 \Delta$	$D_0 e^{-J_2 X}$	ΣD	DO
Creek B	0							2.9
13	5.2	0.901	0.502	0.399	2.05	2.51	4.56	3.3
14	10.8	0.425	0.004	0.421	2.16	0.02	2.18	5.7

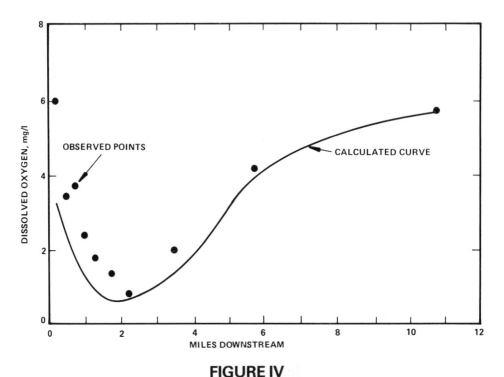

FIGURE IV

Calculated and Observed DO Profile

PROBLEM 3–2

Given the data in table I, determine the oxygen distribution in a portion of the Delaware Estuary.

TABLE I

Station	Mile Pt	Temp. (°C)	DO (mg/l)	BOD$_5$ (mg/l)	Cl (mg/l)
0	0.0	17			
1	4.0	17	3.9	7.5	780
2	9.5	17	5.2	4.0	1350
3	16.0	17	6.7	2.5	2200
4	19.5	17	7.2	1.8	3000
5	24.5	17	7.5	2.2	5000

Flow = 7450 cfs (209 cu. m/sec)
Mean cross-sectional area = 149,000 ft^2 (13,480 m^2)
Mean depth = 15.2 ft.

Estimation of Dispersion Coefficient ε

The chloride data is plotted on semilogarithmic paper as shown in figure I.

$$U = \frac{Q}{A} = \frac{7450\,(3600)\,24}{149,000\,(5280)} = 0.82 \text{ miles/day}$$

$$\frac{U}{\epsilon} = -\frac{1}{X}\,\ln\frac{C_0}{C} = -\frac{1}{20}\,\ln\left(\frac{880}{5200}\right) = 0.09$$

$$\epsilon = \frac{U}{0.09} = \frac{0.82}{0.09} = 9.2 \text{ sq. mile/day}$$

Estimation of Reaeration Coefficient

$$K_2 = \frac{12.9\,V_0^{1/2}}{H^{3/2}}$$

V_0 = Mean average tidal velocity in estuary case

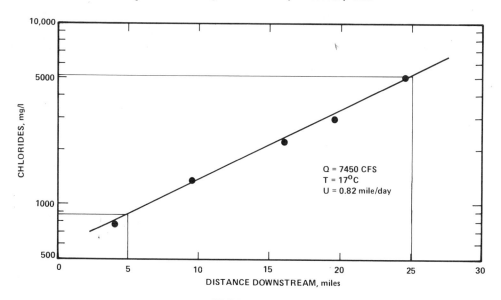

FIGURE I

Evaluation of Dispersion Coefficient

For Delaware River:

Mean Depth = 15.2 ft.

(V_o) Mean tidal velocity = 1.0 fps

General Procedure for V_o

Refer to figure II.

Mean average flood velocity = 0.70 V_F (max)

Mean average ebb velocity = 0.70 V_E (max)

If the flood cycle is equal in duration to ebb cycle, then mean average tidal velocity over complete cycle is:

$$V_0 = \frac{0.70\ V_F\ (\text{max}) + 0.70\ V_E\ (\text{max})}{2}$$

or:

$$V_0 \simeq 0.35\ (V_F\ (\text{max}) + V_E\ (\text{max}))$$

Tidal velocity information is available from *Tidal Current Tables* published yearly by USCGS.

Calculation:

$$K_2 = \frac{12.9\ (1.0)^{1/2}}{(15.2)^{3/2}} = 0.22/\text{day}$$

Estimation of deoxygenation rate coefficient

The deoxygenation rate coefficient j_1 is computed as the slope of a semilogarithmic plot of BOD_5 against distance downstream as shown in figure III.

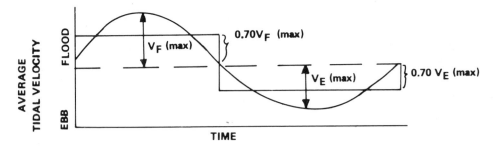

FIGURE II

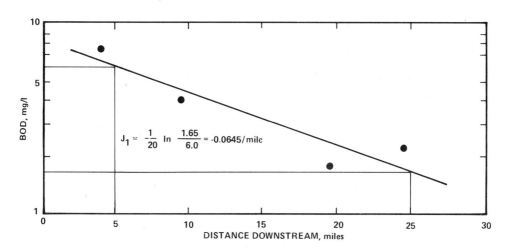

FIGURE III

Evaluation of J_1

Computation of Dissolved Oxygen Profile

Parameters (analysis begins at mile point 4.0) are:

Flow = 7450 cfs, T = 17° C

Cross–sectional area = 149,000 ft^2

U = 0.82 miles per day

Turbulent diffusion coefficient (ε) = 9.2 sq. mile/day

Evaluated from chloride plot (figure I)

Velocity and ε assumed constant.

J_1 = −0.0645/mile at 17° C (−0.0403/km)

(From figure III)

Assume $K_r = K_1$.

Calculation of K_1

$$K_1 = \frac{U^2}{4\epsilon} \left[\left(1 - \frac{2\epsilon J_1}{U}\right)^2 - 1 \right]$$

$$= \frac{(0.82)^2}{4(9.2)} \left\{ \left[1 - \frac{2\,(9.2)\,(-0.0645)}{0.82} \right]^2 - 1 \right\} = 0.09/\text{day at } 17° \text{ C}$$

Calculation of J_2

K_2 = 0.22/day @ 20° C

K_2 = 0.22 $(1.047)^{17-20}$ = 0.19/day @ 17° C

$$J_2 = \frac{0.82}{2\,(9.2)} \left[1 - \sqrt{1 + \frac{4\,(0.19)\,(9.2)}{(0.82)^2}} \right] = -0.1060/\text{mile}$$

Initial BOD_5 @ (MP = 4.0) = 6.3 mg/l (From figure III)
Ratio of ultimate to five-day BOD = 1.30 (From long-term BOD analysis)

L_0 = 1.30 (6.3) = 8.2 mg/l
At T = 17° C, C_s = 9.4 mg/l
Initial DO concentration (C_0) = 3.9 mg/l
D_0 = 9.4 − 3.9 = 5.5 mg/l

$$F = \frac{K_1}{K_2 - K_1} = \frac{0.09}{0.19 - 0.09} = 0.90$$

FL_0 = 0.90 (8.2) = 7.38

Oxygen Balance Calculation

$$D = FL_0(e^{J_1X} - e^{J_2X}) + D_0 e^{J_2X}$$

The results are shown in table II.

TABLE II

X	e^{J_1X}	e^{J_2X}	Δ	$FL_0Δ$	$D_0 e^{J_2X}$	D	DO^b
0^a						5.5	3.9
2	0.879	0.809	0.070	0.52	4.45	4.97	4.4
4	0.773	0.654	0.119	0.88	3.60	4.48	4.9
6	0.679	0.530	0.149	1.10	2.92	4.02	5.4
8	0.597	0.428	0.169	1.25	2.35	3.60	5.8
10	0.524	0.347	0.177	1.31	1.91	3.22	6.2
15	0.380	0.204	0.176	1.30	1.12	2.42	7.0
20	0.275	0.120	0.155	1.14	0.66	1.80	7.6

^{a}X = 0 at Mile Pt = 4.0.
^{b}The calculated DO profile is shown in figure IV.

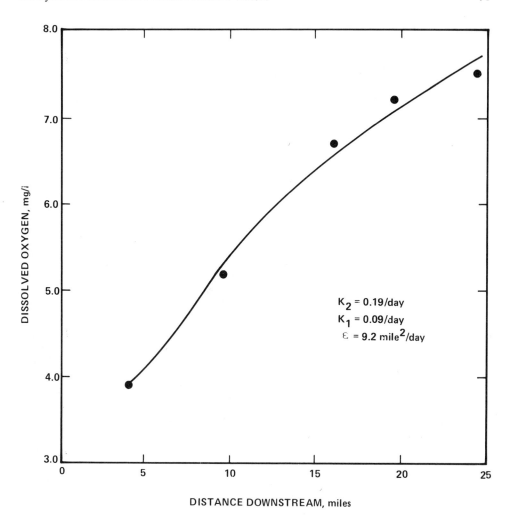

FIGURE IV

DO Profile of Delaware River

REFERENCES

1. O'Connor, D. J. and Dobbins, W. E. 1958. Mechanism of reaeration in natural streams, *Trans. ASCE* 123: 641.
2. Churchill, M. A. *et al.* 1962. The prediction of stream reaeration rates. *J. Sanit. Eng. Div., ASCE,* 88, SA4, 1.
3. Langbein, W. B. and Durum, W. H. 1967. The aeration capacity of streams. *U.S. Geological Survey Circular No. 542* Washington, D.C.
4. Owens, M. *et al.* 1964. Some reaeration studies in streams. *Intl. J. Air Water Poll.* 8: 469.
5. Bansal, M. K. 1973. Atmospheric reaeration in natural streams. *Water Research* 7: 769.
6. Isaacs, W. P. *et al.* 1969. An experimental study of the effect of channel surface roughness on the reaeration rate coefficient. *Proc. 24th Ind. Waste Conf.* Lafayette, Ind.:1 Purdue University.
7. Kothandaraman, V. and Ewing, B. B. 1969. Probabilistic analysis of dissolved oxygen-biochemical oxygen demand relationship in streams. *J. Water Pollution Control Federation* 41, no. 2, R73.
8. Negulescu, M. and Rojanski, V. 1969. Recent research to determine the reaeration coefficient. *Water Research* 3: 189.
9. Thackston, E. L. and Krenkel, P. A. 1969. Reaeration prediction in natural streams. *J. Sanit. Eng. Div., ASCE,* 95, SA1, 65.
10. Kothandaraman, V. 1970. Effects of contaminants on reaeration rates in river water. *Proc. 25th Ind. Waste Conf.,* Lafayette, Ind.: Purdue University.
11. Wilson, G. T. and MacLeod, N. 1974. A critical appraisal of empirical equations and models for the prediction of the coefficient of reaeration of deoxygenated water. *Water Research* 8: 341.

12. Tsivoglou, E. C. and Neal, L. A. 1976. Tracer measurement of reaeration: III. Predicting the reaeration capacity of inland streams. *J. Water Pollution Control Federation* 48: 2669.

13. Edeline, F. and Lambert, G. 1974. A simple simulation method for river self-purification studies. *Water Research* 8: 297.

14. Novotny, V. and Krenkel, P. A. 1975. A waste assimilative capacity model for a shallow, turbulent stream. *Water Research* 9: 233.

15. Bosko, K. 1967. Discussion in *Advances in Water Pollution Research*. Washington, D.C.: Water Pollution Control Federation.

16. Novotny, V. 1969. Boundary layer effect on the course of the self-purification of small streams. *Advances in Water Pollution Research*. Oxford: Pergamon Press.

17. Downing, A. L. *et al.* 1964. Nitrification in the activated sludge process. *J. Proc. Inst. Sewage Purification* 130.

18. Edberg, N. and Hofsten, B. V. 1973. Oxygen uptake of bottom sediments studied in situ and in the laboratory. *Water Research* 7: 1285.

19. Oldaker, W. H. *et al.* 1966. *Report on Pollution of the Merrimack River and Certain Tributaries, Part IV – Pilot Plant Study of Benthic Oxygen Demand,* Lawrence, Mass: U.S. Dept. of the Interior, FWPCA, Northeast Region.

20. Edwards, R. W. and Owens, M. 1965. The oxygen balance of streams. *Ecology and the Industrial Society,* Fifth Symposium of the British Ecological Society, 149. Oxford: Blackwell.

21. Camp, T. R. 1963. *Water and its impurities,* 2nd ed. New York: Reinhold Publishing Co.

22. Thomann, R. V. 1972. *Systems Analysis and Water Quality Management.* New York: McGraw-Hill.

23. Nemerow, N. L. 1974. *Scientific Stream Pollution Analysis.* Washington, D.C.: Scripta Book Co.

24. Department of Scientific and Industrial Research. 1963. *Water Pollution Research.*

25. O'Connor, D. J. and DiToro, D. M. 1967. An analysis of the dissolved oxygen variation in a flowing stream. *Advances in Water Quality Improvement.* Austin: University of Texas Press.

26. Kittrell, F. W. and Kochtitzky, O. W. Jr. 1947. Natural purification characteristics of a shallow turbulent stream. *Sewage Works J.* 19:1 1032.

27. Biguria, G. *et al.* 1969. Distributed parameter model of thermal effects in rivers. *Chemical Engineering Progress Symposium Series 65* no. 97: 86.

28. Leopold, L. B. and Maddock, T. 1953. The hydraulic geometry of stream channels and some physiographic implications. *U.S. Geological Survey Professional Paper 252.* Washington, D.C.

29. Dobbins, W. E. 1964. BOD and oxygen relationships in streams. *J. Sanit. Eng. Div., ASCE,* 90, SA3, 53.

30. O'Connor, D. J. and DiToro, D. M. 1970. Photosynthesis and oxygen balance in streams. *J. Sanit. Eng. Div., ASCE,* 96, SA2, 547.

Characteristics of Municipal Wastewaters

Municipal sewage is composed organic matter present as soluble, colloidal, and suspended solids. The pollutional contributories in sewage are usually expressed as a per capita contribution. A study of data reported by 73 cities in 27 states in the United States [1] during the period 1958–1964 showed a sewage flow of 135 gallon/capita/day (511 liter/capita/day) and a BOD_5 and suspended solids content of 0.20 pound/capita/day (90.7 gram/capita/day) and 0.23 pound/capita/day (104 gram/capita/day), respectively. The average composition of municipal sewage is shown in table 4.1. It should be recognized that the presence of industrial wastes

TABLE 4.1

Average Characteristics of Municipal Sewage[a]

Characteristics	Maximum	Mean	Minimum
pH	7.5	7.2	6.8
Settleable solids, mg/l	6.1	3.3	1.8
Total solids, mg/l	640	453	322
Volatile total solids, mg/l	388	217	118
Suspended solids, mg/l	258	145	83
Volatile suspended solids, mg/l	208	120	62
Chemical oxygen demand, mg/l	436	288	159
Biochemical oxygen demand, mg/l	276	147	75
Chlorides, mg/l	45	35	25

[a]After Hunter and Heukelekian [3].

in the municipal system may radically alter these concentrations. In addition, these concentrations may be expected to vary by about a factor of 3 over a 24-hour period. The chemical characteristics of sewage are summarized in table 4.2.

TABLE 4.2
Chemical Characteristics of Municipal Sewage[a]

Constituent	Type	Concentration
Volatile acids	Formic, acetic, propionic, butyric, and valeric	8.5–20 mg/l
Nonvolatile soluble acids	Glutaric, glycolic, lactic, citric, benzoic, and phenyllactic	Any acid 0.1–1.0 mg/l
Higher fatty acids	Palmitic, stearic, and oleic	⅔ fatty acid content
Proteins and amino acids	At least twenty types	45–50% of the total nitrogen
Carbohydrates	Glucose, sucruse, lactose, some galactose, and fructose	—

[a]After Walter [2].

The general composition of municipal sewage has been reported by Hunter and Heukelekian [3] and is summarized in table 4.3. An average pollution loading and sewage volume is shown in table 4.4. Sewage volumes and BOD from various services are summarized in table 4.5. The distribution of organic parameter contents of sewage is summarized in table 4.6.

TABLE 4.3
Municipal Sewage Composition[a]

Fraction, %	Total Solids	Volatile Solids	Organic Matter[b]	Nitrogenous Matter[c]	COD
Settleable	18	28	30	23	34
Supracolloidal	11	22	19	34	27
Colloidal	7	11	10	11	14
Soluble	64	37	41	22	25

[a]From Hunter and Heukelekian [3].
[b]Volatile solids or organic carbon.
[c]Organic nitrogen data.

TABLE 4.3 (continued)
Municipal Sewage Composition[a]

Strength Parameter	Particulates, %	Solubles, %
Total solids	34.7	65.3
Volatile solids	57.6	42.4
COD	77.3	22.7
Organic nitrogen	80.5	19.5

TABLE 4.4
Average Pollutional Loading and Wastewater Volume from Domestic Household (Four Members)[a]

Wastewater event (1)	Number per day (2)	Water volume per use, in gallons (3)	Total water use, in gallons (4)	BOD$_5$, in pounds per day (5)	Suspended solids, in pounds per day (6)
Toilet	16	5	80	0.208	0.272
Bath/Shower	2	25	50	0.078	0.050
Laundry	1	40	40	0.085	0.065
Dishwashing	2	7	14	0.052	0.026
Garbage disposal	3	2	6	0.272	0.384
Total			190	0.695	0.797

[a]From Ligman *et al.* [5].

TABLE 4.5
Sewage Volume and BOD for Various Services[a]

Type	Volume		5-day BOD	
	gal/capita/day	liter/capita/day	lb/capita/day	g/capita/day
Airports				
Each employee	15	56.8	0.05	22.7
Each passenger	5	18.9	0.02	9
Bars				
Each employee	15	56.8	0.05	22.7
Plus each customer	2	7.6	0.01	4.5
Camps and resorts				
Luxury resorts	100	378	0.17	77.2
Summer camps	50	189	0.15	68.1
Construction camps	50	189	0.15	68.1
Domestic sewage				
Luxury homes	100	378	0.20	90.7
Better subdivisions	90	340	0.20	90.7
Average subdivisions	80	302	0.17	77.2
Low-cost housing	70	265	0.17	77.2
Summer cottages, etc.	50	189	0.17	77.2
Apartment houses	75	284	0.17	77.2
(*Note*: If garbage grinders installed, multiply BOD factors by 1.5.)				
Factories (exclusive of industrial and cafeteria wastes)	15	56.8	0.05	22.7
Hospitals				
Patients plus staff	150–300 (avg 200)	578–1136 (avg 757)	0.30	136

[a]After Goodman and Foster [4].

Type	gal/capita/day	liters/capita/day	g 5-day BOD/capita/day	lb 5-day BOD/capita/day
Hotels, motels, trailer courts, boarding houses (not including restaurants or bars)	50	189	68.1	0.15
Milk plant wastes	100–225 gal/1000 lb of milk	835–1880 l/1000 kg of milk	560–1660 g/1000 kg of milk	0.56 to 1.66/1000 lb of milk
Offices	15	56.8	22.7	0.05
Restaurants				
Each employee	15	56.8	27.2	0.06
Plus each meal served	3 (per meal)	11.4 (per meal)	13.6 (per meal)	0.03 (per meal)
If garbage grinder provided, add	1 (per meal)	3.8 (per meal)	13.6 (per meal)	0.03 (per meal)
Schools	Elem.　High	Elem.　High	Elem.　High	Elem.　High
Day schools				
Each person, student or staff	15　20	56.8　76	18　22.7	0.04　0.05
Add per person if cafeteria has garbage grinder	—	—	4.5　4.5	0.01　0.01
Boarding schools	75	284	77.2	0.17
Swimming pools				
Employees, plus customers	10	38	13.6	0.03
Theatres				
Drive-in theatre per stall	5	19	9	0.02
Movie theatre per seat	5	19	9	0.02

TABLE 4.5 (continued)

TABLE 4.6
Distribution of Organic Strength Parameter Contents of Sewage[a]

Fraction	Total Solids (mg/l)	Total Solids (%)	Volatile Solids (mg/l)	Volatile Solids (%)	TOC (mg/l)	TOC (%)	COD (mg/l)	COD (%)
Settleable	74	15	59	25	29	27	120	29
Supracolloidal	57	11	43	18	22	20	87	21
Colloidal	31	6	23	9	12	11	43	10
Soluble	351	68	116	48	46	42	168	40
Total	513	–	241	–	109	–	418	–

[a]After Rickert and Hunter [6].

REFERENCES

1. Loehr, R. C. 1968. Variation of wastewater parameters, *Public Works*, 99, 81.
2. Walter, L. 1961. Composition of sewage and sewage effluents, Parts 1 and 2, *Water Sewage Works,* 108, Nos. 11 and 12, 428 & 478.
3. Hunter, J. V. and Heukelekian, H. 1965. The composition of domestic sewage fractions, *J. Water Pollution Control Federation*, 37, 1142.
4. Goodman, B. and Foster, J. W. 1969. *Notes on Activated Sludge,* 2nd ed. Lenexa, Kans.: Smith and Loveless Co.
5. Ligman, K. *et al.* 1974. Household wastewater characterization, *J. Environmental Engineering Div.,* ASCE, 100, 201.
6. Rickert, D. A. and Hunter, J. V. 1971. General nature of soluble and particulate organics in sewage and secondary effluent, *Water Research,* 5, 421.

Industrial Wastewaters

THE INDUSTRIAL WASTE SURVEY

One of the first considerations in developing a water quality management program for an industry is to define the methodology and the procedures involved in developing and conducting an industrial waste survey. Considering the industrial plant, wastewater can be attributed to four sources: process waste waters, sanitary sewage, storm runoff and tank cleanings, and other various noncontinuous miscellaneous sources of waste.

Depending on the location of the plant, the sanitary sewage can either be treated with the industrial waste, treated independently of the industrial waste in a separate system, or it can be discharged into a municipal sewer. One problem in treating the sanitary wastewater with the industrial wastewater is the possible requirement for chlorination or disinfection of the total plant effluent. In these cases, depending on sewer segregation costs, it is better to handle the sanitary wastewater and the industrial wastewater separately.

Considering the process wastewaters in the industrial waste survey, the first step is to review the plant process makeup and the wastewater sources. The objective is to understand the industrial process, plant layout, raw material inputs, chemicals and materials input, and products and wastes output. The second step is to define the raw materials makeup and the wastewater characteristics. At this point we establish what parameters should be analyzed. In many cases, this is obvious from a knowledge of the industry, such as a food processing plant. Other industries, such as chemicals and petrochemicals, require consideration of a variety of pollutional parameters that will depend on the chemical inputs, the raw material inputs, and the product outputs. Table 5.1 summarizes typical parameters by industrial category.

TABLE 5.1
Significant Wastewater Parameters for Selected Industrial Classifications

Parameter	Aluminum	Automobile	Beet Sugar	Beverage	Fruits & Vegs.	Livestock Feeding	Dairy	Nitrogen Fertilizer	Phosphate Fertilizer	Flat Glass	Cement/Concrete	Grain Milling	Leather	Meat Products	Metal Finishing	Petroleum Refining	Plastics & Synthetics	Pulp & Paper	Steam Power	Steel	Textiles
Color		X	X	X	X		X						X	X		X		X			X
Suspended Solids	X	X	X	X	X		X	X	X	X	X	X	X	X	X	X	X	X	X	X	X
Oil and Grease	X	X	X	X									X	X	X	X	X	X		X	X
BOD₅		X	X	X	X	X	X			X		X	X	X		X	X	X	X		X
Ammonia Nitrogen		X	X		X	X	X	X	X		X		X	X		X				X	
Phosphorus	X	X		X	X	X	X	X	X	X	X			X		X	X	X			X
Chromium		X								X			X		X	X			X	X	
Cyanide								X							X	X				X	
Copper		X								X					X	X	X		X		
Nickel									X		X				X						
Iron	X	X			X	X	X	X		X	X	X			X	X	X		X	X	
Zinc	X	X								X		X	X		X	X	X		X	X	X
Phenols	X	X				X	X				X		X			X	X			X	X
COD	X									X					X	X					
Chlorides		X						X	X							X				X	
Nitrates								X	X	X							X				
Sulfate		X								X						X	X			X	
Tin		X								X										X	

Parameter	Aluminum	Automobile	Beet Sugar	Beverage	Fruits & Vegs.	Livestock Feeding	Dairy	Nitrogen Fertilizer	Phosphate Fertilizer	Flat Glass	Cement/Concrete	Grain Milling	Leather	Meat Products	Metal Finishing	Petroleum Refining	Plastics & Synthetics	Pulp & Paper	Steam Power	Steel	Textiles
Lead	X	X													X	X					
Cadmium		X													X						
Total Dissolved Solids	X	X	X	X	X	X		X			X	X	X	X	X	X	X	X	X		X
Alkalinity			X								X	X	X								X
Temperature			X	X	X		X				X	X	X			X			X	X	X
Toxic Organics											X	X	X				X		X	X	X
Free Chlorine	X																		X		
Fluoride	X								X												
pH	X	X	X	X	X	X	X				X	X	X	X		X	X	X	X	X	X
Aluminum	X	X											X			X	X		X	X	
Total Coliforms			X	X	X	X									X			X			

TABLE 5.1 (continued)
Significant Wastewater Parameters for Selected
Industrial Classifications

It is important to develop a sewer map. It is surprising how frequently a sewer map is not available, particularly when a plant was built in segments over a long period of time. A tracer study is often required to determine the fate of various wastewater streams. Sampling and measurement stations must be located on the sewer map. An example of a tomato processing plant is shown in figure 5.1.

At the completion of the survey the characteristics and the volume of all the major wastewater discharges considering operating units, operating processes, and operating areas should be defined. The sampling stations should be selected to achieve this objective. When certain process areas are inaccessible, sampling stations should be selected that permit estimation of the wastewater by difference.

A total effluent measurement device should be installed that may be a weir or a Parshall flume. Figure 5.2 illustrates a typical effluent measurement structure. In cases with buried sewer lines other procedures may be necessary.

The analyses to be run are selected, which refers back to step 2 and table 5.1. The types of samples and sampling procedures must be considered if the data is collected for the purpose of discharging to a joint wastewater treatment plant. To provide a basis to define treatment criteria for the plant or to develop effluent rate charges, the samples will usually be 24-hour flow weighted composites. If the data is to be used to design a wastewater treatment plant, then it is necessary to establish a sampling and compositing schedule to reflect the design needs for each treatment operation. Table 5.2 summarizes a suggested sampling schedule for a biological treatment process for the major parameters. Depending on the industrial category (table 5.1) other parameters may also be required.

A reasonably constant output of wastewater from the industrial plant would fall into the category of low variability, while numerous batch processes discharged at different times or clean-up at differing times showing large load changes would be considered of high variability.

In considering BOD, high variability may consist of 4-hour composites and low variability, 12-hour composites. The number of BODs run should depend on what kind of a correlation can be obtained between BOD and COD or TOC. Whether COD, TOC, or BOD is employed depends on the availability of instruments. It is initially desirable to determine whether a correlation between BOD and COD or TOC exists. If so, only 24-hour composite BODs need be run. A useful test is the short-term COD test reported by Jeris. The compositing periods are picked relative to the response in the wastewater treatment process to be designed. For example, variations in BOD in an activated sludge plant are significant because these influence the oxygen input requirements and the effluent quality, which in turn determines pretreatment requirements such as equalization. Suspended solids tend to equalize through the process and longer-term composites are adequate. Alkalinity/acidity and pH involve the potential installation of neutralization, and short-term impacts are required in order to design effective

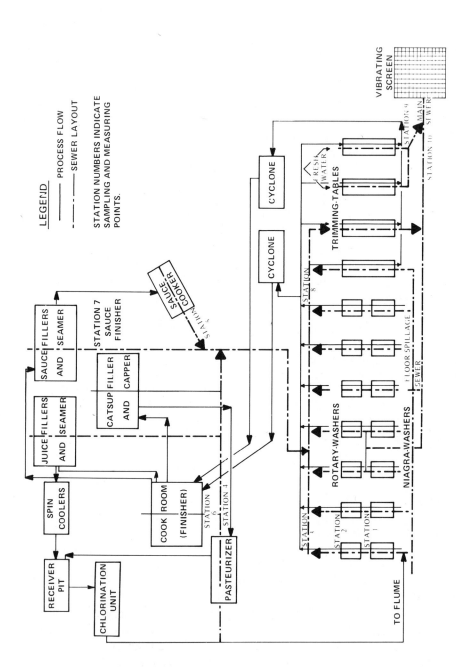

FIGURE 5.1
Flow Diagram of a Tomato Processing Plant

FIGURE 5.2
Typical Effluent Measurement Structure

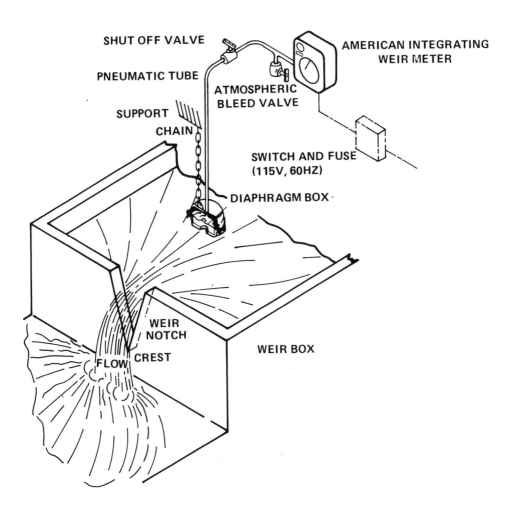

neutralization facilities. Nitrogen and phosphorous, when serving as nutrients, have a reserve capacity in the system and need only be determined on long-term composites. If biological treatment facilities are to be constructed in northern climates, wastewater temperature is an important consideration. Heavy metals and toxic organics, if present, should be regularly monitored.

TABLE 5.2
High Frequency Sampling
(Compositing Schedule)

Parameter	High Variability (hours)	Low Variability (hours)
TOC or COD	2	8
BOD	4	8
Suspended Solids	2	8
Total Dissolved Solids	4	8
Alkalinity or Acidity	4	8
Heavy Metals	4	24
Chlorine	1	4
pH	Continuous	4
Total Nitrogen (depending on plant)	24	24
Total Phosphorus (depending on plant)	24	24
Temperature	Continuous	8
Dissolved Oxygen	Continuous	8

A few of the more common wastewater samplers are shown in figures 5.3 and 5.4. One sampler operates either from a pump or vacuum (figure 5.3) and discharges to a series of sample bottles on a rotating table. A given volume of sample is discharged into a bottle and the table rotates one slot for the next sample. Usually samples are collected each hour for 24 hours. The sample can be adjusted to discharge a sample every 5 minutes in order to get a short-term picture of waste load variability. This device does not flow weight the composite. In order to composite a sample, e.g., 4 hour, 8 hour, etc., it is necessary to obtain the flow record and weight the samples for the composite according to the flow.

The second type of sampler, the Trebler Sampler (figure 5.4), depends upon flow variation over a weir. A curved scoop rotates and collects a quantity of sample proportional to the flow over the weir.

A pump with a timer is probably the most common type of sampler. The pump discharges a given volume into a vessel, for example 100 milliliters every 10 minutes. This does not give a weighted sample.

A sample can be collected off the totalizer of a flow meter. The flow meter records and totalizes flow. An adjustment can be made such that for every 350 gallons that passes through the flow meter, a given volume of sample is taken. The composite is therefore weighted according to the flow. Other samplers are discussed in detail in reference [1].

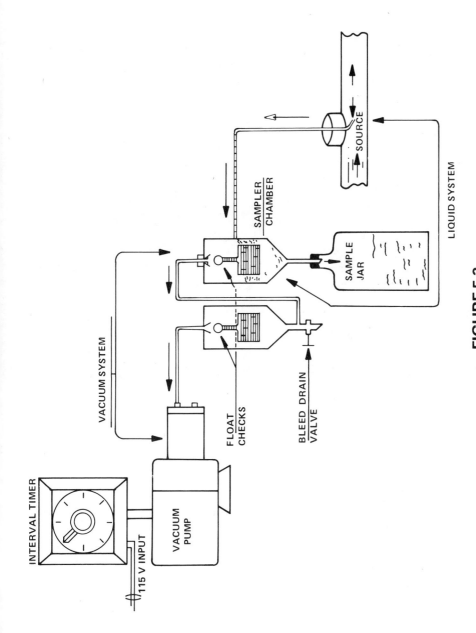

FIGURE 5.3

"CVE" Sampler System Schematic

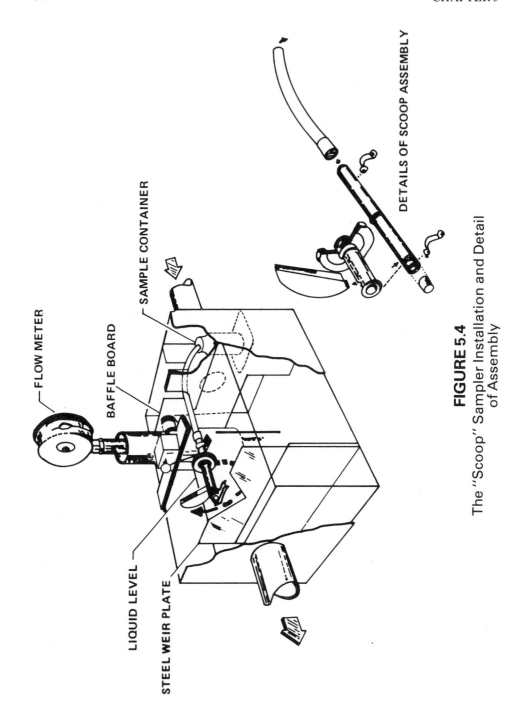

FIGURE 5.4

The "Scoop" Sampler Installation and Detail
of Assembly

An important consideration with industrial wastes as opposed to municipal sewage is that frequently the wastewaters contain greases, oils, and various types of suspended materials. It is important when a sampling program is established that a representative sample is obtained.

The duration of the sampling program depends in part on what the data will be used for. If its purpose is to generate data for a wastewater treatment plant design, the program must be continued to encompass all of the operating ranges and operating variables within the industrial plant. For example, in the paper board industry where repetitive production schedules usually exist, several days' survey will usually provide sufficient data. By contrast, in a multiproduct chemical plant, longer sampling periods would be required and in some cases additional data should be obtained for a change of campaign or products. In any case, the duration of the industrial waste survey is a judgment made on a knowledge of the processing operations. Special cases are those industries, such as food processing, which involve seasonal operation on different products. For example, one canning operation processes peaches for six weeks, tomatoes for eight weeks, and baby food the rest of the year. In this case, the peach processing provides the greatest organic load, the tomatoes the greatest hydraulic load, and the baby food a very low organic and hydraulic load. Careful consideration must be made to develop a wastewater management program and treatment system to effectively encompass all these operating conditions.

There are various methods for measuring flow in the industrial waste survey. In most cases a weir is the simplest method available for measurement of the total plant outfall. A number of industrial operations use gutters instead of pipes to transport wastewater from various parts of the plant to the main sewer. Frequently it is possible to place a small weir in the gutter and get an estimate of the flow rate. A bucket and a stopwatch can be employed for small wastewater flows. A pumping duration and rate determination is particularly applicable for wastewater flows that are inaccessible but pumped either continuously or intermittently. In a sewer a floating object such as an orange can be timed between two points and the flow can be calculated from the continuity equation. The mean velocity of flow will be 0.8 times the velocity of the orange floating on the surface. Plant water use records may give a good estimate of wastewater flow taking into account water losses in the industrial process. A change of level in a tank or reactor may be employed when dealing with batch processes, batch dumps, and batch discharges.

A flow and material balance diagram is developed from the industrial waste survey data (table 5.3). The same diagram for a detergent and cosmetic product manufacturing plant is shown in figure 5.5. The flow and material balance diagram is intended to serve two purposes. First, it shows where the major and minor sources of pollution exist. Second, and perhaps more important, it indicates where recirculation segregation and separate treatment might be considered.

TABLE 5.3
Industrial Waste Survey Data

Waste source	Sampling station	COD, mg/liter	BOD, mg/liter	SS, mg/liter	ABS, mg/liter	Flow, gpm
Liquid soap	D	1,100	565	195	28	300.0
Toilet articles	E	2,680	1,540	810	69	50.0
Soap production	R	29	16	39	2	30.0
ABS production	S	1,440	380	309	600	110.0
Powerhouse	P	66	10	50	0	550.0
Condenser	C	59	21	24	0	1,100.0
Spent caustic	B	30,000	10,000	563	5	2.0
Tank bottoms	A	120,000	150,000	426	20	1.5
Fly ash	F			6,750		10.0
Main sewer		450	260	120	37	2,150.0

From these considerations a new raw waste load is computed for the treatment plant design. In most cases waste load reduction at the source is favored over end-of-pipe treatment. Waste load reduction in this case does not refer to major process changes involving high capital dollars, but rather to segregation or recycling, which reduce the raw waste load with minimum capital expenditure.

Consider the plant shown in figure 5.5(a), which summarizes the waste survey data. The options to redefine the raw waste load for wastewater treatment removes F, which is the fly ash sewer (a noncontaminated stream), high in suspended solids to be discharged directly into a fly ash lagoon. Noncontaminated streams P, which is the powerhouse, and C, which is a condenser, may also be removed from the sewer. Since these two sources have virtually no pollutants present, they can be discharged directly to the river (see figure 5.5b). There is one other option that might be considered. Sources A and B, the spent caustic and the tank bottoms, are both extremely low in volume with extremely high concentrations of BOD and COD. One alternative is to consider a wastewater treatment scheme that would remove these sources and treat them separately. Economics would dictate the feasibility of this option. A possible alternative in this case would be to batch treat the small volumes to a BOD in the order of 300 mg/l and then discharge that effluent with the other streams to general treatment in a smaller aerated lagoon or activated sludge plant. An important consideration is that if biological processes are to be considered then streams that might be toxic should be removed for separate treatment.

Another example of redefining a raw waste load from the industrial waste survey is shown for a corn processing operation in table 5.4 and figure 5.6.

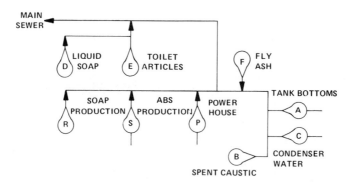

| | LINE | | | | | | | | | MAIN |
	D	E	R	S	P	F	B	C	A	SEWER
COD, lb/day	3950	1600	10	1900	440		720	780	2150	11550
BOD, lb/day	2030	920	6	500	65		240	280	2700	6740
SS, lb/day	700	485	14	410	330	800	14	320	8	3080
ABS, lb/day	100	41		800						941
Flow, gpm	300	50	30	110	550	10	2	1100	1.5	2150

(a) ORIGINAL WASTEWATER FLOW DIAGRAM

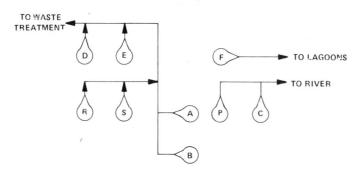

| | LINE | | | | | | TOTAL |
	D	E	R	S	A	B	
COD, lb/day	3950	1600	10	1900	2150	720	10330
BOD, lb/day	2030	920	6	500	2700	240	6396
SS, lb/day	700	485	14	410	8	14	1631
ABS, lb/day	100	41		800			941
Flow, gpm	300	50	30	110	1.5	2	493.5

(b) MODIFIED WASTEWATER FLOW DIAGRAM

FIGURE 5.5

Wastewater Flow Diagram in a Detergent and Cosmetic Product Manufacturing Plant

Unit	Line	Flow, gpm	Water use	Possible change	Estimated new flow, gpm
Rotary ear	1	21.7	Loosens and removes sand, etc.	Screen 7 and reuse in washer	21.7
Cutting machines	2	27.0	Flumes kernels to flotation washer	Screen and re-circulate	0.0
Scavenger reel	3	10.4	Flumes waste separated by flotation washer	Use smaller flow or remove as solid	5.2
Wash reel	4	18.0	Rewashes corn (not necessary)	Remove	8.0
Blancher	5	4.5	Overflow from makeup waste	No change	4.5
Cooling reel	6	24.5	Cool corn after blanching	Use smaller flow	10.0
Holding tank	7	16.9	Overflow from makeup water	Screen and reuse in rotary washer	0.0
Shaker screen	8	2.1	Flumes waste separated by reel	Remove as solid or reuse screened	0.0
Total		125.1			49.4

TABLE 5.4

Unit Waste Flows for Corn Processing Operation and Possible Changes to Reduce Flow

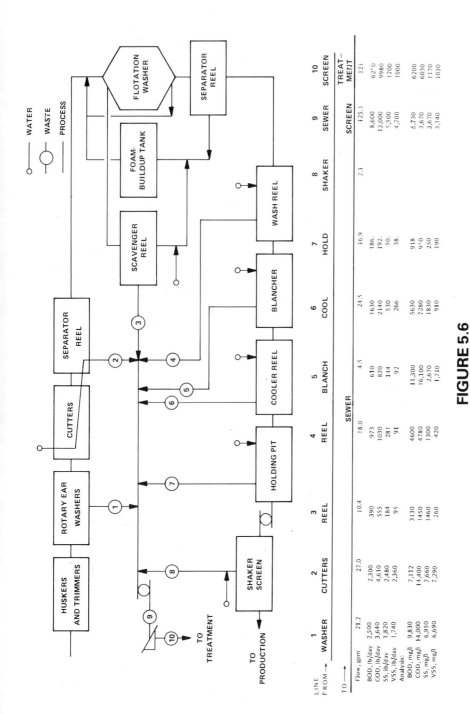

LINE FROM → TO →	1 WASHER	2 CUTTERS	3 REEL	4 REEL	5 BLANCH	6 COOL	7 HOLD	8 SHAKER	9 SEWER SCREEN	10 TREATMENT SCREEN
				SEWER						
Flow, gpm	21.7	27.0	10.4	18.0	4.5	24.5	16.9	2.1	125.1	121
BOD, lb/day	2,500	2,300	390	973	610	1630	186.		8,600	6270
COD, lb/day	3,640	4,610	555	1030	870	2140	192.		13,000	9980
SS, lb/day	1,820	2,480	184	281	144	530	50.		5,500	1700
VSS, lb/day	1,740	2,360	95	91	92	266	38.		4,700	1900
Analysis:										
BOD, mg/l	9,830	7,112	3130	4600	11,300	5630	918		6,730	6200
COD, mg/l	14,000	14,400	1450	4780	16,100	7280	9?0		3,670	6030
SS, mg/l	6,950	7,660	1460	1300	2,670	1830	250		3,670	1170
VSS, mg/l	6,690	7,290	760	420	1,710	910	190		3,140	1030

FIGURE 5.6

Waste-Flow Diagram and Material Balance at
a Corn Plant

The variability in waste load should be statistically defined to provide a basis for the design of the wastewater treatment units. Depending on the nature of the industrial operation a normal or skewed distribution may be obtained.

PROBLEM 5–1

For small amounts of industrial waste survey data (that is, less than 20 datum points), the statistical correlation procedure is as follows:

1. Arrange the data in increasing order of magnitude (first column of table I).
2. In the second column of table I m is the assigned serial number from 1 to n where n is the total number of values.
3. The plotting position is determined by dividing the total number of samples into 100 and assigning the first value as one half this number (third column of table I).

$$\text{Plotting position} = \frac{100}{n} + \text{previous probability}$$

For $m = 5$:

$$\text{Plotting position} = \frac{100}{9} + 38.85$$

$$= 49.95$$

TABLE I
Statistical Correlation of BOD Data

BOD mg/l	m	Plotting Position
200	1	5.55
225	2	16.65
260	3	27.75
315	4	38.85
350	5	49.95
365	6	61.05
430	7	72.15
460	8	83.75
490	9	94.35

4. These data are illustrated in figure I. The standard deviation of these data is calculated by:

FIGURE I
Probability Plot of BOD

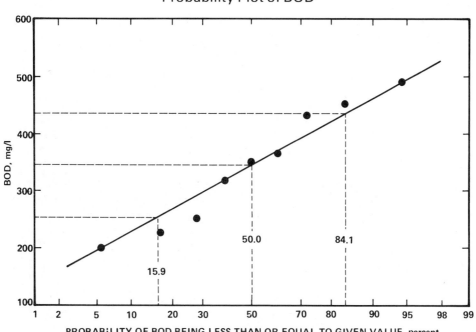

$$\sigma = \frac{X_{84.1\%} - X_{15.9\%}}{2}$$

$$= \frac{436 - 254}{2}$$

$$= 91$$

and mean:

$$\overline{X} = X_{50.0\%}$$

$$= 345$$

   When large numbers of data are to be analyzed it is convenient to group the data for plotting, e.g., 0 to 50, 51 to 100, 101 to 150, etc. The plotting position is determined as $m/(n + 1)$, where m is the cumulative number of points and n the total number of observations. The statistical distribution of data serves several important functions in developing the industrial waste management program.

In some cases an industrial plant that has batch process operations produces high variability due to the discharge schedule of the batch processes. These should be identified separately since it may be desirable to bleed the batch discharges or change the discharge schedule to minimize the variability of clean-up periods that might cause abnormal variability.

STORM WATER CONTROL

In most industrial plants, it is now necessary to contain and control pollutional discharges from storm water. Pollutional discharges can be minimized by providing adequate diking around process areas, storage tanks, and liquid transfer points with drainage into the process sewer. Contaminated storm water is usually collected based on a frequency for the area in question (for example, a ten year storm) in a holding basin. The collected water is then passed through the wastewater treatment plant at a controlled rate. A total storm runoff flow and contaminant loading of a refinery petrochemical installation is shown in figure 5.7.

PROBLEM 5-2

If runoff is to be stored or surged prior to eventual treatment, anticipated rainfall volumes must be estimated. The probable rainfalls for periods of one day to one year are given in table I and figure I for the two-year storm.

The areas and runoff coefficients associated with a refinery complex are shown in table II.

TABLE I

Time Period (days)	Cumulative Rainfall (inches)
1	2.5
5	3.7
10	4.5
30	6.6
60	10.7
120	18.0
240	30.0
360	41.0

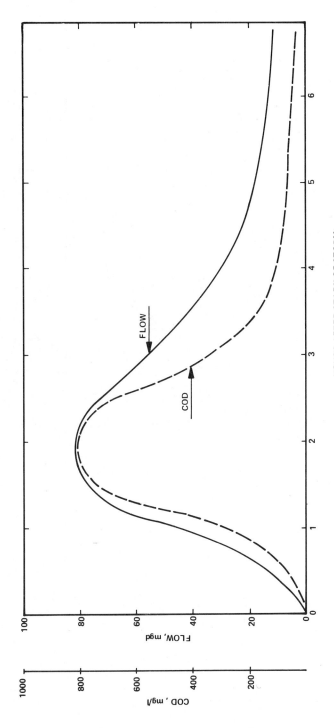

FIGURE 5.7

Total Storm Runoff Flow and Contaminant Loading
Characteristics of Typical
Refinery/Petrochemical Installation

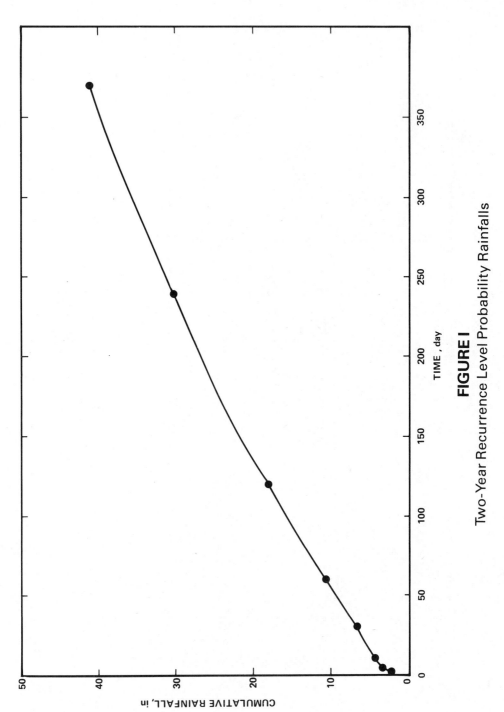

FIGURE I

Two-Year Recurrence Level Probability Rainfalls

TABLE II

Type of Surface	Area (acres)	Runoff Coefficient (C)
Water surface	21.8	1.0
Dirt or gravel	115.4	0.5
Bermed tank farm	166.9	0.7
Grass	176.7	0.5
Asphalt or concrete	55.6	0.9

Determine:

1. The cumulative runoff (MG) as a function of time (days).
2. The required minimum return rate for treatment (MGD) for storage periods of 10, 30, 60, and 120 days.
3. The required storage capacity (MG) for storage periods of 10, 30, 60, and 120 days.

Total effective area = $21.8 \cdot 1.0 + 115.4 \cdot 0.5 + 166.9 \cdot 0.7$

$$+ 176.7 \cdot 0.5 + 55.6 \cdot 0.9$$

$$= 334.7 \text{ acre}$$

Based on the table of cumulative rainfall with various time periods the cumulative runoff as a function of time can be estimated by:

$$q = i(CA)$$

in which CA is effective area and i is rainfall intensity. Example calculation for 10-day period at 4.5 in rainfall is:

$$q = 4.5 \text{ in} \cdot 334.7 \text{ acre} \cdot \frac{43560 \text{ ft}^2}{\text{acre}} \cdot \frac{0.0833 \text{ ft}}{\text{in}} \cdot \frac{7.48 \text{ gal}}{\text{ft}^3}$$

$$= 4.088 \times 10^7 \text{ gal or } 40.9 \text{ MG}$$

The results are tabulated and plotted as follows (see figure II):

Time, days	Runoff, MG
1	22.7
5	33.6
10	40.9
30	60.0
60	97.2
120	163.5
240	272.5
360	372.4

FIGURE II
Cumulative Runoff Curve

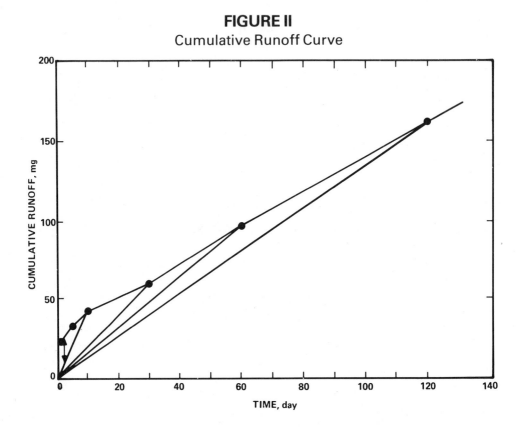

The minimum allowable return rate for treatment from a storage basin is defined by the slope of a straight line connecting the origin and a point on the cumulative runoff curve representing the maximum practical storage period. A sample calculation for 10 days is:

$$\text{Return rate} = \frac{40.9 - 0}{10 - 0} \text{ MG/day}$$

$$= 4.1 \text{ MG/day}$$

The required storage is defined by the maximum vertical distance between the cumulative runoff curve and the straight line depicting rate of removal. Example for 10 days:

$$\text{Required storage} = 18 \text{ mg}$$

Storage Period (days)	Return Rate (MGD)	Required Storage (MG)
10	4.09	18.0
30	2.00	20.9
60	1.62	24.0
120	1.36	28.0

POLLUTION REDUCTION

The 5 major ways of reducing pollution are:

1.   Recirculation. In the paper board industry white water from a paper machine can be put through a saveall to remove the pulp and fibre and recycled to various points in the paper making process.

2.   Segregation. In the soap and detergent case previously discussed clean streams were separated for direct discharge. Concentrated or toxic streams may be separated for separate treatment.

3.   Disposal. In many cases concentrated wastes can be removed in a semi-dry state. In the production of ketchup, the kettle bottoms after cooking and preparation of the product are usually flushed to the sewer. The total discharge BOD and suspended solids can be markedly reduced by removal of this residue in a semi-dry state for disposal. In breweries the secondary storage units have a sludge in the bottom of the vats which contains both BOD and suspended solids. Removal of this as a sludge rather than flushing to the sewer will reduce the organic and solids load to treatment.

4.   Reduction. It is common practice in many industries, such as breweries and dairies, to have hoses continuously running for clean-up purposes. The use of automatic cutoffs can substantially reduce the wastewater volume.

The use of drip pans to catch products, in such cases as in a dairy or ice cream manufacturing plant, instead of flushing this material to the sewer considerably reduces the organic load. A similar case exists in the plating industry where a drip pan placed between the plating bath and the rinse tanks will reduce the metal dragout.

5.   Substitution. The substitution of chemical additives of a lower pollutional effect in processing operations, e.g., substitution of surfactants for soaps in the textile industry.

An example of water reuse and suspended solids or fiber recovery in a board mill is shown in figure 5.8. In this case, fiber recovery may be accomplished using a Waco filter, a disc filter, or a flotation saveall. In some instances a flotation saveall might be followed with a filter to obtain a higher recovery.

FIGURE 5.8

Pulp and Fiber Recovery in a Board Mill

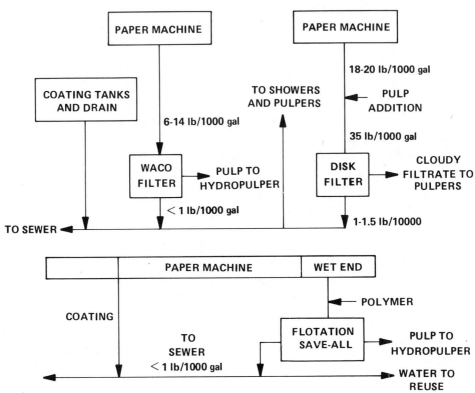

To date the discussion has considered in-plant measures to reduce waste-water volume and strength that do not involve major capital expenditures or process modifications. It is possible in many cases to markedly reduce the raw waste load by in-plant process changes. In general, this involves an economic trade-off between in-plant changes and end-of-pipe treatment. Some examples of this follow.

In the petroleum industry, major pollutant discharges from a petroleum refinery is the sour water or sour condensates. These are condensates from the various processing operations and are particularly high in sulfides and ammonia. A steam stripper accomplishes 96–99% sulfide removal and anywhere from 40 to 95% removal of ammonia depending on how the stripper is operated. Sulfides strip most effectively at low pH while ammonia strips most effectively at high pH. Low nitrogen levels can be achieved by the addition of a second stripper with a caustic leg. The substitution of surface condensers for barometric condensers significantly reduces the pollutional load.

Spent caustic in a refinery is an alkaline waste, containing high concentrations of sulfides, mercaptans, and phenolates. Removal of spent caustic for separate treatment may markedly reduce wastewater treatment costs and can in some cases result in a marketable product.

In the organic chemicals industry, an example might be drawn from the Union Carbide plant at South Charleston, West Virginia. This plant is reasonably representative of what might be expected from a major multiproduct chemicals complex. A detailed study of waste load reduction by in-plant changes showed that the present plant flow of 11.1 million gallons/day and 55,700 pounds of BOD/day could be reduced to a flow of 8.3 million gallons/day and a BOD to 37,100 pounds/day. The ways of achieving these reductions are shown in table 5.5. Equipment revision and additions and unit shutdowns refers to those units that would be replaced. Incineration involves the option of taking the more concentrated wastewater streams and rather than discharging to the main sewer, segregating them with incineration. Reprocessing refers to taking the tank bottoms that still have product present and reprocessing them for further product recovery.

TABLE 5.5
General Analysis of In-Plant Modifications

Type of Modifications	Description	% of total RWL reduction
Equipment revision & additions	Self-explanatory	25
Unit shutdowns	Shutdowns due to the age of the unit or the product. These shutdowns are not a direct result of pollution considerations, but they are somewhat hastened by these considerations.	10
Scrubber replacement	Replacement of scrubbers associated with amine production by burning of the off-vapors.	3
Segregation, collection & incineration	Of specific concentrated wastewater streams.	35
Raw material substitutions	Self-explanatory	3
Reprocessing	Collecting tail streams from specific processes, then putting these streams through an additional processing unit to recover more product and concentrate the final waste stream.	3
Miscellaneous small projects	A variety of modifications which individually do not represent a large reduction in RWL.	21

In the kraft pulp and paper industry various options exist for the major processing modifications as shown in tables 5.6 and 5.7.

WATER REUSE

Up to this point, reductions in wastewater volume and strength have been achieved by in-plant changes. The installation of wastewater treatment can be considered for the purpose of recycling rather than for disposal. This becomes particularly significant in the case of the high effluent discharge standards as contemplated in the future. A simplified diagram is shown in figure 5.9. From this figure a relationship may be developed for determining the daily cost of a water management system.

$$C = C_W (1 - f)Q + C_R fQ + C_D [Q(1 - f) - L]$$

where C = daily cost
Q = production water requirements
C_W = freshwater cost
C_R = recycled water costs
C_D = effluent treatment costs
f = fraction recycled
L = total water losses

TABLE 5.6
Unbleached Kraft Processing Unit Operations and Alternatives

Unit Operation	A	B	C
Wood preparation	Wet debarking	Dry debarking	
Pulping	Batch or continuous		
Washing & screening	Blow pit wash, course & fine screening, decker thickening	Digestor wash, two radial filters, brownstock screening[a]	Multistage counter-current vacuum wash; brownstock screening
Chemical recovery	Evaporation, oxidation, recausticizing	Evaporation, oxidation, recausticizing; steam stripping of condensates; spill collection system	
Papermaking	Conventional system with moderate white water recycle	Conventional system with extensive recycle and fiber recovery	

[a]Baseline technology for mills with continuous digestion.

TABLE 5.7
Unbleached Kraft Unit Operations Raw Waste Load Data

Unit Operation	Units	A	B	C
Wood preparation	$Q \times 10^3$ GPT[a]	3.5	1.0	
	BOD, PPT	3.0	2.0	
	SS, PPT	8.0	5.0	
	Color, PPT[b]	10.0	5.0	
Pulping				
Batch digestion	$Q \times 10^3$ GPT	1.2		
	BOD, PPT	8.0		
	SS, PPT	0.1		
Continuous digestion	$Q \times 10^3$ GPT	0.3		
	BOD, PPT	6.0		
	SS, PPT	0.1		
Washing & screening	$Q \times 10^3$ GPT	12.0	5.0	2.0
	BOD, PPT	21.0	10.0	8.0
	SS, PPT	21.0	15.0	10.0
	Color, PPT	80.0	50.0	50.0
Chemical recovery	$Q \times 10^3$ GPT	4.0	2.8	
	BOD, PPT	11.0	4.0	
	SS, PPT	6.0	4.0	
	Color, PPT	90.0	40.0	
Papermaking	$Q \times 10^3$ GPT	12.0	8.0	
	BOD, PPT	14.0	10.0	
	SS, PPT	16.0	12.0	
	Color, PPT	15.0	12.0	

[a]GPT = gal per ton of product.
[b]PPT = pound per ton of product.

PROBLEM 5–3

A plant uses 10,000 gallons/hour of process water with a maximum contaminant concentration of 1 pound/1000 gallon. The raw water supply has a contaminant concentration of 0.5 pound/1000 gallon (see figure I). Optimize a water-reuse system for this plant based on a raw water cost of 20 cents/1000 gallon. The following conditions apply:

Evaporation and product loss (E) = 1000 gal/hr

Contaminant addition (Y) = 100 lb/hr

Maximum discharge to receiving water= 20 lb/hr

FIGURE I
Water Reuse and Treatment Balance

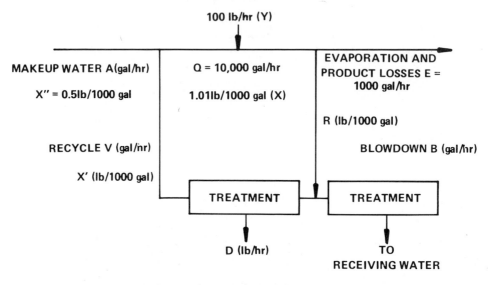

Example calculation: Let $A = 3000$ gallon/hour. Since $A = E + B$, $B = 3000 - 1000 = 2000$ gallon/hour and the recycle, V, is 7000 gallon/hour. By a material balance:

$$R = \frac{(A + V)X + Y}{(A + V) - E} = \frac{(10)(1.0) + 100}{10 - 1} = 12.2 \text{ lb/1000 gal}$$

By a material balance:

$$QX = AX'' + VX'$$

$$X' = \frac{QX - AX''}{V} = \frac{(10)(1.0) - A(0.5)}{V} = \frac{10 - (3)(0.5)}{7} = 1.21$$

The required efficiency of treatment for reuse is:

$$\% \text{ Removal} = \frac{R - X'}{R} \cdot 100 = \frac{12.2 - 1.21}{12.2} \cdot 100 = 90\%$$

The required treatment efficiency for discharge to the river is:

$$\frac{2(12.2) - 20}{2(12.2)} \cdot 100 = 18\%$$

The raw water cost for 3000 gallon/hour is (3) (24) (0.20) = $14.40/day. The cost of effluent treatment for discharge to the river is 5 cents/1000 gallon (see figure II) for a total daily cost of (2) (0.05) (24), or $2.40/day. The cost of treatment for reuse (see figure II) is (7) (0.42) (24) = $70.60/day. The net total cost is $87.40/day. A similar series of calculations is made for freshwater inputs varying from 2000 to 10,000 gallon/hour (no reuse). The total daily water cost can then be plotted versus percent recycle as shown in figure III.

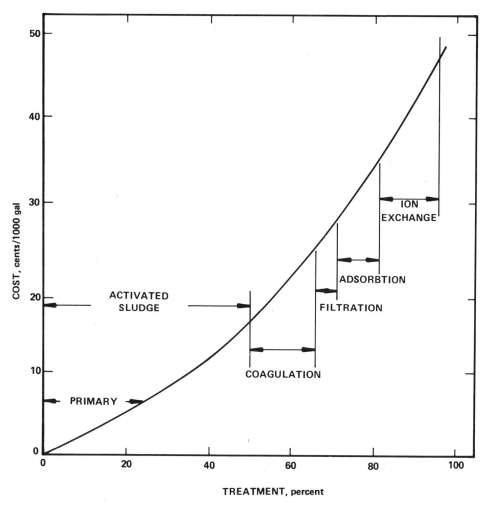

FIGURE II

Relationship Between Total Water Cost and Treatment

FIGURE III
Relationship Between Total Daily Water Cost and Treated Waste Recycle For Reuse

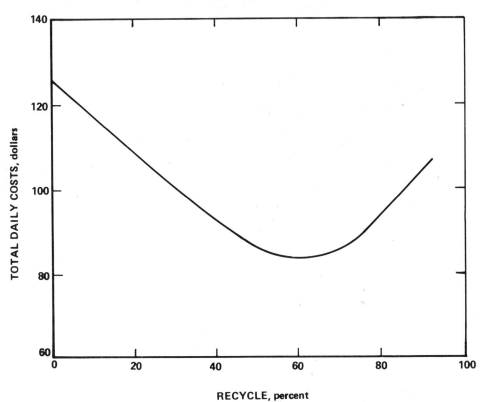

Another study in a bleached kraft mill is shown in figure 5.9. Effluent requirement 1 requires an effluent of 75 percent removal of BOD and suspended solids with no limitation on color. Effluent 2 requires 20 mg/l BOD_5 and suspended solids without color control. Effluent 3 requires 25 mg/l BOD_5 and 30 mg/l SS with 50 units of color. It is obvious from this figure that the optimal recycle will primarily be related to the effluent quality requirement and to the cost of fresh water.

FIGURE 5.9

Optimum Recycle Versus Freshwater Cost
for Varied Effluent Quality [2]

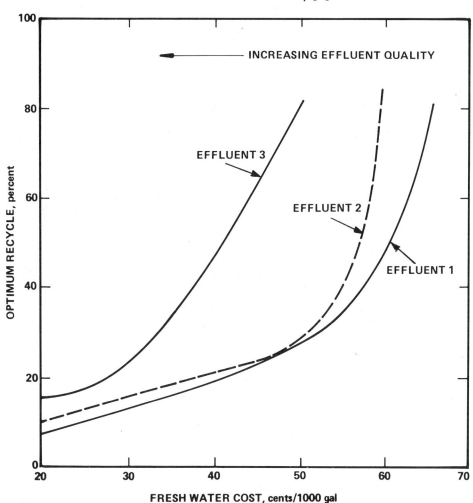

REFERENCES

1. U.S. Environmental Protection Agency. 1973. *Handbook for monitoring industrial wastewater, Technology Transfer.*
2. Lyons, D. N. and Eckenfelder, W. W., Jr. 1971. Optimizing a kraft mill water reuse system, *Water–1970, AIChE, Chem. Eng. Progress Symp. Series,* 67, no. 107:381.

6

Wastewater Treatment Processes

The application of many wastewater treatment processes is related both to the characteristics of the waste and the degree of treatment required. The various processes as a treatment sequence and substitution diagram are shown in figure 6.1. Pretreatment or primary treatment is used for the removal of floating and suspended solids and oils, neutralization and equalization, and to prepare the wastewater for subsequent treatment or discharge to a receiving water. General consideration in pretreatment or primary treatment for secondary biological treatment are summarized in table 6.1. The processes generally applicable to the removal of specific pollutants are shown in table 6.2. The treatment alternatives in most common use today are shown in figure 6.2. The characteristics of the effluent from these processes under optimum operation are summarized in table 6.3.

The selection of a wastewater treatment process or a combination of processes depends upon:

1. the characteristics of the wastewater
2. the required effluent quality
3. the costs and availability of land
4. the future upgrading of water quality standards

For any given wastewater treatment problem several treatment combinations can produce the desired effluent. Only one of these alternatives, however, is the most cost effective. Each process and the applicable design criteria are discussed in detail in subsequent chapters.

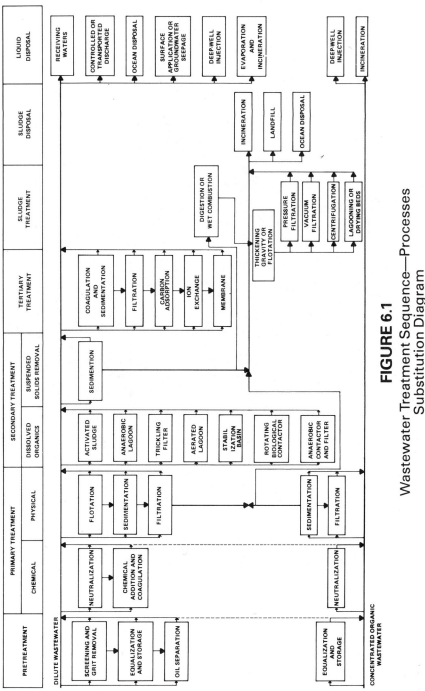

FIGURE 6.1

Wastewater Treatment Sequence—Processes
Substitution Diagram

TABLE 6.1

Pretreatment or Primary Treatment Requirements
for Biological Processes

Characteristics	Treatment
Suspended solids	Lagooning, sedimentation, and flotation
Oil or grease	Skimming tank or separator
Heavy metals	Precipitation or ion exchange
Alkalinity	Neutralization for excessive alkalinity
Acidity	Neutralization
Sulfides	Precipitation or stripping
BOD loading	Equalization

TABLE 6.2

Processes Applicable to Wastewater Treatment

Pollutant	Processes
Biodegradable organics (BOD)	Aerobic biological (activated sludge), aerated lagoons, trickling filters, stabilization basins, anaerobic biological (lagoons, anaerobic contact), deep-well disposal
Suspended solids (SS)	Sedimentation, flotation, screening
Refractory organics (COD, TOC)	Carbon adsorption, deep-well disposal
Nitrogen	Maturation ponds, ammonia stripping, nitrification-denitrification, ion exchange
Phosphorus	Lime precipitation; Al or Fe precipitation, biological coprecipitation, ion exchange
Heavy metals	Ion exchange, chemical precipitation
Dissolved inorganic solids	Ion exchange, reverse osmosis, electrodialysis

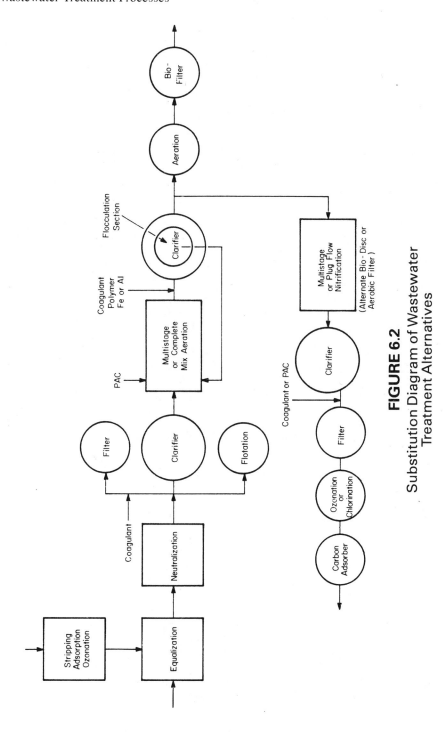

FIGURE 6.2

Substitution Diagram of Wastewater
Treatment Alternatives

Table 6.3. Maximum Effluent Quality Attainable from Waste-Treatment Processes

Process	BOD	COD	SS	N	P	TDS
Sedimentation, % removal	10–30	—	50–90	—	—	—
Flotation,[a] % removal	10–50	[b]	70–95	[c]	[c]	—
Activated sludge, mg/l	<25	—	<20	—	—	—
Aerated lagoons, mg/l	<50	—	>50	—	—	—
Anaerobic ponds, mg/l	>100	—	<100	—	—	—
Deep-well disposal	Total disposal of waste					
Carbon adsorption, mg/l	<2	<10	<1	—	—	—
Ammonia stripping, % removal	—	—	—	>95	—	
Denitrification and nitrification, mg/l	<10	—	—	<5	—	—
Chemical precipitation, mg/l	—	—	<10	—	<1	—
Ion exchange, mg/l	—	—	<1	[d]	[d]	[d]

[a] Higher removals are attained when coagulating chemicals are used.

[b] COD_{inf} − [BOD_u (Removed)/0.9].

[c] N_{inf} − 0.12 (excess biological sludge), lb; P_{inf} − 0.026 (excess biological sludge), lb.

[d] Depends on resin used, molecular state, and efficiency desired.

TABLE 6.3

Maximum Effluent Quality Attainable
from Waste Treatment Processes

Pretreatment and Primary Treatment

Pretreatment and primary treatment are employed primarily for the removal of floating materials and suspended solids and for conditioning the wastewater for discharge to waters of low classification or secondary treatment through neutralization and/or equalization. The various processes used in pretreatment and primary treatment were shown in the previous chapter. The primary treatment requirements for secondary treatment are summarized on page 132.

SCREENING

Screening is used to remove large solids prior to other treatment processes. In municipal sewage treatment, screens are usually provided at the head end of the plant for the removal of coarse materials. Screens for industrial waste treatment are usually of the rotary, vibrating, or eccentric type and are widely used in the canning, brewing, and pulp and paper industries. The screens consist of coarse bars or racks with 1.5- to 2.5-in. (3.8- to 6.4-cm) openings and may be either mechanically or manually cleaned, as shown in figure 7.1.

The types of screens in use today are:

1. bar racks with mechanical or manual cleaning
2. static screens
3. rotary drum screens
4. vibrating screens

Each type of screen is available in a variety of designs, mesh or aperture sizes, and screening surface materials.

FIGURE 7.1
Mechanical Sewage Screen

(Courtesy of Envirex, Inc.)

BAR RACKS

Bar rack screens are widely used in municipal waste treatment plants and may be applied to industrial waste treatment whenever the influent contains a large fraction of coarse debris. The common bar rack consists of parallel bars mounted on a steel frame and placed in a gravity flow channel. The bar rack is placed in the channel in an angle of $0° – 30°$ from the vertical for mechanically cleaned units, and $30° – 45°$ for manually cleaned racks. Bar spacing, which determines the size of debris to be retained by the screen, varies from 0.25 to 3.00 in. (0.64 to 7.62 cm). Finer screens are often preceded by bar racks. The quantity of screenings that may be expected from municipal sewage are discussed in Chapter Eleven.

STATIC SCREENS

With a static screen the wastewater flows by gravity or under pressure over a concave or multiple slope screening surface. The water passes through the screen while the solids are rolled down into a collecting hopper. The device is effective in dewatering slurries containing fatty or sticky fibrous suspended matter. The screening medium is usually made of stainless steel V-bars, positioned horizontally, perpendicular to the flow direction. The bar spacing is usually between 0.01 and 0.10 in. (0.03 and 0.25 cm). Commercially available units have a capacity ranging from 25 to 2,500 gpm (0.1 to 9.5 m^3/min). The loading rate may vary from 2 to 10 gpm/sq ft (0.08 to 0.41 m^3/m^2-min).

ROTARY DRUM SCREENS

Rotary drum screens use metallic or synthetic cloth as screen medium. The cloth is mounted on an open-ended drum that rotates on a horizontal shaft. The waste enters the drum through the end, and passes the screen into a collection well, in which the drum is partially submerged. The water level inside the drum is higher than in the collection well, and this difference in water level provides the necessary driving force for the flow through the screen. The suspended solids that attach to the fabric in the submerged section are removed into a collecting device by spray washers at the upper section of the drum. The screening media may vary in opening size from 0.01 to 0.75 in. (0.03 to 1.91 cm). Similar screens with finer openings (down to 5 μm) are classified as microstrainers. Rotation speed of the drum may vary from 1 to 5 rpm, and may be continuous or intermittent.

VIBRATING SCREENS

The vibrating screen is a structure with means for producing a rapid motion of a perforated or meshed surface. The effectiveness of the screen depends on that motion, which is normally 1,000 to 2,000 rpm in a 0.03125- to 0.125-in. (0.08 to 0.32 cm) orbit. A vibrator may be placed on top of the unit causing a slight rocking action and motion of particles away from the feed end.

In selecting the medium for a vibrating screen, both the opening size and the strength of wire must be considered. When heavy and abrasive solids are present in the wastes, wires should be thicker to reduce the open area. For light and sticky solids a light wire and increased open area are desirable. Vibrating screens are available in a variety of designs and in mesh range of #3 to #200.

Compared to rotary drums, the vibrating screens usually require smaller screen surface area and their initial cost is lower. In many cases the screen is self-cleaned without spray washing, and therefore the screenings are drier.

SEDIMENTATION

Sedimentation is employed for the removal of suspended solids from wastewaters. The process can be considered in three basic classifications, depending on the nature of the solids present in the suspension:

1. Discrete
2. Flocculent
3. Zone settling

In discrete settling, the particle maintains its individuality and does not change in size, shape, or density during the settling process. Discrete settling is observed with suspensions of grit (grit chambers), fly ash, and coal. Flocculent settling occurs when the particles agglomerate during the settling period, resulting in a change in size and settling rate. Examples include domestic sewage and pulp and paper wastes. Zone settling involves a flocculated suspension which forms a lattice structure and settles as a mass, exhibiting a distinct interface during the settling process. Alum flocs and activated sludge usually exhibit zone settling.

Discrete Settling

The settling velocity of discrete particles is related to gravity and viscous forces and is defined by the relationship:

$$v = \sqrt{\frac{4g(\rho_s - \rho_1)D}{3C_D\rho_1}}$$

(7–1)

where ρ_s = density of the particle
       ρ_l = density of the liquid
       D = diameter of the particle
       C_D = drag coefficient

The drag coefficient is related to the Reynolds number as shown in figure 7.2 for spherical particles and encompasses three regions. Many of the solids-separation problems encountered in sewage and waste treatment are defined by Stokes' law.

The ideal tank concept was developed by Hazen [1] and Camp [2] to define relationships applicable to the design of sedimentation basins. This concept is based on the premises that particles entering a tank are uniformly distributed across the influent cross section and that a particle is considered removed when it reaches the bottom of the tank. The settling velocity of any particle can be related to the overflow rate:

$$v_0 = \frac{D}{t} = \frac{DA}{At} = \frac{V/t}{A} = \frac{Q}{A} \qquad (7\text{--}2)$$

FIGURE 7.2
Settling Characteristics of Discrete Particles

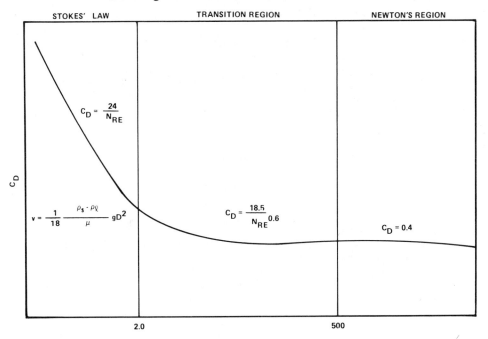

$$N_{RE} = \frac{vD\rho}{\mu}$$

Referring to figure 7.3 and the premises established for the ideal settling tank, all particles with settling velocities greater than v_o are completely removed and particles with settling velocities less than v_o are removed in the ratio v/v_o or h/h_o. The removal of discrete particles is related only to overflow rate. The removal of particles from suspensions with a wide range of particle sizes can be computed by graphical integration from the relationship:

$$\text{Removal} = (1 - C_o) + \frac{1}{v_o} \int_0^{C_o} v\,dC \qquad (7\text{–}3)$$

in which C_o is the fraction of particles with a settling velocity equal to or less than v_o as shown in figure 7.4.

Since settling basins are subject to turbulence, short-circuiting, and velocity gradients, a correction must be made to the ideal settling tank. Dobbins [3] and Camp [2] have developed a correction for turbulence. Alternatively, depending on the hydraulic characteristics of the settling basin, the overflow rate will be decreased by a factor of 1.25 to 1.75 and the detention period increased by a factor of 1.5 to 2.0 [4].

Scour occurs when the flow through velocity is sufficient to resuspend previously settled particles. Scour is usually not a problem in large settling tanks, but can occur in grit chambers and narrow channels.

Flocculent Settling

Flocculent settling occurs when the settling velocity of the particles increases as it settles through the tank depth, owing to coalescence with other particles, thereby increasing the settling rate and yielding a curvilinear settling path, as shown in figure 7.5.

FIGURE 7.3
Discrete Particle Settling

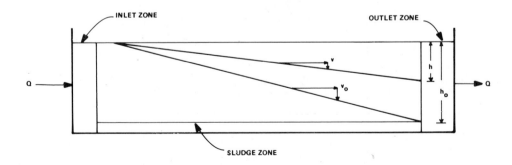

FIGURE 7.4
Settling-Velocity Analysis Curve for Suspension of Nonflocculating Particles

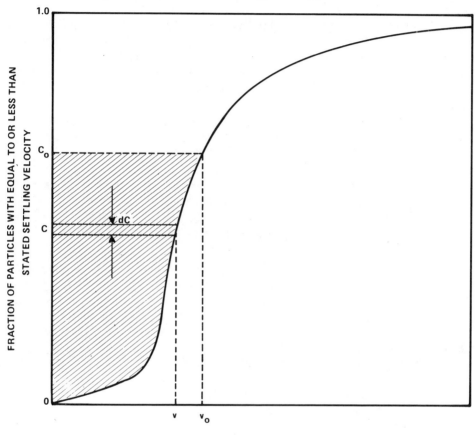

FIGURE 7.5
Flocculent Particle Settling

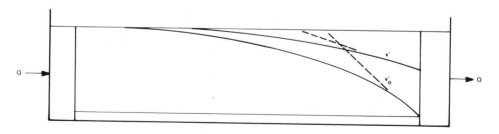

A mathematical analysis is not possible in the case of flocculent suspensions, due to interparticle contact during the settling of flocculent suspensions. Therefore, a laboratory settling analysis is required to establish the necessary design parameters [5].

When flocculation occurs, both overflow rate and detention time become significant design factors. The degree of flocculation is influenced by the initial concentration of suspended solids, so that the anticipated range of suspended solids in the wastewater should be considered in the design. Correction for turbulence, short-circuiting, and other conditions must be applied to the relationships developed from the laboratory study.

Design formula based on quiescent settling columns may not be as accurate as those based on continuous flow tracer laboratory model studies. But, these continuous tracer studies may not be justified as they require excessive attention and planning.

The suspended solids and BOD removal from domestic sewage and from pulp and paper mill wastes by sedimentation are shown in figure 7.6.

Zone Settling

Zone settling is characterized by activated sludge and flocculated chemical suspensions when the concentration of solids exceeds approximately 500 mg/l. The floc particles adhere together and the mass settles as a blanket, forming a distinct interface between the floc and the supernatent. The settling process is distinguished by four zones, as shown in figure 7.7. Initially, all the sludge is at a uniform concentration.

During the initial settling period, the sludge settles at a uniform velocity. The settling rate is a function of the initial solids concentration. As settling proceeds, the collapsed solids on the bottom of the settling unit build up at a constant rate. A zone of transition results, through which the settling velocity decreases as a result of an increasing concentration of solids. The concentration of solids in the zone settling layer remains constant until the settling interface approaches the rising layers of collapsed solids and a transition zone occurs. Through the transition zone, the settling velocity decreases, owing to the increasing density and viscosity of the suspension surrounding the particles. When the rising layer of settled solids reaches the interface, a compression zone occurs.

In the separation of flocculent suspensions, both clarification of the liquid overflow and thickening of the sludge underflow are involved. The overflow rate for clarification requires that the average rise velocity of the liquid overflowing the tank be less than the zone settling velocity of the suspension. The tank-surface-area requirements for thickening the underflow to a desired concentration level are related to the solids loading to the unit and are usually expressed in terms of a mass loading, lb of solids/ft^2/day (kg/m^2/day), or a unit area, ft^2/lb of solids/day (m^2/kg/day).

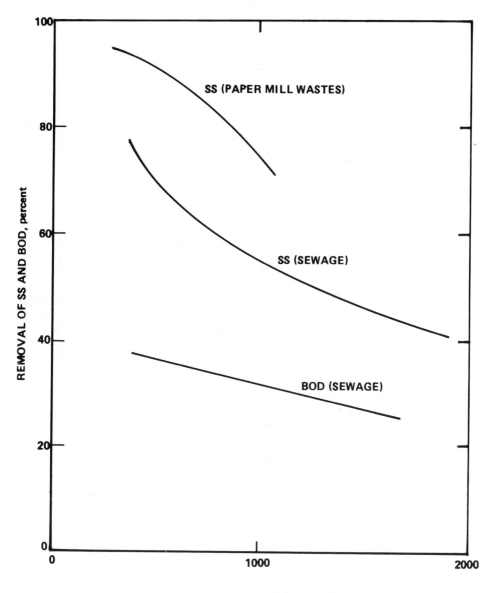

FIGURE 7.6

Suspended Solids and BOD Removal from Domestic
Sewage and Paper Mill Wastes by Sedimentation

FIGURE 7.7
Settling Zones

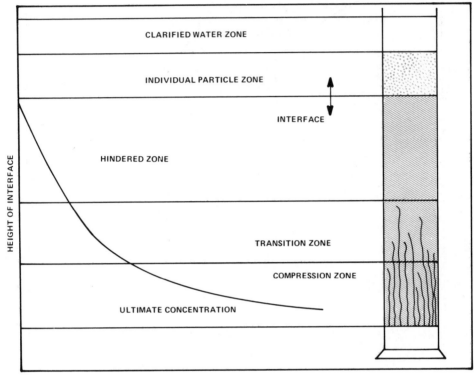

TIME

GRIT REMOVAL

Sand and other heavy particles that settle as discrete particles should be removed early in the treatment train to prevent abrasive wear on pumps and mechanical equipment, clogging of pipes by deposition, and accumulation in sludge holding tanks and digestors. Proper design will prevent the removal of organic material with the grit, thus simplifying grit disposal.

Several systems are available for grit removal. They depend on the amount of grit in the water, size of the plant, and cost. Standard systems include:

1. Channel shaped settling tanks used in tandem—one tank is hand cleaned while the other is in use. Extra volume is provided to allow grit accumulation.

2. Aerated units with hopper bottoms used where the deposition of organics becomes a problem.

3. Clarifier units with mechanical scrapping arms.

Velocity is crucial in grit settling tanks—1.0 fps (0.75 – 1.25 fps) (0.23 – 0.38 m/s) is recommended to allow deposition of grit particles while scouring out organics. An analysis of the prospective waste will help determine the parameters needed in the equation (7–1) that calculates the particle settling velocity. Basin length is determined by horizontal velocity (1 fps) and the settling velocity (see figure 7.8). The cross-sectional area is determined by the flow rate and the requirement that the horizontal velocity be 1 fps (0.3 m/sec). The 1 fps (.3 m/sec) can be maintained over a range of flow rates by the use of a proportional weir that follows the grit chamber.

CLARIFIERS

Clarifiers may either be rectangular or circular. In most rectangular clarifiers, scraper flights extending the width of the tank move the settled sludge toward the inlet end of the tank at a speed of about 1 ft/min (0.3 m/min). Some designs move the sludge toward the effluent end of the tank, corresponding to the direction of flow of the density current. A typical unit is shown in figure 7.9.

Circular clarifiers may employ either a center feed well or a peripheral inlet. The tank can be designed for center sludge withdrawal or vacuum withdrawal over the entire tank bottom.

Circular clarifiers are of three general types. With the center feed type (figure 7.10), the waste is fed into a center well and the effluent is pulled off at the weir along the outside. With a peripheral feed tank (figure 7.11), the effluent is pulled off at the tank center. With a rim-flow clarifier (figure 7.12), the peripheral feed and effluent drainoff are also along the clarifier rim, but this type is usually used for larger clarifiers.

The circular clarifier usually gives the optimal performance. Rectangular tanks may be desired where construction space is limited. In addition a series of rectangular tanks would be cheaper to construct due to the "shared wall" concept.

FIGURE 7.8
Grit Chamber Design

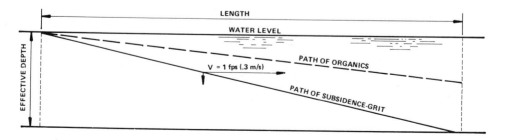

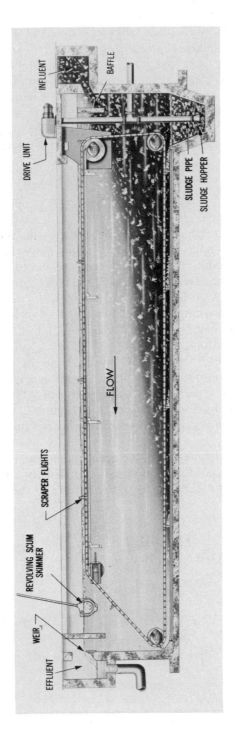

FIGURE 7.9
Rectangular Clarifier
(Courtesy of Envirex, Inc.)

FIGURE 7.10
Circular Clarifier
(Courtesy of Dorr Oliver Incorporated.)

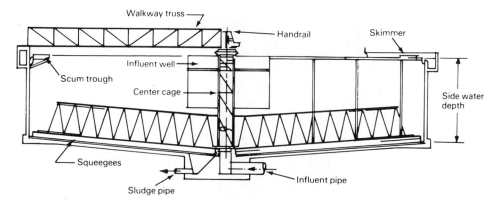

The gravity thickener is also a variation of the clarifier concept. In this case the solids loading is much higher. Higher torque is required along with more heavily constructed rakes.

The reactor-clarifier is another variation where the functions of chemical mixing, flocculation, and clarification are combined in the highly efficient solids-contact unit. This combination achieves the highest overflow rate and the highest effluent quality of all clarifier designs. Figure 7.13 shows an example.

The circular clarifier can be designed for center sludge withdrawal or vacuum withdrawal over the entire tank bottom. Center sludge withdrawal requires a minimum bottom slope of 1 in./ft (8.3 cm/m). The flow of sludge to the center well is largely hydraulically motivated by the collection mechanism, which serves to overcome inertia and avoid sludge adherence to the tank bottom. The vacuum drawoff is particularly adaptable to secondary clarification and thickening of activated sludge. These units are shown in figures 7.10 and 7.11.

The mechanisms can be of the plow type or the rotary-hoe type. The plow-type mechanism employs staggered plows attached to two opposing arms that move about 10 ft/min (3 m/min). The rotary-hoe mechanism consists of a series of short scrapers suspended from a rotating supporting bridge on endless chains that make contact with the tank bottom at the periphery and move to the center of the tank.

An inlet device is designed to distribute the flow across the width and depth of the settling tank. The outlet device is likewise designed to collect the effluent uniformly at the outlet end of the tank. Well-designed inlets and outlets will reduce the short-circuiting characteristics of the tank. Increased weir length can be provided by extending the effluent channels back into the basin or by

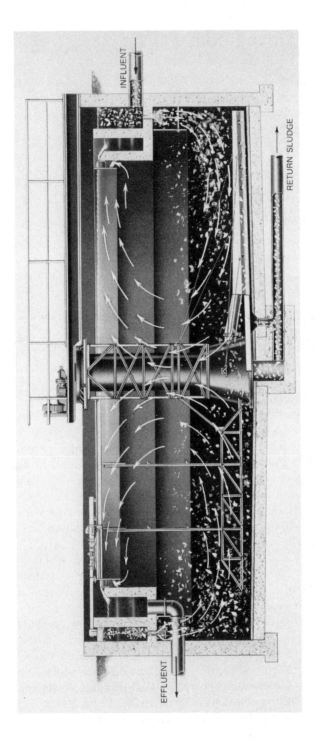

FIGURE 7.11
Clarifier with Suction Sludge Drawoff
(Courtesy of Envirex, Inc.)

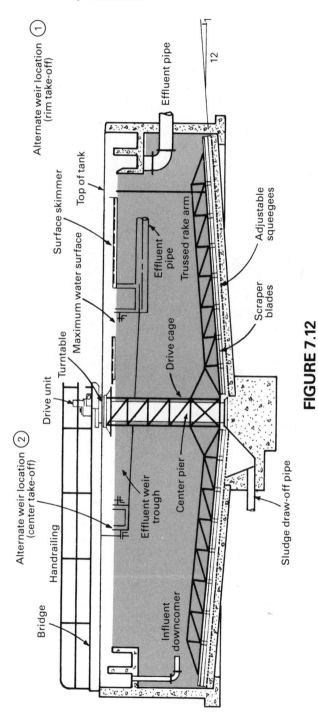

FIGURE 7.12
Rim-Flo Clarifier

(Courtesy of Environmental Equipment Division, FMC Corporation.)

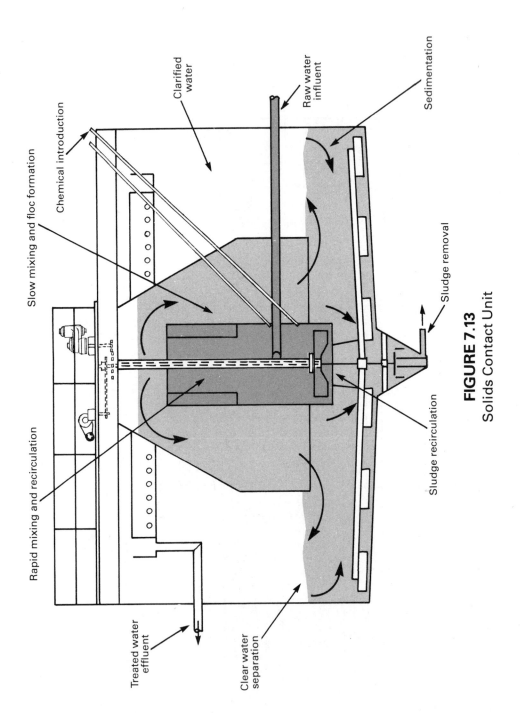

FIGURE 7.13
Solids Contact Unit

providing multiple effluent channels. In circular basins, inboard or radial weirs will ensure low takeoff velocities. Relocation of weirs is sometimes necessary to minimize solids carryover induced by density currents resulting in upwelling swells of sludge at the end of the settling tank.

TUBE CLARIFIERS AND HIGH RATE SEDIMENTATION

Significant developments in sedimentation theory have been very sporadic. They have been limited to the work of Hazen (1904), Camp (1946), and more lately the work of Hansen and Culp (1960s). Hansen and Culp [6] have applied the theories of Hazen [1] in the development of tube clarifiers.

Tube clarifiers offer increased removal efficiency at higher loading rates and lower detention times. An immediate advantage is that modules of inclined tubes constructed of plastic can be installed in existing clarifiers to upgrade performance.

There are two types of tube clarifiers, the slightly inclined and the steeply inclined units. The slightly inclined unit usually has the tubes inclined at the 5° angle. For the removal of discrete particles an inclination of 5° has proven most efficient. As shown in figure 7.14 the slightly inclined unit should be followed by a filter and the unit should be drained and backwashed with the filter.

The steeply inclined unit (figure 7.14) is less efficient in removal of discrete particles but can be operated continuously. When the tubes are inclined greater than 45°, the sludge is deposited and slides back out of the tube forming a countercurrent flow. In practice most wastes are flocculent in nature, and removal efficiency is improved when the tubes are inclined to 60° to take advantage of the increased flocculation that occurs as the solids slide back out the tube. Steeply inclined units are usually used where sedimentation units are being upgraded.

There are several criteria on which to base the physical design of tube settlers:

1. Tube length—2 – 8 ft (0.6 – 2.4 m) with the longest possible length being desired.
2. Tube diameter—0.5 – 4.0 in. (1.27 – 10.16 cm).
3. Tube shape—round, square, rectangular, hexagonal, or chevron shaped. The chevron is recommended for its highest perimeter/area ratio by Ford and Manning [8]. In steeply inclined units it is desirable to choose a shape that can be closely packed with the option to reverse the direction of inclination on alternative alternate layers.
4. Tube inclination—5° to 45–60°.
5. Flow rate—1 to 10 gal/min/ft^2 (41–410 l/min/m^2 of tube end area.
6. Influent temperature—to maintain an even flow distribution over all the tubes the influent temperature should not exceed the bulk liquid temperature of the tank by 1° C (0.2° C is preferable).
7. Polyelectrolyte addition—dependent on waste.

FIGURE 7.14
Two Basic Tube Clarifier Designs (After Culp *et al* [7])

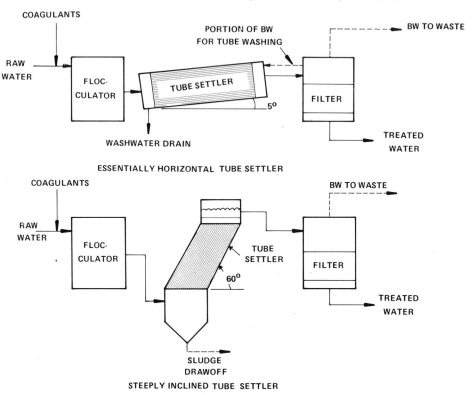

The amount of polyelectrolyte that can cause the maximum flocculation is probably more important than the other design parameters.

Another high rate sedimentation unit is the Lamella separator. This device, developed in Sweden, is also a tilted plate separator but the flow is directed downward between the plates employing concurrent flow of liquid and solids as shown in figure 7.15.

DESIGN PROCEDURE FOR FLOCCULENT SETTLING

1. Develop the settling rate-time relationship curves for at least three different concentrations of suspended solids to cover the expected range of fluctuation in suspended solids concentrations (figure 7.16).

2. The overflow rates and detention times for various percent removals are computed from the curves as follows. The overflow rate or the settling velocity

FIGURE 7.15
Tilted Plate Separator Systems

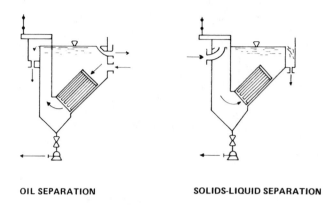

OIL SEPARATION   SOLIDS-LIQUID SEPARATION

TILTED-PLATE SEPARATOR

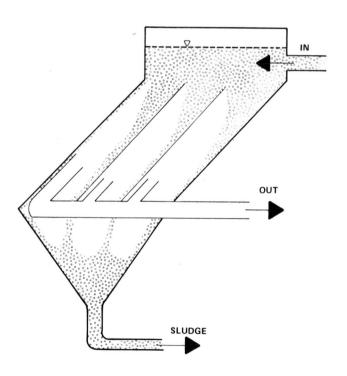

LAMELLA SEPARATOR

FIGURE 7.16

Design Procedure for Flocculent Settling. (a) Percent
SS Removed vs. Time and Depth; (b) Overflow
Rate vs. Percent Removal;
(c) Detention Time vs. Percent Removal

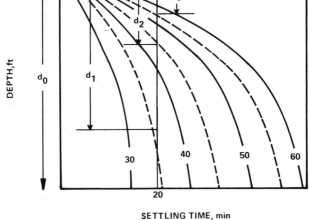

SETTLING TIME, min

(a)

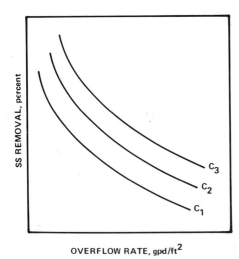

OVERFLOW RATE, gpd/ft^2

(b)

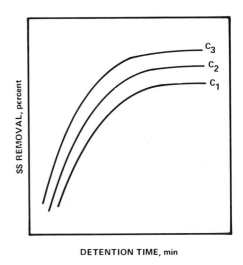

DETENTION TIME, min

(c)

v_o is the effective depth (six feet) divided by the time required for a given percent to settle through the effective depth. All particles having a settling velocity equal to or greater than v_o will be removed in an ideal basin. Particles with a lesser settling velocity, v, are removed in proportion v/v_o. For a given settling depth and a detention period t (min), a certain percentage (χ) of the suspended solids is removed completely (figure 7.16a). Particles in each additional 10 percent range are removed in the proportion v/v_o or in proportion to the average depth settled to the total settling depth. Each subsequent percentage range is computed in a similar manner and the total removal determined as follows:

$$\text{total percent removal} = \chi + \frac{d_1}{d_o}\,(10) + \frac{d_2}{d_o}\,(10) + \frac{d_3}{d_o}\,(10) \qquad (7\text{--}4)$$

Overflow rates [as gal/ft^2/day(m^3/m^2/day)] are computed from settling velocities.

   3.   Overflow rates against percent suspended solids removal (figure 7.16b) and retention time against percent suspended solids removal (figure 7.16c) are developed.

   4.   For a required percent suspended solids removal, the overflow rate and the detention time for a given initial suspended solids concentration are selected.

   5.   For prototype design, the overflow rate is decreased by a factor from 1.25 to 1.75 and the detention time increased by a factor from 1.5 to 2.0 to account for the effects of turbulence, short-circuiting, and inlet and outlet losses.

   6.   The surface area and depth are computed based on the design overflow rate and detention time.

PROBLEM 7–1

   A 3-meter settling column is filled with a discrete-particle suspension. At various times, the samples are taken from sampling points 1.2 and 2.4 m below the surface of the liquid and the suspended solids contents determined. Use the results of the settling column analysis given in table I to de-

TABLE I
Results of Settling Column Test

Time, min	0	15	30	45	60	90	180
mg/l	400	384	324	248	184	92	24
SS at 1.2 m, (%)	(1.00)	(0.96)	(0.81)	(0.62)	(0.46)	(0.23)	(0.06)
mg/l	412	408	400	383	354	288	132
SS at 2.4 m, (%)	(1.00)	(0.99)	(0.97)	(0.93)	(0.86)	(0.70)	(0.32)

termine the percent removal of suspended solids in an ideal horizontal-flow sedimentation basic operating at 1060 gpd/ft² (0.03 m³/m²/min). The figures shown in parentheses in table I are the computed concentration fractions.

The settling velocities of particles in the suspension are calculated by dividing the depth of sampling by the sampling time:

Time, min	0	15	30	45	60	90	180
v at 1.2 m, cm/min	—	8.0	4.0	2.67	2.0	1.33	0.67
v at 2.4 m, cm/min	—	16.0	8.0	5.33	4.0	2.67	1.33

The distribution of settling velocity versus the fraction of particles having equal to or less than stated settling velocity is shown in figure I. From figure I, C_o is obtained as 0.75 with v_o of 3.0 cm/min. The integral part in equation (7–3) is the shaded area in figure I and is evaluated graphically as 1.28 cm/min. Therefore:

$$\text{percent removal} = (1 - 0.75) + \frac{1.28}{3.0} = 67.7\%$$

PROBLEM 7–2

A flocculent suspension containing 500 mg/l suspended solids is filled in an 8-ft-tall column. Samples are taken with time and depth from the settling column, and the suspended solids concentration is determined for each sample as shown in table I.

Problem 7–2 (a)

Determine the relationships between percent removal and overflow rate and detention period for a horizontal-flow sedimentation basin with a 6-ft depth based upon laboratory results.

Problem 7–2 (b)

Calculate the required criteria for the basin design that reduces suspended solids to 150 mg/l. Use scale-up factors of

FIGURE I
Distribution of Settling Velocities

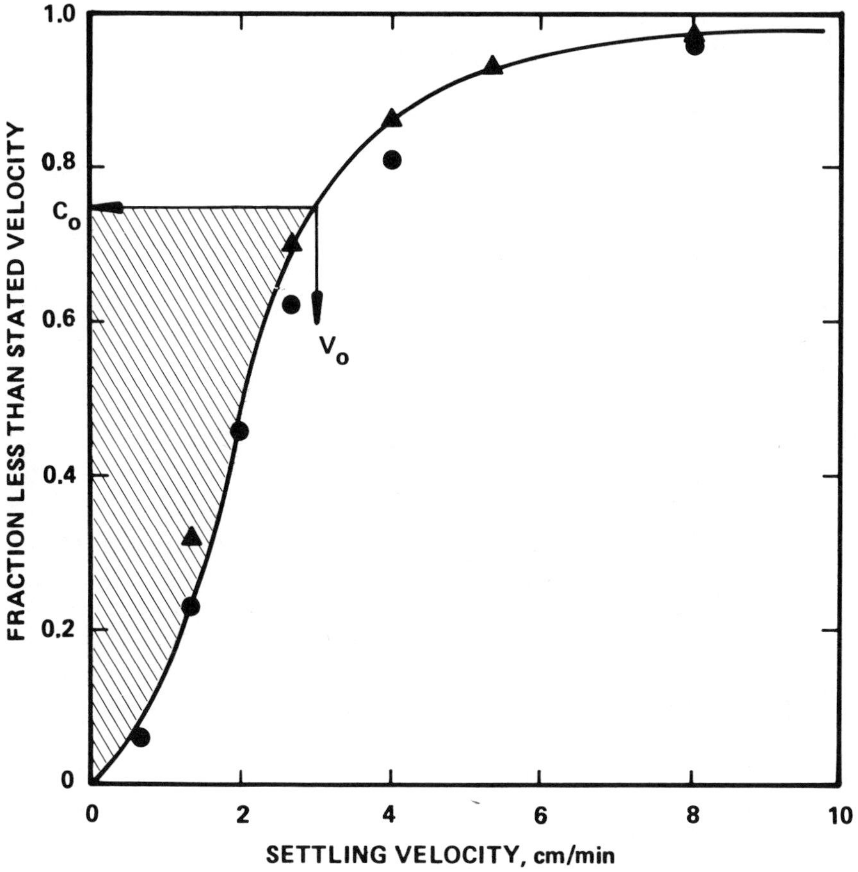

Time, min	0	5	10	20	30	40	50
Sampling Depth, ft			SS concentration, mg/l				
1	490	450	355	170	80	45	5
3	500	490	445	340	235	95	30
5	510	495	475	385	260	110	50
7	500	505	480	380	225	120	70

TABLE I
Results of Settling Column Test

1.5 and 1.75 for overflow rate and detention period, respectively.

SOLUTIONS

Problem 7–2 (a)

1. The percent removal of solids for each sample is calculated to be as follows:

Time, min	0	5	10	20	30	40	50
Sampling Depth, ft				SS Removal, %			
1	—	8	28	65	84	91	99
3	—	2	11	32	53	81	94
5	—	3	7	25	49	78	90
7	—	0	4	24	55	76	86

2. Plot the data as shown in figure I with depth as the ordinate, time as the abscissa, and parameters of percent removal. Estimate the isopercent removal lines by interpolation and/or judgement.

3. The total percent removal in an ideal horizontal-flow basin for any detention period and depth is given by equation (7–4). Figure I also illustrates sample calculation of the percent removal at one value of detention period (35 min) and overflow rate of (6 ft/35 min) = 1840 gpd/ft^2.

$$\text{Total percent removal} = 60 + \frac{4.0}{6}(10) + \frac{2.35}{6}(10) + \frac{1.18}{6}(10)$$

$$= 72.6\%$$

4. Figure II shows the results of several calculation similar to step 3 for different values of detention period and overflow rate.

Problem 7–2 (b)

1. Reduction of suspended solids from 500 to 150 mg/l is:

$$\frac{500 - 150}{500} \times 100 = 70.0 \text{ percent removal}$$

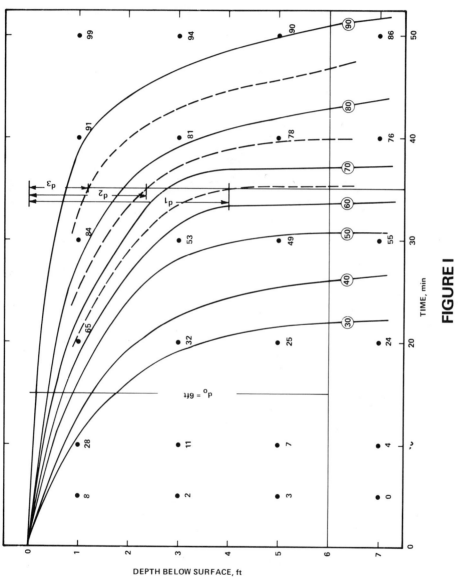

FIGURE I

Results of Settling Column Test

FIGURE II
Detention Period and Overflow Rate
Versus Percent Removal

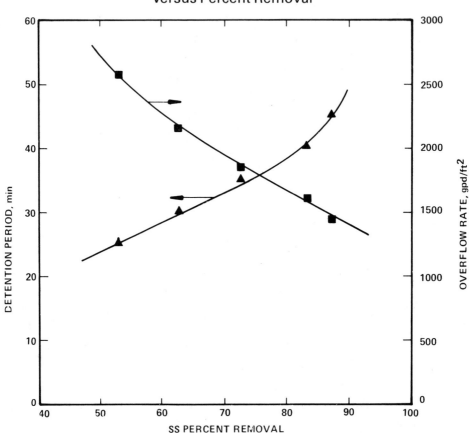

2. Required ideal detention period and overflow rate are 33 min and 1920 gpd/ft^2, respectively, as read from figure II.

3. Applying scale-up factors gives final design bases:

Detention period = 1.75 (33 min) = 58 min

Overflow rate = (1920 gpd/ft^2)/1.5 = 1280 gpd/ft^2

FLOCCULATION AND CHEMICAL ADDITION

Chemical addition, flocculation, and sedimentation are the primary unit operations in water treatment, while chemical addition and flocculation have been largely ignored in wastewater treatment. The efficiency of a clarifier can be significantly improved if increased particle flocculation occurs. This increased flocculation can be fostered by three methods:

1. Mechanical flocculation
2. Chemical addition and flocculation
3. Sludge return

Most municipal wastes and many industrial wastes flocculate under gentle agitation without the addition of chemicals. Typical results show a 10 percent increase in removal efficiency. Mechanical flocculation units vary from small high speed impellers to the low speed paddle units. Air flocculation is also a possibility. It can provide in addition to flocculation some pre-aeration of the waste. Some manufacturers (Dorr-Oliver, Inc. and Envirotech Co.) provide sedimentation units that are combined with flocculation and pre-aeration. One problem here is that a malfunction of the flocculator or aerator will cause the shut down of the sedimentation basin while repairs are made.

SEPARATION TECHNIQUES FOR OILY MATERIALS

Free and floating oily materials are separated from wastewaters by gravity clarification where the oily materials, which have a specific gravity less than water, float to the surface of a separator to be skimmed off for additional recovery or disposal. Successful separation by gravity requires that the oil globules be of sufficient size that they will readily separate in an adequate detention time. If the oily materials are highly dispersed, some form of coalescence might be introduced ahead of the gravity separator, or coagulant aids might be added to coagulate several oily particles to form larger particles that will separate in a specific time period. Occasionally, it is more advantageous to add coagulants and coagulate the oily materials with heavier compounds so that the oil and the coagulants settle to the bottom of a gravity sedimentation tank whereby they are removed. Air flotation is commonly used to enhance the separation of oily materials by lowering the specific gravity of the oily compounds even greater by the attachment of air bubbles to the oil particles. Air flotation is discussed later.

Emulsified oily materials require special treatment to break the emulsions so that the oily materials will be free and can be separated by gravity, coagulation, or air flotation. The breaking of emulsions is a complex art and may require thorough bench-scale and pilot plant investigations prior to developing a final process. The principles of breaking emulsions are highly involved and depend on many factors, and trial-and-error experimentation is essential.

Emulsions may be broken by heating, which reduces the viscosity of the oil phase and in some instances weakens the interfacial films of various emulsifying agents. Coalescence and separation of the oil and water phases can then take place. Heating generally applies to a very concentrated emulsion which may have resulted from a gravity oil-water separator. Adjustment of the pH by caustic or acid may help facilitate the breaking of an emulsion with heat.

Comparatively stable mixtures sometimes may be separated by centrifuging. If there is a difference in specific gravity between the oil and water, centrifugal force may be used to accelerate separation.

Stable W-O emulsions, particularly those stabilized by finely divided solids, can be resolved by filtration. Rotary vacuum precoat filters are sometimes employed for this purpose. The breaking of the emulsion occurs as a result of rupturing the globules in the dispersed phase when passing through the interstices of the filter cake and precoat material and as a result of removing the stabilizing solids.

Some emulsions can be destroyed under the influence of a strong electrical field by application of the principle of the Cottrell precipitator. The emulsion is passed between two electrodes and subjected to a high-potential, pulsating current. Oil globules are attracted electrically and coalesce until the masses are of sufficient size to settle by gravity.

Emulsions can be broken by chemicals that will balance or reverse the interfacial surface tension on each side of the interfacial film, neutralize stabilizing electrical charges, or precipitate emulsifying agents. Anionic and cationic surface-active agents are not compatible and will tend to neutralize one another. When dealing with emulsions stabilized with electrolytes, reactive anions, such as OH^- and PO_4^{---}, will break W-O emulsions. In the same sense, reactive cations, such as H^+, Al^{+++}, and Fe^{+++} will break O-W emulsions. The use of a heavy metal salt to form a flocculant precipitate of a heavy metal hydroxide to break the O-W emulsions is typified by the flocculation processes. Common chemicals that have been utilized to break emulsions or coagulate the colloidal particles include alum, ferrous sulfate, ferric sulfate or chloride, sodium hydroxide, calcium chloride, sulfuric acid, lime, soda, silicate, borax, sodium sulfate, and commercial organic-treating chemicals.

Principles of Gravity Oil-Water Separation

The principles involved in the gravity separation of oil from wastewater can be expressed mathematically, which provides the basis for design. In an oil separator, free oil floats to the surface of the tank, where it is skimmed off. The principles discussed on discrete settling and equation (7–1) apply except that the lighter-water oil globules rise through the liquid. The design of oil separators (see figure 7.17) as developed by the American Petroleum Institute [9] is based on the removal of free oil globules larger than 0.015 cm in diameter. The Reynolds number is less than 0.5, so Stokes' law applies. A procedure that considers short-circuiting and turbulence has been developed [9].

Design improvement such as the addition of inclined parallel plates to serve as collecting surfaces for oil globules and modifications to the inlet structure have improved separator performance. As the wastewater flows between the plates, the lighter oil droplets float upward toward the corrugation, coalesce into larger drops while rising to the top portion of the plate pack, and finally rise to the top of the tank. The plates establish laminar flow conditions through the plate pack and decrease drastically so that oil drops must rise to be collected. A corrugated plate interceptor (CPI) is shown in figure 7.18.

FIGURE 7.17
Corrugated Plate Oil Separator

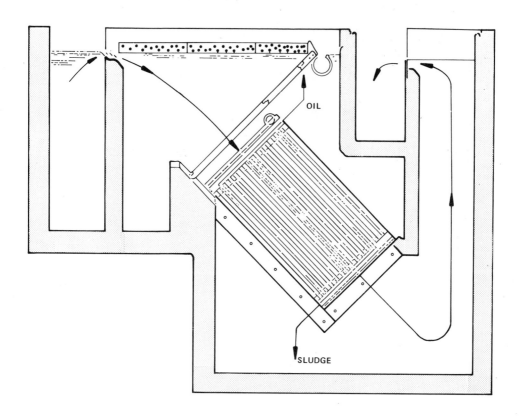

FIGURE 7.18

Example of General Arrangement for API Separator

(Courtesy of the American Petroleum Institute.)

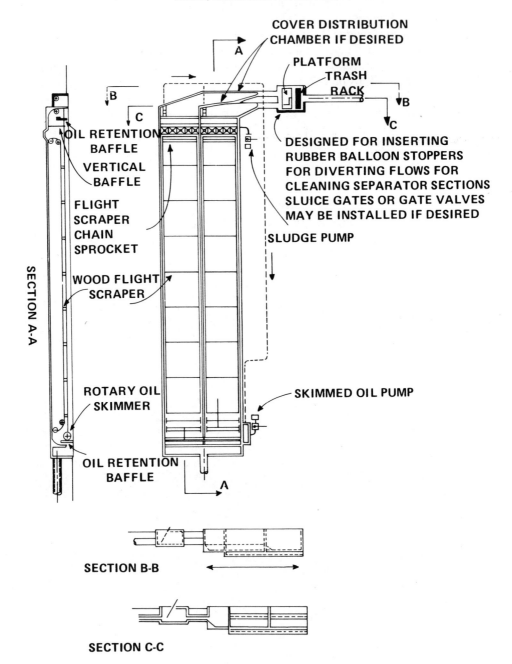

COVER DISTRIBUTION
CHAMBER IF DESIRED

A

PLATFORM
TRASH
RACK

B

C

B

C

OIL RETENTION
BAFFLE

VERTICAL
BAFFLE

FLIGHT
SCRAPER
CHAIN
SPROCKET

WOOD FLIGHT
SCRAPER

ROTARY OIL
SKIMMER

OIL RETENTION
BAFFLE

DESIGNED FOR INSERTING
RUBBER BALLOON STOPPERS
FOR DIVERTING FLOWS FOR
CLEANING SEPARATOR SECTIONS
SLUICE GATES OR GATE VALVES
MAY BE INSTALLED IF DESIRED

SLUDGE PUMP

SKIMMED OIL PUMP

SECTION A-A

A

SECTION B-B

SECTION C-C

FLOTATION

The flotation process is a method of clarification or thickening that is similar to gravity sedimentation except that the materials are removed by floating to the surface where they are skimmed off rather than settling to the bottom of a clarifier for subsequent withdrawal. Air flotation has been used for many years in treating wastewaters by separating suspended solids, oils, greases, fibers, and other low density solids from liquid as well as for thickening of activated sludge and flocculated chemical sludges. In order for particles to float to the surface, it is necessary that the specific gravity of these particles be less than that of water. The flotation process is accomplished by introducing dissolved gases in excess of saturation into solution and releasing these gases at atmospheric pressure. This mechanism reduces the specific gravity of suspended or oily materials by attaching fine gas bubbles to the particulate matter, thereby enhancing separation.

Air Flotation Systems

There are three flotation design schemes:

1. Pressurized full-flow dissolved air flotation (DAF) without recycle
2. Pressurized dissolved air flotation with recycle
3. Induced air flotation

The difference in these schemes depends on the location and method of adding the bubbles to the wastewater. In the first two instances, i.e., dissolved air flotation, bubbles are added to the wastewater by suddenly reducing the pressure on a supersaturated portion of the waste, causing the excess air to come out of the solution as fine bubbles. In the last instance, induced air flotation, the bubbles are added to the wastewater by entraining air in the liquid by mechanical or diffused air aeration. The dissolved air method produced much finer bubbles on the order of 50 to 100 μm as compared to 500 to 1000 μm for the induced air flotation system.

The pressurized full-flow and recycle systems are shown in figure 7.19. A DAF unit is shown in figure 7.20. The full-flow unit has the advantage of providing maximum bubble-particle contact, although the use of pressurized recycle requires a smaller pressurizing pump, utilizes simpler controls, minimizes emulsion formation, and optimized floc formation. The design variables for the dissolved air system include pressure, recycle ratio, overflow rate, and retention period. The solids loading is considered when dissolved air flotation is used for thickening. The pressurized tank is usually maintained at 40–60 psig (approximately 3 to 5 atmospheres); a 30 to 40 percent recycle ratio is normally applied for pretreatment; the overflow rate varies from 1 to 4 gpm/sq ft (0.04 to 0.16 m^3/m^2-min), including recycle, with a retention period of between 20 to 40 minutes. The performance of DAF treating oily wastewaters is shown on table 7.1. Oil removal from refinery wastewaters is shown in table 7.2. The removal of suspended solids from wastewaters is shown in table 7.3. Performance of a DAF unit for oil removal is shown in figure 7.21.

FIGURE 7.19
Pressurized Dissolved Air Flotation Systems
(From Eckenfelder [10])

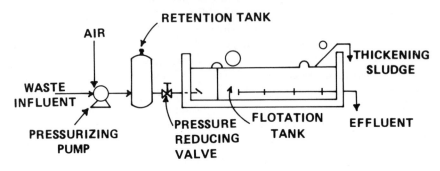

(a). PRESSURIZED FULL FLOW DAF UNIT (WITHOUT RECYCLE)

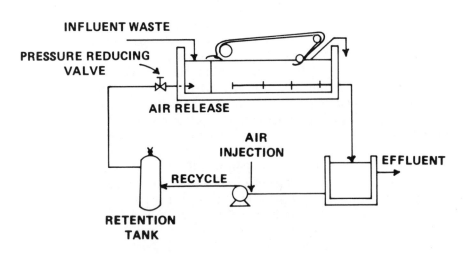

(b). PRESSURIZED RECYCLE DAF UNIT

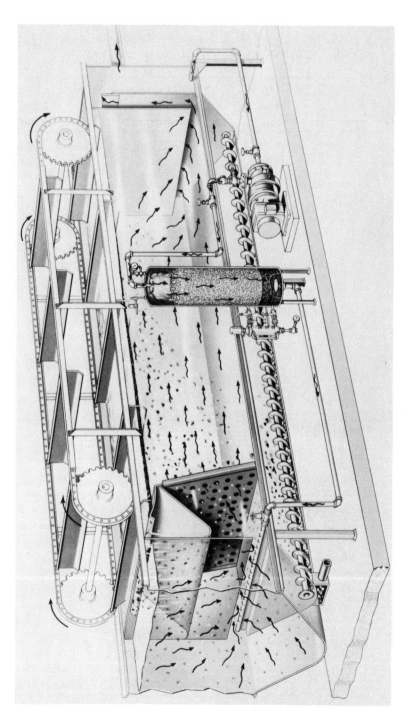

FIGURE 7.20
Dissolved Air Flotation Unit
(Courtesy of Envirex, Inc.)

TABLE 7.1
Air Flotation Treatment of Oily Wastewaters

Industrial Source	Alum, mg/l	Oil Concentration, mg/l		Percent Removal
		Influent	Effluent	
Refinery	0	125	35	72
	100	100	10	92
Oil tanker ballast water	100 (+ 1 mg/l polymer)	133	15	89
Paint manufacture	150 (+ 1 mg/l polymer)	1900	0	100
Aircraft maintenance	30 (+ 10 mg/l activated silica)	250–700	20–50	90+
Meat Packing		3830	270	93
Meat Packing		4360	170	96

TABLE 7.2
DAF Oil Removal Performance

Influent Oil, mg/l	Effluent Oil, mg/l	Removal, %	Chemicals[a]	Configuration
1930 (90%)	128 (90%)	93	Yes	Circular
580 (50%)	68 (50%)	88	Yes	Circular
105 (90%)	26 (90%)	78	Yes	Rectangular
68 (50%)	15 (50%)	75	Yes	Rectangular
170	52	70	No	Circular
125	30	71	Yes	Circular
100	10	90	Yes	Circular
133	15	89	Yes	Circular
94	13	86	Yes	Circular
638	60	91	Yes	Rectangular
153	25	83	Yes	Rectangular
75	13	82	Yes	Rectangular
61	15	75	Yes	Rectangular
360	45	87	Yes	Rectangular

[a]Alum most common, 100–130 mg/l; polyelectrolyte, 1–5 mg/l occasionally added.

TABLE 7.3

DAF Performance as Clarifiers

Type of Application	Influent SS mg/l	Effluent SS mg/l	Removal, %
Domestic sewage	180	63	65
Domestic sewage	145	70	52
Domestic sewage	130	59	55
Domestic sewage	158	93	41
Domestic sewage	171	65	62
Domestic sewage	142	66	53
Domestic sewage	148	54	64
Domestic sewage	103	44	57
Domestic sewage	128	54	58
Domestic sewage	185	80	57
Domestic sewage	177	40	74
Domestic sewage	183	82	56
Refinery	58	14	76
Refinery	59	32	46
Paper mill	5700	400	93
Aircraft mn.	70	25	65
Meat packing	4360	170	96
Meat packing	3830	270	93

The induced air flotation system, shown in figure 7.22, operates on the same principles as a pressurized air DAF unit. The gas, however, is self-induced by a rotor-disperser mechanism. The rotor, the only moving part of the mechanism that is submerged in the liquid, forces the liquid through the disperser openings, thereby creating a negative pressure. It pulls the gas downward into the liquid causing the desired gas-liquid contact. The liquid moves through a series of four cells before leaving the tank, and the float skimmings pass over effluent weirs on each side of the unit. This type of system offers the advantages of significantly lower capital cost and smaller space requirements than the pressurized system, and current performance data indicate that these systems have the capacity to effectively remove free oil and suspended materials. The disadvantages include higher connected power requirements than pressurized system, performance dependent on strict hydraulic control, less chemical addition and flocculation flexibility, and relatively high volumes of float skimmings as a function of liquid throughput (3 to 7 percent of the incoming flow for induced air systems is common compared to less than one percent for pressurized air systems). Performance data for induced air flotation is shown in table 7.4.

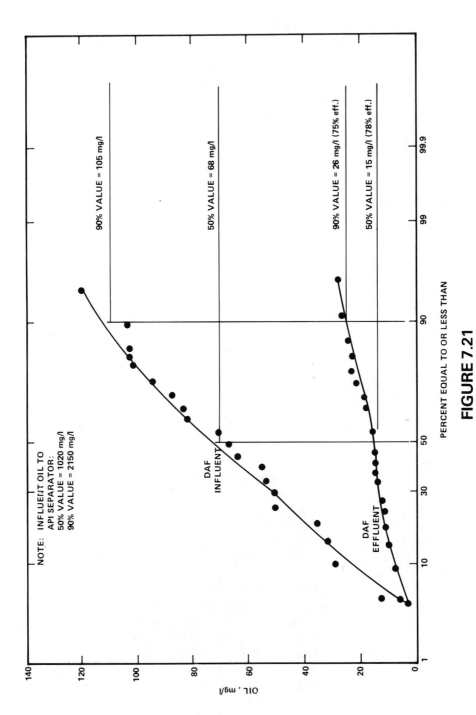

FIGURE 7.21

Oil Removal Efficiency of a Dissolved Air Flotation Unit

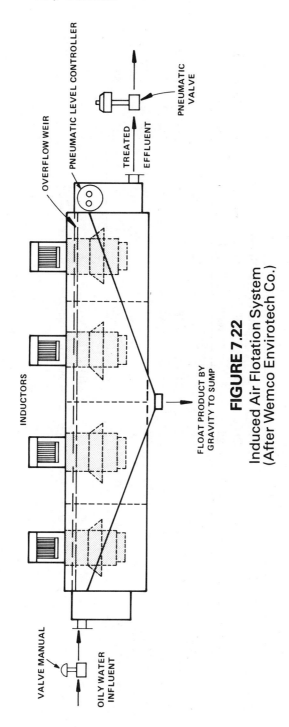

FIGURE 7.22

Induced Air Flotation System
(After Wemco Envirotech Co.)

Chemical Additive	Conc. mg/l	pH	Influent COD mg/l	Removal, %	Influent Oil mg/l	Effluent Oil mg/l	Removal, %	Sulfides mg/l	Removal, %
(Raw Wastewater)	—	9.0	1378	—	111	—	—	318	—
None	0.0	9.0	1060	23	58	30	48	238	25
Polyelectrolyte	25.0	9.0	1171	15	42	16	62	256	20
Polyelectrolyte	6.0	9.0	988	28	41	15	63	256	20
Polyelectrolyte	12.5	9.0	868	37	41	15	63	—	—
Polyelectrolyte	12.5	9.0	892	35	44	17	60	223	30

TABLE 7.4

Performance Data—Induced Air Flotation (Bench Scale)

The flotation process is particularly applicable when the materials to be removed are light and of a low density. For example, the reduction in specific gravity of a particle from 1.01 to 0.95 by the addition of air bubbles will create an upward driving force 5 times the gravity sedimentation driving force. Consequently, the overflow rates for flotation systems are generally much greater than those for gravity sedimentation systems. Overflow rates on the order of 100 to 9000 gal/day/sq ft (4 to 367 m^3/m^2-day) are not uncommon as compared to 500 to 2000 gal/day/sq ft (20.4 to 81.5 m^3/m^2-day) for sedimentation systems.

Air-to-Solids Ratio

The most important parameter in flotation design is the air-to-solids ratio. There is a quantity of air that can effectively be absorbed or entrapped onto wastewater particles. If less than the optimum amount of air is employed, the performance of the flotation system is impaired, such as effluent quality in terms of oil and grease or suspended solids, or the percent solids in the case of thickening. If more than the optimal amount of air is employed, the performance of the system is improved and capital and operating expenditures will be wasted. The air-to-solids ratio is expressed in terms of pounds of air released per pound of solids fed and affects the performance of a flotation system by:

1. the oil or suspended solids content in the effluent
2. the velocity of rise of the floating materials, which will control the overflow rate and the detention time in the basin
3. the degree of thickening in the float solids

These performance indicators are improved by an increase in the air-to-solids ratio up to the optimum level as shown in figures 7.23 and 7.24. The optimum varies for different wastewaters and materials to be removed and must be determined by bench-scale and pilot investigations for each case.

The quantity of gas that will be theoretically released from solution when the pressure is reduced to atmospheric levels can be calculated from the following:

$$A = QC_s\,[f(P/14.7 + 1) - 1]\,8.34 \tag{7-5}$$

where A = gas released at atmospheric pressure, lb/day
   C_s = gas saturation at atmospheric conditions, mg/l
   f = gas solubility in wastewater/gas solubility in water
   P = gauge pressure, psig
   Q = pressurized flow, mgd

The solubility of air in water varying with temperature is shown in table 7.5.

An air-to-solids parameter based on solids in the wastewater has been used to relate process performance. These relationships for no recycle and recycle are:

FIGURE 7.23
Effect of Air-to-Solids Ratio on Float Solids

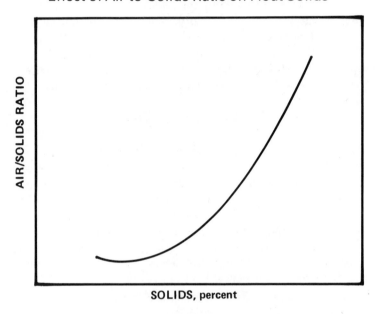

SOLIDS, percent

FIGURE 7.24
Effect of Air-to-Solids Ratio on Effluent
Suspended Solids

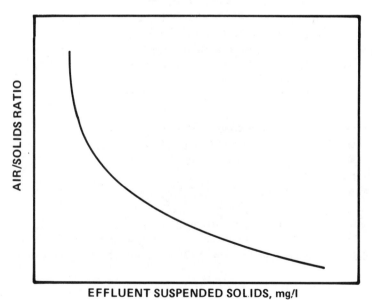

EFFLUENT SUSPENDED SOLIDS, mg/l

TABLE 7.5
Air Characteristics and Solubilities[a]

Temp.		Volume Solubility		Weight Solubility		Density	
°C	°F	ml/l	C.F. / 1000 gal	mg/l	lbs / 1000 gal	g/l	lbs. / C.F.
0	32	28.8	3.86	37.2	0.311	1.293	0.0808
10	50	23.5	3.15	29.3	0.245	1.249	0.0779
20	68	20.1	2.70	24.3	0.203	1.206	0.0752
30	86	17.9	2.40	20.9	0.175	1.166	0.0727
40	104	16.4	2.20	18.5	0.155	1.130	0.0704
50	122	15.6	2.09	17.0	0.142	1.093	0.0682
60	140	15.0	2.01	15.9	0.133	1.061	0.0662
70	158	14.9	2.00	15.3	0.128	1.030	0.0643
80	176	15.0	2.01	15.0	0.125	1.000	0.0625
90	194	15.3	2.05	14.9	0.124	0.974	0.0607
100	212	15.9	2.13	15.0	0.125	0.949	0.0591

[a]Values presented in absence of water vapor and at 14.7 psia pressure.

$$A/S = \frac{C_s\,[f(P/14.7 + 1) - 1]}{X_0} \quad \text{(no recycle)} \tag{7–6}$$

where A/S = air/solids ratio, lb air released/lb solids applied
X_0 = average influent suspended solids, mg/l

and:

$$A/S = \frac{RC_s}{QX_0}\,[f(P/14.7 + 1) - 1] \quad \text{(with recycle)} \tag{7–7}$$

where R = pressurized recycle, mgd
Q = wastewater flow, mgd

The chemical treatment in air flotation is extremely important, particularly when the soluble, colloidal, or emulsified components of the wastestream are to be treated. The effect of chemical addition (coagulants and/or polyelectrolytes) can be assessed using bench-scale jar test procedures. If there is justification for chemical addition, then a rapid mix tank and flocculation basin are incorporated into the design of a pressurized full flow or recycle DAF system. In an induced air flotation unit, chemicals are normally added in the influent line to the unit, sometimes employing a static mixer to enhance contact, then allowing flocculation to occur in the first cell of the induced air system.

PROBLEM 7–3

A wastestream at 150 gpm and a temperature of 103°F contains significant quantities of nonemulsified oil and non-settleable suspended solids. The concentration of oil and grease, as measured by the hexane extractables test, was found to be 120 mg/l and the suspended solids were approximately 130 mg/l. Since the oily materials were of mineral origin and were not relatively biodegradable, it was considered necessary to reduce the levels of oil and grease to less than 20 mg/l and to reduce the effluent SS to 20 mg/l before entering the biological system.

Laboratory studies further showed the following information:

alum dose = 50 mg/l
pressure = 50 psig
sludge = 3% by weight

Calculate:

1. The recycle rate
2. Surface area of the flotation unit
3. Sludge quantities generated

An air-to-solids ratio for effluent oil and grease of 20 mg/l is found from figure I.

A/S = 0.03 lb air released/lb solids applied

At 103° F the weight solubility of air is 18.6 mg/l from table 7.5. The value of f is assumed to be 0.5.

1. The recycle rate:

$$R = \frac{(A/S)QX_0}{C_s\,[f(P/14.7 + 1) - 1]}$$

$$= \frac{0.03 \cdot 150 \cdot 130}{18.6\,[0.5\,(50/14.7 + 1) - 1]}\ \text{gpm}$$

$$= 26.2\ \text{gpm}$$

2. Required surface area

The hydraulic loading is determined from figure II for effluent oil and grease of 20 mg/l,

$$L = 2.6\ \text{gpm/ft}^2$$

$$A = \frac{Q + R}{L}$$

$$= \frac{150 + 26.2}{2.6} \, ft^2$$

$$= 68 \, ft^2$$

FIGURE I
Effect of A/S Ratio on Effluent Quality

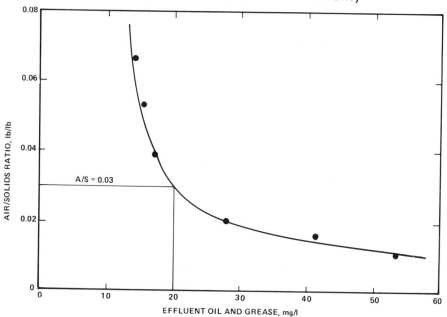

3. Sludge quantities generated

$$SS \ sludge = (130 - 20) \cdot 150 \ gpm \cdot \left(\frac{1440 \ min}{day} \right) \left(\frac{MG}{10^6 \ gal} \right) \cdot 8.34$$

$$= 198 \ lb/day$$

$$Oil \ and \ grease \ sludge = (120 - 20) \cdot 150 \ gpm \cdot \left(\frac{1440 \ min}{day} \right) \left(\frac{MG}{10^6 \ gal} \right) \cdot 8.34$$

$$= 180 \ lb/day$$

$$Alum \ sludge = \left(\frac{0.64 \ mg \ sludge}{mg \ alum} \right) \cdot 50 \ mg/l \ alum \cdot 150 \ gpm \cdot$$

$$\left(\frac{1440 \ min}{day} \right) \left(\frac{MG}{10^6 \ gal} \right) \cdot 8.34$$

$$= 58 \ lb/day$$

$$Total \ sludge = 436 \ lb/day$$

$$\text{Total sludge volume} = \frac{436}{0.03} \text{ lb/day} \cdot \left(\frac{\text{gal}}{8.34 \text{ lb}} \right) \left(\frac{\text{day}}{1440 \text{ min}} \right)$$

$$= 1.21 \text{ gpm}$$

FIGURE II
Effect of Surface Loading Rate on Effluent Quality

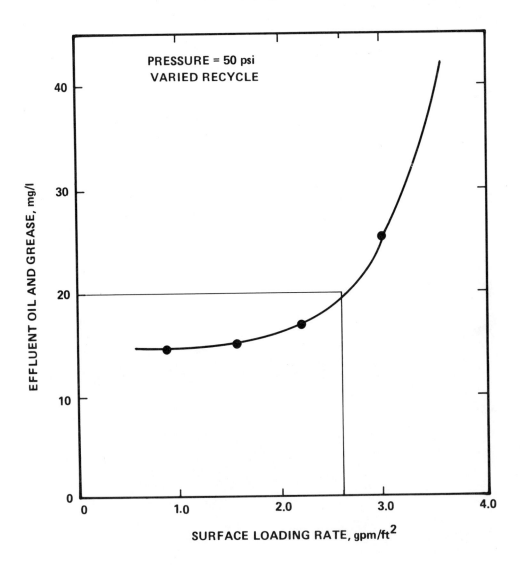

EQUALIZATION

The objective of equalization is to minimize or control fluctuations in wastewater characteristics in order to provide optimum conditions for subsequent treatment processes. The size and type of equalization basin provided varies with the quantity of waste and the variability of the wastewater stream. The basin should be of a sufficient size to adequately absorb waste fluctuations caused by variations in plant production scheduling and to dampen the concentrated batches, periodically dumped or spilled to the sewer.

The purposes of equalization for industrial treatment facilities are:

1. To provide adequate dampening of organic fluctuations in order to prevent shock loading of biological systems.
2. To provide adequate pH control or to minimize the chemical requirements necessary for neutralization.
3. To minimize flow surges to physical-chemical treatment systems and permit chemical feed rates compatible with feeding equipment.
4. To provide continuous feed to biological systems over periods when the manufacturing plant is not operating.
5. To provide capacity for controlled discharge of wastes to municipal systems in order to distribute waste loads more evenly.
6. To prevent high concentrations of toxic materials from entering the biological treatment plant.

Mixing is usually provided to ensure adequate equalization and to prevent settleable solids from depositing in the basin. In addition, the oxidation of reduced compounds present in the waste stream or the reduction of BOD by air stripping may be achieved through mixing and aeration. Methods that have been used for mixing include:

1. Distribution of inlet flow and baffling
2. Turbine mixing
3. Diffused air aeration
4. Mechanical aeration

The most common method is to provide submerged mixers or, in the case of a readily degradable wastewater such as a brewery, to use surface aerators employing a power level of approximately 15–20 hp/MG (0.003–0.004 kw/m^3). Air requirements for diffused air aeration are approximately 0.5 cu ft air/gal (67 l air/m^3) of waste.

The equalization basin may be designed with a variable volume to provide a constant effluent flow (figure 7.25) or with a constant volume and an effluent flow which varies with the influent. The variable volume basin is particularly applicable to the chemical treatment or wastes having a low daily volume. This type basin may also be used for discharge of wastes to municipal sewers. It may be desirable to program the effluent pumping rate to discharge the maximum quantity of waste during periods of normally low flow to the municipal treatment facility.

FIGURE 7.25
Equalization Basin Types

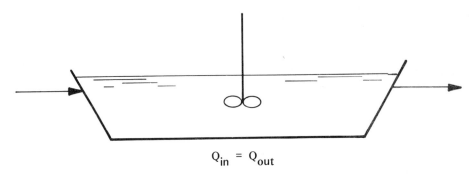

$$Q_{in} = Q_{out}$$

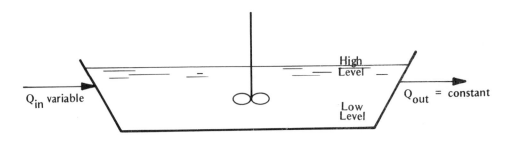

Q_{in} variable

High
Level

Low
Level

Q_{out} = constant

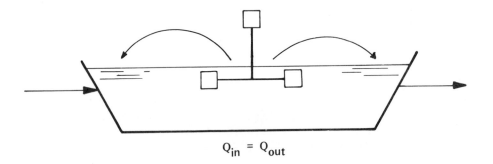

$$Q_{in} = Q_{out}$$

Basins have a constant volume with variable flow discharge that are used primarily to aid neutralization (where both acidic and alkaline waste streams are present) and to equalize concentration fluctuations for chemical or biological treatment. Depending on the topography in individual locations, effluent pumping may not be required for this type of basin whereas it is usually a requirement for constant flow configurations.

When considering equalization basins in which occasional dumps or spills cause unusually high concentrations say 1 percent of the time, a spill basin with an automatic bypass activated by a monitor (such as TOC) may be used as shown in figure 7.26. As shown in this figure the bypass is activated when the TOC exceeds

FIGURE 7.26
Equalization of Waste Composition for Biological Treatment

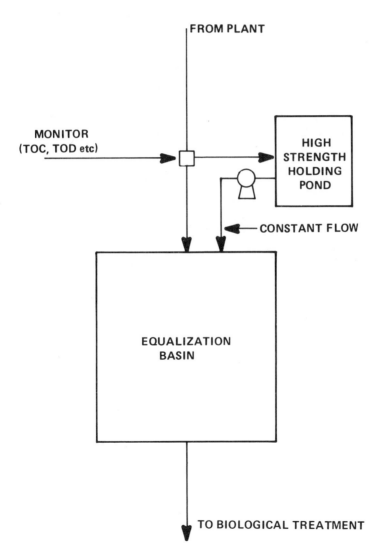

4000 mg/l and continues until the concentration drops below 4000 mg/l at which time the wastewater flow again diverts to the equalization basin. High concentrations are usually the result of an accidental spill and an alarm alerts plant personnel to seek out and correct the cause of the spill.

The equalization basin is usually sized to restrict the discharge to a maximum concentration commensurate with the maximum permissible discharge from subsequent treatment units. For example, if the maximum effluent from an activated sludge unit is 50 mg/l BODs, the maximum effluent from the equalization basin may be computed and thereby provide a basis for sizing the unit.

For the case of near constant wastewater flow and a normal statistical distribution of wastewater composite analyses the required equalization retention time is:

$$t = \frac{\Delta(S_i)}{2(S_e)} \tag{7-8}$$

where Δ = time interval over which samples were composited, hrs
       t = detention time, hrs
       S_i = variance of the influent wastewater concentration (the square of the standard deviation)
       S_e = variance of the effluent concentration at a specified probability (e.g., 99 percent)

Where a completely mixed basin is to be used for treatment, such as in an activated sludge basin or an aerated lagoon, this volume can be considered as part of the equalization volume. For example, if the completely mixed aeration basin volume is 8 hours and the total required equalization volume is 16 hours, then the equalization basin only needs to have a capacity of 8 hours. For more complex cases involving variable flow and concentration, nonnormal distributions and a reaction occurring in the basin reference is made to Novotny and England [11] and DiToro [12].

NEUTRALIZATION

Many industrial wastes contain acidic or alkaline materials that require neutralization prior to discharge to receiving waters or prior to chemical or biological treatment. For biological treatment to natural waters, a pH in the biological system should be maintained between 6.5 to 8.5 to ensure optimum biological activity. The biological process itself provides a neutralization and a buffer capacity as a result of the production of CO_2, which reacts with caustic and acidic materials. The degree of preneutralization required depends, therefore, on the ratio of BOD removed and the causticity or acidity present in the waste. These requirements are discussed in Chapter Nine.

Types of Processes

Mixing acidic and alkaline waste streams. This process requires sufficient equalization capacity to effect desired neutralization. There may be a danger of toxic gases or by-products.

Acid wastes neutralization through limestone beds. These may be downflow or upflow systems. The maximum hydraulic rate for downflow systems is 1 gpm/ft^2 (40 1/min/m^2) to ensure sufficient retention time. The acid concentration should be limited to 0.6 percent H_2SO_4 to avoid coating of limestone with nonreactive $CaSO_4$ and excessive CO_2 evolution, which limits complete neutralization. High dilution or dolomitic limestone requires longer detention periods to effect neutralization. Hydraulic loading rates can be increased with upflow beds, because the products of reaction are swept out before precipitation. A limestone bed system is shown in figure 7.27. The relationship between terminal pH and flow rate for neutralization with calcined magnesite is shown in figure 7.28.

Mixing acid wastes with lime slurries. The neutralization depends on the type of lime used. The magnesium fraction of lime is most reactive in strongly acid solutions and is useful below pH 4.2. Neutralization with lime can be defined by a basicity factor obtained by titration of a one-gram sample with an excess of HCl, boiling 15 minutes, followed by backtitration with 0.5 N NaOH to the phenolphthalein end point.

In lime slaking, the reaction is accelerated by heat and physical agitation. For high reactivity, the lime reaction was complete in 10 minutes. Storage of lime slurry for a few hours before neutralization may be beneficial. Dolomitic

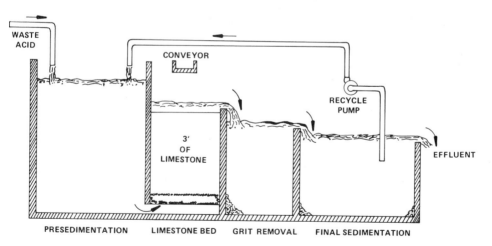

FIGURE 7.27
Simplified Flow Diagram of Limestone Neutralization
System (Adapted from Tully, T.J. [13])

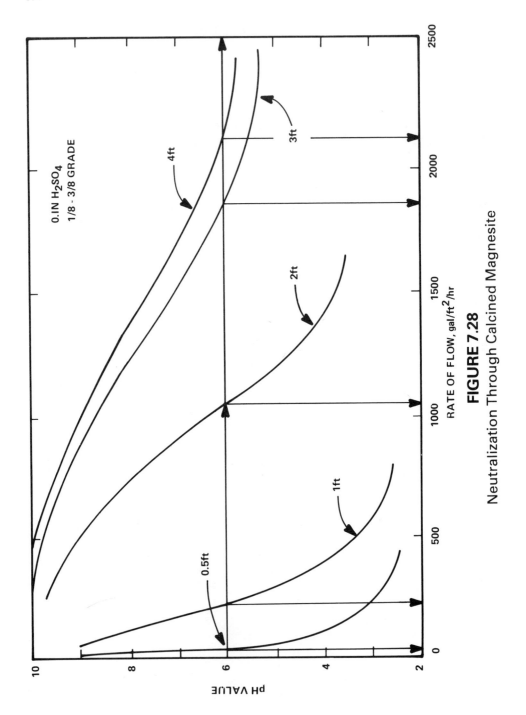

FIGURE 7.28

Neutralization Through Calcined Magnesite

quicklime (only the CaO portion) hydrates except at elevated temperature. Slaked quicklime is used as 8 to 15 percent lime slurry. Neutralization can also be accomplished using NaOH, Na_2CO_3, or NH_4OH.

Basic (Alkaline) Wastes

Any strong acid can be used effectively to neutralize alkaline wastes, but cost considerations usually limit the choice to sulfuric or possibly hydrochloric acid. The reaction rates are practically instantaneous, as with strong bases.

Flue gases that can contain 14 percent CO_2, can be used for neutralization. When bubbled through the waste, the CO_2 forms carbonic acid, which then reacts with the base. The reaction rate is somewhat slower but is sufficient if the pH need not be adjusted to below 7 or 8. Another approach is to use a spray tower in which the stack gases are passed countercurrent to the waste liquid droplets.

All of the above processes usually work better with the stepwise addition of reagents, that is, a staged operation. Two stages, with possibly a third tank to even out any remaining fluctuations, are generally optimum.

Systems

Batch treatment is used for waste flows to 100,000 gpd (380 m³/day). Continuous treatment employs automated pH control—where air is used for mixing, minimum air rate is 1 to 3 cfm/ft² (0.3 to 0.9 m³/min/m²) at 9 ft (3 m) liquid depth. If mechanical mixers are used 0.2–0.4 hp/1000 gal (0.04–0.08 kw/1000 l) is required. A batch system is shown in figure 7.29.

Control of Process

The automatic control of pH for waste streams is one of the most troublesome, for the following reasons:

1. The relation between pH and concentration or reagent flow is highly nonlinear, particularly when close to neutral (pH 7.0). The nature of the titration curve as shown in figure 7.30 favors multi-staging in order to insure close control of the pH.
2. The influent pH can vary at a rate as fast as 1 pH unit per minute.
3. The waste stream flow rates can double in a few minutes.
4. A relatively small amount of reagent must be thoroughly mixed with a large liquid volume in a short time interval.

Advantage is usually gained by the stepwise addition of chemicals (see figure 7.31). In reaction tank 1, the pH may be raised to 3 to 4. Reaction tank 2 raises the pH to 5 to 6 (or any other desired end point). If the waste stream is subject to slugs or spills, a third reaction tank may be desirable to effect complete neutralization.

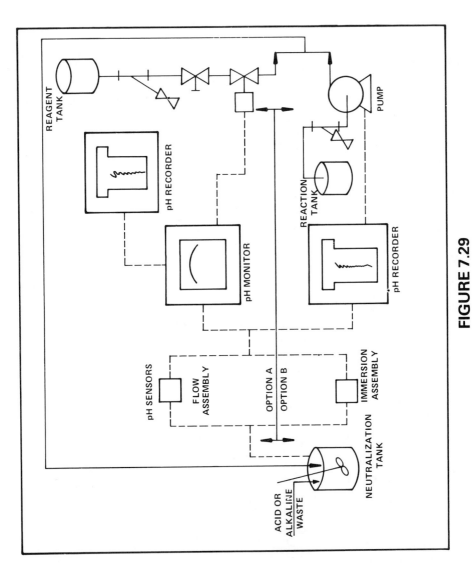

FIGURE 7.29

Schematic of Batch Neutralization System

FIGURE 7.30
Lime-Waste Titration Curve

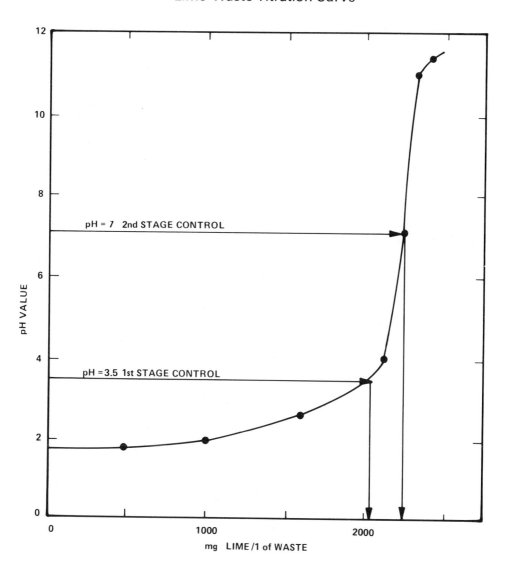

pH = 7 2nd STAGE CONTROL

pH = 3.5 1st STAGE CONTROL

pH VALUE

mg LIME /1 of WASTE

FIGURE 7.31

Schematic of Continuous Neutralization System

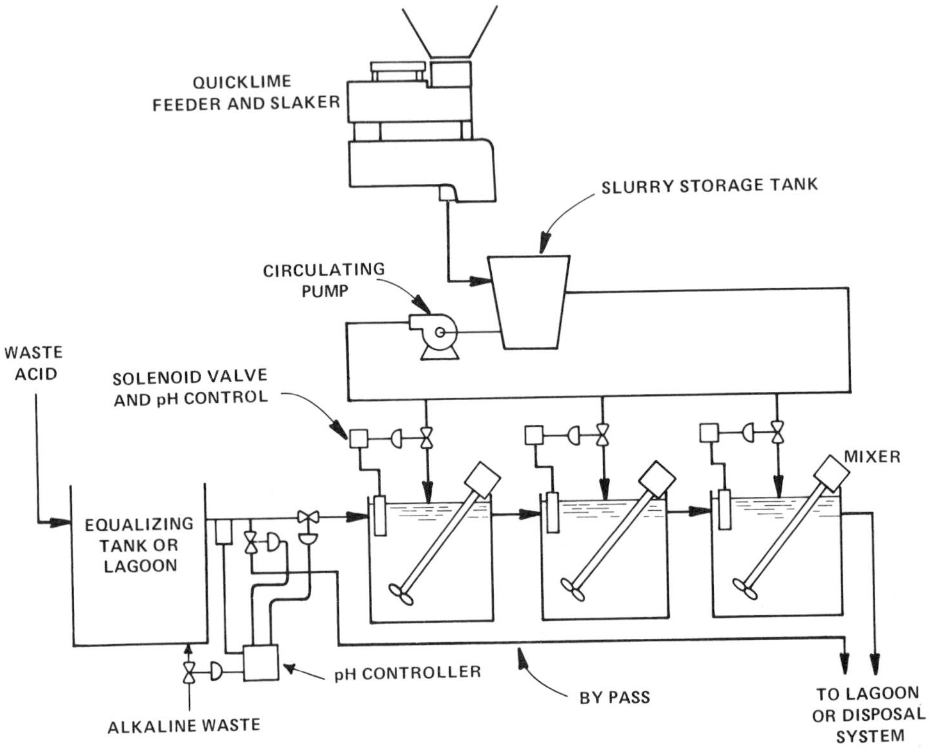

PROBLEM 7–4

An equalization system is to be designed for an industrial wastewater plant that has a highly variable product mix. The data, shown in figure I, consist of grab samples of the raw wastewater taken at two-hour intervals. An analysis of plant flow records indicate that the discharge rate is relatively constant with respect to the grab samples taken for design of the equalization system. The average waste flow is 1.4 mgd.

Problem 7–4 (a)

For constant operation of the plant 24 hr/day, 7 days/week, what equalization volume must be designed to provide a maximum effluent BOD 30 percent greater than the average with a 99.5 percent confidence level?

FIGURE I
Statistical Distribution of Biochemical Oxygen Demand
of Two-Hour Raw Waste Grab Samples

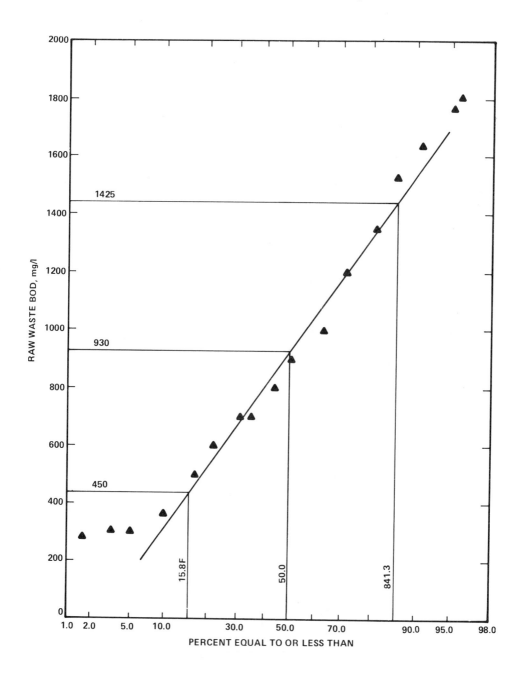

Problem 7–4 (b)

How many times will the desired value be exceeded in a one-year period?

Problem 7–4 (a)

Influent Standard Deviation:

$$\sigma_{INF} = \frac{BOD_{84.13\%} - BOD_{15.87\%}}{2}$$

$$= \frac{1425 - 450}{2}$$

$$= 488 \text{ mg/l}$$

Influent Variance:

$$S'_{INF} = (\sigma_{INF})^2$$

$$= (488 \text{ mg/l})^2$$

$$= 238000 \text{ (mg/l)}^2$$

Influent Mean:

$$\overline{X} \simeq 50 \text{ Percentile Number}$$

$$= 930 \text{ mg/l}$$

Maximum Allowable Effluent Concentration:

$$X_{MAX} = (1.30) \ (930 \text{ mg/l})$$

$$= 1210 \text{ mg/l}$$

Effluent Standard Deviation:

$$\sigma_{EFF} = \frac{X_{MAX} - \overline{X}}{Y}$$

For 99.5%, $Y = 2.575$

$$\sigma_{EFF} = \frac{1210 - 930}{2.575} \text{ mg/l}$$

$$= 109 \text{ mg/l}$$

Effluent Variance:

$$S'_{EFF} = (\sigma_{EFF})^2$$

$$= (109 \text{ mg/l})^2$$

$$= 11900 \text{ (mg/l)}^2$$

Required Detention Time:

$$t = \frac{(\Delta t)\ (S'_{INF})}{2\ (S'_{EFF})}$$

$$= \frac{2 \cdot 23800}{2 \cdot 11900} \text{ hr}$$

$$= 20.0 \text{ hrs or } 0.833 \text{ days}$$

Basin Volume:

$$V = 1.4 \text{ mgd} \cdot 0.833 \text{ days}$$

$$= 1.17 \text{ mg or } 156{,}000 \text{ ft}^3$$

Problem 7–4 (b)

Number of Times Value Exceeded $= 365 - 0.995 \cdot 365$
$= 2$ times/year

PROBLEM 7–5

Pilot plant studies on refinery wastewater showed that in order to maintain the pH of the influent wastewater at a value of 8.5, hydrated lime requirements as $Ca(OH)_2$ ranged from 100 mg/l to 700 mg/l with an average value of 500 mg/l. The maximum lime requirement for a pH $= 9.0$ was determined to be 1200 mg/l with a maximum dosage of 1000 mg/l being expected over an extended period. The mean wastewater flow is 5.0 MGD, and the maximum expected wastewater flow is 250,000 gal/hr. A caustic (NaOH) feed system is to be used as an alternative for a backup lime system. Determine the following:

Problem 7–5 (a)

The lime storage capacity (ft³) based on the controlling requirement of either the mean monthly requirement or the maximum two-week requirement. Assume the use of pebbled quick lime (CaO), which has a bulk density of 65 lb/ft³.

Problem 7–5 (b)

The capacity of the lime slaker and mechanism for transporting the bulk lime based on the maximum estimated requirements.

Problem 7–5 (c)

The mean and maximum slaking water requirements assuming a 10 percent by weight slurry.

Problem 7–5 (d)

The size of the slurry control tank based on a minimum resident time of 5 minutes.

Problem 7–5 (e)

The size of the caustic storage tank based on 24 hours caustic feed at the maximum extended usage equivalent to a hydrated lime requirement of 1000 mg/l. Assume NaOH is available having a purity of 98.9% and a solubility of 2.5 lbs/gal.

Problem 7–5 (f)

The maximum caustic (NaOH) feed rate for backup of the lime system.

Problem 7–5 (a)

Lime Storage Capacity:

Mean monthly requirement of lime (pH = 8.5):

$$Ca(OH)_2 = 5.0 \cdot 500 \cdot 8.34 \text{ lb/day}$$

$$= 20,850 \text{ lb/day}$$

$$CaO = 20850 \cdot \frac{56}{74} \text{ lb/day}$$

$$= 15{,}800 \text{ lb/day or } 237 \text{ ton/month}$$

Maximum two-week requirement of CaO (pH = 9.0):

$$CaO = 5.0 \cdot 1000 \cdot 8.34 \cdot \frac{56}{74} \text{ lb/day}$$

$$= 31{,}600 \text{ lb/day or } 221 \text{ ton/2 weeks}$$

Mean monthly requirements controls:

$$\text{Capacity} = 15800 \cdot 30/65 \text{ ft}^3$$

$$= 7{,}300 \text{ ft}^3$$

Problem 7–5 (b)

Capacity of Lime Slaker and Mechanism for Handling Bulk Lime (CaO):

(Basis: maximum hourly flow rate and maximum dosage)

$$\text{Capacity} = 0.25 \cdot 1200 \cdot 8.34 \cdot \frac{56}{74} \text{ lb/hr of CaO}$$

$$= 1{,}900 \text{ lb/hr of CaO}$$

Problem 7–5 (c)

Water Requirements for Slaking to a 10% By Weight Slurry:

$$\text{Mean slaking water} = 20850 \cdot \frac{0.9}{0.1} \cdot \frac{1}{8.34} \text{ gal/day}$$

$$= 22{,}500 \text{ gal/day or } 15.6 \text{ gpm}$$

$$\text{Maximum slaking water} = 1900 \cdot \frac{74}{56} \cdot \frac{0.9}{0.1} \cdot \frac{1}{8.34} \text{ gal/hr}$$

$$= 2{,}710 \text{ gal/hr or } 45.2 \text{ gpm}$$

Problem 7–5 (d)

Size Slurry Control Tank:

Maximum Rate from Part (3) is 45.2 gpm
Capacity = 5 min × 45.2 gpm = 226 gal

Problem 7–5 (e)

Caustic Storage Tank:

$$\text{Pure NaOH} = 5.0 \cdot 1000 \cdot \frac{40}{37} \cdot 8.34 \text{ lb/day}$$

$$= 45,100 \text{ lb/day}$$

$$\text{Capacity for NaOH} = 45,100 \cdot \frac{1}{0.989} \cdot \frac{1}{2.5} \text{ gal}$$

$$= 18,240 \text{ gal or } 2,440 \text{ ft}^3$$

Problem 7–5 (f)

Maximum Caustic Feed Rate:

(Basis: $Ca(OH)_2$ Dosage = 1200 mg/l and Flow = 250,000 gal/hr)

$$\text{Feed rate} = 0.25 \cdot 1200 \cdot \frac{40}{37} \cdot 8.34 \cdot \frac{1}{0.989} \cdot \frac{1}{2.5} \text{ gal/hr}$$

$$= 1,094 \text{ gal/hr or } 18.2 \text{ gpm}$$

PROBLEM 7–6

A wastewater flow of 100 gpm equivalent to 0.1N H_2SO_4 requires neutralization prior to secondary treatment. This flow is to be neutralized to a pH of 7.0 using a limestone bed. Figure I presents the results of a series of laboratory pilot tests using a one-foot diameter limestone bed. This data was for upflow units with the effluent being aerated to remove residual CO_2. Assume limestone is 60 percent reactive. Design a neutralization system specifying:

Problem 7–6 (a)

Most economical bed depth of limestone

Problem 7–6 (b)

The pounds of acid/day to be neutralized

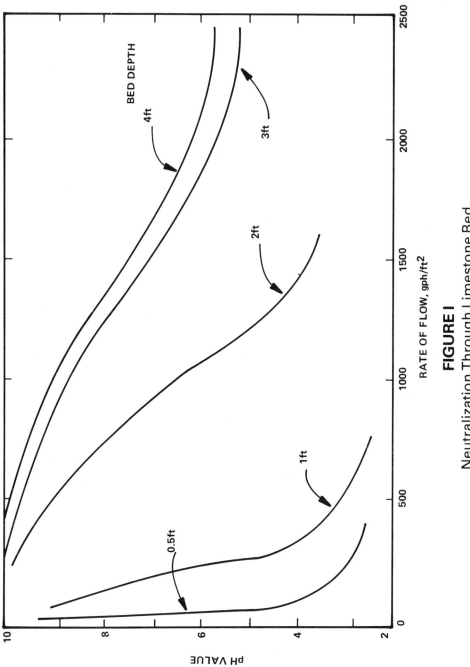

FIGURE I

Neutralization Through Limestone Bed

Problem 7–6 (c)

Limestone requirements on an annual basis

For pH 7.0, the allowable hydraulic loadings to the limestone bed with various depths are estimated from figure I.

The required flow rate per unit limestone value can be calculated by:

$$\text{gal/ft}^3\text{-hr} = \frac{\text{hydraulic loading (gal/ft}^2\text{-hr)}}{\text{bed depth (ft)}}$$

Then the following table can be derived:

Depth, ft	0.5	1.0	2.0	3.0	4.0
Hydraulic loading, gal/ft²-hr	50	200	850	1500	1700
Flow rate per unit limestone volume, gal/ft³-hr	100	200	425	500	425

Problem 7–6 (a)

By plotting the flow rate per unit limestone value against the limestone bed depth, the most economical bed depth of limestone is found to be 3 ft. (see figure II).

Problem 7–6 (b)

The weight of acid to be neutralized per day:

$$= (100 \text{ gpm } 0.1 \text{ N } H_2SO_4) \left(\frac{4900 \text{ mg/l}}{0.1 \text{ N } H_2SO_4} \right) \left(\frac{1440 \text{ min}}{\text{day}} \right) \left(\frac{\text{mg}}{10^6 \text{ gal}} \right) \quad (8.34)$$

$$= 5{,}890 \text{ lb/day}$$

Problem 7–6 (c)

Annual limestone requirements:

$$= 5890 \cdot \frac{50}{49} \cdot 365 \cdot \frac{1}{0.60} \text{ lb/year of } CaCO_3$$

$$= 3{,}660{,}000 \text{ lb/year of } CaCO_3$$

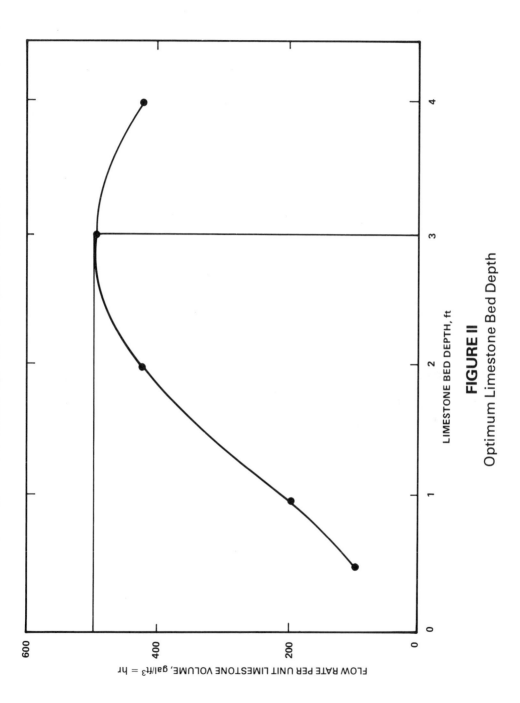

FIGURE II

Optimum Limestone Bed Depth

REFERENCES

1. Hazen, A. 1904. On sedimentation. *Trans. ASCE* 53: 45.
2. Camp, T. R. 1946. Sedimentation and the design of settling tanks. *Trans. ASCE* 111: 895.
3. Dobbins, W. E. 1961. *Advances in sewage treatment design, sanitary engineering division,* New York, NY.: Manhattan College Metropolitan Section.
4. Eckenfelder, W. W. and O'Connor, D. J. 1961. *Biological waste treatment,* Oxford: Pergamon Press.
5. Eckenfelder, W. W. and Ford, D. L. 1969. *Laboratory and design procedures for waste treatment process design.* CRWR Rep. 31. Austin: University of Texas.
6. Hansen, S. P. and Culp, G. L. 1967. Applying shallow depth sedimentation theory. *J. American Water Works Association* 59: 1134.
7. Culp, G. L. *et al.* 1968. High rate sedimentation in water treatment works. *J. American Water Works Association* 60: 681.
8. Ford, D. L. and Manning, F. S. 1977. Oil removal from wastewaters. In *Process Design in Water Quality Engineering: New Concepts and Developments* Nashville: Vanderbilt University.
9. American Petroleum Institute. 1959. *Manual on disposal of refinery wastes,* vol. 1. New York: API.
10. Eckenfelder, W. W. 1966. *Industrial water pollution control,* New York: McGraw-Hill.
11. Novotny, V. and England, A. J. 1974. Equalization design techniques for conservative substances in wastewater treatment systems. *Water Research* 8: 325.
12. DiToro, D. M. 1975. Statistical design of equalization basin *J. Environ. Eng. Div., ASCE* 101: 917.
13. Tully, T. J. 1958. Waste acid neutralization. *Sew. Ind. Wastes* 30: 1385.

Oxygen Transfer and Aeration

The oxygen-transfer process can be considered to occur in three phases. Oxygen molecules are initially transferred to the liquid surface, resulting in a saturation or equilibrium condition at the interface. This rate is very rapid. The liquid interface has a finite thickness with unique properties. The layer of film is composed of water molecules with their negative ends facing the gas phase and is estimated to be at least three molecules thick. During the second phase, the oxygen molecules must pass through this film by molecular diffusion. In the third phase, the oxygen is mixed in the body of liquid by diffusion and convection. It is assumed that at very low mixing levels (laminar flow conditions) the rate of oxygen absorption is controlled by the rate of molecular diffusion through the undisturbed liquid film (phase II). At increased turbulence levels, the surface film is disrupted and renewal of the film is responsible for transfer of oxygen to the body of the liquid [1, 2, 3]. This surface renewal can be considered as the frequency with which fluid with a solute concentration C_L is replacing fluid from the interface with a concentration C_s (figure 8.1).

The oxygen-transfer rate can be expressed

$$N = K_L A (C_s - C_L) \qquad (8-1)$$

where N = mass of oxygen transferred per unit time
K_L = liquid film coefficient (related to surface renewal rate; increased with turbulence or renewal rate)
A = interfacial area for transfer
C_s = saturation concentration of oxygen at interface
C_L = concentration of oxygen in the body of the liquid

Equation (8–1) can be reexpressed in concentration units:

$$\frac{N}{V} = \frac{dC_L}{dt} = K_L \frac{A}{V} (C_s - C_L) = K_L a (C_s - C_L) \qquad (8-2)$$

in which $K_L a$ is the overall coefficient for oxygen transfer and includes both the liquid film coefficient (K_L) and the interfacial area per unit volume (A/V). In aeration practice it is impossible to measure the interfacial area, so the overall coefficient $K_L a$ is used for aeration design. $K_L a$ can be determined from a semilogarithmic plot of $(C_s - C_L)$ against time of aeration as shown in figure 8.2.

PROBLEM 8–1

An air diffuser yielded a $K_L a$ of 6.5/hr at 15°C. The air flow is 50 ml/min, the bubble diameter 0.15 cm and the bubble velocity 28 cm/sec. The depth of the aeration column is 250 cm with a volume of 4000 cm^3.

Problem 8–1 (a)

Compute the liquid film coefficient K_L.

FIGURE 8.1
Mechanism of Oxygen Transfer

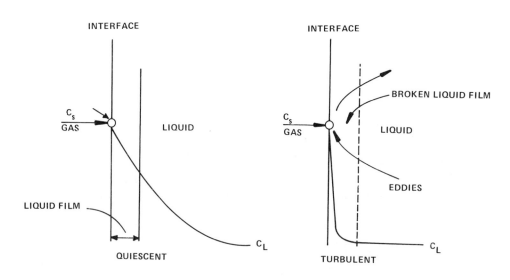

FIGURE 8.2

Determination of $K_L a$ From Experimental Data

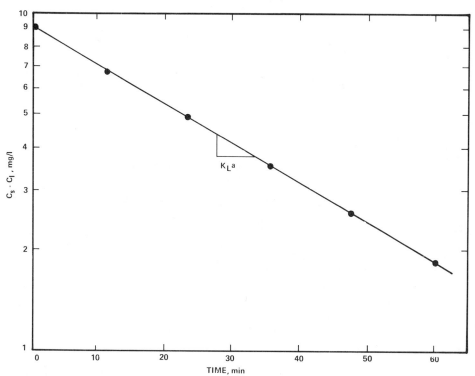

Problem 8–1 (b)

Compute the overall coefficient $K_L a$ at 250°C for $\Theta =$ 1.024.

The interfacial area per unit volume, a, is defined as:

$$a = \frac{A}{V} = \frac{K_L a}{K_L}$$

where A = interfacial area for transfer
V = volume of liquid body

Further, the interfacial area of spherical air bubble in the liquid can be expressed as:

$$A = \frac{6 G_s H}{d_B v_B}$$

where G_s = air flow
      H = liquid depth
      d_B = diameter of air bubble
      v_B = bubble rising velocity

Problem 8–1 (a)

The liquid film coefficient:

$$K_L = \frac{V K_L a}{A} = \frac{d_B\, v_B\, V\, K_L a}{6\, G_s\, H}$$

$$= \frac{0.15 \cdot 28 \cdot 4000 \cdot 6.5 \cdot 1/3600}{6 \cdot 50 \cdot 1/60 \cdot 250} \text{ cm/sec}$$

$$= 0.0243 \text{ cm/sec or } 87.4 \text{ cm/hr}$$

Problem 8–1 (b)

The overall coefficient:

$$K_L a_{(T)} = K_L a_{(15)} \theta^{T-15}$$

$$K_L a_{(25)} = 6.5\,(1.024)^{25-15}$$

$$= 8.24/\text{hr}$$

FACTORS AFFECTING OXYGEN TRANSFER

Several factors affect the performance of aeration devices. Since manufacturers test and rate their equipment under standard conditions in a test tank, (20°C and zero dissolved oxygen), correction must be made for operation in wastewater systems. These corrections are summarized below.

Oxygen Saturation, C_s

The value of oxygen saturation in water is related to the partial pressure of oxygen in the gas phase by Henry's law;

$$C_s = Hp \qquad\qquad (8\text{–}3)$$

where C_s = saturation concentration of oxygen
      H = Henry's constant, available in handbooks
      p = partial pressure of oxygen (for air, .209 × total pressure)

An increase in temperature, dissolved solids, altitude, or waste constituents will affect oxygen saturation. The effect of temperature on oxygen saturation in pure water is shown in table 8.1. Increasing altitude causes a decrease in total pressure and thus oxygen partial pressure.

The effect of dissolved solids and waste concentration on oxygen saturation varies from one waste to another and must be measured in the field. The relationship is usually given by:

$$\beta = \frac{C_s(\text{waste})}{C_s(\text{water})} \tag{8-4}$$

where β is experimentally determined. The effect of total dissolved solids on C_s is also shown in table 8.1.

For submerged aeration devices a correction must be made for increased oxygen partial pressure due to submergence. Traditionally the saturation concentration at the tank middepth, C_{sm}, has been used as the average saturation concentration. C_{sm} is given by:

$$\begin{aligned} C_{sm} &= \frac{C_s}{2}\left(\frac{P_b}{p} + \frac{O_t}{20.9}\right) \\ &= \frac{C_s}{2}\left(\frac{p + 0.433h}{p} + \frac{O_t}{20.9}\right) \end{aligned} \tag{8-5}$$

where C_s = saturation concentration at surface
      P_b = absolute pressure at diffuser depth, psia
      p = basin ambient barometric pressure, lbs/in^2
      h = diffuser submergence, ft
      O_t = oxygen in exit gas, percent by volume

$$= \frac{21(1-E)\,100}{79 + 21(1-E)} \quad (\%)$$

where E = decimal fraction of oxygen transferred to liquid

Although middepth saturation concentration has been traditionally used for average saturation throughout the tank, Schmit et al. [4] from studies with air diffusers found the relationship:

$$C_{s\,\text{ave}} = C_s\left(\frac{14.7 + 0.4335 \times h}{14.7}\right) \tag{8-6}$$

where h = submergence of diffuser, ft
      χ = decimal of submerged depth at which pressure corresponds to measured saturation concentration.

Temperature		Elevation							TDS (at sea level)			
°F	°C	0'	1000'	2000'	3000'	4000'	5000'	6000'	400ppm	800ppm	1500ppm	2500ppm
32.0	0	14.6	14.1	13.6	13.1	12.6	12.1	11.7	–	–	–	–
35.6	2	13.8	13.3	12.8	12.4	11.9	11.5	11.1	13.74	13.68	13.58	13.42
39.2	4	13.1	12.6	12.2	11.8	11.4	10.9	10.5	13.04	12.98	12.89	12.75
42.8	6	12.5	12.0	11.6	11.2	10.8	10.4	10.0	12.44	12.38	12.29	12.15
46.4	8	11.9	11.4	11.0	10.6	10.2	9.9	9.5	11.85	11.80	11.70	11.58
50.0	10	11.3	10.9	10.5	10.1	9.8	9.4	9.1	11.25	11.20	11.12	11.00
53.6	12	10.8	10.4	10.1	9.7	9.4	9.0	8.6	10.76	10.71	10.64	10.52
57.2	14	10.4	10.0	9.6	9.3	8.9	8.6	8.3	10.36	10.32	10.25	10.15
60.8	16	10.0	9.6	9.2	8.9	8.6	8.3	8.0	9.96	9.92	9.85	9.75
64.4	18	9.5	9.2	8.9	8.5	8.2	7.9	7.6	9.46	9.43	9.36	9.27
68.0	20	9.2	8.8	8.5	8.2	7.9	7.6	7.3	9.16	9.13	9.06	8.97
71.6	22	8.8	8.5	8.2	7.9	7.6	7.3	7.1	8.77	8.73	8.68	8.60
75.2	24	8.5	8.2	7.9	7.6	7.3	7.1	6.8	8.47	8.43	8.38	8.30
78.8	26	8.2	7.9	7.6	7.3	7.1	6.8	6.6	8.17	8.13	8.08	8.00
82.4	28	7.9	7.6	7.4	7.1	6.8	6.6	6.3	7.87	7.83	7.78	7.70
86.0	30	7.6	7.4	7.1	6.9	6.6	6.4	6.1	7.57	7.53	7.48	7.40
89.6	32	7.4	7.1	6.9	6.6	6.4	6.2	5.9	7.4	–	–	–
93.2	34	7.2	6.9	6.7	6.4	6.2	6.0	5.8	7.2	–	–	–
96.8	36	7.0	6.7	6.5	6.3	6.0	5.8	5.6	7.0	–	–	–
100.4	38	6.8	6.6	6.3	6.1	5.9	5.6	5.4	6.8	–	–	–
104.0	40	6.6	6.4	6.1	5.9	5.7	5.5	5.3	6.6	–	–	–

TABLE 8.1

Solubility of Oxygen (mg/l) at Various Temperatures,
Elevations, and Total Dissolved Solid Levels

χ varied from 0.22 to 0.33 over a range of depths of diffuser submergence varying from 11.0 to 21.0 feet.

Oxygen Transfer Coefficient, $K_L a$

The liquid film coefficient K_L and the interfacial area per unit volume, A/V, which constitute the oxygen transfer coefficient, are both affected by the type and turbulence level of aeration equipment.

The overall mass transfer coefficient, $K_L a$, includes the effects of changes in the liquid film coefficient, K_L, and the interfacial area per unit volume, A/V. The liquid film coefficient in aeration processes as defined by Danckwertz [2] is the square root product of the molecular diffusion coefficient, D_L, and the rate of surface renewal:

$$K_L = \sqrt{D_L r} \qquad (8\text{--}7)$$

The surface renewal rate, r, is the average frequency with which the interfacial film is replaced with liquid from the body of solution. In the turbulent regime which prevails in most aerobic biological systems, the oxygen transfer rate is a function of surface renewal. K_L is dependent on the surface tension and molecular characteristics which prevail at the gas-liquid interface of a given fluidized system, and A/V, which depends on the turbulence and bubble patterns in an aeration system.

One of the most significant factors that affects $K_L a$, based on diffusivity and viscosity, is temperature. This temperature effect can be defined by the relationship:

$$K_L a_{(T)} = K_L a_{(20°)} \theta^{(T\text{-}20)} \qquad (8\text{--}8)$$

The Θ value has been reported to vary from 1.016 to 1.037. Values of 1.020 and 1.028 are normally used for bubble systems, while a value of 1.024 has been used for mechanical aeration systems.

Values of $(K_L a C_s)$ for stream aeration along with various types of aeration equipment have recently been correlated with temperature by Imhoff and Albrecht [5] as shown in figure 8.3. This correlation combines the temperature effect of both $K_L a$ and C_s at 10° C as the base temperature. Figure 8.3 indicates that the level of turbulence in a system significantly affects the temperature dependency of $(K_L a C_s)$. For low turbulence, diffused aeration systems molecular diffusion has a minor influence. This conclusion agrees well with the data of Metzger and Dobbins [6] who have determined the average Θ values for the liquid film coefficient, K_L, to be 1.032 for low intensity mixing and 1.006 for high intensity mixing.

When oxygen is supplied to fluidized systems treating wastewaters via aerobic biological oxidation, it is necessary to define a correction factor that relates the oxygen transfer to the nature of the waste. Using the transfer of oxygen

FIGURE 8.3

Effect of Temperature on Relative DO Absorption
$(K_L a_{(T)} C_{s(T)} / K_L a_{(10)} C_{s(10)})$ [5]

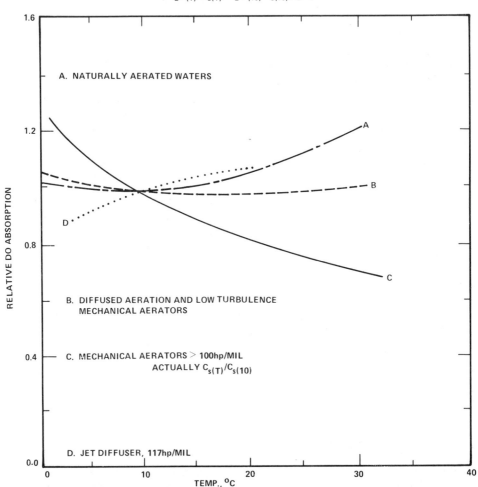

to tap water as the datum, the parameter, α, serves as this correction factor. Specifically, it relates the overall mass transfer coefficient $(K_L a)$ of the wastewater to that of tap water:

$$\alpha = \frac{K_L a(\text{wastewater})}{K_L a(\text{tap water})} \qquad (8\text{–}9)$$

Although α is represented in the mathematical relationships describing diffused, turbine, and surface aeration system performance, there are many variables that affect its magnitude. These include:

1. Temperature of the mixed liquor
2. Nature of dissolved organic and mineral constituents
3. Characteristics of the aeration equipment (diffused or mechanical)
4. The liquid mixing intensity that affects the surface renewal rate
5. The liquid depth and geometry of the aeration basin

Effect of Waste Characteristics on Oxygen Transfer

The presence of surface-active agents have a marked effect on the oxygen-transfer rate as they affect both the liquid film coefficient K_L and the A/V ratio and hence $K_L a$. This effect will be reflected by changes in concentration of surfactant and by changes in the turbulence and mixing in the system.

A surfactant concentrates at an interface such that the interfacial concentration is greater than that in the body of the liquid. As a result, a "film" of adsorbed surfactant molecules is concentrated at the interface, which provides a barrier to diffusion of oxygen across the interface.

The changes in transfer rate in the presence of surface-active materials is defined by the coefficient, α. Increasing the concentration of surfactant will decrease α until the interfacial surface is saturated. Further increases will not affect α (figure 8.4). In bubble aeration, the presence of surfactants will markedly decrease the bubble size and hence increase A/V. Under these conditions, it is possible for α to increase in some cases to values greater than 1.0 (figure 8.4), because the increased effect of A/V exceeds the decrease in K_L caused by the surface barrier.

The degree of turbulence in the system also affects α.

Under laminar conditions (approaching a stagnant film surface) there is substantially no effect on α because the resistance in the bulk of solution to oxygen transport exceeds the combined interfacial resistance [6]. This condition would rarely be encountered in aeration practice. Under moderately turbulent conditions, a maximum depression occurs because the interfacial resistance to molecular diffusion by the adsorbed surfactant molecules controls the transfer rate.

At high degrees of turbulence, α approaches unity as a result of the high surface-renewal rates, resulting in an inability to establish an adsorption equilibrium at the interface (figure 8.5). As a result of the increased A/V values associated with turbulent conditions at the surface and bubble entrainment (figure 8.5), α often exceeds unity.

The coefficient α may be expected to increase or decrease and approach unity during the course of biooxidation when the substances affecting the transfer rate are being removed in biological process as shown in (figure 8.6). Alpha (α) values reported for various wastewaters are shown in table 8.2.

FIGURE 8.4
Effect of Concentration of Surface Active Agent on Oxygen Transfer

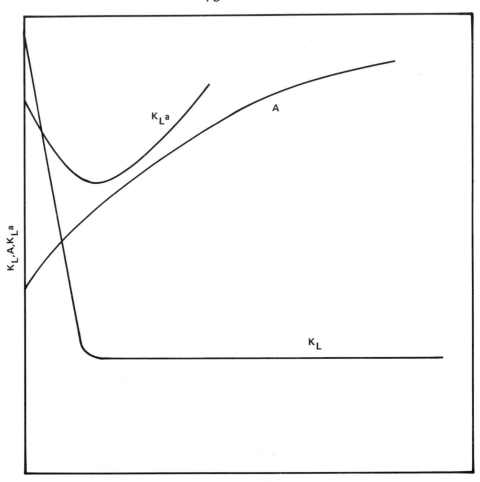

CONCENTRATION OF SURFACTANTS

Dissolved Oxygen Concentration in Bulk Liquid, C_L

According to equation (8–2) the lower the C_L value, the greater the rate of oxygen transfer. In setting a C_L value, two factors must be considered:

1. Minimum dissolved oxygen concentration required by the biological floc particles to maintain their maximum oxygen utilization rate

FIGURE 8.5
Effect of Turbulence on Oxygen Transfer

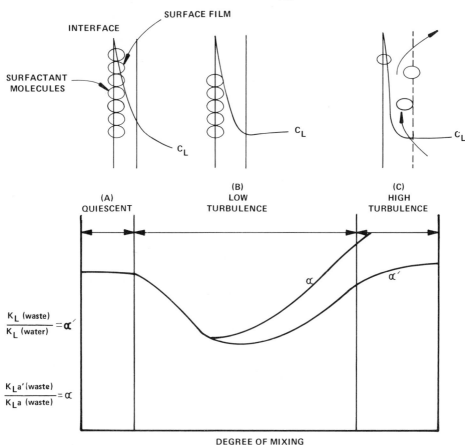

2. Varying oxygen demands due to flow and organic load variations

Data by Borowski and Johnson [8] indicate that if dispersed cells are used for biological oxidation, a very low oxygen concentration (0.04 mg/l) is required in solution. However, due to the nature of activated sludge, the cells are clumped together to form a floc particle allowing them to settle from solution. Thus, to reach the active sites at the bacterial cell membranes, oxygen must penetrate the liquid film surrounding the floc particles and diffuse through the floc matrix to the active sites. For the typical size of activated sludge particles (20 to 115 μm), a dissolved oxygen concentration in the order of 1.0 mg/l should be maintained in extended aeration systems and 2.0–2.5 mg/l for conventionally loaded systems. When nitrification is to be achieved, the dissolved oxygen level should be in excess of 2.0 mg/l.

FIGURE 8.6

Effect of Wastewater Treatment Efficiency on α

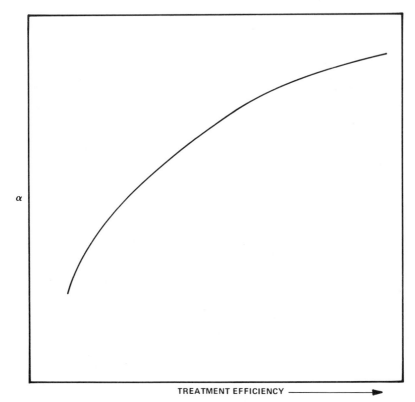

TREATMENT EFFICIENCY ⟶

TABLE 8.2

Alpha Values for Various Wastes[a]

Type of Waste	Alpha Value
Water	1.0 (by definition)
Sewage	0.8–0.95
Synthetic sewage	0.8–0.85
Synthetic fiber	0.45–0.65
Apple processing waste	0.8
Soap and detergent production	0.6 (raw) 0.85 (treated)
Oils and essence production	0.5 (raw) 0.90 (treated)
Paper (reworked)	0.65–0.90
Pulp and paper (integrated mill)	0.7–0.8

[a]From [7].

FIGURE 8.7

Determination of K_La Under Nonsteady State
Conditions in Biological Systems

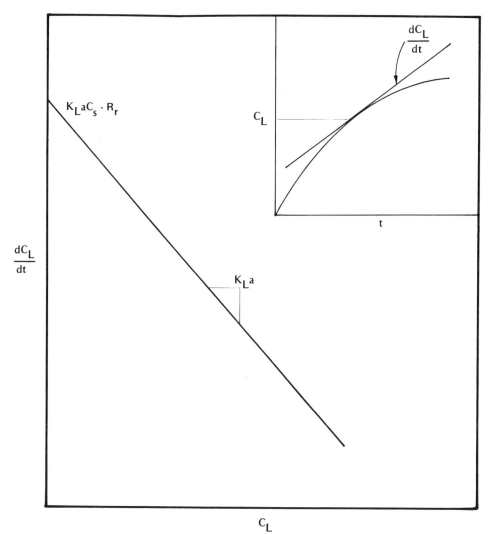

In the presence of biomass in which oxygen is being removed from the system equation (8–2) becomes:

$$dC_L/dt = K_La\,(C_s - C_L) - R_r \tag{8–10}$$

in which R_r is the oxygen uptake rate in mg O_2/l/hr. Under steady state conditions K_La can be computed:

$$K_La = \frac{R_r}{C_s - C_L} \tag{8–10a}$$

Under non-steady state conditions, equation (8–10a) can be modified:

$$dC_L/dt = (K_La\,C_s - R_r) - K_La\,C_L \tag{8–10b}$$

and K_La determined from a plot of dC_L/dt against C_L as shown in figure 8.7 (see page 211).

PROBLEM 8–2

The following parameters are measured at steady state in an activated sludge plant:

DO at steady state = 2.8 mg/l

R_r = 0.835 mg/1-min

Temp. = 22° C

Compute K_La.
  At steady state:

$$K_La = \frac{R_r}{C_s - C_L}$$

At 22° C, C_s = 8.78 mg/l and:

$$K_La = \frac{0.835}{8.78 - 2.8}\ \text{min}^{-1}$$

$$= 0.140\ \text{min}^{-1}\ \text{or}\ 8.38\ \text{hr}^{-1}$$

PROBLEM 8–3

The following data was obtained for nonsteady state aeration in an activated sludge basin:

Time (min)	C_L (mg/l)
0.0	0.52
0.5	0.70
1.0	0.93
2.0	1.23
3.0	1.55
4.0	1.80
5.0	2.00
10.0	2.20

Determine the $K_L a$.
Unsteady state equation is:

$$\frac{dC_L}{dt} = K_L a \, (C_s - C_L) - R_r$$

Rearrangement yields:

$$\frac{dC_L}{dt} = (K_L a \, C_s - R_r) - K_L a \, C_L$$

By plotting dC_L/dt against C_L the negative slope of the resulting straight line is $K_L a$. The dC_L/dt values are calculated and plotted in table I and figures I and II (see pages 214 and 215).

DESIGN RELATIONSHIPS— MANUFACTURERS SPECIFICATIONS

Three types of aeration equipment are generally employed in wastewater treatment; diffused, turbine, and surface. A manufacturer generally designates the oxygen transfer capability of his equipment in terms of either (1) the rated capacity (also called standard transfer efficiency), N_o, the pounds of O_2 transferred per hour per unit horsepower, or (2) the oxygen absorption efficiency, E_o, the pounds of oxygen transferred to solution per aeration unit. The latter term applies only to turbine and diffused aeration where a known quantity of oxygen is supplied by blower.

In evaluating the capacity of their equipment, manufacturers use the following standard conditions:

Medium = tap water
Temperature = 20° C
C_L = O, thus yielding the maximum driving force

TABLE I

t (min)	C_L (mg/l)	dC_L/dt
0	0.51	0.43
1	0.91	0.37
2	1.25	0.32
3	1.55	0.27
4	1.80	0.22
5	1.99	0.17

FIGURE I

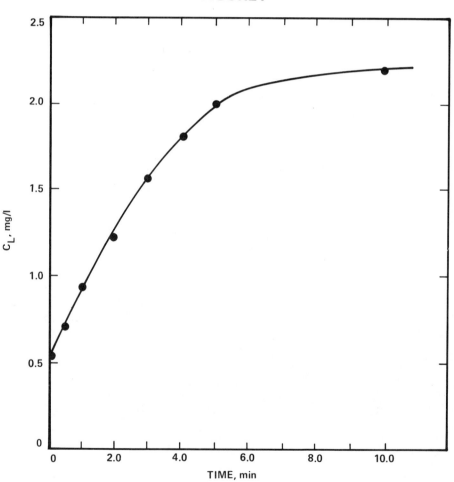

FIGURE II

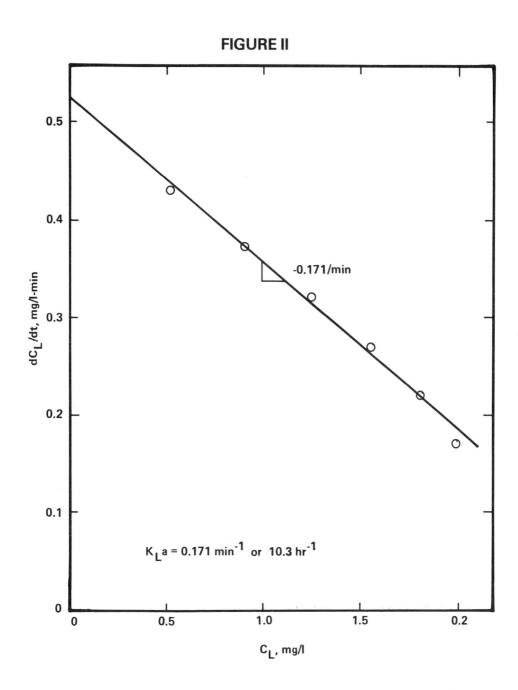

By dividing equation (8–1) by the horsepower per unit volume and applying the above standard conditions, the rated capacity is given by:

$$N_o = K_L a_{(20)} \frac{V}{hp} C_{s(20)} {}^{(8.34)} \tag{8–11}$$

where N_o = lb O_2l/hp-hr
$C_s(20)$ = 9.2 mg/l (see *Standard Methods*) for surface aerators
$C_s(20)$ = $C_{sm}(20)$ for diffused and turbine aeration
hp   = aeration equipment horsepower
V = aeration tank volume, MG

For diffused and turbine aeration equipment, rated capacity may also be given in terms of lb O_2/hr—aeration unit. The oxygen absorption efficiency for these types of equipment is given by:

$$E_o = \frac{\text{lb } O_2/\text{hr absorbed}}{\text{lb } O_2/\text{hr supplied}} = \frac{K_L a_{(20)} \, C_{sm\,(20)} \, V(8.34)}{1.05 \, G_o} \tag{8–12}$$

in which G_o is air flow rate, scfm (standard condition, for blower or compressor, is normally dry air at 1 atmosphere pressure and 70° F. The value 1.05 is the conversion factor (0.075 lb/ft³ air × 0.232 lb O_2/lb air × 60 min/hr).

To design aeration systems, the manufacturers' specifications must be corrected to account for the actual operating conditions in the aeration tank as follows:

$$\frac{N}{N_o} = \frac{E}{E_o} = \frac{(\beta C_s - C_L) \, \alpha \, \theta^{(T-20)}}{C_{s\,(20)}} \tag{8–13}$$

where N and E are the rated capacity and absorption efficiency under design conditions.

AERATION EQUIPMENT

The aeration equipment commonly employed in the wastewater treatment field consists of air diffusion units; turbine aeration systems in which air is released below the rotating blades of a submerged impeller; static aerators where air bubbles are released under submerged vertically mounted cylindrical tubes that contain baffles; and surface aeration units in which oxygen transfer is accomplished by high surface turbulence and liquid spray. The common types of aeration devices are shown in figure 8.8.

Diffused Aeration

Diffused aeration equipment is basically of two types. One type produces a small bubble from a porous media and the other uses a large orifice or a hydraulic shear device to produce larger air bubbles.

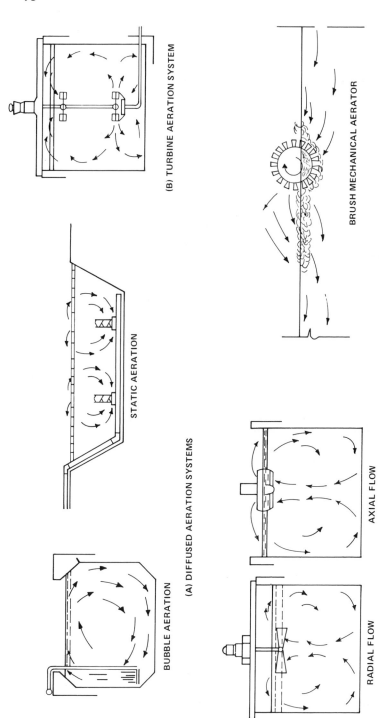

(A) DIFFUSED AERATION SYSTEMS

(B) TURBINE AERATION SYSTEM

(C) SURFACE AERATION SYSTEMS

FIGURE 8.8

Schematics of Selected Aeration Equipment

Porous media are either tubes or plates constructed of carborundum or tightly wrapped plastic wrap or nylon. Tubes are placed at the side wall of the aeration tank perpendicular to the wall two to three feet from the bottom and they generate a rolling motion to maintain mixing. Minimum and maximum spacings are required to maintain solids in suspension and to avoid bubble coalescence, respectively. The size of bubbles released from this type of diffuser range from 2.0 to 2.5 mm.

To maintain adequate mixing, the maximum width of aeration tank is approximately twice the depth. This width can be doubled by placing a line of diffusion units along the center line of the aeration tank.

Large-bubble air diffusion units will not yield the oxygen-transfer efficiency of fine-bubble diffusers, because the interfacial area for transfer is considerably less. These units have the advantage, however, of not requiring air filters and of generally requiring less maintenance. Large-bubble diffusers are placed along the side wall of an aeration tank in a manner similar to porous diffusers. These units generally operate over a wider range of air flows per unit.

The variables affecting the performance of diffused aeration units are air flow rate, tank liquid depth, and tank width. These are shown for a sparger unit in figure 8.9.

The performance of all air diffusion units can be expressed by the relationship:

$$N = C G_s^n \frac{H^m}{W^p} (\beta C_{sm} - C_L) \cdot 1.02^{(T-20)} \cdot a \qquad (8\text{--}14)$$

where N = lb of O_2/hr/aeration unit
      C = constant for the aeration unit
      G_s = air flow, scfm/aeration unit
      H = liquid depth, ft
      W = aeration tank width, ft
      C_L = dissolved oxygen concentration in liquid, mg/l
      T = temperature, °C
      α = oxygen-transfer coefficient ratio of the waste to water
    n,m,p = exponents characteristic of the aeration device

Aeration performance characteristics are summarized in table 8.3. An air diffusion system is shown in figure 8.10.

Turbine Aeration

In turbine aeration, air is discharged from a pipe or sparge ring beneath the rotating blades of an impeller. The air is broken into bubbles and dispersed throughout the tank contents. Present commercial units employ one or more submerged impellers and may employ an additional impeller near the liquid sur-

FIGURE 8.9

Oxygen Transfer Characteristics of a Sparger Unit

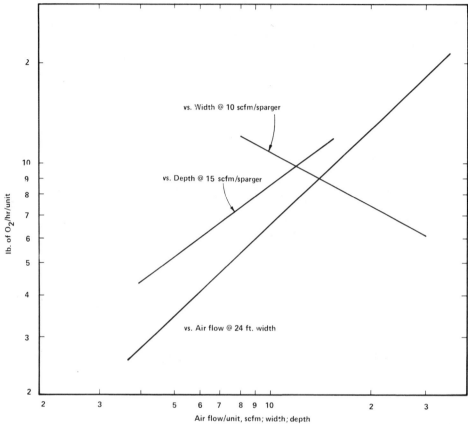

face for oxygenation from induced surface aeration. A typical turbine unit is shown in figure 8.11.

In addition to air flow, both the diameter and the speed of the impeller will affect the bubble size and velocity, thus influencing the overall transfer coefficient, K_La. The performance of turbine aerators can generally be defined by the relationship:

$$N = CR^x G_s^n d^y (\beta C_{sm} - C_L) \cdot 1.02^{(T-20)} \cdot \alpha$$

(8–15)

where N = lb O_2 transferred/hp-hr,
C = constant characteristic of aeration device,
R = impeller peripheral speed, ft/sec,
d = impeller diameter, ft
x, y, n = exponents characteristic of the aeration device.

TABLE 8.3

Diffused Aeration-Performance Characteristics

Unit	C	n	Conditions
Saran tubes	0.0160	0.90	9-in. spacing, wide band; 14.4 ft depth, 24-ft width
Saran tubes	0.0170	0.81	9-in. spacing, narrow band; 14.4-ft depth, 24-ft width
Saran tubes	0.0150	0.92	9-in. spacing, narrow band; 14.4-ft depth, 24-ft width
Spargers	0.0081	1.02	24-in. spacing, wide band; 14.8-ft depth, 24-ft width
Spargers	0.0062	1.02	9-in. spacing, narrow band; 14.8-ft depth, 24-ft width
Spargers	0.0064	1.02	9/32-in. orifice; 25-ft width 15-ft depth
Spargers	0.0068	1.02	13/64-in. orifice; 25-ft width 15-ft depth
Plate tubes	0.0350	0.49	Single row; 25-ft width, 15-ft depth
Plate tubes	0.0200	0.80	Double row; 25-ft width, 15-ft depth
INKA system	0.0036	0.95	6.8-ft width, 6-ft depth, 2.6-ft submergence

All other variables are as previously assigned for equation (8–14).

In practice, C, x, y, and n will be unknown and the designer must rely on the value of N_o supplied by the manufacturer.

When air is applied beneath the impeller, the actual power drawn is reduced because a less dense mixture is being pumped by the turbine. Quirk [9] developed two significant correlations between the oxygen-transfer efficiency and the power supplied to the system from the rotor (HP_r) and the compressor (HP_c), as shown in figure 8.12.

In most cases the optimum power split (power split, $P_d = HP_r/HP_c$) = 1.0. (This implies an equal power expenditure by the turbine and the blower.) At extremely high air rates, ($P_d < 1.0$), large bubbles and flooding of the impeller yield poor oxygenation efficiencies, while at very low rates ($P_d > 1.0$) too much turbine horsepower is being expended in fluid mixing. Variation in oxygen input to the system can most easily be adjusted by varying the air rate under the impeller. This in turn will change P_d. The anticipated range of operation should cover the maximum range of operation efficiency as related to P_d.

Available data indicate that the oxygenation efficiency of turbine aerators in water should vary from 2.5 to 3.0 lb of O_2/hp-hr (1.52 to 1.82 kg of O_2/kwh)

FIGURE 8.10
Air-Diffusion System
(Courtesy of Chicago Pump Company.)

FIGURE 8.11
Turbine Aeration Unit
(Courtesy of Mixing Equipment Company.)

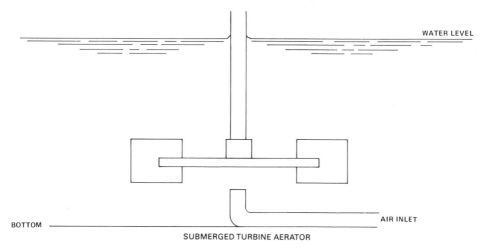

FIGURE 8.12
Design Relationships for Turbine Aeration

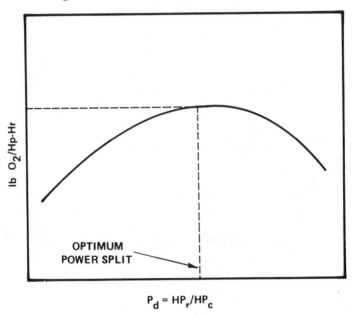

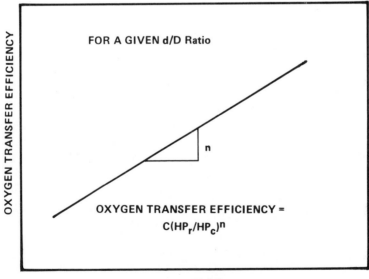

including motor and reducer losses. The ratio of turbine diameter (d) to tank diameter (D) should be between 0.1 and 0.2. Baffles are normally required to eliminate swirling and vortexing in circular tanks. Four baffles placed at equal intervals around the circumference of a round tank usually suffice, while two baffles on opposite walls are best for a square tank. In a rectangular tank with a length to width ratio greater than 1.5, no baffles are required.

Static Aeration

Static aeration systems consist of vertical cylindrical tubes placed at specified intervals in an aeration basin and containing fixed internal elements. A central compressor supplies an air source through a sparger at the bottom of the tubes, and an air-water mixture is forced through the cylinder. There is an intimate air-water contact as the mixture travels up through the elements within the static mixer cylinder. Most of the oxygen transfer occurs within the cylinder, although there is additional transfer at the surface turbulent area where the air-water mixture is discharged from the cylinder. The transfer efficiency of static aeration systems is reported to be higher than that obtained from a conventional diffused air-sparger or orifice system [10]. The exact transfer obtained using a static aeration system is a function of several design features:

1. The bottom sparger design
2. Diameter of the cylinder
3. Length of the cylinder
4. The air flow rate
5. The liquid submergence
6. The imported liquid velocity and mixing level
7. The design of the fixed elements and associated pressure drop through the cylinder

The air flow per mixer cylinder normally varies from 0.34 to 0.35 m³/min with a delivery air pressure in the range of 0.7 to 1 kg/cm² (gauge).

To date, static aeration systems have been used primarily in aerated lagoons carrying relatively low concentrations of biological suspended solids. They have several advantages in this application, including low annual costs and relatively high transfer efficiencies. They have also been applied as mixers in small mechanical or neutralization rapid mix tanks. A static aerator is shown in figure 8.13. A layout is shown in figure 8.14. The oxygen transfer characteristics of static aerators is shown in figure 8.15.

Surface Aerators

Surface aerators have become increasingly popular in recent years, particularly in industrial treatment applications. There are three basic classifications of surface aerators; specifically, the radial flow slow speed aerator, the axial flow

FIGURE 8.13
The Static Tube Aerator

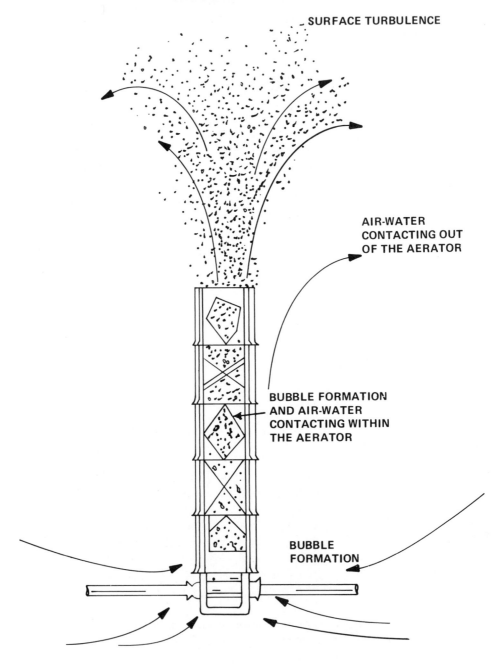

SURFACE TURBULENCE

AIR-WATER
CONTACTING OUT
OF THE AERATOR

BUBBLE FORMATION
AND AIR-WATER
CONTACTING WITHIN
THE AERATOR

BUBBLE
FORMATION

FIGURE 8.14
Static Aerator Layout

FIGURE 8.15

Observed Transfer Rate for Tubes at Fifteen-Foot
Centers [11]

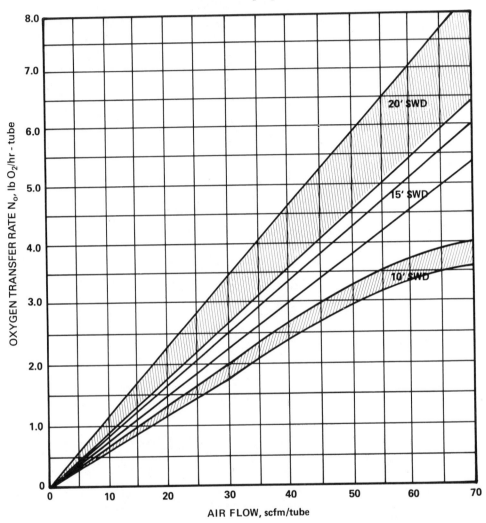

high speed unit, and the brush mechanical aerator. A brief discussion of each
follows.

The radial flow, slow speed aerator is essentially a low head, high volume
pump. Assuming an exit liquid velocity of 8 fps and impeller submergence of 6
inches, for example, the total dynamic head would be only 1.5 feet. The volume
pumped per unit horsepower (nameplate) is a function of the motor size and

ranges from 2 to 15 cfs/hp (0.08 to 0.57 m³/hp-sec). Oxygen transfer is accomplished primarily through entrainment associated with an induced hydraulic jump. The primary component of these units are the motor, gear reducer, and impeller. The impellers vary in diameter from 3 to 12 ft (0.9 to 3.7 m), and motor sizes range from 3 to 150 hp (2.2 to 110 kw). The output speed ranges from 30 to 60 rpm. These aerators are generally capable of mixing the contents of large aeration basins varying in depth from 3 to 18 ft (0.9 to 5.5 m). This depth can be increased by utilizing draft tubes or increasing the impeller submergence depth. These units may be bridge mounted, pier mounted or float mounted (figures 8.16 and 8.17).

The axial flow, high speed aerator is widely used and consists of a motor and propeller assembly which is usually float-mounted (figure 8.18). The major components are a nonsubmersible motor, riser tube, fiberglass or stainless steel float, and propeller deflector. They have a lower pumping capacity per unit horsepower than the radial flow units, but impart a higher velocity to the liquid (12 to 18 fps) (3.6 to 5.5 m/sec). As the radial flow speed units, they are low head, high volume pumps working at a total dynamic head of 3 to 6 ft (0.9 to 1.8 m). Most of the oxygen transfer occurs in the spray pattern, although transfer also occurs in the turbulent area in the outer peripheral area of the spray. These units range in

FIGURE 8.16
Surface Aeration Unit
(Courtesy of Mixing Equipment Company.)

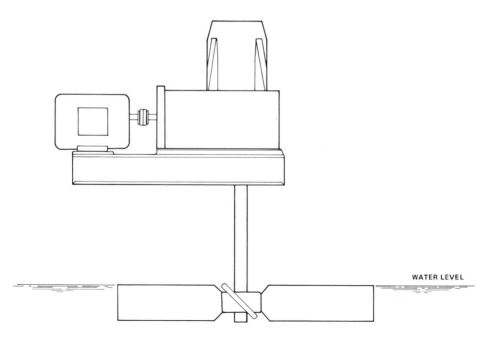

WATER LEVEL

FIGURE 8.17
Floating Surface Aeration Unit
(Courtesy of Mixing Equipment Company.)

FIGURE 8.18
Floating Surface Aeration Unit
(Courtesy of Peabody Wells, Inc.)

size from one to 150 hp (0.8 to 110 kw), and the rotational speed is in the 900 to 1400 rpm range. The liquid depths required for proper operation vary from 3 to 15 ft (0.9 to 4.6 m) depending on the aerator size.

The brush aerator is most frequently used in oxidation ditches. It consists of a rotating cylindrical wire brush mounted at the water surface as shown in figure 8.19. Guiding walls and/or baffles are sometimes installed in front of or behind the brushes to insure turbulent velocity patterns. Mixing is accomplished by an induced velocity below the rotating element. Efficiencies of 3.0–3.5 lbs O_2/hp-hr (1.8–2.1 kg O_2/kwh) have been observed [12].

Oxygen transfer in most types of surface aerators may be considered to occur in two ways:

1. Transfer to droplets and thin sheets of liquid sprayed from the blades of the unit

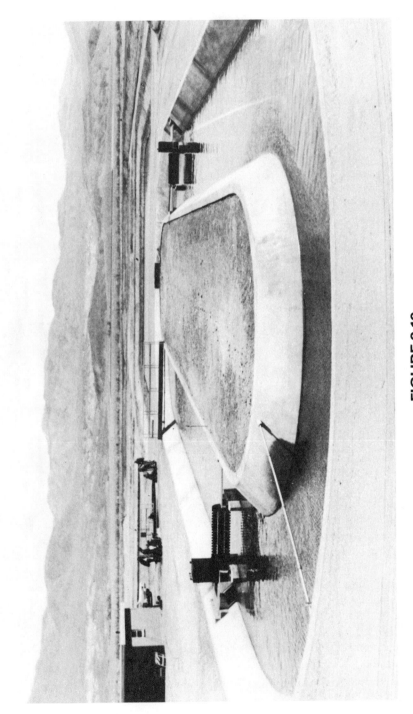

FIGURE 8.19

Brush Aerator System

(Courtesy of Lakeside Equipment Co.)

2. Transfer at the turbulent liquid surface and from entrained air bubbles where the spray strikes the surface of the liquid.

The amount of transfer that occurs in the spray is a function of the liquid pumping rate and the degree of saturation attained in the spray. The oxygen transfer rate per unit power of a high-speed aerator is lower than a low-speed one due to the lower volume of liquid pumped and exposed to the air. The capital cost of a high-speed aerator is less than a low-speed aerator. Performance data based on aerators from 5 to 75 hp (3.7 to 56 kw) and liquid volumes varying from 30,000 to 325,000 gal (114 to 1,230 m³) are shown in figure 8.20.

Kormanik [14] has shown a relationship between oxygen transfer rate and power per unit surface area. The power level dependency is greater for high-speed aerators than for low-speed aerators as shown in figure 8.21. A mathematical relationship can be established for a given type and size of aerator:

$$N_o = K_a P_a + N_s \tag{8--16}$$

where N_o = total oxygen transferred under standard conditions per unit horsepower

P_a = horsepower/ft^2 or watts/m^2 of basin surface area

N_s = oxygen transferred from the spray

K_a = constant characteristic of the aeration device

It should be recognized that aerators serve two functions in the biological process:

1. The transfer of the required oxygen and including sufficient mixing to maintain uniform oxygen throughout the basin, as in the case of the aerobic-facultative lagoon

2. Keeping the biological solids in suspension, as in the aerobic lagoons and the activated sludge process.

For conventional or high-rate organic loadings, the power required for oxygen transfer is usually considerably in excess of that required for mixing. In large aerobic-facultative lagoons or extended aeration systems, however, power for mixing may control the aerator design. Figure 8.22 shows the power relationships for dispersion of oxygen and for maintenance of activated sludge in suspension for low speed surface aerators [15]. Experiments on one type of surface aerator indicated a minimum power level for oxygen dispersion of 6 to 8 hp/MG [1.18 to 1.58 kw/m³ (10³)] of basin volume. (This power level is defined as that which will maintain a dissolved oxygen concentration ± 0.4 mg/l throughout the basin.) Experiments in the United States and Germany have shown that a velocity of 0.4 to 0.5 fps (12.2 to 15.2 cm/sec) may be necessary to maintain normal activated sludge in suspension. Sludges with a high inorganic content may require higher velocities.

The maintenance of solids in suspension limits the depth of aeration basin with some types of aeration units. Data indicates that units without a draft tube

FIGURE 8.20
Surface Aerator Performance Characteristics [13]

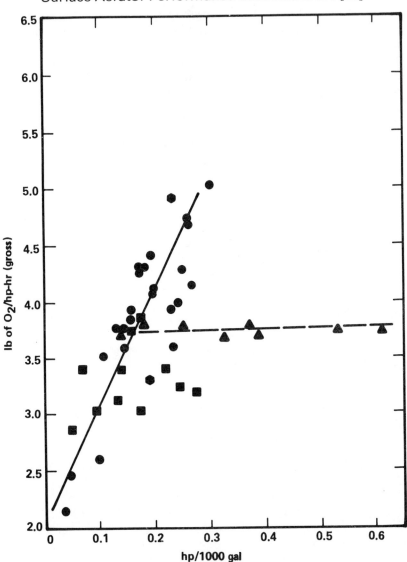

have a maximum depth of 12 ft (3.65 m) (unless supplementary mixing is provided) and units with a draft tube to 15 to 17 ft (4.57 to 5.18 m). Problems also arise with low surface tensions of less than 50 dynes/cm caused by the presence of surface active agents. In such cases the flow velocities in the aeration basin fall below 15

FIGURE 8.21

Comparative Effect of Surface Area on High- and Low-Speed Surface Aerators [14]

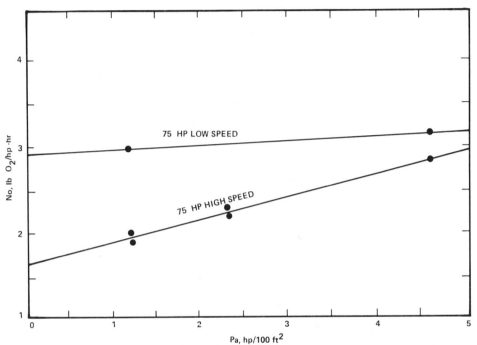

cm/sec as shown in figure 8.23. Modifications to the aerator such as the use of a bottom mixing impeller are necessary in these cases. The jet aerator supplies oxygen through an aspirator into liquid pumped around the aeration basin as shown in figure 8.24.

Data has been obtained from field operating units to indicate that increased transfer rates are obtained at increased oxygen uptake rates as shown in figure 8.25. This increase has been postulated to result from direct transfer of oxygen from the gas phase to the biological floc [16]. The general characteristics of available aeration equipment are summarized in table 8.4.

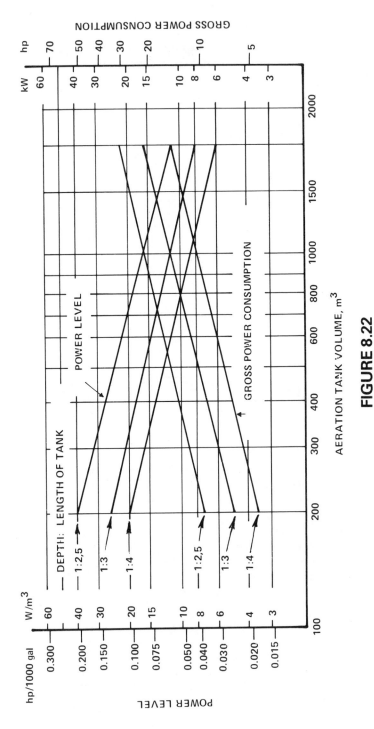

FIGURE 8.22

Power Requirements for Aeration Tank Mixing (Surface
Tension > 60 Dynes/cm) [15]

FIGURE 8.23
Influence of Detergent Concentration on Velocity Along
Tank Floor [15]

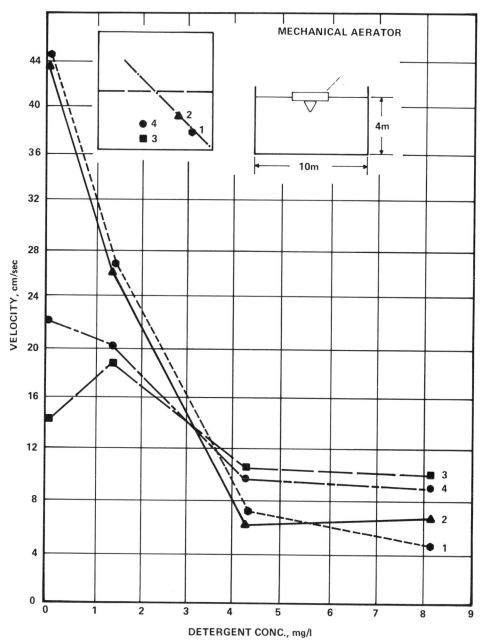

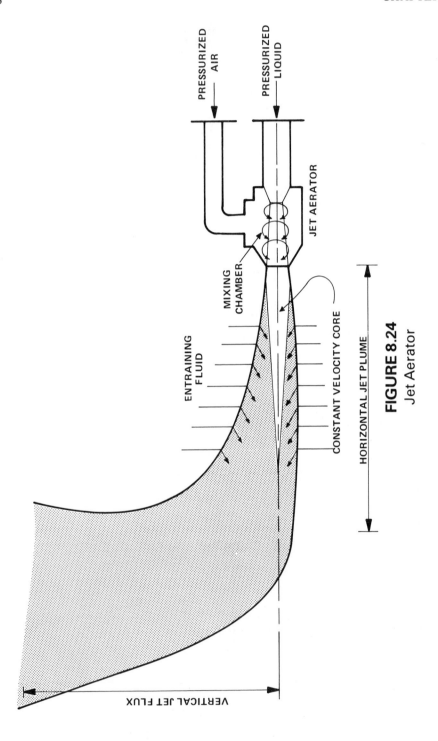

FIGURE 8.24
Jet Aerator

FIGURE 8.25
Field Oxygen Transfer Experiences

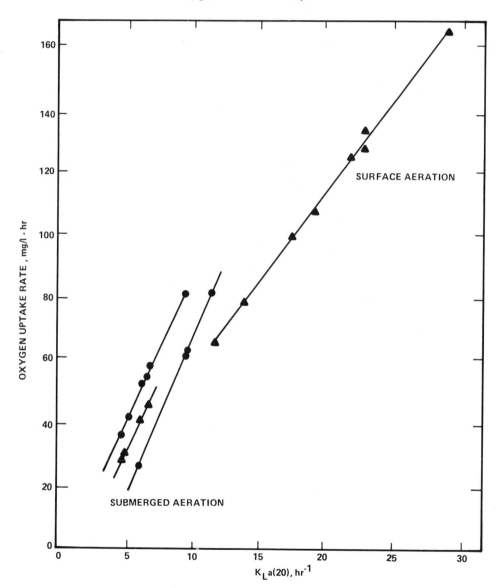

Equipment type	Equipment characteristics	Processes where used	Advantages	Disadvantages	Reported transfer efficiency[b] kg O_2/k Wh
Diffused aeration (bubbler)					
Porous diffusers	Produce fine or small bubbles. Made of ceramic plates or tubes, plastic-wrapped or plastic-cloth tube or bag.	Large, conventional, activated sludge process.	High oxygen transfer efficiency; good mixing; maintain high liquid temperature. Varying air flow provides good operational flexibility.	High initial and maintenance costs; tendency to clog; not suitable for complete mixing.	0.9–1.5
Nonporous diffusers	Made in nozzle, valve, orifice or shear types, they produce coarse or large bubbles. Some made of plastic with check-valve design.	All sizes of conventional activated sludge process.	Nonclogging; maintain high liquid temperature; low maintenance cost.	High initial cost; low oxygen transfer efficiency; high power cost. Clogging occurs.	
(Static)	Produces high shear and entrainment as water-air mixture is forced through vertical cylinder containing static mixing elements. Cylinder construction is metal, plastic, or polyethylene.	Primarily aerated lagoon applications.	Economically attractive; low maintenance; high transfer efficiencies for diffused air systems. Well suited for aerated lagoon application.	Ability to adequately mix aeration basin contents is questionable. Application for use in high rate biological systems unconfirmed.	2.0–3.0

Mechanical aeration:					
radial flow (slow speed)	Low output speed; large diameter turbine, usually fixed-bridge or platform mounted. Used with gear reducer.	All sizes of conventional, activated sludge and aerated lagoon processes.	High oxygen transfer efficiency; tank design flexibility; good transfer efficiency. High pumping capacity.	Some icing in cold climates. Initial cost higher than axial flow aerators. Gear reducer often causes maintenance problems.	2.5–3.5
Axial flow (high speed)	High output speed. Small diameter propeller. They are direct, motor-driven units mounted on floating structure.	Aerated lagoons and activated sludge processes.	Low initial cost; simple to install and operate; good transfer efficiency; adjust to varying water level. Flexible operation.	Some icing in cold climates; poor maintenance accessibility.	
Brush aeration	Low output speed; used with gear reducer.	Oxidation ditch applied either as an aerated lagoon or as an activated sludge process.	Relatively low initial cost, easy to install and operate, good maintenance accessibility, moderate transfer efficiency.	Subject to operational variables which may affect efficiency.	
Turbine aeration	Units contain a low speed turbine and provide compressed air on sparge ring; fixed-bridge application.	Conventional, activated sludge process.	Good mixing; high capacity input per unit volume; deep tank application; moderate efficiency; wide oxygen input range; operational flexibility.	Require both gear reducer and compressor; tendency to foam; high total power requirements.	1.7–2.4

TABLE 8.4

Characteristics of Available Aeration Equipment[a]

[a]From [17].
[b]Test conditions and procedures not documented.

PROBLEM 8–4

For an oxygen requirement of 1375 lbs O_2/hr in an aeration basin of 2.87 MG, design the following four aeration systems. Common input to each system is as follows:

$$C_L = 1.50 \text{ mg/l}$$
$$\text{Temp} = 27.5° \text{ C}$$
$$\beta = 0.90$$

Problem 8 – 4 (a) Diffused Aerators:

Design a sparger system
$G_s = 7.0$ scfm/unit
$\alpha = 0.80$
$E_o = 10.3\%$ (oxygen transfer efficiency)
$n = 1.02$
$m = 0.72$
$p = 0.35$
$C = 0.0081$

Problem 8–4 (b) Turbine Aerators:

$d/D = 0.12$
$P = 0.95$ (optimum power split)
$E_o = 20\%$ (oxygen transfer efficiency)
$\alpha = 0.85$

Problem 8–4 (c) High-Speed Surface Aerators:

$N_o = 0.26 P_a + 1.65$
where $N_o = $ lb O_2/hp-hr
$P_a = $ hp/100 ft^2
$\alpha = 0.90$

Problem 8–4 (d) Static Aerators:

Design for a 20-ft side wall depth (SWD)
$\alpha = 0.85$
The static aerator performance curve is shown in figure I.

Problem 8–4 (a) Diffused Aerators:

(1) Select depth = 16 ft, width = 30 ft
(2) Calculate C_{sm}
At 27.5° C, $C_s = 8.0$ mg/l

$$O_t = \frac{21 \cdot (1 - E) \cdot 100}{79 + 21 \cdot (1 - E)}$$

$$= \frac{21 \cdot (1 - 0.103) \cdot 100}{79 + 21 \cdot (1 - 0.103)} \%$$

$$= 19.25\%$$

$$C_{sm} = \frac{C_s}{2} \left(\frac{p + 0.433 \, h}{p} + \frac{O_t}{20.9} \right)$$

FIGURE I

Observed Transfer Rate for Tubes at Seven and One-Half Foot Centers

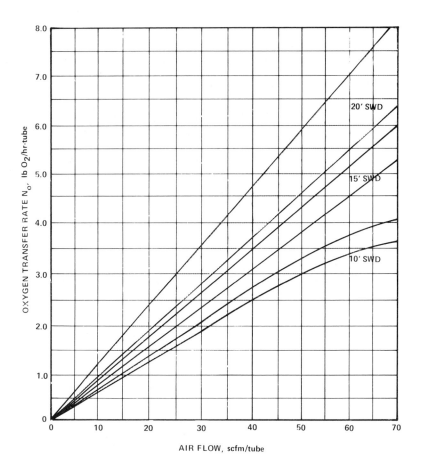

$$= \frac{8.0}{2} \left(\frac{14.7 + 0.433 \cdot 16}{14.7} + \frac{19.25}{20.9} \right) \text{mg/l}$$

$$= 9.57 \text{ mg/l}$$

(3) Calculate air diffuser performance

$$N = CG_s{}^n \frac{Hm}{Wp} (\beta C_{sm} - C_L) \cdot 1.02^{(T-20)} \cdot a$$

$$= 0.0081 \cdot 7.0^{1.02} \frac{16^{0.72}}{300^{0.35}} (0.90 \cdot 9.57 - 1.50) \cdot 1.02^{(27.5-20)} \cdot 0.80 \text{ lb/unit-hr}$$

$$= 0.871 \text{ lb/unit-hr}$$

No. of unit required

$$= \frac{1375}{0.871} = 1580 \text{ units}$$

Length of aeration basin

$$= \frac{2.87 \times 10^6 \cdot 0.134}{16 \cdot 30} = 800 \text{ ft}$$

Spacing between units

$$= \frac{800}{1580} = 0.51 \text{ ft or 6 in}$$

(4) Calculate air flow rate of the blower

$$G_s = 7.0 \cdot 1580 = 11,060 \text{ scfm}$$

(5) Calculate horsepower of the blower required

Efficiency = 70%, $\Delta P = 8$ psi

$$\text{Total hp} = \frac{G_s \cdot \Delta P \cdot 144}{E \cdot 33000}$$

$$= \frac{11060 \cdot 8 \cdot 144}{0.70 \cdot 33000}$$

$$= 552 \text{ hp}$$

Problem 8–4 (b) Turbine Aerators

(1) Select depth = 16 ft
(2) Calculate C_{sm}

$$O_t = \frac{21 \cdot (1 - 0.20) \cdot 100}{79 + 21 \cdot (1 - 0.20)} \%$$

$$= 17.54\%$$

$$C_{sm} = \frac{8.0}{2} \left(\frac{14.7 + 0.433 \cdot 16}{14.7} + \frac{17.54}{20.9} \right) \text{mg/l}$$

$$= 9.24 \text{ mg/l}$$

(3) Calculate air flow rate
The oxygen transfer efficiency is defined as:

$$E_o = \frac{R_r \text{ (lb/hr)}}{1.05 \, G_s \text{ (scfm)}}$$

$$G_s = \frac{1375}{1.05 \cdot 0.20}$$

$$= 6,550 \text{ scfm}$$

(4) Calculate required horsepower of the compressor
Efficiency $= 70\%$, $\Delta P = 8$ psi

$$\text{Total hp} = \frac{6550 \cdot 8 \cdot 144}{0.70 \cdot 33000}$$

$$= 327 \text{ hp}$$

(5) Calculate required horsepower of the rotor

$$HP_r = P_d \cdot HP_c$$

$$= 0.95 \cdot 327$$

$$= 311 \text{ hp}$$

(6) Calculate the oxygenation capacity

$$N = \frac{R_r}{HP_c + HP_r}$$

$$= \frac{1375}{327 + 311} \text{ lb O}_2\text{/hp-hr}$$

$$= 2.16 \text{ lb O}_2\text{/hp-hr}$$

(7) Calculate the oxygenation capacity under standard conditions

$$C_{s\,(20)} = 9.2 \cdot \frac{9.24}{8.0} = 10.63 \text{ mg/l}$$

$$N_o = \frac{NC_{s\,(20)}}{(\beta C_{sm} - C_L)\,\theta^{(T-20)}\,a}$$

$$= \frac{2.16 \cdot 10.63}{(0.90 \cdot 9.24 - 1.50) \cdot 1.02^{(27.5-20)} \cdot 0.85} \text{lb } O_2/\text{hp–hr}$$

$$= 3.42 \text{ lb } O_2/\text{hp–hr}$$

Problem 8–4 (c) High-Speed Surface Aerators

(1) Select depth $= 10$ ft

$$\text{Surface area} = \frac{2.87 \times 10^6 \times 0.134}{10} \text{ ft}^2$$

$$= 38,400 \text{ ft}^2$$

(2) Pick a value of P_a to calculate the required horsepower for mixing and keeping the solids in suspension.

$$HP_{mixing} = \frac{38400}{100 \cdot P_a} \text{ hp}$$

(3) Calculate N_o

$$N_o = 0.26\,P_a + 1.65$$

(4) Calculate N

$$N = N_o \frac{(0.90 \cdot 8.0 - 1.50) \cdot 1.02^{(27.5-20)} \cdot 0.90}{9.2} \text{ lb/hp–hr}$$

(5) Required horsepower for oxygenation

$$HP_{oxygenation} = \frac{1375}{N} \text{ hp}$$

(6) Repeat steps 2 through 5 for various P_a values and the results are plotted in figure II. The intercept of two curves gives the total power requirement of 930 hp and P_a of 2.42.

FIGURE II
Power Requirements for Mixing and Oxygenation

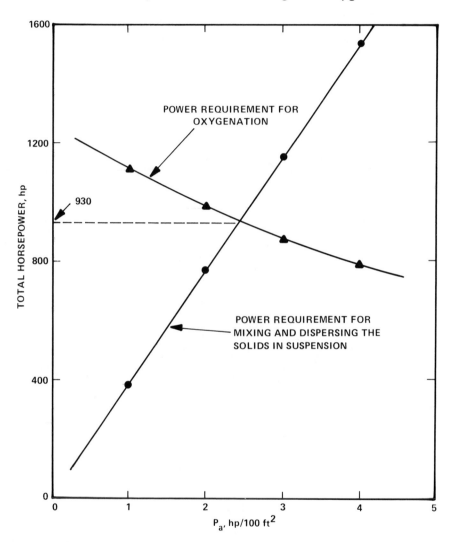

P_a	HP_{mixing}	N_o	N	$HP_{oxygenation}$
1	384	1.91	1.24	1109
2	768	2.17	1.40	982
3	1152	2.43	1.57	875
4	1536	2.69	1.74	790

(7) Calculate required N_o

$$N_o = 0.26 \cdot 2.42 + 1.65 \text{ lb/hp-hr}$$
$$= 2.28 \text{ lb/hp-hr}$$

Problem 8–4 (d) Static Aerators:

(1) For tube spacing of 7.5 ft the required number of tubes is:

$$\text{tube numbers} = \frac{\text{surface area}}{(\text{tube spacing})^2}$$

$$= \frac{2.87 \times 10^6 \cdot 0.134}{20 \cdot (7.5)^2}$$

$$= 342 \text{ tubes}$$

(2) Calculate N

$$N = \frac{1375}{342}$$

$$= 4.02 \text{ lb/hr–tube}$$

(3) Calculate N_o

Assume $E_O = 8\%$

$$O_t = \frac{21 \cdot (1 - 0.08) \cdot 100}{79 + 21 \cdot (1 - 0.08)}$$

$$= 19.65\%$$

$$C_{sm} = \frac{8.0}{2} \left(\frac{14.7 + 0.433 \cdot 20}{14.7} + \frac{19.65}{20.9} \right)$$

$$= 10.12 \text{ mg/l}$$

$$C_{s(20)} = 9.2 \cdot \frac{10.12}{8.0} = 11.64 \text{ mg/l}$$

$$N_o = \frac{4.02 \cdot 11.64}{(0.90 \cdot 10.12 - 1.50) \cdot 1.02^{(27.5-20)} \cdot 0.85}$$

$$= 6.24 \text{ lb/hr-tube}$$

(4) From the static aerator performance curve is obtained,

Air flow = 60 scfm/tube

(5) Calculate required air flow rate:

$G_s = 60 \cdot 342$

$= 20,500$ scfm

(6) Check oxygen transfer efficiency:

$$E_o = \frac{R_r}{1.05 \, G_s}$$

$$= \frac{1375}{1.05 \cdot 20500}$$

$= 0.064$ or 6.4%, it is less than the assumed value of 8% in step 3.

(7) Repeat steps 3 through 6 until the calculated efficiency is very close to the assumed values. The final results are:

$E_o = 6.4\%$

$N_o = 6.23$ lb/hr-tube

$G_s = 20,500$ scfm

(8) Calculate required horsepower of the blower:
Efficiency = 70%, $\Delta P = 8$ psi

$$\text{Total hp} = \frac{20500 \cdot 8 \cdot 144}{0.70 \cdot 33000}$$

$$= 1,020 \text{ hp}$$

REFERENCES

1. Lewis, W. K. and Whitman, W. G. 1924. Principles of gas absorption. *Ind. Eng. Chem.* 16: 1215.
2. Danckwertz, P. V. 1951. Significance of liquid film coefficients in gas absorption. *Ind. Eng. Chem.* 43: 1460.
3. Dobbins, W. E. 1964. Mechanism of gas absorption by turbulent liquids. In *Advances in Water Pollution Research* vol. 2. Oxford: Pergamon Press.
4. Schmit, F. L. *et al.* 1975. Diffused air in deep tank aeration. *Proc. 30th Ind. Waste Conf.* Lafayette, Ind.: Purdue Univ.
5. Imhoff, K. R. and Albrecht, D. 1972. Influence of temperature and turbulence on the oxygen transfer in water. *Proc. 6th Int'l Conf. Water Poll. Res.* Oxford, England: Pergamon Press, Jerusalem.
6. Metzger, J. and Dobbins, W. E. 1967. Role of fluid properties in gas transfer. *Environ. Sci. Tech.* 1: 57.
7. Barnhart, E. L. 1966. *Factors affecting the transfer of oxygen in aqueous solution.* M.S. Thesis, Sanitary Engr. Prog., Manhattan College, Bronx, N.Y.
8. Borkowski, J. D. and Johnson, M. J. 1967. Experimental evaluation of liquid film resistance in oxygen transport to microbial cells. *Applied Microbiology* 15: 1483.
9. Quirk, T. P. "Optimization of gas-liquid contacting systems," unpublished report (1962).
10. Epstein, A. C. and Glover, C. "Full scale aeration studies of static aeration systems," Permutit Co., Report of Kenics Co. Houston, TX (1971).
11. Bennett, C. and Shell, G. 1976. Submerged static aerators: what are they all about? *Water Wastes Eng.*, 13, 5: 37.
12. Stalzer, W. and von der Emde, W. 1972. Tanks with turbulent flow generated by mammoth rotors. *Water Research* 6: 417.
13. Eckenfelder, W. W. and Ford, D. L. 1967. Engineering aspects of surface aeration design. *Proc. 22nd Ind. Waste Conf.* Lafayette, Ind.: Purdue Univ.
14. Kormanik, R. 1976. How does tank geometry affect the oxygen transfer rate of mechanical surface aerators? *Water Sew. Works* 123, 1: 64.
15. Kalbskopf, K. H. 1972. Flow velocities in aeration tanks with mechanical aerators. *Water Research* 6: 413.
16. Albertson, O. E. and DiGregorio, D. 1975. Biologically mediated inconsistencies in aeration equipment performance. *J. Water Pollution Control Federation* 47 : 976.
17. Nogaj, R. J. 1972. Selecting wastewater aeration equipment. *Chem. Engr.* 79, 8: 95.

Biological Waste Treatment

Many types of microorganisms are active in the breakdown of organic matter and the resulting stabilization of organic wastes. These microorganisms may be broadly classified as aerobic, facultative, or anaerobic. Aerobic organisms require molecular oxygen for their metabolic processes. Anaerobic organisms function in the absence of oxygen and obtain their energy from organic compounds. Facultative organisms can function aerobically in the presence of oxygen or anaerobically in the absence of oxygen. A majority of the organisms found in biological wastewater treatment processes are of the facultative type. Most biological systems treating organic wastes depend upon heterotrophic organisms that utilize organic carbon as an energy source and as a carbon source for cell synthesis.

Autotrophic organisms, on the other hand, do not require an organic carbon source, but rather use an inorganic carbon source such as CO_2 or bicarbonate. Chemosynthetic autotrophs obtain energy from the oxidation of inorganic compounds such as nitrogen or sulfur. Photosynthetic autotrophs utilize solar energy for the synthesis of carbon dioxide to cellular protoplasm and produce molecular oxygen as a by-product.

AEROBIC SYSTEMS

In aerobic biological treatment systems the reactions occurring are:

$$\begin{array}{c}\text{organics} \\ \text{(BOD, COD, TOC)}\end{array} + (a')\, O_2 + N + P \xrightarrow{k} (a)\ \text{cells} + CO_2 + H_2O + \begin{array}{c}\text{nondegradable} \\ \text{soluble residue}\end{array} \qquad (9\text{--}1)$$

$$\text{cells} + O_2 \xrightarrow{b} CO_2 + H_2O + N + P + \begin{array}{c}\text{nondegradable cellular} \\ \text{residue}\end{array} \qquad (9\text{--}2)$$

249

(It should be noted that these reactions also occur in streams and natural waters and in the BOD bottle.) Biological treatment design involves the balancing of these equations for the wastewater in question.

The pertinent process parameters a, a', b, and k are shown in equations (9–1) and (9–2). For a known single compound such as glucose or phenol it is possible from available information to balance these equations. For the complex mixed wastewaters usually found in practice, however, it is not possible to balance these equations and a laboratory or pilot plant study is needed to define the parameters. These may be defined:

a — fraction of substrate removed (as BOD, COD, or TOC) converted to cells as VSS (volatile suspended solids)

a' — fraction of substrate oxidized for energy (as BOD, COD, or TOC)

b — fraction per day of degradable cell mass as VSS oxidized

k — substrate removal rate coefficient

In the stabilization of an organic substrate, a portion of the energy obtained from the reaction is used for biological synthesis and the remainder to satisfy the energy requirements for growth. A small portion of the energy is used for cellular maintenance (figure 9.1). Equations (9–1) and (9–2) are schematically shown in figure 9.2. The symbols and parameters used in the development of design relationships are also shown in figure 9.2. As can be seen from figure 9.1

FIGURE 9.1
The Mechanism of Aerobic Biological Oxidation

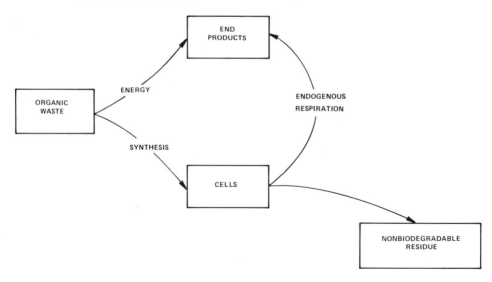

and equation (9–1) all but a small portion of the substrate removed is either converted to cell mass as VSS or oxidized for energy. Therefore, on a COD basis $1.4a + a' \simeq 1$. It should be noted that this relationship does not apply when using BOD_5 since the BOD_5 is a fraction of the COD or ultimate BOD. To illustrate the breakdown of organic matter by aerobic oxidation a hypothetical example using 1 pound COD is shown in problem 9–1. It is assumed in this example that $a' = 0.5$ on a COD basis.

FIGURE 9.2

Schematic of Aerobic Biooxidation Process

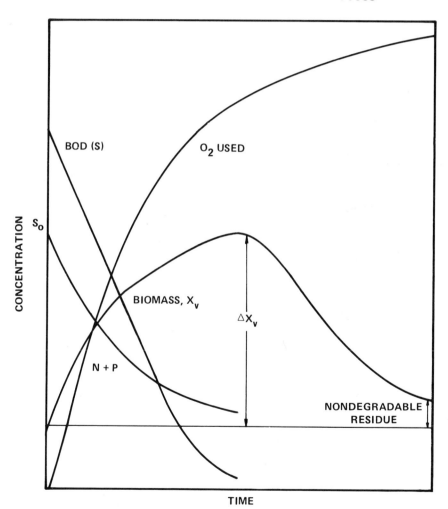

BOD (S)

O_2 USED

CONCENTRATION

S_o

BIOMASS, X_v

$\triangle X_v$

N + P

NONDEGRADABLE RESIDUE

TIME

PROBLEM 9–1

In autotrophic growth, considerable energy must be expended to convert CO_2 to an intermediate for cell synthesis. As a result, yield coefficients of less than 0.1 are usually obtained.

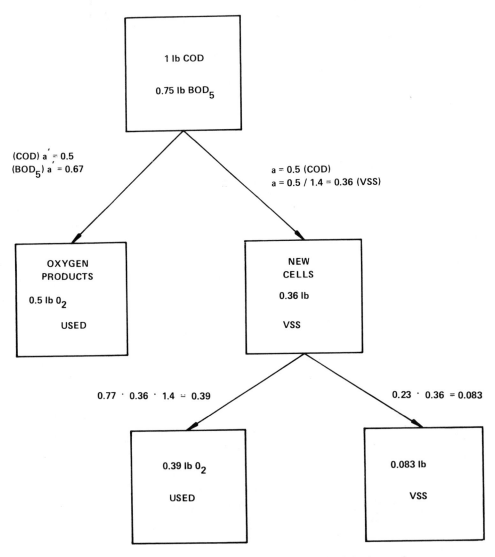

Hypothetical Estimation of Aerobic Oxidation of
1 lb COD

The measured yield coefficients for various systems are summarized in table 9.1 and include the influence of volatile suspended solids originally present in the waste and, hence, denoted as $\bar{a}$.

It has been shown by McCarty [1] that the similarity of the biochemistry of synthesis of all microorganisms under a wide variation in environment permits calculation of cellular yields from thermodynamic considerations when the composition of the substrate is known.

CELL YIELD

In aerobic growth, energy is released from the conversion of organic carbon, resulting in considerable energy being available for synthesis, and, hence, a relatively high yield coefficient, a. McCarty [1] obtained values for a that vary from 0.30 to 0.51 for glucose, analine, lactate, and acetate. Servizi and Bogan [2] showed a value of 0.39 for a variety of compounds. Eckenfelder and O'Connor [3] showed yield coefficients varying from 0.37 to 0.46 for several readily degradable organic wastewaters.

In anaerobic systems, less energy is obtained from the organic conversion, and, hence, the growth yield is much less than for aerobic systems. Yield coefficients varying from 0.032 to 0.27 were found depending on substrate. A detailed study by Andrews *et al.* [4] on a synthetic soluble substrate showed a yield coefficient of 0.14. It should be realized that most of these growth yields include both the acid formers (first stage) and the methane organisms.

TABLE 9.1
Aerobic Biological Waste Treatment Parameters[a]

Waste	$\bar{a}$ (BOD$_5$ basis)	a' (BOD$_5$ basis)	b, day^{-1} (VSS basis)
Domestic	0.73	0.52	0.075
Refinery	0.49–0.62	0.40–0.77	0.10–0.160
Chemical and petrochemical	0.31–0.72	0.31–0.76	0.05–0.180
	0.56		
Brewery	0.56	0.48	0.100
Pharmaceutical	0.72–0.77	0.46	–
Kraft pulping and bleaching	0.50	0.65–0.80	0.080

[a]All parameters include the effect of influent suspended solids.

Porges [5] showed that activated sludge from the treatment of dairy wastewaters has an average composition of $C_5H_7NO_2$. Other investigators have shown a similar composition for sludge treating other wastewaters. The degradable portion of the biomass [equation (9–2)] has been reported as 77 percent with 23 percent of the volatile suspended solids as nondegradable residue. This is considered in the time frame of the process since very long aeration periods will result in a reduction of the nondegradable volatile suspended solids.

It can be seen from equations (9–1) and (9–2) that biological volatile solids are generated from organic removal and oxidized by endogenous metabolism. In addition if the wastewater contains volatile suspended solids in the influent wastewater (as might be expected in domestic wastewater or pulp and fiber in pulp and paper mill wastewaters) the nondegradable portion of these solids will contribute to the accumulated volatile suspended solids in the system.

The sludge yield for a biological system can be estimated from the relationship:

$$\Delta X_v = fs_i + aS_r - bxX_v \qquad (9-3)$$

where a = cell yield coefficient
       b = cell endogenous rate, day^{-1}
       x = biodegradable fraction of the mixed-liquor volatile suspended solids
       X_v = mixed-liquor volatile suspended solids, lb
       s_i = influent volatile suspended solids, lb/day
       S_r = BOD removed, lb/day
       f = fraction of influent volatile suspended solids not degraded

The coefficient f can be related to sludge age since the longer the solids remain in the aeration system the greater will be the rate of degradation. A typical correlation of equation (9–3) is shown in figure 9.3.

PROBLEM 9–2

Data from the treatment of a wastewater is shown in table I. Correlate the results and determine the coefficients a and b.

The data are correlated in accordance with equation (9–3) and the resulting equation is:

$$\Delta X_v = 0.41 \, S_r - 0.06 \, xX_v$$

considering that a is 0.41 and b is 0.06 day^{-1} (see figure I).

FIGURE I
Evaluation of Coefficients *a* and *b*

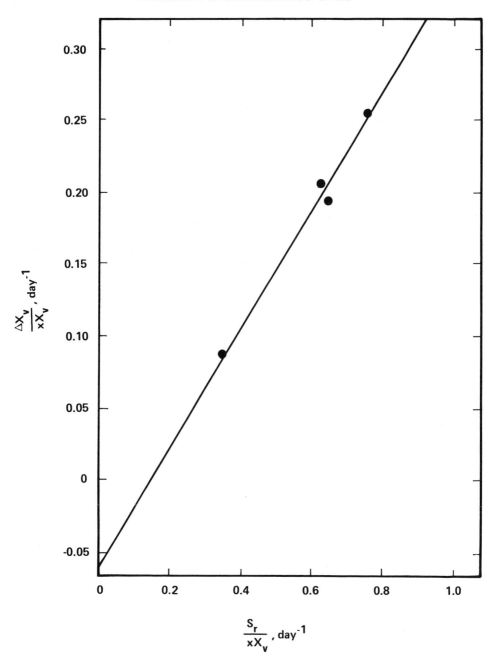

TABLE I

Unit	1	2	3	4
F/M, day^{-1}	0.133	0.298	0.410	0.279
Sludge age, days	37.0	11.5	7.5	11.4
MLVSS, mg/l	2020	2210	2310	2410
Influent BOD, mg/l	158	158	158	158
Effluent BOD, mg/l	7	5	5	7
x	0.31	0.45	0.53	0.43
t, days	0.69	0.24	0.167	0.235

SLUDGE AGE

Sludge age is defined as the average length of time the biomass is under aeration. In a flow through system, sludge age is the reciprocal of the dilution rate. For growth to occur and to effect BOD removal, the growth rate becomes:

$$\mu = \frac{1}{\theta_c} = R_\mu$$

where θ_c = sludge age
R_μ = dilution rate

In a system with sludge recycle and a wastage of excess sludge, sludge age is defined as:

$$\theta_c = \frac{X_v}{\Delta X_v} = \frac{X_v}{aS_r - bxX_v} \text{ (for soluble wastewaters)}$$

where X_v and ΔX_v are expressed in kg and kg/day (lb and lb/day), respectively.

This equation also applies relative to the limiting growth rate for specific organisms. If the sludge age is less than the reciprocal of the growth rate of the organism in question it will be washed out of the system. Examples are nitrifying organisms and methane organisms in anaerobic processes.

For a soluble wastewater the degradable fraction, x, will be related to the sludge age or the length of time the biomass is under aeration since as the aeration time is increased the percentage of nondegradables also increases. This is shown in figure 9.4. For a soluble wastewater, x can be computed from the relationships:

$$x = \frac{aS_r + bX_v - \sqrt{(aS_r + bX_v)^2 - (4bX_v)(0.77\, aS_r)}}{2bX_v} \tag{9-4}$$

FIGURE 9.3
Determination of Sludge Production Coefficients

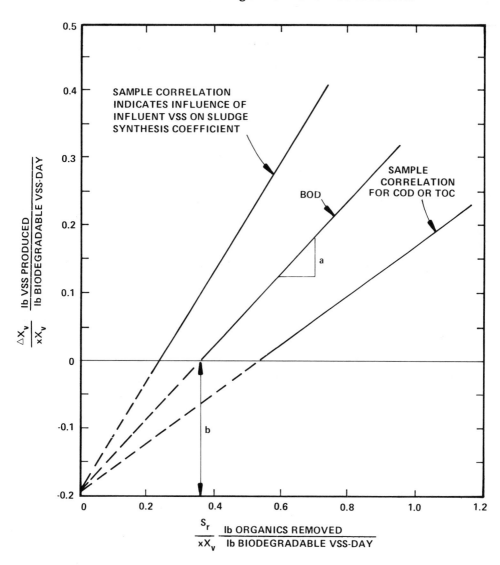

when the wastewater contains volatile suspended solids equation (9–4) may be modified to:

$$x = \frac{aS_r + bX_v + fs_i - \sqrt{(aS_r + bX_v + fs_i)^2 - (4bX_v)(0.77\,aS_r)}}{2bX_v}$$

(9–5)

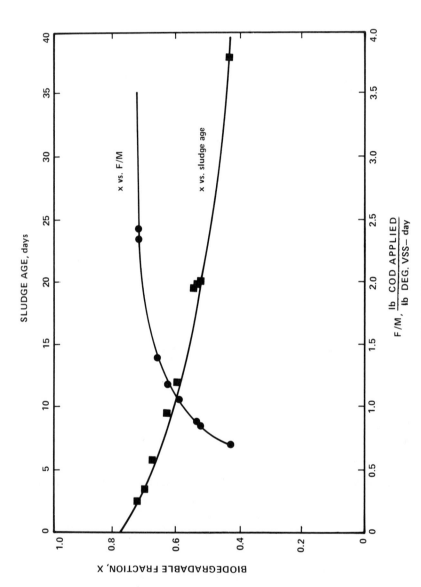

FIGURE 9.4

Variation of Degradable Fraction with Sludge Age
and with F/M (food/microorganisms ratio) for
Synthetic Sewage

The degradable fraction x can also be computed from the relationship:

$$x = \frac{aS_r}{bX_v} - \frac{1}{b\theta_c}$$ (9–5a)

In the case of soluble wastewaters the active portion of the biomass has been considered to be the degradable fraction, x, divided by 0.77 as calculated by equation (9–4). A number of other procedures have been employed to determine the active fraction of the biomass. These are adenosine triphosphate (ATP), dehydrogenase enzyme (TTC), plate counts, and oxygen uptake rate. While these procedures are useful for experimental investigation and for plant operational control they are less applicable for process design and evaluation. The computed degradable fraction, x, reasonably correlates with these other parameters as shown in figure 9.5 and should be applicable to design procedures. In the case of activated sludge where the loadings are high and a major portion of the sludge is active biomass equation (9–3) can be approximated as:

$$\Delta X_v = aS_r + fs_i - bX_v$$

It should be emphasized that in many cases it is not feasible to divide the sludge yield ΔX_v into the contribution by microbial synthesis and the accumulation of volatile suspended solids originally present in the waste. The experimental coefficient $\bar{a}$ is then used for engineering design (table 9.1), and equation (9–3) becomes:

$$\Delta X_v = \bar{a}S_r - bxX_v$$ (9–6)

It should be noted that for a soluble waste, $\bar{a}$ is approximately equal to a. Sludge production from the treatment of domestic sewage by the activated sludge process is shown in figure 9.6.

In a flowthrough system, without clarifier and recycle, the concentration of solids in the effluent will be equal to the concentration in the reactor, and the sludge yield is equal to the solids lost in the effluent. Equation (9–3) can be re-expressed for the flowthrough system as:

$$X_v Q = s_i + aS_r - bxX_v V$$

when the influent volatile suspended solids are not degraded. Dividing by the volumetric flow rate, Q, and rearranging:

$$X_v = \frac{s_o + a(S_o - S_e)}{1 + bxt}$$ (9–7)

in which X_v is the concentration of volatile suspended solids maintained in the reactor; s_o is the influent VSS concentration; S_o and S_e are influent and effluent BOD concentrations, respectively; and t is the hydraulic detention time. Equation (9–7) applies to aerobic lagoons.

FIGURE 9.5
Variation of Viable Cell Parameters (Fraction) with
Sludge Age

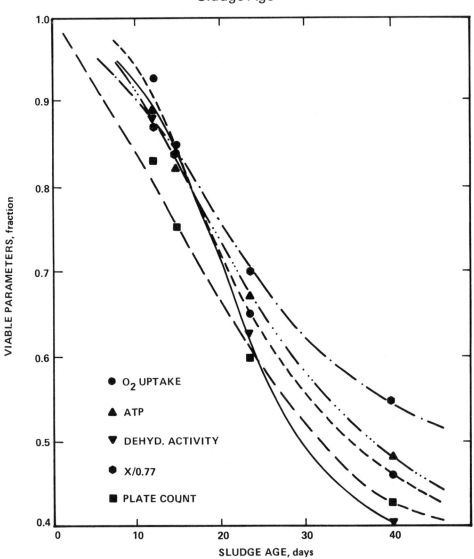

FIGURE 9.6
Sludge Production in the Activated Sludge Process
Treating Domestic Sewage
(After Wuhrmann [8])

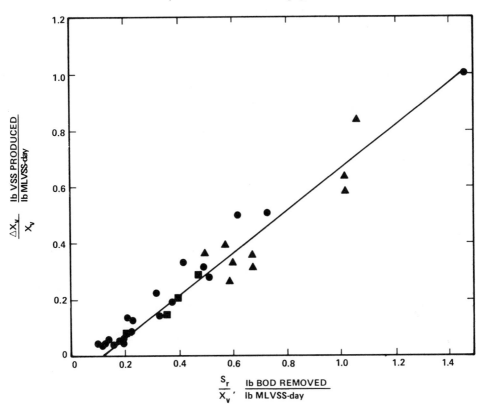

ENDOGENOUS METABOLISM

Endogenous metabolism occurs in all cells in which energy is utilized for cellular maintenance. Endogenous metabolism may be defined by the coefficient, b, which has the units of reciprocal time; that is, there is a fractional decrease in cell mass per day. It should be noted the coefficient, b, applies to the degradable cellular solids, particularly in the case of total oxidation systems [3]. Endogenous values reported for several systems considering total volatile suspended solids are summarized in table 9.1. The coefficient, b, considering only degradable solids will usually vary from 0.1 to 0.2 at 20° C. The temperature effect on this coefficient is discussed under temperature effects.

When high aeration solids are carried, the accumulation of nonbiodegradable mass reduces the percentage of active organisms present in the system based on total volatile suspended solids. For example, the data of Wuhrmann [6] would indicate 59 percent active organisms at 6000 mg/l relative to substantially total activity at less than 500 mg/l aeration solids at the same applied loading.

OXYGEN UTILIZATION
IN AEROBIC SYSTEMS

In aerobic systems, the portion of the substrate not utilized for cellular synthesis uses oxygen for energy. In addition, oxygen is used for cellular maintenance (endogenous respiration) as shown in equations (9–1) and (9–2). The resulting relationship is:

$$R_r = a'S_r + b'xX_v \tag{9-8}$$

and is shown in figure 9.7. When volatile suspended solids are undergoing slow degradation in the aeration system, a' will reflect this oxygen usage. Also, nitrification will vastly increase the value of a'. The coefficient b' reflects the oxygen used for endogenous respiration. Assuming a mean cellular composition of $C_5H_7NO_2$ [5] the oxygen requirements can be computed:

$$C_5H_7NO_2 + 5O_2 \longrightarrow 5CO_2 + 2H_2O + NH_3$$

and

$$\frac{5O_2}{C_5H_7NO_2} = \frac{160}{113} = 1.42$$

This coefficient has been found to vary in practice from 1.32 to 1.54. When a material balance is based on COD or BOD_u, $1.4a + a' \simeq 1$, because the organic substrate carbon results in the production of CO_2 or the formation of biological cells. When based on BOD_5, the following conversion must be made:

$$a'(BOD_5) = \frac{1}{BOD_5/BOD_u} - 1.42a(BOD_5)$$

The oxygen uptake rate in a biological process can be determined in several ways, namely by oxygen uptake rate measurement, by off-gas analysis or by a COD balance. Oxygen uptake rate is measured by withdrawing a sample of sludge, aerating to raise the dissolved oxygen and measuring by means of a dissolved oxygen probe the decrease in dissolved oxygen. This is shown in figure 9.8. Caution must be exercised in interpreting the results of this test. Since a sample is withdrawn from an aeration basin receiving wastewater and no wastewater is added during the course of the test, results will usually read lower than the actual uptake rate. The error increases as the oxygen uptake rate increases. It is possible

FIGURE 9.7
Determination of Oxygen Utilization Coefficients

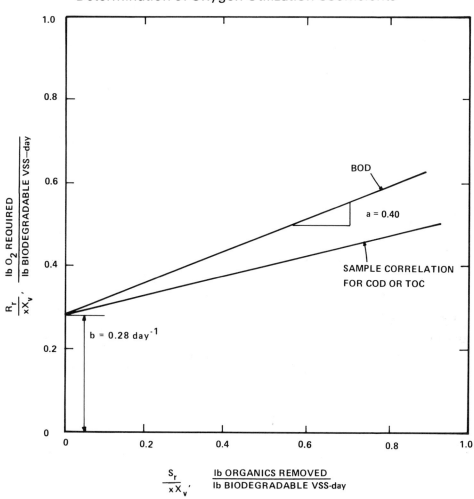

to correct for this error by adding wastewater to the test cell at the same rate as it is being added to the aeration basin.

Off-gas analysis has been employed to measure oxygen uptake rate in high purity oxygen systems that employ covered aeration basins but it is not feasible for air or mechanical aerators. In cases where strippable volatile constituents are not present in the wastewater a COD balance should yield the most accurate measure of total oxygen requirements. In this case the oxygen utilized is:

$$R_r = COD_{INF} - COD_{EFF} - COD_{VSS\ generated}$$

FIGURE 9.8
Dissolved Oxygen Probe

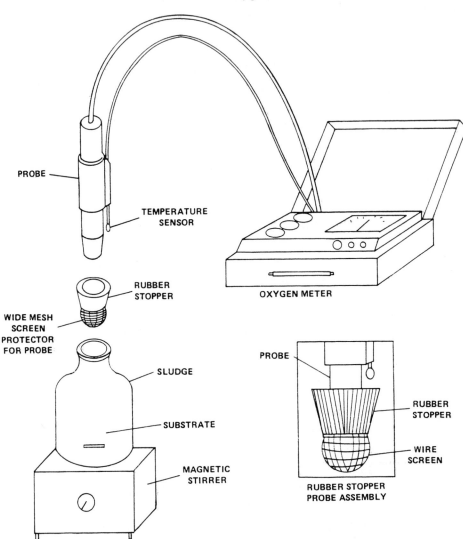

The oxygen usage can then be computed by a COD balance over a reasonable time period.

In the direct method by manometric analysis, a sludge sample is respired in a closed oxygen or air atmosphere at constant temperature. The sample is agitated and the oxygen utilized is measured with respect to time by noting the decrease in gas volume or pressure. The Sierp and Warburg assemblies are typical examples. The Warburg apparatus consists of a reaction flask connected to a manometer in a constant temperature bath as shown in figure 9.9. The sludge-waste mixture is kept agitated by a shaker assembly. As oxygen is biologically utilized it is replenished from the gaseous phase above the sample. The system is maintained at constant volume by adjusting the manometer columns before read-

FIGURE 9.9
Complete Warburg Apparatus

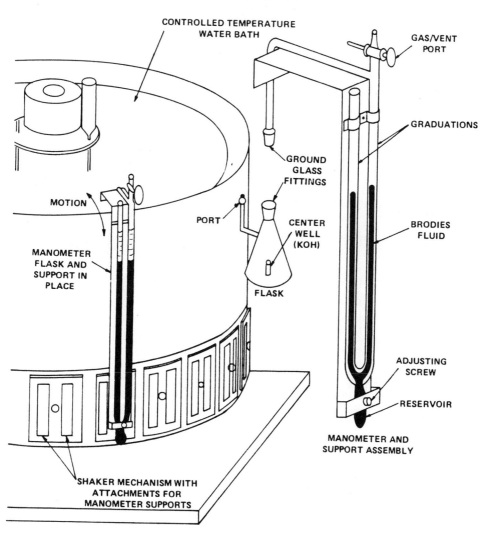

ing. The CO_2 evolved is eliminated by absorption in KOH in a small center well of the test flask as shown in figure 9.10. The removal of oxygen from the gas phase creates a pressure difference which may be read on a manometer. A control flask is employed to correct for barometric and temperature variations. One drawback to this procedure lies in the fact that the measured rate is due to a decrease in oxygen partial pressure in the gas phase rather than of oxygen in solution. With high uptake rates, oxygen absorption from gas to liquid may limit the respiration rather than the characteristics of the culture itself. The formulae and constants necessary to compute oxygen utilization from Warburg data are summarized below:

$$\text{mg/l } O_2 \text{ utilized} = \frac{1000}{V} \frac{32}{22.4} hk' = 1430 \frac{hk'}{V}$$

where V = volume of sample, ml
   h = manometer pressure change, cm
   k' = a flask constant determined for each flask, sample, volume, and temperature by the following formula:

$$k' = \frac{V_g \left(\frac{273}{T} \right) + V_f d}{P_o}$$

where V_g = gas volume in closed system, ml
   V_f = liquid volume in flask, ml
   d = $\dfrac{0.0325 \text{ ml gas}}{\text{ml liquid}}$ = mg O_2 per ml liquid, where total pressure (the sum of the partial pressure of the gas plus the aqueous tension 20° C) is 760 mm Hg and temperature is 20° C
   P_o = 1.001 cm (1 atm of Brodie's solution)
   T = absolute temperature, °K

It should be emphasized that the Warburg test yields a summation of oxygen utilization. The instantaneous utilization rate is obtained from the slope of the oxygen-time curve.

Manometric data is not applicable to the determination of oxygen requirements in continuous completely mixed systems under varying loading conditions. However, it is useful to estimate the coefficient a', to compare the biodegradation rates of various wastewater components, and to determine toxicity in industrial wastewaters.

A typical Warburg curve showing cumulative oxygen consumed in the presence of substrate and for the endogenous blank is shown in figure 9.11. After most of the substrate is depleted the substrate remaining (as BOD, COD, or TOC) is measured in the flask. The coefficient a' is estimated as the total oxygen consumed minus the endogenous blank divided by the substrate removed equation

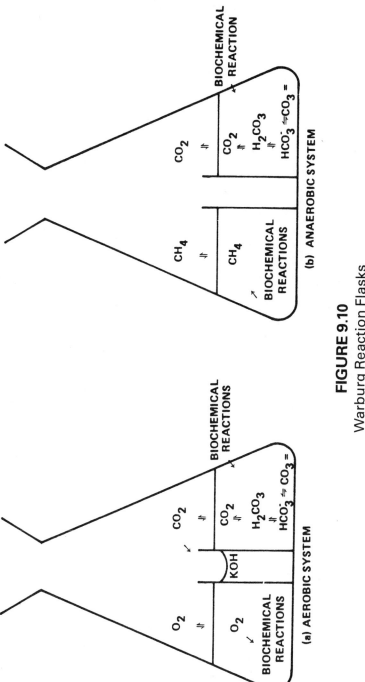

FIGURE 9.10
Warburg Reaction Flasks

FIGURE 9.11
Typical Curves of Warburg Analysis

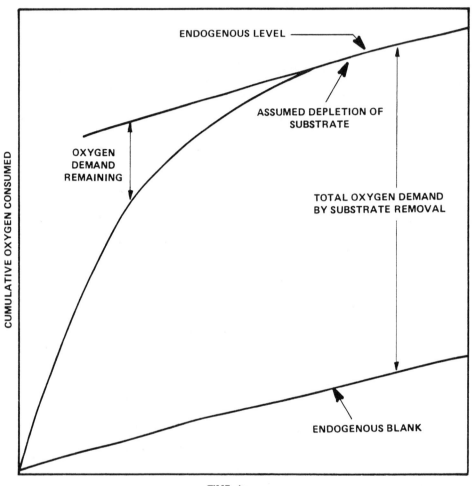

(9–8). This calculation assumes that the endogenous rate remains constant for both flasks, while in fact biomass production in the substrate flask increases the endogenous rate. In most cases, however, this increase is small and does not introduce much error in the estimation of a'.

The biodegradability of various components of a wastewater can be estimated from a Warburg analysis. Using acclimated seed, Warburg curves for the various wastewater components are generated and plotted as shown in figure 9.11. The difference in total oxygen consumed and the endogenous blank are tabulated until the substrate flask approaches the oxygen consumption rate of the endogenous blank at which point the substrate is assumed depleted. These differences are plotted on a semilogarithmic paper and the k computed as the slope of the plot. Note that the corrected oxygen consumed should be directly proportional to the substrate removed as shown by equation (9–8). Data for two waste streams are shown in figure 9.12.

FIGURE 9.12
Manometric Results of Two Waste Streams

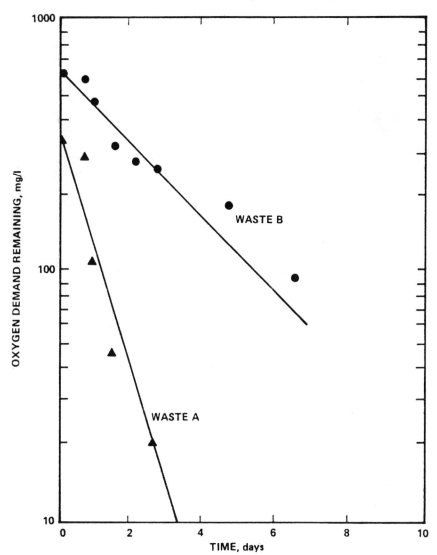

In the determination of toxicity in the biooxidation process various dilutions of the component in question is added to flasks containing acclimated seed. As shown in figure 9.13, a reduction in oxygen uptake rate is evidenced when inhibition or toxicity occurs. Note that inhibitory substances which are biodegradable at low concentration will undergo continuous biodegradation in completely mixed aeration processes. In this case the permissible influent concentration is greater than that shown by the Warburg test by the amount of biooxidation that occurs in the basin. For example if the biooxidation rate is 100 mg/l and the toxicity threshold is 100 mg/l, the influent concentration should not exceed 200 mg/l.

KINETICS OF ORGANIC REMOVAL

The mechanism of organic removal from soluble mixed substrates is a complex phenomena. Grau *et al.* [7] have indicated that the mechanism can generally be described as a sequence of three complex processes, namely, contact of a cell with a molecular substrate, transport of the molecule into the cell and intermediate metabolism of the substrate. Large molecules must first be broken down externally before they can be classified into three main groups:

1. Single component substrates that are directly transportable into the cell.
2. Multicomponent substrates that are represented by a mixture of several single substrates.
3. Complex substrates that have to be changed externally prior to transport into the cell.

The removal of single component substrates has been described by linear removal kinetics or a zero order reaction by Wuhrmann [8] and by Tischler and Eckenfelder [9]. The linear removal concept states that substrate removal will follow zero order kinetics to very low concentration levels. Data reported by Tischler and Eckenfelder are shown in figure 9.14.

Single substrate removal has also been defined by the Monod equation:

$$\frac{-dS}{dt} = \frac{\bar{\mu} X_v}{a} \frac{S}{K_s + S} \tag{9-9}$$

or more commonly as:

$$\mu = \bar{\mu} \frac{S}{K_s + S}$$

where S = substrate concentration
$\bar{\mu}$ = maximum growth rate of organisms
μ = specific growth rate of organisms
a = biomass yield coefficient
X_v = biomass concentration

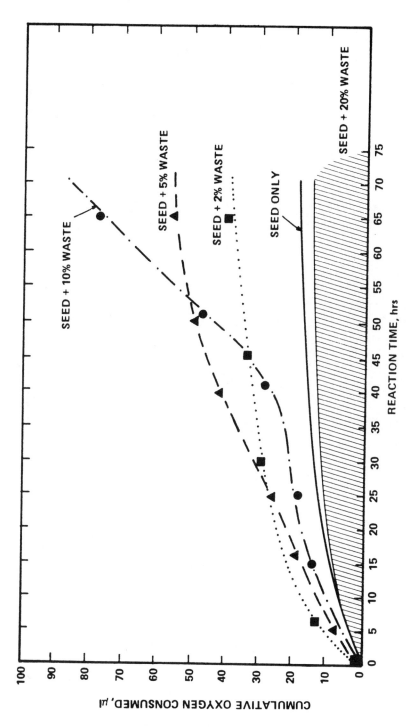

FIGURE 9.13

Dilution Effect on Respiration Rates

FIGURE 9.14
Linear Substrate Removal

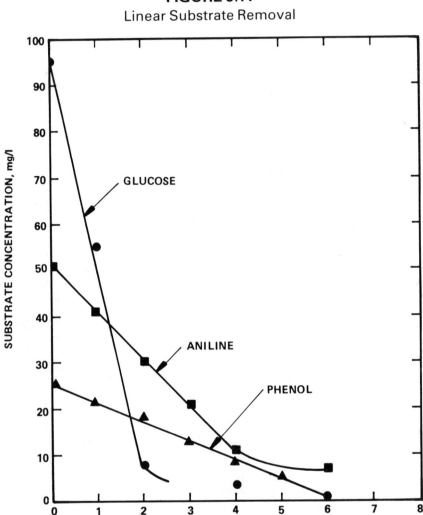

K_s = Monod's constant which is defined as the substrate concentration when the rate is one-half the maximum rate.

Incorporating this into a substrate balance around the aeration basin and clarifier in order to eliminate recycle effects, the Monod model becomes:

$$\frac{S_o - S_e}{X_v t} = \frac{\bar{\mu}}{a} \frac{S_e}{K_s + S_e} \qquad (9\text{--}10)$$

Manipulation into a linear form yields:

$$\frac{X_v t}{S_o - S_e} = \frac{K_s}{\mu}\frac{a}{S_e} + \frac{a}{\bar{\mu}} \qquad (9\text{–}11)$$

It should be noted that a can previously be determined from the combined substrate and material balances about the system, and that the Monod expression does not account for varying initial substrate values. Relationships for K_s and $\bar{\mu}$ as some functions of S_o must be found experimentally and incorporated into the final expression.

At high concentrations of S_e when substrate concentration is not limiting microbial growth, $S_e > K_s$ and equation (9–10) reduces to a zero order expression:

$$\frac{S_o - S_e}{X_v t} = \frac{\bar{\mu}}{a} \qquad (9\text{–}12)$$

At very low concentrations of S_e, $K_s > S_e$ and equation (9–10) reduces to a first-order expression:

$$\frac{S_o - S_e}{X_v t} = \frac{\bar{\mu}}{aK_s} S_e \qquad (9\text{–}13)$$

The linear removal concept and the monod kinetic relationship are compatible since at low K_s values, the Monod equation closely approximates a zero order model. Monod's studies indicated that, in fact, K_s values are often very small and that growth rates observed are commonly independent of substrate concentration to very low levels. Monod reported a K_s value of 4 mg/l for a pure culture of E. coli growing on glucose and Wuhrmann has reported a value of 0.2 mg/l for a mixed culture activated sludge utilizing the same substrate.

It is rare in treating industrial wastewaters that a single component is considered, rather, many components exist in the wastewater and undergo biological removal. Tischler and Eckenfelder [9] developed a mathematical model for multicomponent removal of organics. This model was based on the fact that when considering an acclimated mixed culture, organics would be removed simultaneously, all at a zero order rate to very low concentrations. This is shown in figure 15. Assume a wastewater contains three components, A, B, and C. With an acclimated culture all components are removed simultaneously at different rates as shown in figure 9.15(a). When an overall organic parameter such as COD is used, the total COD is the sum of all components remaining. The removal rate is constant until time t_1 when component A is substantially removed. The rate then reduces reflecting components B and C until time t_2 when component B is substantially removed. The rate then reflects only component C. In most wastewaters there are many components and figure 9.15(b) appears as a smooth curve. The curve can usually be linearized on a semilogarithmic plot as shown in figure 9.15(c).

FIGURE 9.15
Schematic Representation of Organic Removal

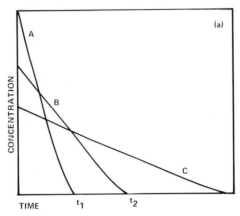

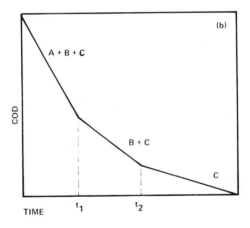

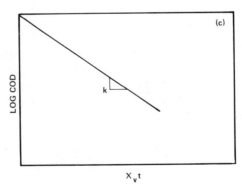

Sequential growth and removal have also been described in some cases for mixed cultures [10]. It was further shown by Tischler and Eckenfelder that when considering total substrate removal such as BOD, COD, or TOC, the overall removal relationship for a batch reaction can be described by a first-order or second-order equation. A second order reaction applies when one or more components have a very low removal rate such as domestic wastewater.

$$\frac{S}{S_o} = e^{-k_1 X_v t} \tag{9-14}$$

or

$$\frac{S}{S_o} = \frac{1}{1 + k_2 X_v t} \tag{9-15}$$

The effect of initial substrate concentration in batch oxidation was studied by Tischler [11], and Grau et al. [7]. Through a mathematical analysis, Tischler showed that the removal rate coefficient is inversely proportional to the initial substrate concentration and that equations (9–14) and (9–15) can be modified:

$$\frac{S}{S_o} = e^{-K_1 X_v t / S_o} \tag{9-16}$$

and

$$\frac{S}{S_o} = \frac{1}{1 + K_2 X_v t / S_o} \tag{9-17}$$

Through similar reasoning, Grau et al. arrived at the same conclusion. When considering continuous completely mixed reactors, a similar analysis can be made. Adams et al. [12] through a series of continuous activated sludge studies on peptone and an organic chemical wastewater showed that the kinetics of organic removal considering influent concentration follow the relationship as shown in figure 9.16 and defined by equation (9–18).

$$\frac{S_o - S_e}{X_v t} = K \frac{S_e}{S_o} \tag{9-18}$$

or

$$S_e = S_o / \left(1 + \frac{K}{F/M}\right) \tag{9-19}$$

where F/M is organic loading defined as $S_o / X_v t$.

The implication of equation (9–19) is that at a constant organic loading, F/M, expressed as kg BOD applied/day/kg VSS, the effluent soluble BOD is directly proportional to the influent BOD. For example, the influent BOD is reduced from 400 mg/l to 20 mg/l in 0.5 days at a F/M of 0.27 day^{-1}. If the influent

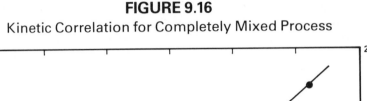

FIGURE 9.16

Kinetic Correlation for Completely Mixed Process

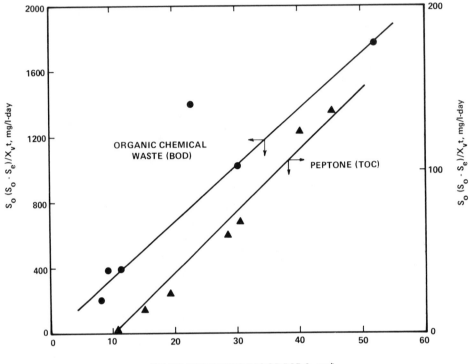

SOLUBLE EFFLUENT TOC OR BOD S_e mg/l

BOD is increased to 800 mg/l, the F/M must be reduced to 0.13 day^{-1} and the retention time increased to two days in order to maintain the same effluent quality of 20 mg/l. If the same percent removal is desired, the F/M remains the same and increasing the influent BOD from 400 to 800 mg/l requires an increase in retention time from 0.5 to 1.0 days. The relationship between influent and effluent BOD for various organic loading levels is shown in figure 9.17.

Organic compounds will have different rates of biodegradation depending on their molecular structure. Table 9.2 shows the relative biodegradability of certain organic compounds. In a comprehensive study Pitter [13] determined the rate and degree of biodegradation of a wide variety of organics as shown in table 9.3.

FIGURE 9.17
Relationship Between Influent and Effluent BOD at Various Loading Levels

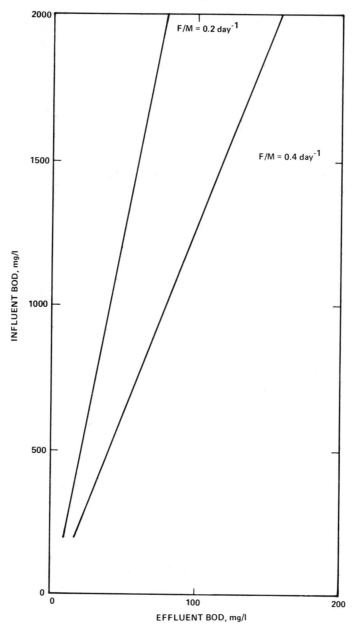

TABLE 9.2
Relative Biodegradability of Certain Organic Compounds

Biodegradable Organic Compounds[a]	Compounds Generally Resistant to Biological Degradation
Acrylic acid	Ethers
Aliphatic acids	Ethylene chlorohydrin
Aliphatic alcohols	Isoprene
(normal, iso, secondary)	Methyl vinyl ketone
Aliphatic aldehydes	Morpholine
Aliphatic esters	Oil
Alkyl benzene sulfonates with	Polymeric compounds
exception of propylene-	Polypropylene benzene
based benzaldehyde	sulfonates
Aromatic amines	Selected hydrocarbons
Dichlorophenols	Aliphatics
Ethanolamines	Aromatics
Glycols	Alkyl-aryl groups
Ketones	Tertiary aliphatic alcohols
Methacrylic acid	Tertiary benzene sulfonates
Methyl methacrylate	Trichlorophenols
Monochlorophenols	
Nitriles	
Phenols	
Primary aliphatic amines	
Styrene	
Vinyl acetate	

[a]Some compounds can be degraded biologically only after extended periods of seed acclimation.

TABLE 9.3
Biological Degradability of Organic Substances[a]

Biological degradability of aliphatic compounds		
Compound	**Percent removed (based upon COD)**	**Rate of biodegradation (mg COD/g VSS-hr)**
Ammonium oxalate	92.5	9.3
n-butanol	98.8	84.0
Sec. butanol	98.5	55.0
Tert. butanol	98.5	30.0
1,4-Butanediol	98.7	40.0
Diethylene glycol	95.0	13.7
Diethanolamine	97.0	19.5
Ethylene diamine	97.5	9.8
Ethylene glycol	96.8	41.7
Glycerol	98.7	85.0
Glucose	98.5	180.0
n-Propanol	98.8	71.0
Iso-Propanol	99.0	52.0
Triethylene glycol	97.7	27.5

Biological degradability of cycloaliphatic compounds		
Compound	**Percent removed (based upon COD)**	**Rate of biodegradation (mg COD/g VSS-hr)**
Borneol	90.3	8.9
Caprolaciam	94.3	16.0
Cyclohexanol	96.0	28.0
Cyclopentanol	97.0	55.0
Cyclohexanone	96.0	30.0
Cyclopentanone	95.4	57.0
Cyclohexanolone	92.4	51.5
1,2-Cyclohexanediol	95.0	66.0
Dimethylcyclohexanol	92.3	21.6
4-Methylcyclohexanol	94.0	40.0
4-Methylcyclohexanone	96.7	62.5
Menthol	95.1	17.7
Tetrahydrofurfuryl alcohol	96.1	40.0
Tetrahydrophthalimide	0	–
Tetrahydrophthalic acid	0	–

TABLE 9.3 (Continued)

Biological degradability of aromatic compounds

Compound	Percent removed (based upon COD)	Rate of biodegradation (mg COD/g VSS-hr)
Aniline	94.5	19.0
Aminophenolsulphonic acid	64.6	7.1
Acctanilide	94.5	14.7
p-Aminoacetanilide	93.0	11.3
o-Aminotoluene	97.7	15.1
m-Aminotoluene	97.7	30.0
p-Aminotoluene	97.7	20.0
o-Aminobenzoic acid	97.5	27.1
m-Aminobenzoic acid	97.5	7.0
p-Aminobenzoic acid	96.2	12.5
o-Aminophenol	95.0	21.1
m-Aminophenol	90.5	10.6
p-Aminophenol	87.0	16.7
Benzenesulphonic acid	98.5	10.6
m-Benzenedisulphonic acid	63.5	3.4
Benzaldehyde	99.0	119.0
Benzoic acid	99.0	88.5
o-Cresol	95.0	54.0
m-Cresol	95.5	55.0
p-Cresol	96.0	55.0
D-Chloramphenicol	86.2	3.3
o-Chlorophenol	95.6	25.0
p-Chlorophenol	96.0	11.0
o-Chloroaniline	98.0	16.7
m-Chloroaniline	97.2	6.2
p-Chloroaniline	96.5	5.7
2-Chloro-4-nitrophenol	71.5	5.3
2,4-Dichlorophenol	98.0	10.5
1,3-Dinitrobenzene	0	–
1,4-Dinitrobenzene	0	–
2,3-Dimethylphenol	95.5	35.0
2,4-Dimethylphenol	94.5	28.2
3,4-Dimethylphenol	97.5	13.4
3,5-Dimethylphenol	89.3	11.1
2,5-Dimethylphenol	94.5	10.6
2,6-Dimethylphenol	94.3	9.0
3,4-Dimethylaniline	76.0	30.0

Biological degradability of aromatic compounds

Compound	Percent removed (based upon COD)	Rate of biodegradation (mg COD/g VSS-hr)
2,3-Dimethylaniline	96.5	12.7
2,5-Dimethylaniline	96.5	3.6
2,4-Diaminophenol	83.0	12.0
2,5-Dinitrophenol	(see note 1)	
2,6-Dinitrophenol	(see note 1)	
2,4-Dinitrophenol	85.0	6.0
3,5-Dinitrobenzoic acid	50.0	–
3,5-Dinitrosalicylic acid	0	–
Furfuryl alcohol	97.3	41.0
Furfurylaldehyde	96.3	37.0
Gallic acid	90.5	20.0
Gentisic acid	97.6	80.0
p-Hydroxybenzoic acid	98.7	100.0
Hydroquinone	90.0	54.2
Isophthalic acid	95.0	76.0
Metol	59.4	0.8
Naphtoic acid	90.2	15.5
1-Naphthol	92.1	38.4
1-Naphthylamine	0	0
1-Naphthalenesulfonic acid	90.5	18.0
1-Naphthol-2-sulphonic acid	91.0	18.0
1-Naphthylamine-6-sulphonic acid	0	0
2-Naphthol	89.0	39.2
p-Nitroacetophenone	98.8	5.2
Nitrobenzene	98.0	14.0
o-Nitrophenol	97.0	14.0
m-Nitrophenol	95.0	17.5
p-Nitrophenol	95.0	17.5
o-Nitroluene	98.0	32.5
m-Nitrotoluene	98.5	21.0
p-Nitrotoluene	98.0	32.5
o-Nitrobenzaldehyde	97.0	13.8
m-Nitrobenzaldehyde	94.0	10.0
p-Nitrobenzaldehyde	97.0	13.8
o-Nitrobenzoic acid	93.4	20.0
m-Nitrobenzoic acid	93.4	7.0
p-Nitrobenzoic acid	92.0	19.7
o-Nitroaniline (see note 2)	0	–
m-Nitroaniline (see note 2)	0	–

TABLE 9.3 (Continued)

Biological degradability of aromatic compounds		
Compound	Percent removed (based upon COD)	Rate of biodegradation (mg COD/g VSS-hr)
p-Nitroaniline (see note 2)	0	–
Phthalimide	96.2	20.8
Phthalic acid	96.8	78.4
Phenol	98.5	80.0
Phloroglucinol	92.5	22.1
n-Phenylanthranilic acid	28.0	–
o-Phenylendiamine (see note 3)	33.0	–
m-Phenylendiamine (see note 3)	60.0	–
p-Phenylendiamine (see note 3)	80.0	–
Pyrocatechol	96.0	55.5
Pyrogallol	40.0	–
Resorcinol	90.0	57.5
Salicylic acid	98.8	95.0
Sulphosalicylic acid	98.5	11.3
Sulphanilic acid	95.0	4.0
Thymol	94.6	15.6
p-Toluenesulphonic acid	98.7	8.4
2,4,6-Trinitrophenol	0	–

[a] From [13].

Note 1. 2,5- and 2,6-dinitrophenol were at higher concentrations not degraded. 2,6-dinitrophenol was at lower concentrations decomposed with long adapted activated sludge (40 days). 2,5-Dinitrophenol was biochemically stable.

Note 2. The degradation of nitroanilines was determined photometrically in the concentration range from 25 to 30 mg/l.

Note 3. The degradation of phenylenediamines was determined photometrically in the concentration range from 25 to 30 mg/l; p-Phenylenediamine was comparatively well degradable.

PROBLEM 9–3

The following data was obtained on the treatment of a wastewater in a completely mixed aeration unit. Correlate the data and develop a kinetic model.

Test 1:

Detention time, days	0.82	0.84	0.97	1.02
MLVSS, mg/l	2380	2630	2710	3370
Influent TOC, mg/l	87.8	215	410	730
Effluent TOC, mg/l	11.0	14.4	27.8	41.0

Test 2:

Detention time, days	0.18	0.38	0.66	1.26
MLVSS, mg/l	2530	2800	2790	2750
Influent TOC, mg/l	75.3	186	370	760
Effluent TOC, mg/l	12.8	18.8	29.5	43.7

The data is tabulated as shown below:

S_o (mg/l)	S_e (mg/l)	X_v (mg/l)	t (days)	$\dfrac{S_o(S_o\text{-}S_e)}{X_v t}$ (mg/l-day)
87.7	11	2380	0.82	3.91
215	14.4	2630	0.84	19.52
401	27.8	2710	0.97	56.93
730	41.0	2370	1.02	146.32
75.3	12.8	2530	0.18	10.33
186	18.8	2800	0.38	29.23
370	29.5	2790	0.66	68.42
760	47.5	2750	1.26	156.32

The results are plotted as shown in figure 9.16 (curve for peptone) and the resulting kinetic equation is:

$$\frac{S_o(S_o - S_e)}{X_v t} = 4.36 \, (S_e - 11.25)$$

in which the specific reaction rate coefficient K is 4.36 day^{-1}.

Consideration of equation (9–18) leads to the conclusion that the reaction rate coefficient K is dependent on the composition of the wastewater. Values computed for several industrial wastewaters are shown in table 9.4. While K might be expected to be relatively constant in wastewaters of constant composition such as a dairy waste, a variable wastewater composition resulting from changes in product mix as is encountered in the organic chemicals industry results in a variable K. In order to define an effluent quality and its variability from a biological wastewater treatment facility for this type of wastewater, the reaction rate K must be treated as a statistical variable.

BOD removal from domestic sewage presents a unique case. In a domestic sewage the major part (75–90 percent) of the BOD is present either in colloidal or suspended form. When treated by the activated sludge process these organics are rapidly removed by adsorption and enmeshment in the biological floc. It is for this reason that contact periods of less than 15 minutes result in BOD removals in excess of 80 percent. This phenomena provides the basis for the contact stabilization process and the physical-chemical process in which suspended and colloidal BOD is removed by coagulation followed by soluble BOD removal on activated carbon. Figure 9.18 illustrates this phenomena. The mechanism of BOD removal from domestic wastewater can therefore be considered as a two phase reaction, an initial rapid removal of BOD followed by a slow removal of soluble BOD. It has been shown that this removal process can be modeled as a second order reaction

TABLE 9.4
Reaction Rate Coefficients for Organic Wastewaters

Wastewater	k days^{-1}	Temp. °C
Potato processing	36.0	20
Peptone	4.03	22
Sulfite paper mill	5.0	18
Vinyl acetate monamer	5.3	20
Polyester fiber	14.0	21
Formaldehyde, propanol, methanol	19.0	20
Cellulose acetate	2.6	20
AZO dyes, epoxy, optical brighteners	2.2	18
Petroleum refinery	9.1	20
Vegetable tannery	1.2	20
Organic phosphates	5.0	21
High nitrogen organics	22.2	22
Organic intermediates	20.6	26
	5.8	8
Viscose rayon and nylon	8.2	19
	6.7	11
Soluble fraction of domestic sewage	8.0	20

FIGURE 9.18

BOD Removal from Sewage and Industrial Waste in the Activated Sludge Process

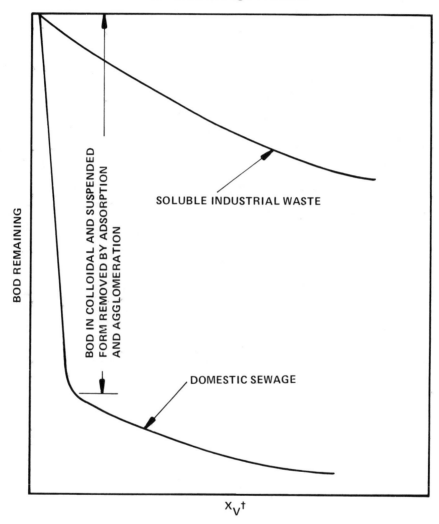

or as a composite exponential which considered two or more reactions. These are shown in figure 9.19. McCauley [14] concentrated the soluble fraction of domestic sewage and ran BOD removal studies as shown in figure 9.20.

     The kinetic relationships assume that all the organics removed in the process undergo biological oxidation and synthesis. Some wastewaters contain volatile organics that will undergo simultaneous biological oxidation and air strip-

FIGURE 9.19
BOD Removal Characteristics from Domestic Sewage

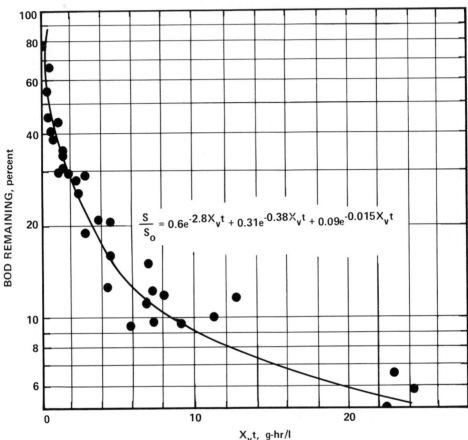

$$\frac{S}{S_o} = 0.6e^{-2.8X_v t} + 0.31e^{-0.38X_v t} + 0.09e^{-0.015X_v t}$$

ping in the aeration process. In these cases it is necessary to operate a sterile test unit under the same aeration conditions to determine the air stripping rate coefficient. The rate coefficient from biological unit that includes both biooxidation and stripping is then corrected for stripping to reflect only biological oxidation. When scaling to full-scale design, the stripping effect should be estimated for the equipment and turbulence to be used.

The kinetic relationships previously developed relate to soluble effluents. The BOD contributed by the effluent suspended solids in a process must therefore be added to the soluble BOD as a terminal step in a process design calculation:

$$BOD_{total} = BOD_{soluble} + f' \cdot SS$$

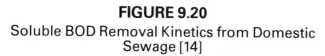

FIGURE 9.20
Soluble BOD Removal Kinetics from Domestic Sewage [14]

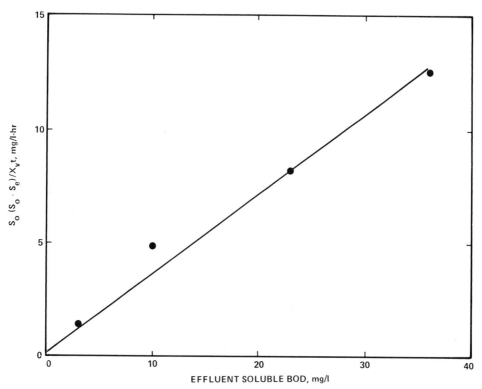

The coefficient, f', is related to the influent waste characteristics (primarily the nature of its suspended solids content, if any) and to the sludge age, since increasing sludge age increases the inert content of the sludge. A typical correlation is shown in figure 9.21.

It should also be noted that small residual BOD or COD will remain even after long periods of aeration, because autooxidation of the sludge results in resolubilization of cellular material that is subsequently used for synthesis as shown in figure 9.22.

It should be recognized that in any one plant as the composition of the waste changes, the overall removal rate may also change, owing to variation in removal rate of particular constituents and their initial concentration. This results in a variation in the coefficient K. As seen in figure 9.23, a wastewater with a constant composition such as a dairy waste yields a near constant K, while a varying composition of chemical wastewater shows a high variability in K.

FIGURE 9.21
BOD Characteristics of Biological Sludge

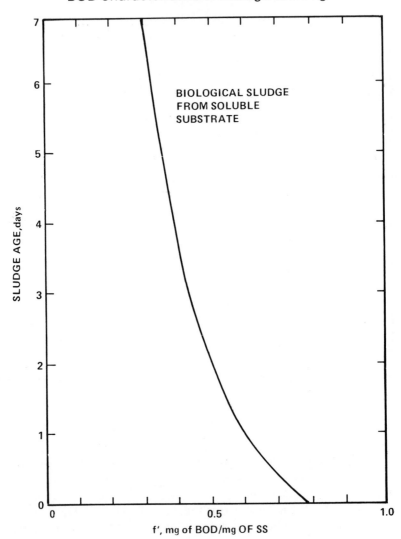

BIOLOGICAL SLUDGE
FROM SOLUBLE
SUBSTRATE

SLUDGE AGE, days

f', mg of BOD/mg OF SS

FIGURE 9.22
Effluent Levels During Biological Oxidation

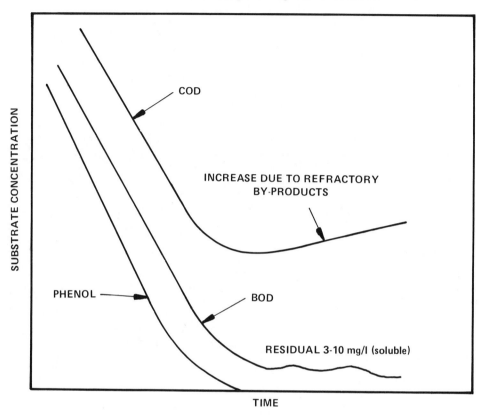

RELATIONSHIPS BETWEEN SLUDGE AGE, ORGANIC LOADING (F/M) AND REACTION RATE

Removal rate for a mixture of organics can be expressed by the relationship:

$$\frac{S_o - S_e}{X_v t} = K \frac{S_e}{S_o} \qquad (9\text{--}20)$$

Sludge age is expressed:

$$\theta_c = \frac{X_v t}{a S_r - b x X_v t} \qquad (9\text{--}21)$$

FIGURE 9.23
Variability in K as Related to Wastewater Composition

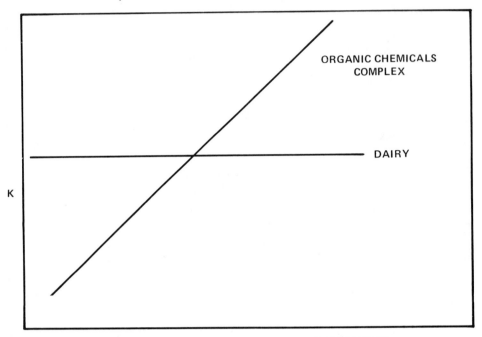

ORGANIC CHEMICALS
COMPLEX

DAIRY

K

PERCENT OF TIME VALUE IS EQUAL TO OR LESS THAN

and organic loading as:

$$F/M = \frac{S_o}{X_v t} \qquad (9\text{--}22)$$

in which S_o, S_e, X_v and S_r are all in concentration units, e.g., mg/l. Substituting equation (9–22) into equation (9–21):

$$\frac{1}{\theta_c} = a\,F/M\,\frac{S_r}{S_o} - bx$$

For high removal levels $S_r \rightarrow S_o$ and:

$$\frac{1}{\theta_c} = a\,F/M - bx \qquad (9\text{--}23)$$

It is noticed that sludge age is inversely proportional to F/M. Substituting equation (9–20) in equation (9–21)

$$\frac{1}{\theta_c} = aK\,\frac{S_e}{S_o} - bx \qquad (9\text{--}24)$$

If the kinetic equation is based on only active biomass, xX_v, the equation (9–24) can be modified:

$$\frac{1}{\theta_c} = axK_A \frac{S_e}{S_o} - bx \qquad (9\text{–}24a)$$

For long sludge ages as $\Theta_c \rightarrow \infty$, equation (9–24) becomes:

$$\frac{S_e}{S_o} = \frac{bx}{aK} \qquad (9\text{–}25)$$

As the sludge age approaches ∞, $x \rightarrow O$ and S_e/S_o also approaches zero. If the kinetic equation is based on only active biomass, then equation (9–25) becomes:

$$\frac{S_e}{S_o} = \frac{b}{aK_A}$$

or the fraction of organic remaining approaches a finite value. Results from studies on three substrate mixtures with varying reaction rates are shown in figure 9.24.

FIGURE 9.24
Kinetic Relationships for Various Substrate Mixtures

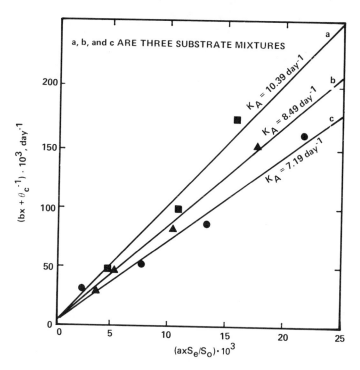

CHARACTER OF BIOLOGICAL SLUDGE

The microbial sludge or biological floc employed in biooxidation processes is a miscellaneous collection of microorganisms such as bacteria, yeasts, molds, protozoa, rotifers, worms and insect larvae in a gelatinous mass. Algae will also be present in these areas exposed to sunlight.

The bacteria are primarily nonnitrifying aerobic spore formers, many of which are of the *B. subtilis* group. Nitrifying bacteria are primarily *Nitrosomonas* and *Nitrobacter*. In most activated sludge processes the sludge appears as zoogleal masses intermixed with filamentous bacteria. One of the principal forms in the zoogleal mass is *Zooglea, ramigera* which has been defined as a gram-negative, non-spore-forming, motile, capsulated rod. Since most bacteria under proper conditions can flocculate, *Zooglea ramigera* may not be a true species, but rather a growth form of many species. Other common forms of bacteria found in activated sludge include *Flavobacterium, Pseudomonas,* and filamentous organisms, the most common of which is *Sphaerotilus natans*. Fungi are more common in trickling filters than in activated sludge. These forms generally exist in the presence of low oxygen tension, low pH or low nitrogen content. Of the protozoa, stalked ciliates are the most common, including *Vorticella, Opercularia,* and *Epistylis*. Free swimming types include *Paramecium, Lionotus,* and *Trichoda*. Some forms of *Flagellata* and *Rhizopodea* are also found. The relationship between the type of protozoa that predominates and the bacterial population seems to depend on the degree of flocculation and the nature of the wastewater. In a well flocculated sludge, stalked ciliates and attached forms are common since they feed on the zoogleal mass. With low flocculation, free swimming forms dominate. The interrelationship between bacteria and protozoa on treatment efficiency is not well defined. There is also a competition for food between the secondary feeders. In a solution with high bacterial populations free swimming protozoa are dominant, but when food becomes scarce stalked protozoa increase in numbers. Stalked protozoa do not require as much energy as free swimming protozoa and can therefore compete more effectively in a system with low bacterial concentrations. It is generally conceded, however, the protozoa aid in clarification. Engelbrecht and McKinney [15] found that sludge developed on structurally related compounds possessed similar morphological characteristics and produced similar biochemical changes. For example, they found that dense flocs were produced from the pentose sugars, xylose and arabinose and that filamentous floc was produced from the hexose sugars, glucose, and fructose.

A wide variation in microorganism population between the assimilative and the endogenous phases was reported by Jasewicz and Porges [16] on dairy waste studies. They found that during the assimilative phase, 74 percent of the organisms were of the genera *Bacillus* or *Bacterium* while only 8 percent of the endogenous sludge was composed of these organisms. The endogenous sludge contained 42 percent of the proteolytic organisms, *Pseudomonas* and *Alcaligenes,* and 48 percent of the saccharolytic organisms, *Flavobacterium* and *Micrococcus.*

Protozoan forms were similar in both sludges. The distribution of microorganisms in trickling filters, likewise, varies with depth depending on the food supply and the growth conditions. In general, algae forms are found in the surface layers and a predominance of nitrifying forms in the lower depths. Activated sludge from two biooxidation processes is shown in figure 9.25.

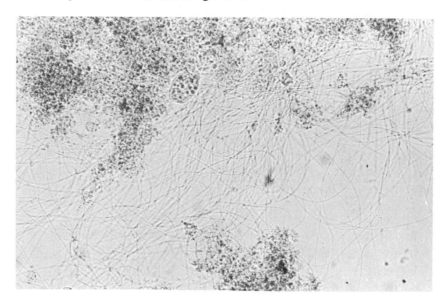

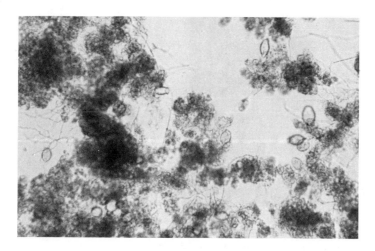

FIGURE 9.25
Activated Sludge from Two Biooxidation Processes

The chemical content of microorganisms depends upon the quantity of water absorbed by the cell and is a function of the pH and the environment. Capsulated organisms retain more moisture than those that do not have a capsule. The average moisture content of bacteria is 80 percent (73–88 percent). The moisture content of yeasts averages 75 percent and molds 85 percent. Phosphorus, potassium, magnesium, and other trace minerals are present in the ash or mineral content and are partly organically bound in proteins, nucleic acids, carbohydrates, lipids, pigments, etc. A high proportion of the total ash consists of potassium and phosphorus. The variation in phosphorus content for most microorganisms is 2.5 to 5.0 percent. The ash content of most biological sludges is low (2–15 percent). When the ash content exceeds this, it is generally due to the presence of non-biological inert substances in the sludge. Nitrogen is present in protoplasm as proteins and amino acids (product of the breakdown of proteins and nucleo-proteins). Nitrogen content will vary from 8 to 15 percent for most bacteria. Yeasts and molds will possess a lower nitrogen content. The nitrogen content is measured by the Kjeldahl distillation. Protein content may be computed as the total organic nitrogen measured above times 6.25. This factor is based on an average nitrogen value of 16 percent for protein. The carbon content of cells is present as complex carbohydrates, etc. and is measured by converting the dry mass to CO_2 and H_2O in which the carbon is equal to 3/11 of the CO_2. The hydrogen content is present as 1/9 of the H_2O produced by the conversion. The carbon content of most microorganisms ranges from 45 to 55 percent. Lipoids (fat) are also present to a varying degree in the cell mass. Table 9.5 shows the chemical formulation of several microorganisms.

TABLE 9.5
Analyses and Empirical Formulas of Microorganisms

	Yeast[a]	Bacteria[b]	Zooglea
Carbon	47·0	47·7	44·9
Hydrogen	6·0	5·7	–
Oxygen	32·5	27·0	–
Nitrogen	8·5	11·3	9·9
Ash	6·0	8·3	–
Empirical formula	$C_{13}H_{20}N_2O_7$	$C_5H_7NO_2$	–
Carbon-nitrogen ratio	5·6/1	4·3/1	4·5/1

[a] From [17].
[b] From [18].

OXYGEN PENETRATION INTO BIOLOGICAL FLOCS

Numerous investigators have shown that under the mixing and aeration conditions in the activated sludge process as conventionally employed, the biological floc may be composed of aerobic surface layers with an anaerobic center. The activity of the floc results from the diffusion of oxygen and nutrients into its mass. If we assume that only the aerobic portion of the floc is effective in organic removal, then on a weight basis, increasing the percentage of the floc, which is aerobic, will result in an increase in the overall organic removal efficiency.

The ideal spherical floc (figure 9.26) can be treated as a porous medium with the oxygen flux being described by the relationship:

$$N_{O_2} = -D_{O_2} \frac{dC_{O_2}}{dr} \tag{9-26}$$

Considering a spherical configuration of surface area $4\pi r^2$, the mass rate of oxygen moving into the floc can be given by:

$$M_{O_2} = -4\pi r^2 D_{O_2} \frac{dC_{O_2}}{dr} \tag{9-26a}$$

The oxygen uptake by microorganisms independent of the dissolved oxygen concentration above some critical value and:

$$-r_{O_2} = -k_r$$

The volume of a sphere is given by $(4/3)\pi r^3$, which allows the mass rate of oxygen utilized by the microorganisms in the floc to be:

$$M_{O_2} = -\frac{4}{3} \pi r^3 k_r \tag{9-27}$$

Equating equations and integrating yields:

$$C_i - C = \frac{k_r}{6D_{O_2}} (R^2 - r^2) \tag{9-28}$$

The fully aerobic diameter of floc can be computed from:

$$C_i - C_c = \frac{d^2}{24D_{O_2}} k_r \tag{9-28a}$$

where C_i = oxygen concentration at the floc surface
C_c = critical oxygen concentration to support a maximum uptake rate, assumed as 0.1 mg/l
k_r = oxygen utilization rate, mg O_2/cm^3 of floc-sec
d = floc diameter, cm

FIGURE 9.26
Representation of Floc

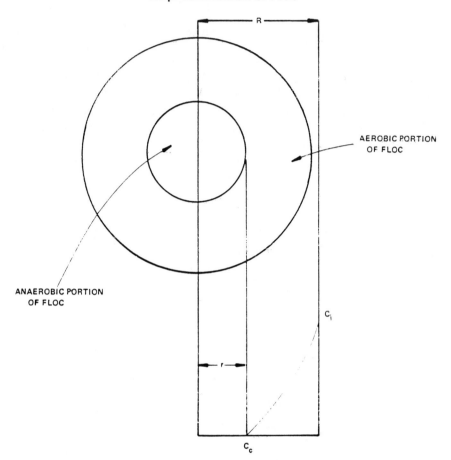

D_{O_2} = oxygen diffusivity in floc material, cm²/sec, assumed as 5
   $\times 10^{-6}$ cm²/sec at 15° C

To calculate k_r, the following relationship is used:

$$k_r = r_r \left(\frac{\text{VSS}}{\text{SS}} \right) \rho_{\text{SS}}$$

where r_r = oxygen utilization rate (R_r/X_v), mg O₂/g VSS-sec
       VSS/SS = volatile fraction of suspended solids
       ρ_{SS} = suspended solids content per unit floc volume ranging
                 from 0.15 to 0.30 g SS/cm³ of floc

The aerobic portion of the floc will depend upon the oxygen gradient across the surface film, ΔC, the floc size, d, and the specific oxygen uptake rate k_r. In any system with a defined organic loading, the aerobic portion of the floc can be increased either by increasing the power level (hp/1000 gal) and thereby reducing the floc size or by increasing the dissolved oxygen driving force, ΔC, which is accomplished by increasing the dissolved oxygen level in the process. As the loading to the process is reduced and the oxygen uptake rate decreases it will take less power and/or oxygen driving force to maintain the floc aerobic since the oxygen uptake rate has decreased. The oxygen penetration into biological films (for example in trickling filters or rotating biological contactors) can be similarly computed from the relationship

$$d = \sqrt{\frac{2D_{O_2}\,(C_i - C_c)}{k_r}} \qquad (9\text{-}29)$$

The endogenous rate coefficient, b, is related to the autooxidation of the biomass, that is the lbs VSS oxidized/day/lb VSS under aeration. As more of the floc becomes aerobic it is reasonable to expect that b will increase since more aerobic activity will result.

Rickard and Gaudy [19] showed the relationship between power level, organic removal, and oxygen uptake rate and the endogenous rate, b. Busch [20] has recomputed their data and showed that increasing the hydraulic shear intensity from 300 to 1000 units resulted in an increase of organic removal rate from 0.3 g COD/g MLSS/hr to 0.7 g COD/g MLSS/hr and an increase in the endogenous coefficient, b, of 0.14 to 0.20 day^{-1}. It is not possible to relate, however, experimental results to common field operating conditions, since the hydraulic shear rate does not readily relate to the power level in aeration basins expressed as hp/1000 gal.

These observations are consistent with the results of Pasveer [21]. Mueller, et al. [22], who developed relationships between floc size and the dissolved oxygen concentration necessary to render the floc completely aerobic, assuming a floc size of 115 μm. Their results showed that at an oxygen uptake rate of 80 mg/l/hr, a dissolved oxygen level of 6.0 mg/l was required to maintain an aerobic floc, 4.0 mg/l/hr required a dissolved oxygen level of 0.6 mg/l. Unfortunately, Mueller et al., did not relate their results to power level, so it is not possible to define floc size in the aeration basins under various turbulence levels. Figure 9.27 shows calculated results from equation (9-28a) using the data of Mueller et al., relating floc size to maintain fully aerobic conditions to dissolved oxygen concentrations. It becomes apparent that the higher the organic loading (F/M) the higher must be the dissolved oxygen to maintain aerobic conditions. For dispersed cells, the critical oxygen level has been reported as 0.1–0.3 mg/l with values as low as 0.001 mg/l for some cells. In the case of biological flocs, oxygen must diffuse into the floc in order to reach the organisms within the floc. The floc particle size will be affected by the basin power input and the basin size.

FIGURE 9.27

Fully Aerobic Floc Diameters Under Various DO Concentration and Oxygen Levels

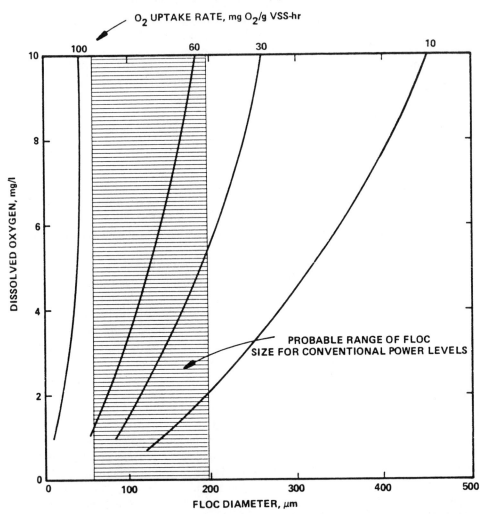

Decreasing the floc size by increased power levels decreases the required dissolved oxygen concentration to maintain aerobic conditions. Zahradka [23] conducted pilot plant studies on Prague sewage in which he varied the organic loading and the power level. He showed that activated sludge under conditions of conventional loading and mixing intensity has considerable removal potential. Increasing the mixing intensity resulted in increased removal as related to loading (F/M). Figure 9.27 would indicate that at endogenous levels of 10 mg O_2/g/hr, 1.0–2.0 mg/l of O_2 are required for fully aerobic conditions.

SLUDGE SETTLING

Sludge settling and compaction characteristics are a primary requisite to successful operation of the activated sludge process. With a poor settling sludge, solids carryover in the effluent will contribute to the BOD (due to endogenous respiration of the activated sludge solids in the BOD bottle). The poor compaction results in a low concentration of return sludge solids, which in turn limits the mixed-liquor suspended solids level. The sludge settling rate relates to the characteristics of the wastewater, for example, nonsettled sewage contains inert materials that enhance the biological sludge settling rate. A poor settling or bulking sludge can be the result of the propagation of filamentous organisms (e.g., *Sphaerotilus*) or from dispersed bacterial growth due to nutrient limitations.

Many filamentous growths are obligatory aerobes that flourish in the presence of an available carbon source such as glucose. The most common form of filamentous bacteria associated with bulking sludge is *Sphaerotilus natans* (figure 9.25), which is characterized by the presence of a prominent contiguous close fitting sheath that endorses a chain of rod-shaped cells. The sheath is of particular importance because it serves as a means of attachment to solid surfaces. *Sphaerotilus* is a strict aerobe, but can survive in systems with dissolved oxygen levels as low as 0.1 mg/l. Other organisms, such as the fungi *Geotrichum candidum*, have shapes very similar to *Sphaerotilus*. *Beggiatoa* is commonly associated with systems containing hydrogen sulfide. The bacteria grow in the form of unattached long cylindrical colorless filaments ranging in length from 80 to over 1500 μm. Some strains of *Beggiatoa* grow as floc composed of long intertwined filaments. *Beggiatoa* is a strict aerobe. Thiothrix oxidizes sulfide and deposits sulfur in extensive intracellular granules. At low concentrations of dissolved oxygen in the mixed liquor (< 0.5 mg/l) there is little oxygen penetration into the biological floc and only a small fraction of the bacterial mass will exhibit aerobic growth [21,24]. The filaments on the other hand, have a very high surface-area-to-volume ratio and will quickly outgrow the bacterial population [20]. High oxygen tension favors the growth of floc forming bacteria. When the carbon source is exhausted, the filaments tend to disappear from the system. In general the propagation of filamentous organisms is favored by higher substrate concentrations in solution and low dissolved oxygen and low nutrient levels. When the floc is completely aerobic, flocculating bacteria can better compete for the available substrate. Fungi, generally filamentous in nature, are formed at a low pH.

Many filamentous organisms are aerobic and can be destroyed by prolonged periods of anaerobiasis. Most of the bacteria, on the other hand, are facultative and can exist for extended periods without oxygen. Although available data are somewhat contradictory, it would appear that at least 12 hours under anaerobic conditions are necessary to restrict the growth of these filaments.

In practice, culture control can frequently be achieved by holding the mixed liquor in the final settling tank for a period sufficient to eliminate the filaments. Filamentous bulking can also be controlled by:

1. The addition of chlorine or hydrogen peroxide [25] to the return
    sludge. Hydrogen peroxide dosages are in the order of 20–50
    mg/l. Chlorine dosages may vary from 8–10 lbs Cl_2/day/1000 lbs
    MLSS in cases of severe bulking to 1–2 lbs Cl_2/day/1000 lbs MLSS
    in cases of moderate bulking.
2. H_2O_2 appears to attack the filamentous cell wall resulting in disin-
    tegration of the organisms. The effect of H_2O_2 addition on SVI is
    shown in figure 9.28.
3. The addition of polyelectrolytes to flocculate and settle the
    filamentous growths.

Filamentous growths are suppressed at high oxygen levels such as is carried in the
high purity oxygen process since the floc is maintained aerobic.

Endogenous sludges can be maintained under anaerobic conditions for
long periods because the exogenous food supply has been exhausted. At high
loadings (F/M), however, gasification due to anaerobic activity may occur, resulting in rising sludge. At very low loadings denitrification may also cause rising
sludge.

Various investigators have related sludge bulking to organic loading. Indications have shown that bulking becomes progressively more severe as the
loading exceeds 0.5 kg of BOD/day/kg of MLSS (0.5 lb of BOD/day/lb of MLSS)
[26]. Von der Emde [27], however, reported excellent settling characteristics with
loading in excess of 2.0 kg of BOD/day/kg of MLSS (2.0 lb of BOD/day/lb of
MLSS). This apparent contradiction can possibly be explained by considering the
cause of bulking and its relationship to the operation of the process.

Sludge bulking can be related to the growth rate or metabolic activity of
the sludge, which in turn is related to the food/microorganisms ratio, F/M. At very
high F/M ratios, the organisms have a maximum growth rate (log growth phase)
and flocculation does not occur [26]. At very low loadings, unoxidized fragments
of floc remain in suspension, resulting in poorer settling.

A number of theories have been advanced to explain the mechanism of
biological flocculation. The differences in the various theories lie in which of the
surface properties is considered to be most important and how this parameter is
affected by nutritional and environmental variations. Forster (28) related zeta
potential to the Sludge Volume Index. The nature of the surface polymers on the
biomass controls the magnitude of the zeta potential. A wide variety of biopoly-
mers, including lipids, proteins, and nucleic acids, have been reported to occur at
bacterial sludge surfaces although the most common type are polysaccharides.
These are produced as a slime or capsule layer by a considerable number of
bacterial species under nutritional limitation. While the mechanisms of biological
flocculation are still in question, there is little doubt that both the nature of the
organic and the organic loading to the process are primary factors influencing the
degree of flocculation in the process. Some filaments present in the sludge can
provide a backbone system with the floc and hence enhance flocculation.

The relationship between process variables and filamentous bulking is

FIGURE 9.28
Hydrogen Peroxide Treatment of Mixed
Municipal/Cannery Waste, Activated Sludge

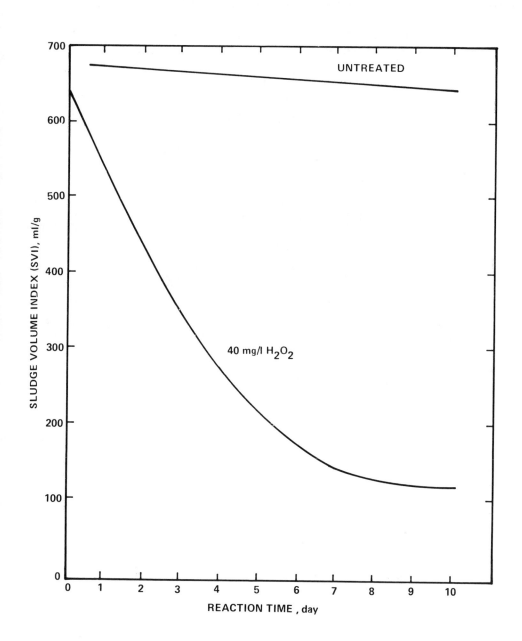

shown in figure 9.29. At low loadings, the flocs become nutrient limited and the growth of dispersed flocs can occur. Under these conditions, Chudoba *et al.* [29] indicated that batch or multiple completely mixed reactors produced a denser sludge than a parallel completely mixed system. The rationale indicated that a greater organic driving force in the initial reactors selectively favored the growth of flocculated bacteria. They showed, for example, in the case of lactose that filamentous organisms will have a higher growth rate at low concentration levels while zoogleal microorganisms will have a higher growth rate at high concentration levels. In this case, a completely mixed aeration tank with low substrate concentration and high purification efficiency will favor filamentous growth while

FIGURE 9.29

Flocculation and Settling Characteristics of Activated Sludge as Related to Organic Loading

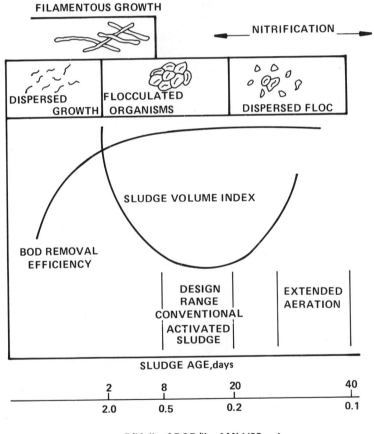

a plug flow or multistage system with high substrate concentration gradients in the initial stages will suppress filamentous microorganisms to a certain extent. Chudoba *et al.* related SVI to the reciprocal of the dispersion number in the aeration basin.

When considering the effect of the F/M ratio on metabolic activity, the availability of the substrate must be considered. Soluble sugars, for example, are immediately available to the organism and consequently yield an immediate growth response. Suspended organics, on the other hand, must undergo sequential breakdown to simpler substrates before being available for synthesis. The growth response is therefore much slower, even at high F/M ratios. At high loading levels, oxygen becomes limiting and higher dissolved oxygen levels are needed to maintain an aerobic floc and hence favor the growth of zoogleal forms over filamentous forms. This would explain why the high purity oxygen process effectively functions at high loadings (F/M > 0.6 day^{-1}) without the propagation of filamentous growths.

In summary, floc characteristics as related to growth rate will be influenced both by the availability of the substrate, the mode of introduction of the waste to the system, and the dissolved oxygen and turbulence in the system. Typical results for an industrial waste in a completely mixed system are found in figure 9.30.

NUTRIENT REQUIREMENTS

Aerobic organisms require minimum quantities of nitrogen and phosphorus and other trace elements for optimal activity. The trace elements required are usually present in sufficient quantity in the carrier water (one exception is when the wastewater emanates from distilled or deionized water). The trace elements required are shown in table 9.6. Nitrogen and phosphorus are frequently deficient in industrial wastewaters and must be added as a nutrient supplement. (In combined treatment of domestic and industrial wastewaters, the excess nitrogen and phosphorus in the sewage may supply the requirements for the industrial waste as shown in figure 9.31.) The quantity of nitrogen and phosphorus required relates to the composition of the biomass. Active biomass contains approximately 12.3 percent nitrogen and 2.6 percent phosphorus. The cellular residue after endogenous metabolism has been reported to contain 7 percent nitrogen and 1 percent phosphorus. Nitrogen and phosphorus will be lost from the process in the excess sludge.

The nitrogen lost will be that present in the active mass plus that present in the nondegradable residue. The minimum quantities of nitrogen required can then be computed:

$$N \text{ (kg/day) (lbs/day)} = 0.123 \frac{x}{0.77} \Delta X_v + 0.07 \frac{(0.77 - x)}{0.77} \Delta X_v \qquad (9\text{–}30)$$

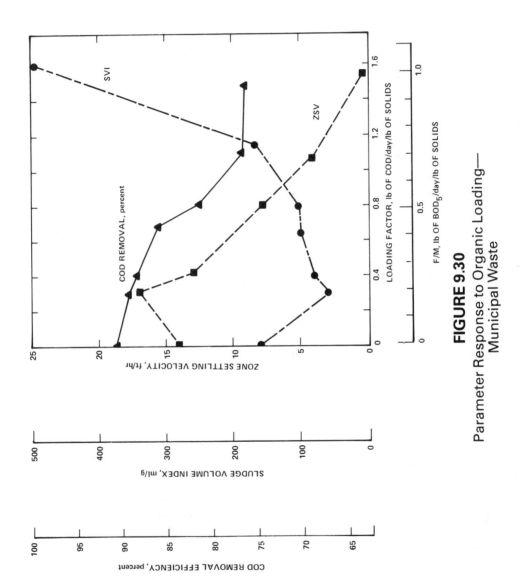

FIGURE 9.30

Parameter Response to Organic Loading—
Municipal Waste

TABLE 9.6
Trace Nutrient Requirements for Biological Oxidation

	(mg/mg BOD)
Mn	10×10^{-5}
Cu	14.6×10^{-5}
Zn	16×10^{-5}
Mo	43×10^{-5}
Se	14×10^{-10}
Mg	30×10^{-4}
Co	13×10^{-5}
Ca	62×10^{-4}
Na	5×10^{-5}
K	45×10^{-4}
Fe	12×10^{-3}
CO_3	27×10^{-4}

TABLE 9.7
Effect of Nutrients on Reaction Rate Coefficient

Waste	Rate – K, day^{-1}	
	Without Nutrients	With Nutrients
Kraft paper	0.35	1.33
Board mill	0.70	3.20
Hardboard	0.34	1.66
Kraft paper	0.26	1.50

In like manner the minimum phosphorus can be computed:

$$P \text{ (kg/day) (lbs/day)} = 0.026 \frac{x}{0.77} \Delta X_v + 0.01 \frac{(0.77 - x)}{0.77} \Delta X_v \qquad (9\text{--}31)$$

In order for nitrogen to be consumed in the process it must be available for assimilation by the organisms. Ammonia nitrogen and nitrate nitrogen are available. Organic nitrogen must first be hydrolyzed in the process in order to be available to the biomass. Depending on the nature of the wastewater a variable portion of the organic nitrogen may become available for synthesis.

In many cases, in aerated lagoons treating pulp and paper mill wastewaters, nitrogen and phosphorus have not been added, but rather the retention time increased. A comparison of operation of an aerated lagoon with and without nutrients is shown in figure 9.32. Calculated reaction rate coefficients are shown in table 9.7. High BOD reductions at minimal nutrients can be achieved if sufficient retention time is provided.

FIGURE 9.31

Nitrogen Requirements

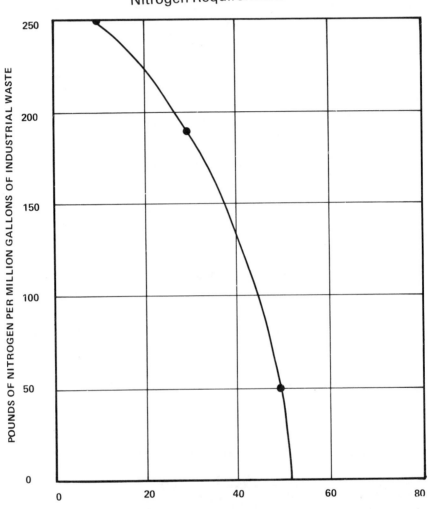

FIGURE 9.32
Effect of Nutrient Addition on BOD Removal in Aerated Lagoons

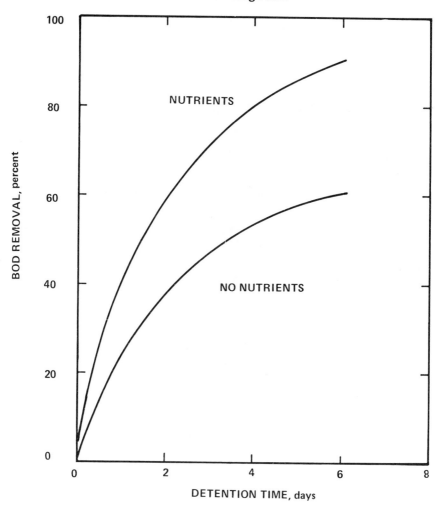

PROBLEM 9–4

Three MGD of an industrial wastewater with a BOD_5 of 2300 mg/l and a nitrogen content of 1.5 mg/l is to be treated in the activated sludge process with 5 MGD domestic sewage with a BOD_5 of 190 mg/l and a nitrogen content of 35 mg/l. Assuming the process with a of 0.52 and b of 0.1 day^{-1}, compute the additional nitrogen required using an F/M of 0.4 day^{-1} and 92 percent BOD removal.

The BOD$_5$ of the influent is:

$$\frac{3 \text{ MGD} \cdot 2300 \text{ mg/l} + 5 \text{ MGD} \cdot 190 \text{ mg/l}}{8 \text{ MGD}} = 981 \text{ mg/l}$$

The nitrogen in the influent is:

$$\frac{3 \text{ MGD} \cdot 1.5 \text{ mg/l} + 5 \text{ MGD} \cdot 35 \text{ mg/l}}{8 \text{ MGD}} = 22.4 \text{ mg/l}$$

The degradable fraction, x, is computed from equation (9–4) as 0.65. The excess volatile suspended solids is:

$\Delta X_v = aS_r - bxX_v$

$= 0.52 \, (0.92 \cdot 981 \text{ mg/l} \cdot 8 \text{ MGD} \cdot 8.34) - 0.1 \cdot 0.65 \cdot 981 \text{ mg/l} \cdot 8 \text{ MGD} \cdot 8.34/0.4$

$= 20680 \text{ lb/day}$

$N_{required} = 0.123 \cdot \dfrac{x}{0.77} \, \Delta X_v + 0.07 \, \dfrac{(0.77 - x)}{0.77} \, \Delta X_v$

$= 0.123 \cdot \dfrac{0.65}{0.77} \cdot 20680 + 0.07 \, \dfrac{(0.77 - 0.65)}{0.77} \, 20680$

$= 2373 \text{ lbs/day}$

$N_{available} = 22.4 \text{ mg/l} \cdot 8 \text{ MGD} \cdot 8.34 = 1495 \text{ lbs/day}$

Supplementary nitrogen = 2373 - 1495 = 878 lbs/day

Effect of Temperature

One of the significant variables in the selection of type of process is the effect of temperature on process performance. The effect of temperature on biological activity is shown in figure 9.33.

The temperature effect on the reaction rate can be expressed by the relationship:

$$K_T = K_{20°C} \, \theta^{T-20} \qquad\qquad (9\text{–}32)$$

Equation (9–32) generally applies over the range of 4–30° C. This range is defined as the mesophilic range. Biological activity can also occur in the ther-

FIGURE 9.33

Effect of Temperature on the Biological Reaction Rate

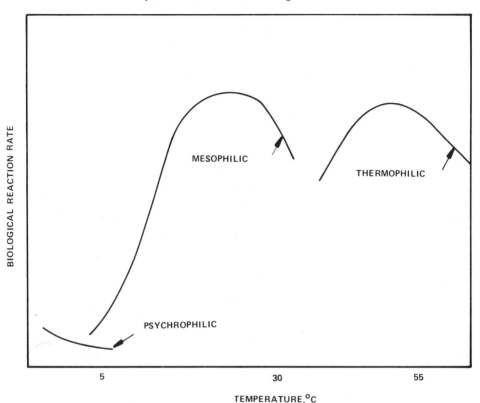

mophilic range, which has an optimum temperature of 55° C. At low temperatures (< 4° C) psychrophilic organisms predominate. The temperature effect on aerobic biological activity is shown in table 9.8.

One explanation for the wide variation in Θ can be rationalized by considering the nature of the process. In the activated sludge process at low loadings, BOD removal and oxidation depend on diffusion of oxygen into the biological flocs, where it is subsequently utilized by the organisms. At low power levels and

TABLE 9.8

Temperature Dependence of Rate Constant
for Activated Sludge

Process	Temperature Range	Θ
Removal Rate of	0–10° C	1.042
Acetate	20–25° C	1.214
Removal Rate of	0–10° C	1.131
Phenol	10–20° C	1.056
Mixed Organic Chem.		1.055
Effluent		

large floc sizes and low temperatures, a low oxygen utilization rate permits greater diffusion of oxygen, and consequently a large portion of the floc is aerobic. At high temperatures, the high respiration rate depletes the oxygen rapidly, and only a small portion of the floc is aerobic. It can be assumed that a large mass of organisms at a low respiration rate (winter) achieves the same degree of oxidation as a small mass at a high respiration rate (summer), and hence, the coefficient Θ is low.

By contrast, at high organic loadings, the floc tends to become dispersed (bulking sludge), and each organism is more directly affected by changes in temperature. The coefficient, Θ, therefore increases. In the case of domestic sewage, where a major portion of the BOD is in suspended or colloidal form, removal in the presence of flocculated sludge is largely physical and, hence, relatively independent of temperature. In aerated lagoons, at a low solids level, the organisms are more dispersed, and the temperature coefficient is higher. This is also true for the BOD bottle and the stream. Trickling filters are analogous to activated sludge except that oxygen diffusion is uniplaner into the film. Similar calculations as for activated sludge lead to a coefficient Θ of 1.035 for trickling filters. Marked improvement in aerated lagoon operation during the winter months can be achieved by adding a recycle and increasing the solids level in the basin.

In many cases the effluent suspended solids increases with decreasing wastewater temperatures. The maximum temperature for effective aerobic biological activity is about 38° C (100° F). At higher temperatures the reaction rate decreases and the effluent suspended solids increases. The effluent suspended solids during winter and summer operation for an activated sludge plant treating an organic chemicals wastewater is shown in figure 9.34.

Since a major portion of BOD removal from domestic sewage is by bioflocculation, which is insensitive to temperature, Θ is very low as compared to soluble organic wastewaters. This comparison is shown in figure 9.35. It should be noted that the endogenous rate coefficient, k_d or b, is also temperature dependent. A coefficient Θ of 1.04 has been reported.

FIGURE 9.34
Variation in BOD and Suspended Solids During Winter and Summer Operation

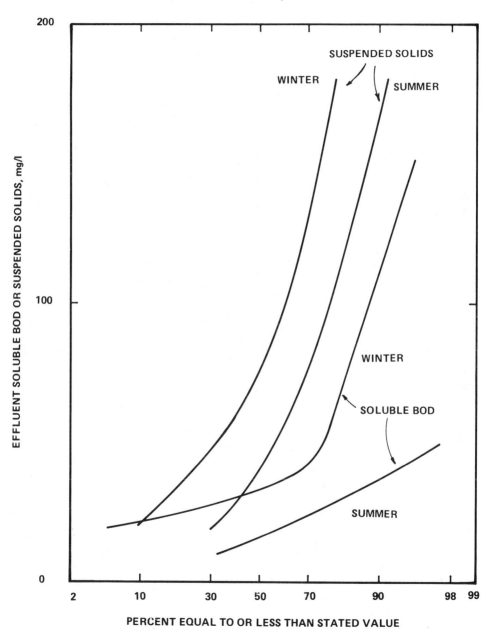

FIGURE 9.35

Temperature Effect on BOD Removal in the Activated Sludge Process for Sewage and Industrial Wastewaters

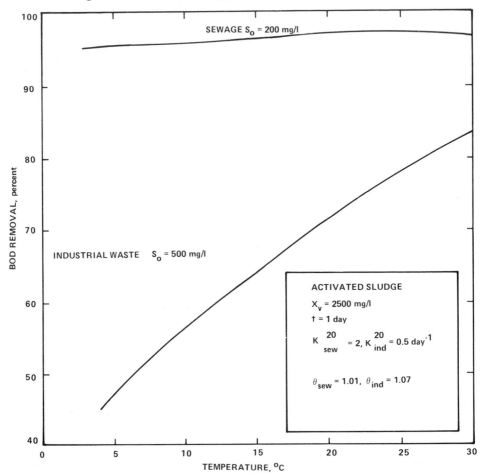

Effects of pH

Most biooxidation systems treating organic wastes have an optimum pH range of pH 6.5–8.5. The efficiency of the process falls off both above and below this range. It is important to note, however, that the pH of significance is that in the aeration basin rather than that of the influent wastewater since bacterial action modifies the pH in the basin. Several cases can be considered: caustic alkalinity present in the wastewater reacts with the carbon dioxide produced by microbial respiration to produce bicarbonate that buffers the process near pH 8.0. Approximately 0.5 lb OH$^-$ as CaCO$_3$/lb BOD$_5$ removed is neutralized in the process.

Nitrification requires sufficient alkalinity to react with the hydrogen ions produced in the reaction.

In the case of organic acids, biooxidation converts the acids to CO_2 and H_2O, thereby increasing the pH. No preneutralization is required if the effluent BOD from a completely mixed reactor is less than about 25 mg/l.

High concentrations of salts of organic acids, such as sodium acetate, reacts to produce sodium carbonate. Organics containing sulfonates can yield sulfuric acid as a reaction product requiring neutralization. Free mineral acidity requires external neutralization.

PROBLEM 9–5

Given the following data treating a fruit processing wastewater:

k, l/mg-day	T,°C
0.0015	3.5
0.0025	7.5
0.0072	15.0
0.0090	17.0
0.0170	20.0

Determine the temperature coefficient Θ.
From the Van't Hoff–Arrhenius equation:

$$\frac{d\,(\ln k)}{dT} = \frac{E}{RT^2}$$

Integration yields:

$$\ln \frac{k_2}{k_1} = \frac{E}{R}\frac{(T_2 - T_1)}{T_1 T_2}$$

Since most wastewater treatment systems are conducted at or near ambient temperature, the relationship can be simplified to:

$$\ln\left(\frac{k_2}{k_1}\right) = (T_2 - T_1)\ln\theta$$

where $\ln\theta = E/(RT_1 T_2)$

By using reference @ 20° C:

$k_1 = 0.017$ l/mg-day
$T_1 = 20°$ C

the following table and graph (figure I) can be derived:

k_2	$\ln\left(\dfrac{k_2}{k_1}\right)$	$T_2 - T_1$
0.0015	−2.43	−16.5
0.0025	−1.92	−12.5
0.0072	−0.86	− 5.0
0.009	−0.64	− 3.0
0.017	0	0

FIGURE I

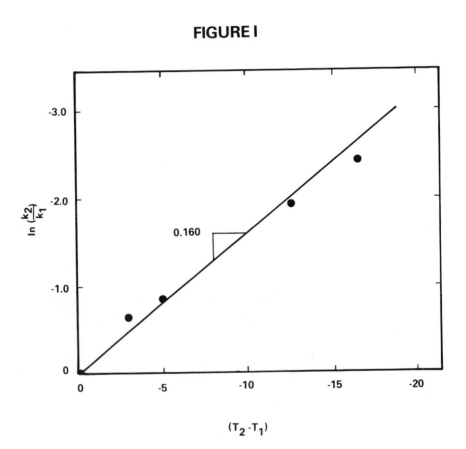

$(T_2 - T_1)$

Toxicity

Inhibition or toxicity to the biological process can be considered in several categories:

1. Organics that are toxic in high concentration but are biodegradable in low concentration, such as phenol. The concentration of significance is that in the basin in contact with the biomass. Therefore, the degradation rate must be sufficient to reduce the concentration of the substance in the basin to a value less than the inhibiting or toxic threshold.

2. Most heavy metals exhibit a toxicity to aerobic and anaerobic biological systems. After acclimation, higher concentrations of metal ion can be tolerated, a portion of which is removed by the biomass in the process. Metal removal by the activated sludge process treating a refinery wastewater is shown in table 9.9. Reported toxicities are shown in tables 9.10 and 9.11. Note that many factors affect heavy metal toxicity. The removal of heavy metals has been attributed to complexing in the cellular slime layer. Most metals are actively complexed by the sludge. Nickel and chromium, however, are poorly complexed. This tends to be related to sludge age. The hardness of the carrier water has an effect. The presence of complexing agents in the wastewater and the state of the metal affects the allowable toxic concentration level. For example, the presence of ammonia influences the toxic level of copper due to copper complexing with the ammonia. It is, therefore, practically impossible to predict heavy metal toxicity for a specific wastewater and experimental studies are usually necessary.

3. Salts in high concentrations inhibit biological activity. The limiting concentration of ammonia has been reported as 1600 mg/l at pH 7.0. A limiting concentration of chloride of 16000 mg/l has been reported. Organics in saline water ($Cl^- > 3\%$) have been successfully treated using marine organisms.

High salt concentration may result in poor flocculation and dispersion. This result is a high concentration of effluent suspended solids as shown in figure 9.36 (see page 518). Note that the increase in suspended solids are dispersed and nonsettleable.

TABLE 9.9
Heavy Metal Removal in the Activated Sludge Process

	Bio Plant Influent (mg/l)	Bio Plant Effluent (mg/l)
Cr	2.2	0.9
Cu	0.5	0.1
Zn	0.7	0.4

TABLE 9.10

Information on Materials that Inhibit Biological
Treatment Processes[a]

Pollutant	Concentration[c]		mg/l
	Aerobic Processes	**Anaerobic Digestion**	**Nitrification**
Copper	1.0	1.0	0.5
Zinc	5.0	5.0	0.5
Chromium (hexavalent)	2.0	5.0	2.0
Chromium (trivalent)	2.0	2000≥	[b]
Total chromium	5.0	5.0	[b]
Nickel	1.0	2.0	0.5
Lead	0.1	[b]	0.5
Boron	1.0	[b]	[b]
Cadmium	[b]	0.02[d]	[b]
Silver	0.03	[b]	[b]
Vanadium	10	[b]	[b]
Sulfides (S)	[b]	100[d]	[b]
Sulfates ($SO_4^=$)	[b]	500	[b]
Ammonia	[b]	1500[d]	[b]
Sodium (Na^+)	[b]	3500	[b]
Potassium (K^+)	[b]	2500	[b]
Calcium (Ca^{++})	[b]	2500	[b]
Magnesium (Mg^{++})	[b]	1000	[b]
Acrylonitrite	[b]	5.0[d]	[b]
Benzene	[b]	50[d]	[b]
Carbon tetrachloride	[b]	10[d]	[b]
Chloroform	18.0	0.1[d]	[b]
Methylene chloride	[b]	1.0	[b]
Pentachlorophenol	[b]	0.4	[b]
1,1,1-Trichloroethane	[b]	1.0[d]	[b]
Trichlorofluoromethane	[b]	0.7	[b]
Trichlorotrifluoroethane	[b]	5.0[d]	[b]
Cyanide (HCN)	[b]	1.0	2.0
Total oil (petroleum origin)[e]	50	50	50

[a] From Federal Guidelines: Pretreatment of Pollutants Introduced into Publicly Owned Treatment Works, Office of Water Program Operations, USEPA, Washington, D.C. (October 1973).

[b] Insufficient data.

[c] Concentrations refer to those present in raw wastewater unless otherwise indicated.

[d] Concentrations apply to the digester influent only. Lower values may be required for protection of other treatment process units.

[e] Petroleum-based oil concentration measured according to the API Method 733-58 for determining volatile and nonvolatile oily materials. The inhibitory level does not apply to oil of direct animal or vegetable origin.

TABLE 9.11

Selected Compounds That Are Toxic Above Certain Levels

Toxic Constituent	Limiting Concentration (mg/l)
Total dissolved solids	16,000
Chloride	15,000
Heavy metals	< 2
Cyanides	60
Nitrites—nitrogen	36
Phenols	140
Ammonia—nitrogen	1,600
Sulfides	100 (bacteria)
	7 (algae)

Special Factors	Limiting Range
Temperature	37° C (100° F)
pH	6.5–8.5
Low dissolved oxygen	< 1 mg/l

NITRIFICATION AND DENITRIFICATION

Nitrification

The most work on nitrification in recent years has been reported by Downing [30] and Wuhrmann [31] and Wong-Chong and Loehr [32]. Nitrification results from the oxidation of ammonia by *Nitrosomonas* to nitrite and the subsequent oxidation of the nitrite to nitrate by *Nitrobacter*. Since a buildup of nitrite is rarely observed, it can be concluded that the rate of conversion to nitrite controls the rate of overall reaction as shown in figure 9.37.

The reactions that occur are as follows. Ammonia is oxidized to nitrite by *Nitrosomonas:*

$$2NH_4^+ + 3O_2 \longrightarrow 2NO_2^- + 2H_2O + 4H^+ + \text{new cells}$$

Other organisms capable of oxidizing ammonia to nitrite are *nitrosococcus, nitrosospima, nitrosocystis,* and *nitrosogloea,* although these are probably of secondary importance in wastewater oxidation systems. The nitrite is then oxidized to nitrate by *Nitrobacter:*

$$2NO_2^- + O_2 \longrightarrow 2NO_3^- + \text{new cells}$$

FIGURE 9.36

Effect of Influent Total Dissolved Solids on Effluent
Suspended Solids

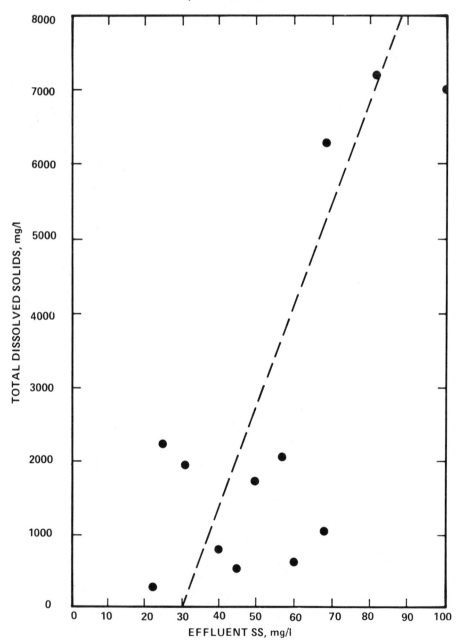

FIGURE 9.37
Nitrification Kinetics of Organic Nitrogen
(After Wong-Chong and Loehr [32])

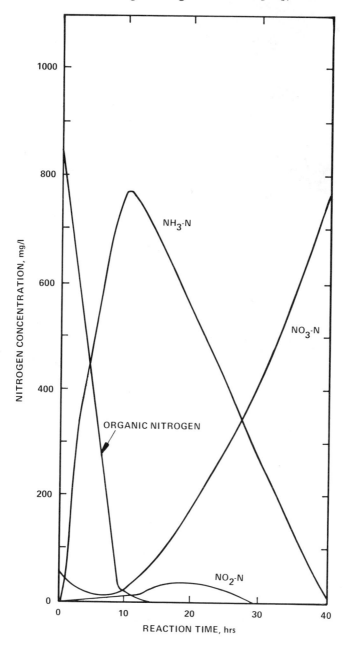

The nitrifiers are autotrophic organisms and use CO_2 or HCO_3^- as a carbon source. The organisms can survive with an initial lag period under anaerobic conditions for at least four hours. The oxygen requirements for nitrification have been formulated by McCarty [33].

Nitrosomonas:

$$55NH_4^+ + 5CO_2 + 76O_2 \longrightarrow C_5H_7NO_2 + 52H_2O + 109H^+ + 54NO_2^-$$

Nitrobacter:

$$400NO_2^- + 5CO_2 + NH_4^+ + 195O_2 + 2H_2O \longrightarrow C_5H_7NO_2 + 400NO_3^- + H^+$$

On this basis 3.22 mg/l O_2 is required for oxidation of 1 mg/l $NH_4^+ - N$ to NO_2^- and 1.11 mg/l O_2 for oxidation of nitrite to nitrate. The total O_2 required would be 4.33 mg/l O_2/mg/l $NH_4^+ - N$ oxidized. The cell yield for *Nitrosomonas* has been reported as 0.05–0.29 and for *Nitrobacter* 0.02–0.08. A value of 0.15 mg VSS/mg $NH_3 - N$ is usually used for design purposes. The nitrification reaction produces acidity and requires 7.14 mg/l alkalinity per mg/l $NH_3 - N$ oxidized.

For effective nitrification to occur, the sludge retention period or sludge age, Θ_c, must be greater than the reciprocal of the growth rate of the nitrifying organisms. Shorter sludge ages result in a washout of these organisms. The results of Downing and Wuhrmann show that these resultant retention periods are usually sufficient to effect substantially complete nitrification. Several investigators have reported nitrification to occur over the temperature range of 5–45° C with the optimum range being 25–35° C. The critical sludge age for nitrification can be computed from the relationship:

$$\theta_c = 2.13e^{0.098(15-T)} \tag{9–33}$$

where Θ_c is sludge age in day and T is temperature in °C. It is customary to apply a safety factor in the order of 2.5 to the Θ_c calculated from equation (9–33) for design purposes.

Nitrification as related to sludge age is shown in figure 9.38. Percent nitrification as defined in these figures is the percentage of total nitrate formed after complete oxidation. Nitrification is a zero order reaction to very low concentration levels in the order of 1.0 mg/l. The pH has a significant effect on the growth rate of both *Nitrosomonas* and *Nitrobacter*. Nitrification proceeds over a pH range of pH 6.0–9.0. The effect of pH on nitrification is shown in figure 9.39. Since nitric acid is produced in the oxidation, it is necessary to provide 7.14 mg/l alkalinity per mg/l $NH_3 - N$ oxidized. The rate of nitrification has been shown to be dependent on the dissolved oxygen level at concentration less than 2.0 mg/l.

The rate of nitrification has been reported by Wong-Chong and Loehr [32] that it is essentially linear with biomass concentration (as VSS) over the range of 0 to 1500 mg/l and decreases above that level. From these data the nitrification rate at 20° C can be estimated as 1.04 mg $NH_3 - N$ oxidized/mg VSS-day. The overall rate of nitrification can then be expressed at 20° C as:

$$R_N = 1.04\, f_N X_v$$

in which R_N is the nitrification rate in mg/l-day; f_N is the fraction of nitrifiers in the mixed liquor; X_v is the mixed liquor volatile suspended solids concentration in mg/l.

The fraction of nitrifiers in the activated sludge system can be estimated by the following relationship:

$$f_N = \frac{a_N N_r}{a_N N_r + a S_r}$$

where a_N = nitrifiers yield coefficient
        a = heterotrophic microorganisms yield coefficient
        N_r = ammonia nitrogen removal by nitrifiers, mg/l
        S_r = BOD removal by heterotrophs, mg/l

Like all biochemical reactions the rate of nitrification is also temperature dependent as shown by equation (9–32) with Θ value in the range of 1.05 to 1.15.

FIGURE 9.38
Relationship Between Nitrification and Sludge Age in the Activated Sludge Process

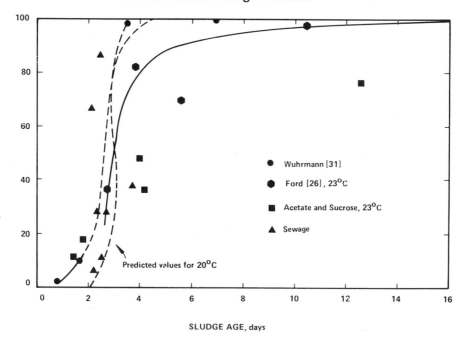

FIGURE 9.39
Effect of pH on Ammonia Oxidation [32]

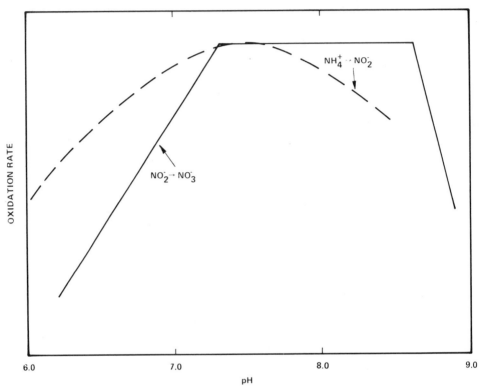

Inhibition of Nitrification

Nitrifying organisms are subject to inhibition by various organic and inorganic compounds. Hockenbury and Grady [34] have summarized data for a variety of organic compounds as shown in tables 9.12 and 9.13. Nitrification is inhibited by high concentrations of ammonia and nitrite. It has been shown that the inhibition results from free ammonia (FA) and free nitrous acid (FNA). The ranges of FA concentrations that begin to inhibit the nitrifying organisms are FA inhibition to *Nitrosomonas,* 10 to 150 mg/l and FA inhibition to *Nitrobacter* 0.1 to 1.0 mg/l. Nitrifying organisms were inhibited at concentrations of FNA between 0.22 and 2.8 mg/l. A chart summarizing inhibition is shown in figure 9.40. The limits shown in figure 9.40 are the lowest reported levels for inhibition.

TABLE 9.12
Organic Compounds That Inhibit Activated Sludge Nitrification

Compound	Concentration[a] (mg/l)
Acetone[b]	2000.000
Allyl alcohol	19.500
Allyl chloride	180.000
Allyl isothiocyanate	1.900
Benzothiazole disulfide	38.000
Carbon disulfide[b]	35.000
Chloroform[b]	18.000
o-Cresol	12.800
Di-allyl ether	100.000
Dicyandiamide	250.000
Diguanide	50.000
2,4-Dinitrophenol	460.000
Dithio-oxamide	1.100
Ethanol[b]	2400.000
Guanidine carbonate	16.500
Hydrazine	58.000
8-Hydroxyquinoline	72.500
Mercaptobenzothiazole	3.000
Methylamine hydrochloride	1550.000
Methyl isothiocyanate	0.800
Methyl thiuronium sulfate	6.500
Phenol[b]	5.600
Potassium thiocyanate	300.000
Skatol	7.000
Sodium dimethyl dithiocarbamate	13.600
Sodium methyl dithiocarbamate	0.900
Tetramethyl thiuram disulfide	30.000
Thioacetamide	0.530
Thiosemicarbazide	0.180
Thiourea	0.076
Trimethylamine	118.000

[a]Concentration giving approximately 75% inhibition.
[b]In the list of industrially significant chemicals.

TABLE 9.13

Effects of Concentrations of Organic Compounds on
Degree of Inhibition of Ammonia Oxidation Exhibited

Compound	Degree of Inhibition at Concentration Indicated (%)				Estimated Concentration Giving 50% Inhibition, (mg/l)
	100 (mg/l)	50 (mg/l)	10 (mg/l)	As Noted (mg/l)	
Dodecylamine	96	95	–	66[b]	< 1
Aniline[a]	86	–	–	76,[c] 89[d]	< 1
n-Methylaniline	90	83	71	–	< 1
1-Naphthylamine	81	81	45	–	15
Ethylenediamine[a]	73	–	41	61[e]	17
Napthylethylenediamine diHCl	93	79	29	–	23
2,2'-Bipyridine	91	81	23	–	23
p-Nitroaniline	64	52	46	–	31
p-Aminopropiophenone	80	56	22	–	43
Benzidine diHCl	84	56	12	–	45
p-Phenylazoaniline	54	47	0	–	72
Hexamethylene diamine[a]	52	45	27	–	85
p-Nitrobenzaldehyde	76	32	29	–	87
Triethylamine	35	–	–	63[f]	127
Ninhydrin	30	26	31	–	> 100
Benzocaine	30	27	0	–	> 100
Dimethylgloxime	30	9	–	56[f]	140
Benzylamine	26	10	0	–	> 100
Tannic acid	20	7	–	22[f]	> 150
Monoethanolamine[a]	16	–	–	20[g]	> 200

[a] From the list of industrially significant chemicals.
[b] 1 mg/l.
[c] 2.5 mg/l.
[d] 5 mg/l.
[e] 30 mg/l.
[f] 150 mg/l.
[g] 200 mg/l.

Heavy metals such as Cr, Ni, and Zn are toxic at low concentrations. Some organic carbon compounds may be toxic at a high concentration level. It has been shown that the rate of nitrification is dependent on the dissolved oxygen level at concentration less than 2.0 mg/l. Hence, a DO level in excess of 2.0 mg/l is recommended in order to insure maximum nitrification rates.

FIGURE 9.40
Ammonia and Nitrite Inhibition to Nitrification [35]

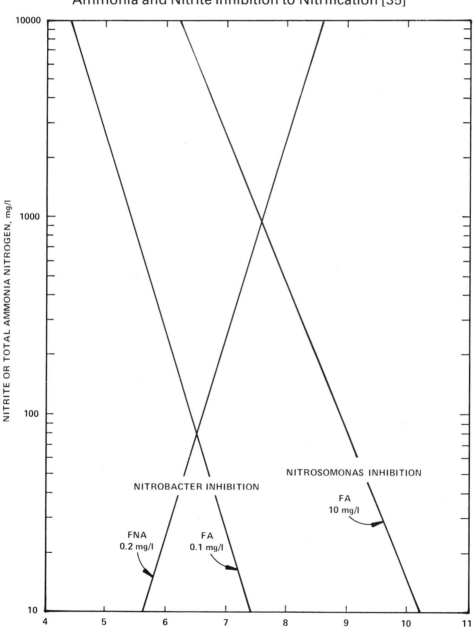

Denitrification

The most exhaustive study on denitrification has been reported by Wuhrmann [31] and McCarty [33]. Many of the heterotrophic bacteria present in activated sludge are facultative and can reduce nitrate. Nitrate under these conditions reduces to N_2 and small quantities of N_2O.

$$NO_3^- + BOD \longrightarrow N_2 + CO_2 + H_2O + OH^- + cells$$

The pH of the mixture profoundly affects the rates of denitrification relative to the presence of dissolved oxygen. pH values in the acid range permit active denitrification in the presence of dissolved oxygen, whereas strict anaerobic conditions should be maintained to promote denitrification under alkaline conditions.

The denitrifying organisms are heterotrophic and require an organic carbon source for growth. Typical facultative heterotrophic bacteria found in activated sludge including *Pseudomonas, Achromobacter, Bacillus,* and *Micrococcus* can cause denitrification. It is possible, however, to use the endogenous byproducts as a food supply for the denitrifiers. Denitrification rates would increase in the presence of available carbon such as supplied by untreated sewage or an industrial wastewater.

The most common external carbon source studied is methanol. The reactions using methanol have been postulated:

First stage:

$$NO_3^- + 1/3 \, CH_3OH \longrightarrow NO_2^- + 1/3 \, CO_2 + 2/3 \, H_2O$$

Second stage:

$$NO_2^- + 1/2 \, CH_3OH \longrightarrow 1/2 \, N_2 + 1/2 \, H_2O + OH^- + 1/2 \, CO_2$$

The overall reaction is:

$$NO_3^- + 5/6 \, CH_3OH \longrightarrow 1/2 \, N_2 + 5/6 \, CO_2 + 7/6 \, H_2O + OH^-$$

74 mg/l methanol is required to reduce mg/l NO_3^- $-N$. Twenty-five to thirty percent is required for cell synthesis. In practice the total methanol requirement is 2.5–3.0 lb methanol/lb of nitrate removed. Denitrification rates using various carbon sources are shown in table 9.14.

Nitrate reduction can be related to oxygen consumption or BOD removal as shown in figure 9.41. The temperature effect on denitrification is shown in figure 9.42.

TABLE 9.14
Denitrification Rates with Various Carbon Sources

Carbon Source	lbs NO$_3$−N/day/lb VSS	Temp., °C
Brewery Waste	0.22–0.25	20
Methanol	0.18	19–24
Volatile Acids	0.36	20
Molasses	0.10	10
	0.036	16

FIGURE 9.41 (a)
Correlation Between Oxygen Uptake and Denitrification Rate

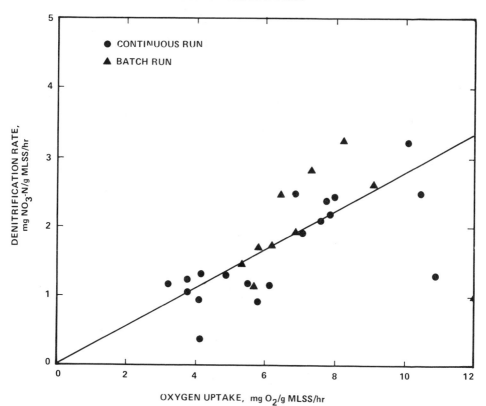

OXYGEN UPTAKE, mg O$_2$/g MLSS/hr

FIGURE 9.41 (b)

(b) NO_3-N/BOD Removal Ratio in Denitrification

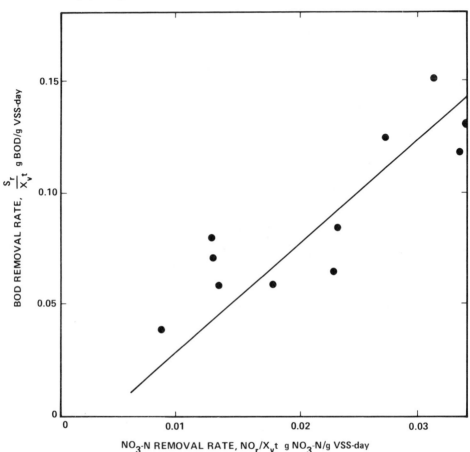

BOD REMOVAL RATE, $\frac{S_r}{X_v t}$ g BOD/g VSS-day

NO_3-N REMOVAL RATE, $NO_r/X_v t$ g NO_3-N/g VSS-day

PROBLEM 9–6a

A 1.0 MGD sewage treatment plant is to achieve nitrification at 15° C. The BOD is 150 mg/l and the NH_3-N is 25
mg/l. The alkalinity is 280 mg/l. Design the facility for the following conditions:

BOD removal = 90%
MLVSS   = 2500 mg/l
a = 0.55
b = 0.1 day^{-1}

FIGURE 9.42
Denitrification Rate: Temperature Effect
(No Nitrate Limited)

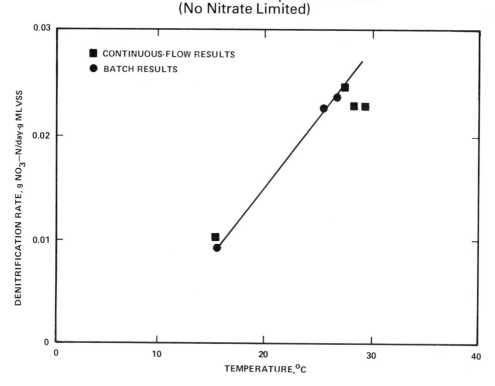

$$x = 0.65$$
$$a_N = 0.15$$
$$\Theta = 1.05$$

1. The critical sludge age at 15° C is computed by equation (9–33) as

$$\theta_c = 2.13 \text{ day}$$

Using a safety factor of 2.5 to account for variability the minimum design sludge age is:

$$2.13 \cdot 2.5 = 5.33 \text{ day}$$

2. For a MLVSS level of 2500 mg/l the required retention time for organic removal is computed by the following relationship:

$$t = \frac{\theta_c a\,(S_o - S_e)}{X_v\,(1 + bx\theta_c)}$$

$$= \frac{5.33 \cdot 0.55 \cdot (150 - 15)}{2500\,(1 + 0.1 \cdot 0.65 \cdot 5.33)}$$

$$= 0.118 \text{ day or } 2.82 \text{ hours}$$

3. Check nitrification rate
    a) Calculate sludge yield by organic removal

$$\Delta X_v = aS_r - bxX_v$$

$$= 0.55 \cdot (135 \cdot 1.0 \cdot 8.34) - 0.1 \cdot 0.65 \cdot 2500 \cdot 0.118 \cdot 1.0 \cdot 8.34$$

$$= 459 \text{ lb/day}$$

  b) Calculate NH_3-N removal by synthesis

$$NH_3 - N \text{ synthesized} = \frac{0.123\,\Delta X_v}{Q \cdot 8.34}$$

$$= \frac{0.123 \cdot 459}{1.0 \cdot 8.34}$$

$$= 6.8 \text{ mg/l}$$

  c) Calculate NH_3-N removal by nitrification

$$NH_3 - N \text{ to be oxidized} = 25 - 6.8$$

$$= 18.2 \text{ mg/l}$$

  d) NH_3-N oxidation rate at 20° C is 1.04 mg NH_3-N oxidized/mg VSS-day. The nitrifier fraction in the mixed liquor can be estimated as:

$$f_N = \frac{a_N N_r}{a_N N_r + aS_r}$$

$$= \frac{0.15 \cdot 18.2}{0.15 \cdot 18.2 + 0.55 \cdot 135}$$

$$= 0.036$$

$$R_N = 1.04\,\theta^{(T-20)}\,f_N X_v$$

$$= 1.04\,(1.05)^{(15-20)} \cdot 0.036 \cdot 2500$$

$$= 73 \text{ mg/l-day}$$

e) Calculate required detention time for nitrification

$$t = \frac{18.2}{73}$$

$$= 0.249 \text{ day or } 6.0 \text{ hours}$$

Therefore, it is nitrification rate controlling.

f) Repeat the calculation of steps (a) to (e) until the adequate detention time is obtained.

Say $t = 0.25$ day or 6.0 hours

$$\Delta X_v = 0.55 \cdot (135 \cdot 1.0 \cdot 8.34) - 0.1 \cdot 0.65 \cdot 2500 \cdot 0.25 \cdot 1.0 \cdot 8.34$$

$$= 280 \text{ lb/day}$$

$$NH_3 - N \text{ synthesized} = \frac{0.123 \cdot 280}{1.0 \cdot 8.34}$$

$$= 4.1 \text{ mg/l}$$

$$NH_3 - N \text{ to be oxidized} = 25 - 4.1$$

$$= 20.9 \text{ mg/l}$$

$$f_N = \frac{0.15 \cdot 20.9}{0.15 \cdot 20.9 - 0.55 \cdot 135}$$

$$= 0.041$$

$$R_N = 1.04 \, (1.05)^{(15-20)} \cdot 0.041 \cdot 2500$$

$$= 83.5 \text{ mg/l-day}$$

detention time for nitrification

$$t = \frac{20.9}{83.5}$$

$$= 0.25 \text{ day or } 6.0 \text{ hours}$$

4. Oxygen Requirement

$$O_2 \text{ for nitrification} = 20.9 \cdot 1.0 \cdot 8.34 \cdot 4.33$$

$$= 755 \text{ lb/day}$$

5. Alkalinity Requirement

The alkalinity remaining after nitrification will be:

$$280 - 7.14 \, (20.9) = 131 \text{ mg/l}$$

PROBLEM 9–6b

Design an aerobic-anoxic process for nitrification/ denitrification of sewage with the characteristics given in Problem 6a and the following conditions:

denitrification rate = 0.06 lb NO_3 – N/lb VSS–day at 20° C

MLVSS = 2500 mg/l

θ = 1.10

K = 6.4 day^{-1} at 15° C

1. Calculate the aerobic volume requirement for nitrification. The aerobic requirement for nitrification is 6.0 hours from Problem 6a. This does not consider a reduction in nitrification rate due to decreased oxygen concentrations in the aerobic zone. If a reduction of 50 percent in nitrification rate be applied, the required aerobic volume will be:

$$V = Qt$$

$$= 1.0 \cdot \frac{6.0 \cdot 2}{24}$$

$$= 0.5 \text{ mg}$$

or

$$t = 6.0 \cdot 2$$

$$= 12 \text{ hours}$$

2. Calculate the anoxic volume requirement for denitrification

$$R_{DN} = 0.06 \, (1.10)^{(15-20)} \cdot 2500$$

$$= 93.1 \text{ mg/l–day}$$

$$t = \frac{20.9}{93.1}$$

$$= 0.224 \text{ day or } 5.4 \text{ hours}$$

$$V = Qt$$

$$= 1.0 \cdot 0.224$$

$$= 0.224 \text{ mg}$$

3. Calculate the BOD removal in anoxic zone
    From Figure 41b it is found that the maximum BOD removal in denitrification system is 4.8 mg BOD removed/mg NO_3-N denitrified. The BOD removal potential in the anoxic zone is:

$$4.8 \cdot 93.1 \cdot 0.224 = 100 \text{ mg/l}$$

4. Calculate the BOD removal in aerobic zone

$$S_e = \frac{S_o{}^2}{S_o + KX_v t}$$

$$= \frac{(150)^2}{150 + 6.4 \cdot 2500 \cdot 0.5}$$

$$= 2.8 \text{ mg/l}$$

There is adequate retention for BOD removal through the aerobic and anoxic zones. However, the distribution of BOD removal between the zones cannot be determined from these calculations.

PROBLEM 9–7

1.0 MGD of an industrial waste containing 500 mg/l NH_3-N is to be oxidized in a biological system at 20° C. Design the system using 2500 mg/l MLVSS. The operating dissolved oxygen level is 2.0 mg/l.

$$a_N = 0.15$$

$$b_N = 0.05 \text{ day}^{-1}$$

$$x = 0.65$$

1. Calculate required detention time:

$$R_N = 1.04 \, X_v$$

$$= 1.04 \cdot 2500$$

$$= 2600 \text{ mg/l–day}$$

$$t = \frac{500}{2600}$$

$$= 0.192 \text{ day or 4.6 hours}$$

2. Calculate required sludge age

$$\theta_c = \frac{X_v t}{aS_r - bxX_v t}$$

$$= \frac{2500 \cdot 0.192}{0.15 \cdot 500 - 0.05 \cdot 0.65 \cdot 2500 \cdot 0.192}$$

$$= 8.1 \text{ day}$$

3. Check the critical sludge age for nitrification

$$\theta_c = 2.13 e^{0.098(15-T)}$$

$$= 2.13 e^{0.098(15-20)}$$

$$= 1.3 \text{ day}$$

4. Calculate solids buildup

$$\Delta X_v = aS_r - bxX_v$$

$$= 0.15 \cdot 500 \cdot 1.0 \cdot 8.34 - 0.05 \cdot 0.65 \cdot 2500 \cdot 1.0 \cdot 0.192 \cdot 8.34$$

$$= 495 \text{ lb/day}$$

5. Oxygen requirement

$$R_r = 500 \cdot 1.0 \cdot 8.34 \cdot 4.33$$

$$= 18100 \text{ lb/day}$$

6. Alkalinity requirement

$$500 \cdot 1.0 \cdot 8.34 \cdot 7.14 = 29800 \text{ lb/day}$$

KINETICS OF ANAEROBIC TREATMENT

Anaerobic treatment is employed for the degradation and breakdown of organic solids or for the breakdown of soluble organics to gaseous end products as shown in figure 9.43. The volatile acids formed in the fermentation are acetic, propionic, and butyric. For the most of the longer-chain volatile acids, a given species of methane organisms converts the acid to methane, carbon dioxide, and a second volatile acid having a shorter carbon chain. The second volatile acid is then fermented in a similar fashion, as shown in figure 9.44. Acetic acid is directly converted to CO_2 and CH_4. Thus the overall conversion is the result of two or more reactions (figure 9.45).

FIGURE 9.43
Mechanism of Anaerobic Sludge Digestion

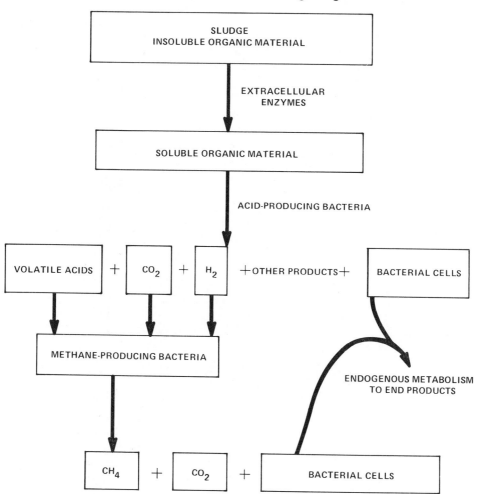

Successful anaerobic treatment depends upon maintaining a balance between the various rates of reaction occurring in the reactor. Since the rate of methane fermentation must control the overall rate to avoid process failure, further consideration of the rate of this fermentation is important.

To effect methane fermentation, sufficient time must be available in the reactor to permit growth of the organisms or they will be washed out of the system. In a completely mixed flowthrough anaerobic reactor, this means that the detention time in the unit must be greater than the reciprocal of the growth rate of

FIGURE 9.44
Methane Fermentation From Volatile Acids

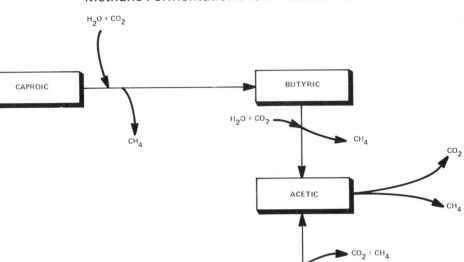

the methane organisms. It is significant that there are several species of methane organisms active in an anaerobic system, all having different growth rates. Andrews *et al.* [4] have shown that some organisms with a high growth rate ($\Theta_c < 2$ days) can produce methane, probably from the fermentation of formate, methanol, CO_2 and H_2, and possibly some volatile acid fermentation. Other organisms require residence times of up to 20 days. Although data are limited, some results have been reported relative to the growth rate of methane organisms. These data are summarized in table 9.15.

As shown in figure 9.45, at low residence times there will be volatile solids reduction as a result of the liquefaction of the solids and the subsequent conversion to volatile acids by acidification. During this period a small amount of methane fermentation may occur (depending on environmental conditions such as pH) primarily due to reduction of formate, methanol, CO_2, and H_2. There will be a decrease in pH and a corresponding increase in the volatile acid concentration. However, there will be very little COD reduction, because the organics have

FIGURE 9.45

Mechanism of Continuous Mixed Anaerobic Digestion

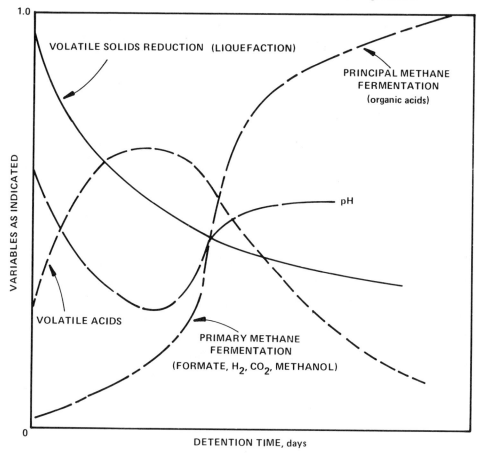

merely been converted from a solid form to a soluble form in the supernatant liquor. When the detention time in the reactor exceeds the reciprocal of the growth rate of the principal methane organisms, there will be a rapid increase in methane production with a corresponding decrease in volatile acid concentration and an increase in pH. There are probably several methane organisms responsible for the volatile acid conversion, each of which will have a different generation time or growth rate; the methane production curve is relatively flat, as shown in figure 9.45.

The acetic acid at higher residence times are obtained from two sources; direct fermentation, and the breakdown of higher carbon acids to acetic. The major part of the methane production comes from acetic acid fermentation, although some is generated from the breakdown of the higher acids. At long resi-

TABLE 9.15

Growth Rate of Methane Organisms

Substrate	Temp. (°C)	Residence Time (days)	References
Methanol	35	2	[36]
Formate	35	3	[36]
Acetate	35	5	[36]
Propionate	35	7.5	[36]
Primary and Activated Sludge	37	3.2	[37]
Acetate	35	2–4.2	[38]
Acetate	25	4.2	[38]
Propionate	25	2.8	[38]
Butyrate	35	2.7	[38]

dence periods, substantially all of the volatile acids are converted to methane and carbon dioxide. It should be noted that since all of the volatile solids present are not degradable in the anaerobic unit, a portion will remain even after long periods of retention. For the digestion of domestic sewage sludge, this fraction is approximately 40 percent.

Gas Production

The major part of the gas produced in breakdown of volatile acids. Some gas is produced by the early stages of methane fermentation of CO_2 and H_2, methanol, etc., but this contribution is probably very small. The gas is composed of CH_4 and depends in large measure on the residence time, the percentage of CO_2 being higher at the lower residence times with corresponding lesser numbers of methane bacteria. Lawrence and McCarty [38] have shown from theoretical considerations supported by experimental evidence that 0.351 m³ of methane gas is produced per kg of COD reduced (5.62 ft³/lb). The reported gas production for volatile solids reduction in a well operating anaerobic digestion tank is 1.06 to 1.25 m³/kg (17 to 20 ft³/lb) of VSS destroyed with a methane content of about 65 percent. This is about equivalent to 0.31 to 0.44 m³ CH_4/kg (5 to 7 ft³ of CH_4/lb) of COD destroyed, which is close to the value reported by Lawrence and McCarty [38]. Note that these values are a maximum, assuming complete conversion of the solids to methane. Volatile solids reduction can, of course, occur by liquefaction and conversion to volatile acids without any COD reduction. Under these conditions, the methane yield per unit of volatile solids reduction may be very low.

All methane bacteria are strictly anaerobic and therefore function in the absence of oxygen and at a low oxidation reduction potential (ORP). The result of

Dirasian, *et al.* [39] indicated that optimum anaerobic sludge digestion occurred at ORP values between -520 and -530 mv. The functional range was -490 to -550 mv with activity decreasing rapidly at the extreme ends of these ranges.

The pH is also critical to optimum methane fermentation. The optimum range of pH is 6.8 to 7.4 with an extreme range of 6.4 to 7.8.

High concentrations of inorganic salts may result in temporary or permanent inhibition of the fermentation process. Heavy metals such as copper, zinc, nickel, and chromium may cause inhibition depending on the state of the material, its solubility, and possible precipitation in the process by combination with sulfide.

The optimum conditions to maintain maximum rates of methane fermentation are summarized in table 9.16.

TABLE 9.16
Environmental Conditions for Methane Fermentation

Variable	Optimum	Extreme
pH	6.8 to 7.4	6.4 to 7.8
Oxidation Reduction Potential, mv	-520 to -530	-490 to -550
Volatile Acids (mg/l as acetic)	50 to 500	$>2,000$
Total Alkalinity (mg/l as $CaCO_3$)	1,500 to 2,000	1,000 to 3,000
Salts:		
NH_4^+ (mg/l as N)		3,000
Na (mg/l)		3,500 to 5,500
K (mg/l)		2,500 to 4,500
Ca (mg/l)		2,500 to 4,500
Mg (mg/l)		1,000 to 1,500
Gas Production, (ft³/lb VSS Destroyed)	17 to 20	–
Gas Composition, % CH_4	65 to 70	–
Temperature, °F	90 to 100	–

AEROBIC BIOLOGICAL TREATMENT PROCESSES

The various biological treatment processes are summarized below:

1. Activated sludge should provide an effluent with a soluble BOD_5 of less than 10–15 mg/l and a total BOD_5, including carryover suspended solids, of less than 30 mg/l. The process requires treatment and disposal of excess sludge

and would generally be considered where high effluent quality is required, available land area is limited, and waste flows exceed 0.1 MGD.

    2.   Extended aeration or total oxidation provides an effluent with a soluble BOD_5 of less than 10 to 15 mg/l and a total BOD_5 of less than 40 mg/l. The suspended solids carryover may run as high as 50 mg/l [high clarity (low solids) effluent usually requires posttreatment by filtration, coagulation, etc.]. This process is usually considered for waste flows less than 1 MGD.

    3.   Contact stabilization is applicable where a major portion of the BOD is present in colloidal or suspended form. As a general rule, the process should be considered when 85 percent of the BOD_5 is removed after 15 minutes of contact with aerated activated sludge. The effluent suspended solids are of the same order as those obtained from activated sludge.

    4.   An aerobic lagoon is only applicable where partial treatment (approximately 50 to 60 percent BOD_5 reduction) and high effluent suspended solids are permissible. This process should be considered as a stage development that can be converted into an extended aeration plant at some future date by the addition of a clarifier, return sludge pump, and additional aeration equipment or for pretreatment of an industrial wastewater prior to discharge to a municipal treatment plant.

    5.   An aerated lagoon will provide effluent soluble BOD_5 of less than 25 mg/l with a total BOD_5 of less than 50 mg/l, depending on the operating temperature. The effluent suspended solids may exceed 100 mg/l. The system is temperature sensitive and treatment efficiencies decrease during winter operation. Posttreatment is necessary if a highly clarified effluent is desired. Large land areas are required for the process.

    6.   High-rate trickling filters will provide 85 percent reduction of BOD_5 for domestic sewage. Roughing filters at high loadings provide 50–60 percent BOD_5 reduction from soluble organic industrial wastes.

    7.   Anaerobic and facultative ponds for industrial waste treatment should only be considered if odors will not cause a nuisance. If high degree treatment is required, these ponds must be followed by an aerobic treatment (aerated lagoons, activated sludge, etc.).

    The various biological treatment processes and their constraints are summarized in table 9.17.

Ponds and Stabilization Basins

    Ponds and stabilization basins are the most common method of treatment where large land areas are available and high quality effluent is not required at all times. Ponds are also the most common method of treatment in developing countries.

    Ponds can be divided into two general classifications, the impounding and absorption basin, and the flowthrough basin. In the impounding and absorption

TABLE 9.17

Design Criteria for Biological Processes

Process	Detention Time, days	Depth, ft	X_v, mg/l	BOD₅ Reduction, %	Conditions
Anaerobic ponds	5–50	8–15	<25	50–80	Loading 250–4000 lbs BOD/acre/day SO₄⁼< 100 mg/l
Facultative ponds	7–50	3–8	<25	70–95	Loading 20–50 lbs BOD/acre/day, 10–50 mg/l algae
Aerobic algal ponds	2–6	0.6–1.0	<50	80–95	Loading 100–200 lbs BOD/acre/day, 100 mg/l algae; periodically mixed; velocity 1–1.5 fps
Aerated lagoons (aerobic)	0.5–3.0	8–16	2.5 BOD removal	50–70	hp/MG > 14–20 fully mixed conditions
Aerated lagoons (facultative)	3–10	8–16	50–100	80–95	hp/MG > 4, < 10
Activated sludge	0.16–0.33	12–16	2000–3000	85–95	F/M for flocculation
Extended aeration	0.5–1.0	12–16	3000–4000	85–95	F/M < 0.2 day⁻¹

basin, either there is no overflow or there is intermittent discharge during periods of high stream flow. In the first case the volumetric capacity of the basin is equal to total waste load less losses by evaporation and percolation. If there is intermittent discharge, required capacity is related to the stream flow characteristics. In view of the large area requirements, impounding basins are usually limited to industries discharging low daily volumes of wastewater in high evaporation geographical areas, to seasonal operations such as the canning industry, and to cases where the total inorganic dissolved solids level must be regulated for discharge to the receiving water.

The flow-through basins can be considered in three classifications depending on the nature of the biological activity involved.

1. *Aerobic Algae Ponds*—The aerobic algae ponds depend on algae to provide sufficient oxygen to satisfy the BOD applied to the pond. Since sunlight and light penetration is essential to oxygen production by photosynthesis, the depth of the pond is limited to that through which light will penetrate. Most treatment systems of this type do not exceed a depth of 18 in. (46 cm). To maintain aerobic conditions in the pond, and to provide uniformity of oxygen, mixing of the pond contents for a few hours each day is essential. Separation of the algae from the effluent is sometimes required to minimize the oxygen demand on receiving waters. The aerobic pond is limited to those waters that are not toxic to algal growth and usually to cases where the algae can be harvested for productive use.

2. *Facultative Ponds*—The facultative pond is separated by the magnitude of the organic loading and thermal stratification into an aerobic surface layer and an anaerobic bottom layer. The aerobic surface layer has a diurnal oxygen variation, decreasing in oxygen content during the night as shown in figure 9.46. Sludge deposited on the bottom of the pond undergoes anaerobic decomposition producing methane and other anaerobic products. These diffuse into the aerobic layers where they undergo oxidation. Odors are produced if an aerobic layer is not maintained. The depth of facultative ponds varies from 3 to 8 ft (0.9 to 2.4 m).

3. *Anaerobic Ponds*—Anaerobic ponds are loaded to such a level that anaerobic conditions exist throughout the liquid volume. The biological process is the same as that occurring in anaerobic processes being primarily organic acid formation followed by methane fermentation.

Design and Operation of Ponds

In aerobic ponds, the amount of oxygen produced by photosynthesis can be estimated from:

$$O_2 = 0.25fS \qquad\qquad (9\text{--}34)$$

where O_2 = oxygen production, lbs/acre/day
       f = light conversion efficiency %
       S = light intensity, calories/cm²/day

FIGURE 9.46

Waste Stabilization Pond, Facultative Type [40]

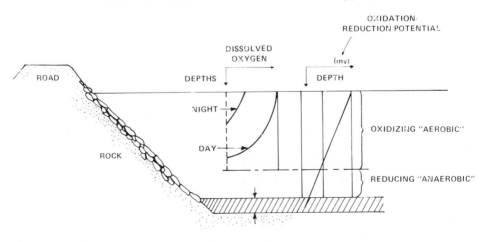

MINIMUM SLUDGE

If the light conversion efficiency is estimated as 4 percent [41], then $O_2 = S$. S is a function of latitude and the month of the year. S may be expected to vary from 100 to 300 calories/cm²/day during winter and summer for latitude 30°. This, in turn, would imply maximum loadings of 100 to 300 lbs BOD_u/acre/day (0.011 to 0.034 kg/m²-day) in order to maintain any aerobic activity in the pond.

The depth of oxygen penetration in a facultative pond has been estimated as a function of surface loading as shown in figure 9.47. It should be noted that the data for figure 9.47 was developed from oxidation ponds treating domestic wastewater in California [41]. Appropriate adjustments to this curve would have to be made for other types of wastewaters being treated in other climatic conditions.

FIGURE 9.47

Depth of Oxygen Penetration in Facultative Ponds [41]

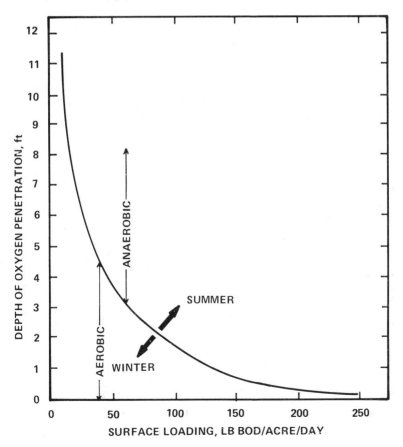

The typical green algae in waste stabilization basins are *Chlamydomonas*, *Chlorella*, and *Euglena*. Common blue-green algae are *Oscillatoria*, *Phormidium*, *Anacystis*, and *Anabaena*. Algae types in a pond will vary seasonally as shown in figure 9.48.

In the treatment of highly colored or turbid wastewaters such as kraft pulp and paper mills, light penetration will be minimal and oxygen input will generate primarily from surface reaeration. Gellman and Berger [42] estimated the oxygen input from reaeration at 45 lbs O_2/acre/day (0.005 kg/m^2-day). The performance of stabilization basins in the pulp and paper industry from their data is shown in figure 9.49.

FIGURE 9.48
Algae Population Distribution with Temperature

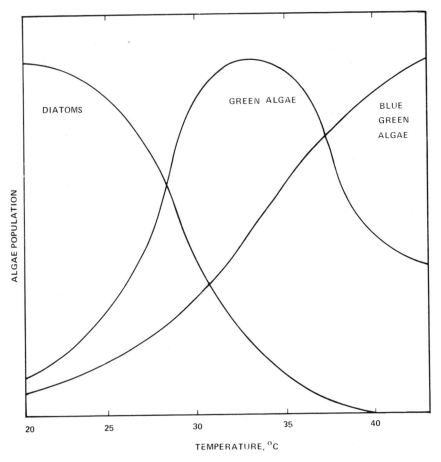

FIGURE 9.49
Waste Stabilization Pond Performance in the Pulp and Paper Industry

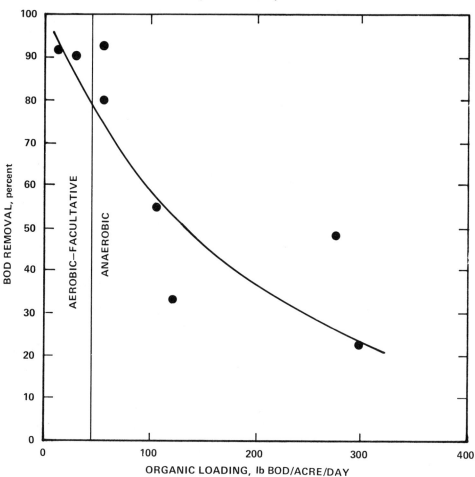

Several concepts have been employed for the design of facultative and anaerobic ponds [43,44]. Functional equations for anaerobic and facultative ponds have been developed as follows:
for a single pond:

$$\frac{S_e}{S_o} = \frac{1}{1+kt} \tag{9-35}$$

and for multiple ponds:

$$\frac{S_e}{S_o} = \frac{1}{(1 + kt_1)(1 + kt_2) \ldots (1 + kt_n)} \qquad (9\text{--}36)$$

for an infinite number of ponds:

$$\frac{S_e}{S_o} = e^{-k_n t_n} \qquad (9\text{--}37)$$

While variable influent concentrations are considered equation (9–35) can be modified to:

$$\frac{S_e}{S_o} = \frac{1}{1 + Kt/S_o} \qquad (9\text{--}38)$$

The equation is functionally similar to that employed for aerated lagoons and activated sludge, except that the rate coefficient k or K includes the effect of biomass concentration. This is because it is generally impractical or impossible to effectively measure the biomass concentration (VSS) in waste stabilization basins.

When multiple ponds are to be considered [equation (9–36)], k is assumed to be the same for all ponds. For complex wastewaters, this is probably not true since the more readily degradable compounds will be removed in the initial ponds. For these cases, an experimental study would need to be conducted for the wastewater in question to define the change in k.

In the treatment of domestic wastewater, Marais and Shaw [44] found k to be 0.23/day and used a design value of 0.17/day [equation (9–35)]. Data from an organic chemicals plant showed k to be 0.05/day under anaerobic conditions at 20° C and 0.5/day under aerobic conditions.

As in all biological processes, the biological activity in the pond will be a function of temperature and the rate coefficient k can be corrected through the application of equation (9–32). Hermann and Gloyna [43] found a coefficient Θ of 1.085 for domestic sewage treatment. Evaluation of 30-day average performance for a kraft pulp and paper mill showed a Θ value of 1.053 as shown in figure 9.50. During winter operation in colder climates, the pond will ice over resulting in anaerobic conditions and reduce performance. (Note that an ice cover will act as an insulator maintaining higher temperatures in the liquid.) Effluent quality from pond performance in Canada is shown in figure 9.51.

The general design criteria for ponds are shown in table 9.18. Average performance data for ponds treating various industrial wastewaters are summarized in table 9.19.

FIGURE 9.50

Temperature Effect on 30-Day Average Performances of Pond Treating a Pulp and Paper Mill Effluent

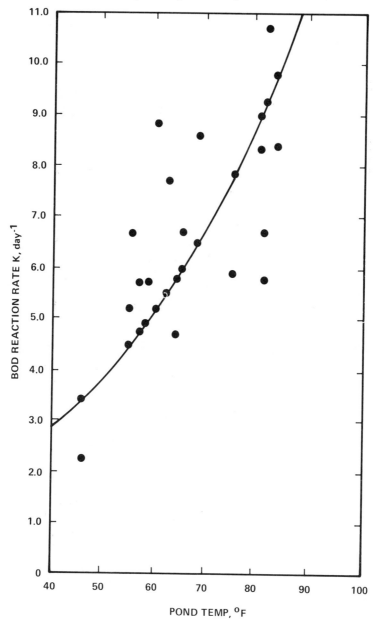

FIGURE 9.51

Typical Seasonal Variation of Effluent BOD in Waste Stabilization Ponds (Canada) [45]

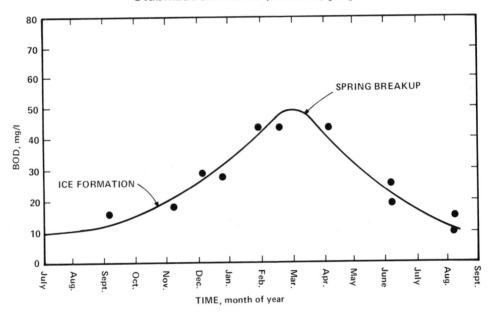

TABLE 9.18

Design Criteria for Lagoons and Stabilization Basins

	Aerobic[a]	Facultative	Anaerobic
Depth, ft	0.6–1.0	3–8	8–15
(Depth, m)	(0.2–0.3)	(1–2.5)	(2.5–5)
Detention, days	2–6	7–50	5–50
BOD Loading			
lb/acre/day	100–200	20–50	250–4000
kg/ha/day	111–222	22–55	280–4500
Percent BOD removal	80–95	70–95	50–80
Algae Conc., mg/l	100	10–50	Nil

[a]Must be periodically mixed; velocity 1 to 1.5 fps.

TABLE 9.19
Summary of Average Performance Data From Stabilization Basins

(a) Anaerobic Ponds

Industry	Area, acres	Depth, feet	Detention Time, days	Loading, lb/acre -day	BOD Removal, percent
Canning	2.5	6.0	15	392	51
Meat and poultry	1.0	7.3	16	1260	80
Chemical	0.14	3.5	65	54	89
Paper	71	6.0	18.4	347	50
Textile	2.2	5.8	3.5	1433	44
Sugar	35	7.0	50	240	61
Wine	3.7	4.0	8.8	–	–
Rendering	1.0	6.0	245	160	37
Leather	2.6	4.2	6.2	3000	68
Potato	10	4.0	3.9	–	–
Average value				860	60

(b) Facultative-Aerobic Ponds

Industry	Area, acres	Depth, feet	Detention Time, days	Loading, lb/acre -day	BOD Removal, percent
Meat and poultry	1.3	3.0	70	72	80
Canning	6.9	5.8	37.5	139	98
Chemical	31	5.0	10	157	87
Paper	84	5.0	30	105	80
Petroleum	15.5	5.0	25	28	76
Petrochemical	–	–	–	100	95
Wine	7	1.5	25	221	–
Dairy	7.5	5.0	98	22	95
Textile	3.1	4.0	14	165	45
Sugar	20	1.5	2	86	67
Rendering	2.2	4.2	48	36	76
Hog feeding	0.6	3.0	8	356	–
Laundry	0.2	3.0	94	52	–
Miscellaneous	15	4.0	88	56	95
Potato	25.3	5.0	105	111	–
Average value				114	81

TABLE 9.19 (Continued)

(c) Combined Anaerobic-Aerobic Ponds

Industry	Area, acres	Depth, feet	Detention Time, days	Loading, lb/acre -day	BOD Removal, percent
Canning	5.5	5.0	22	617	91
Meat and poultry	0.8	4.0	43	267	94
Paper	2520	5.5	136	28	94
Leather	4.6	4.0	152	50	92
Miscellaneous industrial wastes	140	4.1	66	128	–
Resins, alcohols amines, esters styrene, ethylene	760	–	–	41	99
Olefins, glycols, aldehydes, polyolefins	470	–	240	41	99
Average value				100	97

Ponds for treating wastewaters in which aerobic conditions are not maintained can frequently emit odors and provide a breeding ground for insects. The odor problem can frequently be controlled by adding sodium nitrate to maintain the oxidation reduction potential at a level above that at which sulfate reduction occurs. Typical data are shown in figure 9.52. Surface sprays can be used to reduce the fly and insect nuisance and in some cases odor.

A modification of the stabilization basin is the Air Aqua Lagoon System in which air is introduced into the basin through small plastic tubing stretched across the bottom of the basin as shown in figure 9.53. This has the advantage of increasing the vertical mixing of oxygen in the basin as well as adding more oxygen under highly loaded conditions.

High organic loading (500 to 2000 lbs of BOD/acre/day) (0.057 to 0.227 kg of BOD/m²-day) results in removal efficiency in the order of 70 percent. Since this degree of efficiency is usually not adequate for discharge into a receiving water, anaerobic ponds are usually followed by aerobic ponds. Deep depths in anaerobic ponds are desirable to provide maximum heat retention during cold weather operation. Certain design considerations are important to the successful operation of stabilization basins. These have been discussed by Gloyna [40]. Embankments should be constructed of impervious material with maximum-minimum slopes of 3 to 4:1 and 6:1, respectively. A minimum freeboard of three feet (0.91 m) should be maintained in the basin. Provision should be made for bank protection from erosion. Wind action is important for pond mixing and is effective with a fetch of 650 feet (198 m) in a pond with a depth of three feet (0.91 m).

FIGURE 9.52

ORP Control of Nitrate Addition to Pond Treatment of Tomato Wastewaters

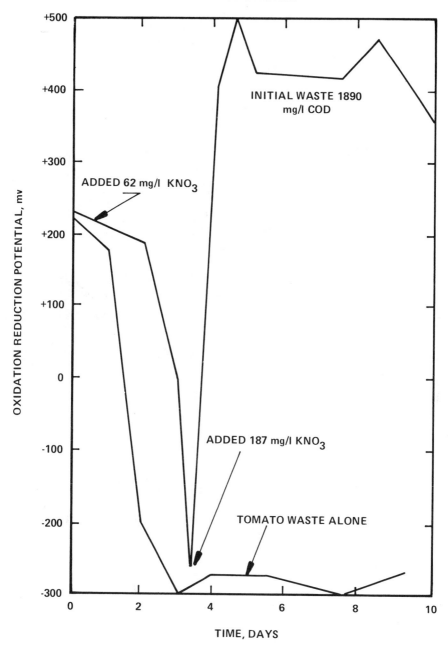

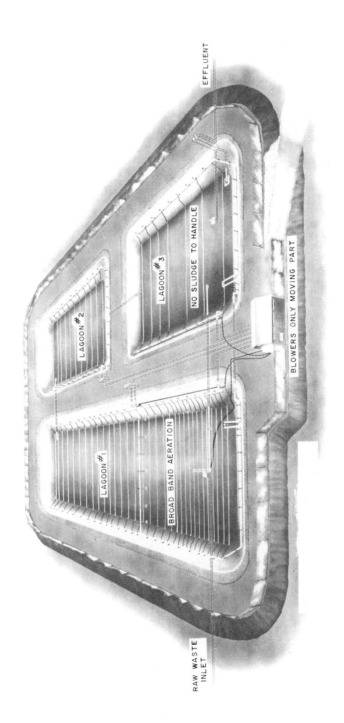

FIGURE 9.53
Air Aqua Lagoon System
(Courtesy of Hinde Engineering Co.)

PROBLEM 9–8

An industrial wastewater with a BOD of 500 mg/l is to be treated in a pond or series of ponds with a total retention time of 50 days with a depth of 6 feet. The anaerobic k at 20° C is 0.05/day and the aerobic k is 0.5/day. Assume the oxygen relationships as shown in figure 9.47 apply and the pond temperature is 20° C

1. For one pond the applied loading is:

$$\text{loading (lbs BOD/acre/day)} = \frac{2.7\, DS_o}{t}$$

$$= \frac{2.7\,(6)\,(500)}{50}$$

$$= 162 \text{ lbs BOD/acre/day}$$

Dissolved oxygen will exist to a depth of 0.8 ft. The mean k can be computed:

$$k = \frac{(0.8)\,(0.5) + 5.2\,(0.05)}{6} = 0.11 \text{ day}^{-1}$$

The effluent BOD is:

$$S_e = \frac{S_o}{1 + kt}$$

$$= \frac{500}{1 + 0.11\,(50)}$$

$$= 77 \text{ mg/l}$$

2. For four ponds in series, the retention time in each pond will be 12.5 days.
   The loading to the first pond is:

$$\frac{2.7\,(6)\,(500)}{12.5} = 648 \text{ lbs BOD/acre/day}$$

and will be anaerobic.
The effluent from the first pond is:

$$S_e = \frac{500}{1 + 0.05\,(12.5)} = 308 \text{ mg/l}$$

The loading to the second pond is:

$$\frac{2.7 \ (6) \ (308)}{500} = 399 \ \text{lbs BOD/acre/day}$$

$$S_e = \frac{308}{1 + (0.05) \ (12.5)} = 190 \ \text{mg/l}$$

The loading to the third pond is 246 lbs BOD/acre/day and the effluent is 117 mg/l.

The loading to the fourth pond is 152 lbs BOD/acre/day and the aerobic depth is 1 ft.

The adjusted k is:

$$\frac{0.5 \ (1) + 0.05 \ (5)}{6} = 0.125 \ \text{day}^{-1}$$

The effluent is:

$$S_e = \frac{117}{1 + (0.125) \ (12.5)} = 45.7 \ \text{mg/l}$$

Four ponds in series with the same total retention time will produce a superior effluent. This assumes the reaction rate k does not change through the series of basins. This is probably not true in many cases and would have to be determined experimentally.

PROBLEM 9–9

A stabilization basin is to be designed to treat a domestic sewage to an effluent BOD_5 of 25 mg/l at 20° C. The influent BOD is 250 mg/l and the k is 0.17/day for a pond depth of 4.5 ft. The wastewater flow is 1 MGD. Compute the effluent BOD under winter conditions with a pond temperature of 10° C.

1. Compute the required retention time:

$$\frac{S_e}{S_o} = \frac{1}{1 + kt}$$

or

$$t = \frac{S_o - S_e}{kS_e}$$

$$= \frac{250 - 25}{0.17\,(25)}$$

$$= 53 \text{ days}$$

2. The area of the pond is:

$$\frac{(1 \text{ MGD}) \ (53 \text{ days}) \times 10^6}{(7.48)(4.5)\,(43560)} = 36 \text{ acres}$$

3. The organic loading to the pond is:

$$\frac{(1 \text{ MGD}) \ (250 \text{ mg/l}) \ (8.34)}{36} = 58 \text{ lbs BOD/acre/day}$$

This should produce an aerobic depth of approximately 3 feet.

4. During winter conditions:

$$k_{10°C} = k_{20°C} \ (1.085)^{(T-20)}$$

$$= 0.075 \text{ day}^{-1}$$

$$S_e = \frac{S_o}{1 + kt} = \frac{250}{1 + 0.075\,(53)} = 50 \text{ mg/l}$$

AERATED LAGOONS

An aerated lagoon is a basin of significant depth 8 to 16 ft (2.4 to 4.8 m) in which oxygenation is accomplished by mechanical or diffused aeration units and through induced surface aeration.

There are two types of aerated lagoons:

1. The aerobic lagoon in which dissolved oxygen and suspended solids are maintained uniformly throughout the basin.
2. The aerobic-anaerobic or facultative lagoon in which oxygen is maintained in the upper liquid layers of the basin, but only a portion of the suspended solids is maintained in suspension. These basin types are shown in figure 9.54. A typical aerated lagoon is shown in figure 9.55.

In the aerobic lagoon, all solids are maintained in suspension and this system may be thought of as a "flowthrough" activated sludge system without solids recycle. Thus, the effluent suspended solids concentration will be equal to the aeration basin solids concentration. In most cases, separate sludge settling and disposal facilities are required. In some cases, the aerobic lagoon can be modified to an extended aeration activated sludge process by the addition of a clarifier and sludge recirculation facilities.

FIGURE 9.54

Comparison of Alternative Aerated Lagoon Systems

a) AEROBIC LAGOON

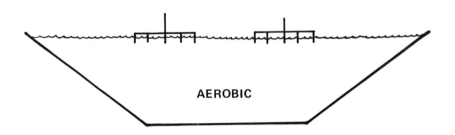

b) AEROBIC–ANAEROBIC LAGOON (FACULTATIVE LAGOON)

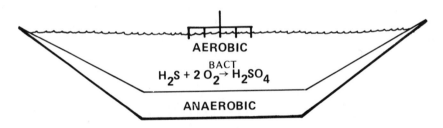

In the facultative lagoon, a portion of the suspended solids settle to the bottom of the basin where they undergo anaerobic decomposition. The anaerobic by-products are subsequently oxidized in the upper aerobic layers of the basin. The facultative lagoon can also be modified to yield a more highly clarified effluent by the inclusion of a separate postsettling pond or a baffled settling compartment.

Aerobic and facultative lagoons are primarily differentiated by the power level employed in the basin. In the aerobic lagoon the power level is sufficiently high to maintain all solids in suspension and may vary from 14 to 20 horsepower per million gallons (2.3 to 3.9 W/m^3) of basin volume, depending on the nature of the suspended solids. Field data demonstrated that 14 horsepower per million gallons (2.8 W/m^3) was generally sufficient to maintain solids in suspension in pulp and paper mill effluents, while 20 horsepower per million gallons (3.9 W/m^3) was required for domestic wastewater treatment.

In the facultative lagoon, the power level employed is only sufficient to maintain dispersion and mixing of the dissolved oxygen. Experience in the pulp and paper industry showed that a minimum power level employing low-speed

FIGURE 9.55
Typical Aerated Lagoon
(Courtesy Degremont, Paris, France)

mechanical surface aerators is 4 horsepower per million gallons (0.79 W/m^3). The use of other kinds of aeration equipment might require different power levels to maintain uniform dissolved oxygen in the basin.

Aerobic Lagoons

At a constant basin detention time, the equilibrium biological solids concentration and the overall rate of organic removal can be expected to increase as the influent organic concentration increases. For a soluble industrial wastewater, the equilibrium biological solids concentration, X_v, can be predicted from the relationship:

$$X_v = \frac{aS_r}{1 + bt} \qquad (9\text{--}39)$$

When nondegradable volatile suspended solids are present in the wastewater, equation (9–39) becomes:

$$X_v = \frac{X_i + aS_r}{1 + bt} \qquad (9\text{--}39a)$$

in which X_i = influent volatile suspended solids. Combining equation (9–39) with the kinetic relationship, equation (9–18), the effluent soluble organic concentration can be computed from equation (9–40):

$$\frac{S_e}{S_o} = \frac{1 + bt}{aKt} \qquad (9\text{--}40)$$

Equation (9–40) can be linearized as shown in equation (9–40a):

$$\frac{S_e}{S_o} = \frac{1}{aKt} + \frac{b}{aK} \qquad (9\text{--}40a)$$

Typical results are plotted in figure 9.56. From equation (9–40) it can be concluded that the fraction of effluent soluble organic concentration remaining is independent of the influent organic concentration. For lagoons with fixed detention times this conclusion is justified, because higher influent organic concentrations result in higher equilibrium biological solids levels and therefore, higher overall BOD removal rates.

From equation (9–40) it can also be shown that the minimum detention time can be determined by employing equation (9–41):

$$t_m = \frac{1}{aK - b} \qquad (9\text{--}41)$$

where t_m equals the minimum detention time for organic removal, days.

FIGURE 9.56

Kinetic Relationships in Aerated Lagoons

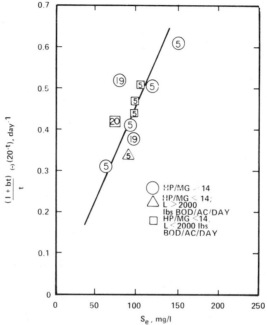

(a) BOD Removal as a Function of Effluent
Concentration

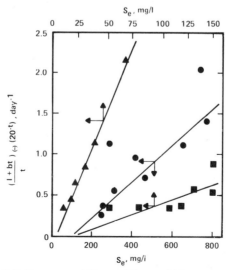

(b) BOD Removal From Three Wastewaters

The specific organic reaction rate coefficient K is temperature dependent, and can be corrected for temperature employing equation (9–32). The oxygen requirements for an aerobic lagoon are computed using the same relationship as employed for activated sludge [equation (9–8)]. Aerobic lagoons are employed for pretreatment of high strength industrial wastewaters prior to discharge to a joint or municipal treatment system, or as the first basin in a two basin aerated lagoon series, followed by a facultative basin. It should be noted that while in an aerobic lagoon, the soluble organic content is reduced, there is an increase in the effluent suspended solids through synthesis. The relationship obtained for an aerobic lagoon pretreating a brewery wastewater is shown in figure 9.57.

Horsepower requirements should generally be designed for summer operation, since the rates of organic removal and benthal feedback will be the greatest during this period of operation.

Temperature Effects in Aerated Lagoons

The performance of aerated lagoons is significantly influenced by changes in basin temperature. In turn, the basin temperatures are influenced by temperature of the influent wastewater and the ambient air temperature. Although heat is lost through evaporation, convection and radiation, it is gained by solar radiation. While several formulas have been developed to estimate the temperature in aerated lagoons, equation (9–42) will usually give a reasonable estimate for engineering design purposes.

$$\frac{t}{D} = \frac{(T_i - T_w)}{f(T_w - T_a)} \tag{9–42}$$

where t = the basin detention time, day

D = the basin depth, ft

T_i = the influent wastewater temperature, °F

T_a = mean air temperature, °F (usually taken as the mean weekly temperature)

T_w = the basin temperature, °F

The coefficient f is a proportionality factor containing the heat transfer coefficients, the surface area increase from aeration equipment, and wind and humidity effects. f has an approximate value of 90 for most aerated lagoons employing surface aeration equipment.

Recently a general temperature model has been developed by Argaman and Adams [46], which considers an overall heat balance in the basin including heat gained from solar radiation, mechanical energy input, biochemical reaction and heat lost by long wave radiation, evaporation from the basin surface, conduc-

FIGURE 9.57
COD Removal From Brewery Wastewater Through an Aerobic Lagoon

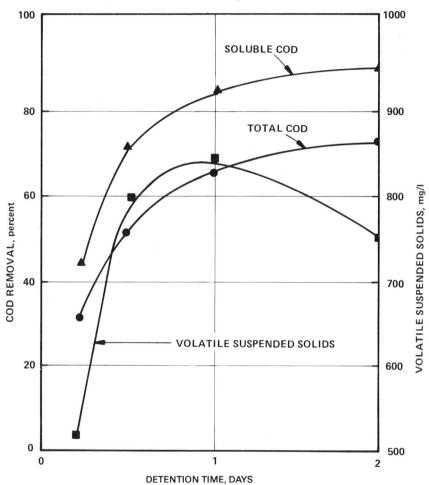

tion from the basin surface, evaporation and conduction from the aerator spray and conduction through the basin walls. Their final equation is:

$$T_w = T_a + \left[\frac{Q}{A} (T_i - T_a) + 10^{-6} (1 - 0.0071 C_c^2) H_{s,o} + 6.95 (\beta - 1) \right.$$

$$\left. + 0.102 (\beta - 1) T_a - e^{0.0604 T_a} \left(1 - \frac{f_a}{100} \right) (1.145 A^{-0.05} + \frac{126 NF V_w}{A} \right.$$

(9–43)

$$+ \frac{10^{-6} H_m}{A} + \frac{1.8 S_r}{A} \Bigg] \Bigg/ \Bigg[\frac{Q}{A} + 0.102 + (0.0686\, e^{0.0604 T_a} + 0.118)\, A^{-0.05}\, V_w$$

$$+ \frac{4.32 N F V_w}{A}\, (3.0 + 1.75 e^{0.0604 T_a}) + \frac{10^{-6} U A_w}{A} \Bigg]$$

where T_w = basin water temperature, °C
        T_a = air temperature, °C

FIGURE 9.58
Cooling Effect of Surface Aerators and Diffused Air System

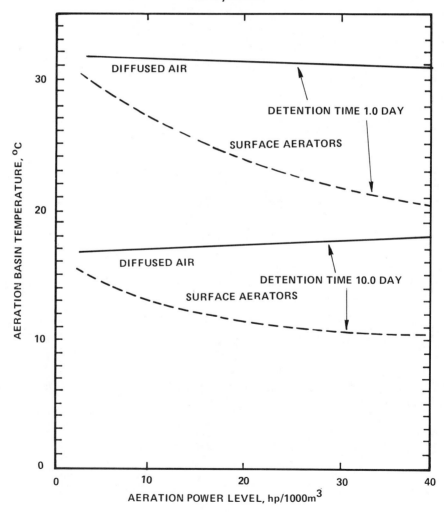

T_i = influent waste temperature, °C
Q = flow rate, m³/day
A = surface area, m²
C_c = average cloud cover, tenths
$H_{s,o}$= average daily absorbed solar radiation under clear sky conditions, cal/m²/day

U = heat transfer coefficient, cal/m²-day-°C
β = atmospheric radiation factor
f_a = relative humidity, percent
N = number of aerators
F = aerator spray vertical cross-section area, m²
V_w = wind speed at tree top, m/sec
H_m = 15.2 × 10⁶ p and p = aeration power, hp
S_r = organic removal rate, kg COD removed/day
A_w = effective wall area, m²

Equation (9–43) can be used to predict the temperature of diffused air systems by substituting:

$$NFV_w = 2Q_A$$

where Q_A = air flow, m³/sec

The temperature in aerated lagoons during winter operation is markedly affected by the type of aeration equipment. Results comparing diffused and surface aerators are shown in figure 9.58. It is noted that the difference in cooling effect between the two aeration methods increases with the aeration power level and is more pronounced at lower detention times.

PROBLEM 9–10

A brewery wastewater is to be pretreated in an aerobic lagoon with a retention period of one day. Estimate the effluent BOD and suspended solids.

BOD$_5$ = 2500 mg/l
K = 40/day
a = 0.5
Inf SS = 500 mg/l (nonvolatile)
b = 0.1/day

$$\frac{S_e}{S_o} = \frac{1 + bt}{aKt}$$

$$S_e = \frac{1 + 0.1 \cdot 1}{0.5 \cdot 40 \cdot 1} (2500) \text{ mg/l}$$

$$= 138 \text{ mg/l as BOD}_5$$

$$X_v = \frac{aS_r}{1 + bt} = \frac{(0.5)(2500 - 138)}{1 + 0.1 \cdot 1} = 1074 \text{ mg/l as biological VSS}$$

Effluent BOD $= 138 + 0.3 \dfrac{\text{mg BOD}}{\text{mg VSS}} (1074 \text{ mg/l VSS}) = 460 \text{ mg/l BOD}$

Effluent SS $= \dfrac{1074}{0.8} + 500 = 1843 \text{ mg/l (Assume 80\% volatile)}$

PROBLEM 9–11

Design a two-stage aerated lagoon system at 12 ft depth to treat a 8.5 MGD industrial wastewater with the following characteristics:

Influent: $BOD_5 = 425$ mg/l
Temp $= 85°$ F
$SS = 0$ mg/l
Ambient Temperature: summer $= 70°$ F
winter $= 34°$ F
Kinetic Variables:
$K = 6.3$ day^{-1} at 20° C
$a = 0.5$
$b = 0.2$ day^{-1}
$a' = 0.52$
$b' = 0.28$ day^{-1}
$\Theta = 1.035$ for BOD reaction rate
$\Theta = 1.024$ for oxygen transfer efficiency
$F = 1.0$ (winter)
1.4 (summer)
$F' = 1.5$ (summer)
$N_o = 3.2$ lb O_2/hp$-$hr
$\alpha = 0.85$
$\beta = 0.90$
$C_L = 1.0$ mg/l

The final effluent should have a maximum soluble BOD_5 of 20 mg/l for the summer and 30 mg/l for the winter.

In general, the required detention time of the lagoon system for reaching a terminal BOD_5 is controlled by winter temperatures while the oxygen requirements and power level will usually be controlled by summer conditions.

I. Design of Basin I Based on Minimum Detention Time

    1. Compute basin volume by assuming a detention time
        For $t = 2$ day

$$V = Qt$$

$$= 8.5 \cdot 2$$

$$= 17 \text{ MG}$$

    2. Calculate winter water temperature in basin

$$\frac{t}{D} = \frac{T_i - T_w}{f(T_w - T_a)}$$

or

$$T_w = \frac{DT_i + ftT_a}{D + ft}$$

$$= \frac{12 \cdot 85 + 1.6 \cdot 2 \cdot 34}{12 + 1.6 \cdot 2}$$

$$= 74.3° \text{ F or } 23.5° \text{ C}$$

    3. Correct the BOD reaction rate for winter conditions

$$K_{(T_2)} = K_{(T_1)} \, \theta^{(T_2 - T_1)}$$

$$K_{23.5} = 6.3 \cdot (1.035)^{(23.5-20)}$$

$$= 7.11 \text{ day}^{-1}$$

    4. Calculate the winter effluent soluble BOD_5

$$\frac{S_e}{S_o} = \frac{1 + bt}{aKt}$$

$$S_e = \frac{1 + 0.2 \cdot 2}{0.5 \cdot 7.11 \cdot 2} \cdot 425$$

$$= 83.7 \text{ mg/l}$$

    5. Repeat steps 2–4 for summer conditions

$$T_w = \frac{12 \cdot 85 + 1.6 \cdot 2 \cdot 70}{12 + 1.6 \cdot 2}$$

$$= 81.8^\circ \text{ F or } 27.7^\circ \text{ C}$$

$$K_{27.7} = 6.3 \cdot (1.035)^{(27.7-20)}$$

$$= 8.21 \text{ day}^{-1}$$

$$S_e = \frac{1 + 0.2 \cdot 2}{0.5 \cdot 8.21 \cdot 2} \cdot 425$$

$$= 72.5 \text{ mg/l}$$

6. Calculate average volatile suspended solids concentration under summer conditions

$$X_v = \frac{aS_r}{1 + bt}$$

$$= \frac{0.5 \cdot (425 - 72.5)}{1 + 0.2 \cdot 2}$$

$$= 126 \text{ mg/l}$$

7. Calculate oxygen requirement

$$R_r = a'S_r + b'X_v$$

$$= 0.52 \cdot (425 - 72.5) \cdot 8.5 \cdot 8.34 + 0.28 \cdot 126 \cdot 8.5 \cdot 2 \cdot 8.34$$

$$= 17996 \text{ lb/day}$$

8. Compute horsepower requirement

$$N = N_o \frac{(\beta C_s - C_L)}{C_{s(20)}} a\theta^{(T-20)}$$

$$= 3.2 \cdot \frac{(0.90 \cdot 7.96 - 1.0)}{9.2} \cdot 0.85 \cdot (1.024)^{(27.7-20)}$$

$$= 2.19 \text{ lb/hp-hr}$$

$$\text{hp} = \frac{R_r}{N}$$

$$= \frac{17996}{2.19 \cdot 24}$$

$$= 342 \text{ hp}$$

9. Check power level

$$P.L. = \frac{hp}{V}$$

$$= \frac{342}{17}$$

$$= 20.1 \text{ hp/mg}$$

For a conservative design, the minimum power level should be 14 hp/MG. If the power level is significantly less than 14 hp/MG, then 14 hp/MG should be used.

10. Repeat steps 1 through 9 using different detention times. The results can be tabulated as shown in Table I.

II. Design of Basin II on Minimum Detention Time

It is assumed that the minimum power level in a facultative lagoon should be 4 hp/MG to maintain 50 mg/l of volatile suspended solids in suspension. The design of Basin II is based on selecting an effluent BOD concentration from the first basin at given detention times to be the influent to the second basin.

TABLE I

Design Summary of Basin I (Aerobic Lagoon)

t (day)	winter or summer	T_w (°F)	(°C)	K (day^{-1})	S_e (mg/l)	X_v (mg/l)	R_r (lb/day)	hp required	P.L. (hp/MG)
1.0	w	79.0	26.1	7.77	131				
	s	83.2	28.4	8.41	121	127	13727	261	30.7
2.0	w	74.3	23.5	7.11	83.7				
	s	81.8	27.7	8.21	72.5	126	17996	342	20.1
3.0	w	70.4	21.3	6.59	68.8				
	s	80.7	27.1	8.04	56.4	115	20436	389	15.3
4.0	w	67.3	19.6	6.21	61.6				
	s	79.8	26.6	7.91	48.4	105	22219	476(423)*	14(12.4)*
5.0	w	64.6	18.1	5.90	57.6				
	s	79.0	26.1	7.77	43.8	95	23480	595(449)	14(10.6)
6.0	w	62.3	16.8	5.64	55.3				
	s	78.3	25.7	7.66	40.7	87	24528	714(469)	14(9.2)

*minimum power level applied

1. Calculate the temperature in Basin II and adjust the BOD reaction rate by assuming a detention time for Basin II.

   For example, the effluent BOD concentration at a 2-day detention time for Basin I are 83.7 mg/l and 72.5 mg/l for winter and summer, respectively.

   Basin temperature

$$T_w = \frac{DT_i + ftT_a}{D + ft}$$

Corrected influent concentration

$$S_o' = FS_o$$

Corrected BOD reaction rate

$$K_{20} = \frac{6.3}{425} S_o'$$

Assume $t = 5$ day

Winter: $T_w = \dfrac{12 \cdot 74.3 + 1.6 \cdot 5 \cdot 34}{12 + 1.6 \cdot 5}$

$$= 58.2° \text{ F or } 14.6° \text{ C}$$

$$K_{20} = \frac{6.3}{425} \cdot 1.0 \cdot 83.7$$

$$= 1.24 \text{ day}^{-1}$$

$$K_{14.6} = 1.24 \cdot (1.035)^{(14.6-20)}$$

$$= 1.03 \text{ day}^{-1}$$

Summer: $T_w = \dfrac{12 \cdot 81.8 + 1.6 \cdot 5 \cdot 70}{12 + 1.6 \cdot 5}$

$$= 77.1° \text{ F or } 25.1° \text{ C}$$

$$K_{20} = \frac{6.3}{425} \cdot 1.4 \cdot 72.5$$

$$= 1.50 \text{ day}^{-1}$$

$$K_{25.1} = 1.50 \cdot (1.035)^{(25.1-20)}$$

$$= 1.79 \text{ day}^{-1}$$

2. Compute the detention time required to reduce the soluble BOD to the prescribed level.

$$S_o' = FS_o$$

$$t = \frac{S_o'(S_o' - S_e)}{KX_vS_e}$$

Winter: $S_o' = 83.7$ mg/l

$$t = \frac{83.7 \cdot (83.7 - 30)}{1.03 \cdot 50 \cdot 30}$$

$$= 2.91 \text{ day}$$

Summer: $S_o' = 1.4 \cdot 72.5$

$$= 102 \text{ mg/l}$$

$$t = \frac{102 \cdot (102 - 20)}{1.79 \cdot 50 \cdot 20}$$

$$= 4.67 \text{ day}$$

Now it is summer conditions controlling.

3. Repeat steps 1 and 2 until the computed detention time in step 2 is close enough to the assumed value in step 1. Final results are as follows:

$$t = 4.67 \text{ day}, V = 39.7 \text{ mg}$$

Winter: $T_w = 58.8° $ F or $14.9°$ C

$$K_{14.9} = 1.04 \text{ day}^{-1}$$

$$S_e = 21.5 \text{ mg/l}$$

Summer: $T_w = 77.3°$ F or $25.2°$ C

$$K_{25.2} = 1.79 \text{ day}^{-1}$$

$$S_e = 20.0 \text{ mg/l}$$

4. Calculate oxygen requirement

$$R_r = F'S_r$$

$$= 1.5 \cdot (1.4 \cdot 72.5 - 20.0) \cdot 8.5 \cdot 8.34$$

$$= 8666 \text{ lb/day}$$

5. Calculate horsepower requirement

$$N = N_o \frac{(\beta C_s - C_L)}{C_s \, (20)} \, a\theta^{(T-20)}$$

$$= 3.2 \cdot \frac{(0.90 \cdot 8.36 - 1.0)}{9.2} \cdot 0.85 \cdot (1.024)^{(25.2-20)}$$

$$= 2.18 \text{ lb/hp–hr}$$

$$hp = \frac{R_r}{N}$$

$$= \frac{8666}{2.18 \cdot 24}$$

$$= 166 \text{ hp}$$

6. Check power level

$$P.L. = \frac{hp}{V}$$

$$= \frac{166}{39.7}$$

$$= 4.2 \text{ hp/MG}$$

TABLE II
Design Summary of Basin II (Facultative Lagoon)

S_o mg/l	T_w (°F)	(°C)	K (day^{-1})	t (day)	S_e (mg/l)	X_v (mg/l)	R_r (lb/day)	hp required	P.L. (hp/MG)
131	55.0	12.8	1.51	8.57	22.1	50			
121	76.2	24.6	2.94	8.57	20.0	50	10740	291(205)*	4.0(2.8)*
83.7	58.8	14.9	1.04	4.67	21.5	50			
72.5	77.3	25.2	1.79	4.67	20.0	50	8666	166	4.2
68.8	59.2	15.1	0.862	3.33	22.3	50			
56.4	77.4	25.2	1.40	3.33	20.0	50	6274	120	4.2
61.6	58.4	14.7	0.761	2.72	23.0	50			
48.4	77.2	25.1	1.19	2.72	20.0	50	5083	97	4.2
57.6	57.3	14.1	0.697	2.34	23.8	50			
43.8	76.9	24.9	1.08	2.34	20.0	50	4392	84	4.2
55.3	56.1	13.4	0.653	2.12	24.6	50			
40.7	76.5	24.7	0.993	2.12	20.0	50	3934	75	4.2

*minimum power level applied

7. Repeat steps 1 through 6 using different detention times in Basin I with the adequate detention time in Basin II to meet the desired effluent quality. The results are tabulated in Table II.

III. Optimum Lagoon System

This two-stage lagoon system can be optimized to minimize either the total detention time or the total horsepower to be installed.

Table III and Figure I show the summarized design results of the two-stage lagoon system. The minimum total required detention time is 6.33 day, with a corresponding total installed horsepower of 509 hp. The minimum total horsepower is 508 hp with a total detention time of 6.67 day. The second alternative may be the optimum system since the increased basin size will be more tolerable to the influent fluctuation in both flow rate and constituent concentration. It will also be justified by an economic gain realized by operating with less power input although the difference in the horsepower installed is minimal. In this alternative, the detention time in Basin I is 2 days and the detention time in Basin II is 4.67 days.

TABLE III

Design Summary of Two-Stage Lagoon System

t, day			hp installed		
Basin I	Basin II	Total	Basin I	Basin II	Total
1.0	8.57	9.57	261	291	552
2.0	4.67	6.67	342	166	508
3.0	3.33	6.33	389	120	509
4.0	2.72	6.72	476	97	573
5.0	2.34	7.34	595	84	679
6.0	2.12	8.12	714	75	789

FIGURE I
Effect of Detention Time in Basin I on Total Required
Time and Horsepower

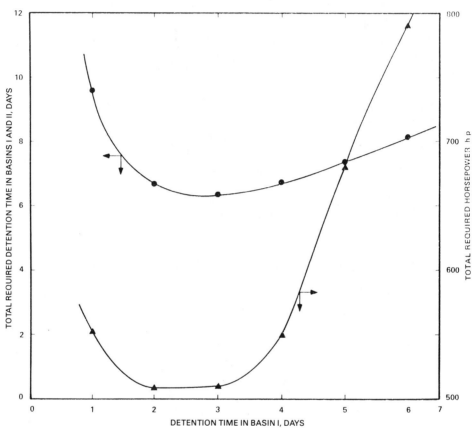

Facultative Lagoons

In a facultative lagoon the biological solids level maintained in suspension is a function of the power level employed in the basin. At low power levels the volatile suspended solids level in the basin may vary from 50 to 100 mg/l. Solids deposited in the bottom of the facultative lagoon will undergo anaerobic degradation, which results in a feedback of soluble organics to the upper aerobic layers. Under these conditions equation (9–40) should be modified to equation (9–44):

$$\frac{S_e}{S_o} = \frac{S_o\,(1+F)}{S_o + KX_v t} \qquad (9\text{–}44)$$

in which F is a coefficient that accounts for organic feedback due to anaerobic activity in the deposited sludge layers. The degree of anaerobic activity is highly temperature dependent and the coefficient F may be expected to vary from 1.0 to 1.4 under winter and summer conditions, respectively, depending on the geographical location of the plant.

In a facultative lagoon biological solids are maintained at a lower level than in an aerobic lagoon, and soluble organics are fed back to the liquid as anaerobic degradation products. It is, therefore, not possible to directly compute the oxygen requirements employing equation (9–8). In this case, the oxygen requirements can be related empirically to organic removal and estimated from equation (9–45):

$$R_r = F'S_r \qquad\qquad (9\text{–}45)$$

in which F' is an overall oxygen utilization coefficient for facultative lagoons. Results obtained for various industrial wastewaters would indicate that the coefficient F' is a function of the degree of organic feedback, which in turn is a function of temperature. In general, depending on the geographic location of the plant, F' can be estimated to vary from 0.8 to 1.1 during winter operation when anaerobic activity in the basin is low, and from 1.2 to 1.5 during summer operation when anaerobic activity in the bottom of the basin is at a maximum. The value selected will depend on the geographical location of the plant.

Aerated Lagoon Systems

Multiple basins may be most effectively employed in aerated lagoon systems under proper conditions. There may be little advantage to series basin operation at a given temperature in terms of effluent organic concentration, but there are two considerations which favor series operation. First, series operation may be desired where land availability is a concern. When considering a thermal balance, a minimum total basin volume can be obtained by employing two basins in series. The first basin volume is minimized to maintain a high temperature, a high biological solids level, and a resulting high BOD reaction rate in an aerobic lagoon. The second basin is a facultative basin at low power (mixing) levels that permit solids settling and decomposition in the bottom of the basin. An optimization procedure [47] can be employed to determine the smallest total basin volume and the lowest aeration horsepower for a specified effluent quality. Where low effluent suspended solids are desired, a final settling basin can be employed. The settling basin should consider four major constraints.

1. A sufficiently long detention period to effect the desired suspended solids removal
2. Adequate volume for sludge storage
3. Minimal algal growth
4. Minimal odors from anaerobic activity

Unfortunately, these design objectives are not always compatible. Frequently short retention times are required to inhibit algal growths, which are too short for proper settling. Also, adequate volume must remain above the sludge deposits at all times to prevent the escape of the odorous gases of decomposition.

In order to achieve these objectives, a minimum detention time of one day is usually required to settle the majority of the settleable suspended solids. Where algal growths pose potential problems, a maximum detention time of three to four days is recommended. For odor control, a minimum water level of three feet should be maintained above the sludge deposits at all times.

Nutrient requirements in aerated lagoons are similarly computed to the activated sludge process. In the case of an aerobic lagoon, 12.3 percent nitrogen and 2.6 percent phosphorus is required for the biomass generated in the process. In the case of a facultative lagoon, however, anaerobic decomposition of sludge deposited in the bottom of the basin will feed back nitrogen and phosphorus. This is usually sufficient for organic removal occurring in this basin, and no additional nitrogen and phosphorus need be added.

TRICKLING FILTERS

A trickling filter is a packed bed covered with slime growth over which wastewater is passed. As the wastewater is passed through the filter, organic matter present in the wastewater is removed by the biological slime. As the waste passes through the filter nutrients and oxygen diffuse into the slimes where assimilation occurs and by-products and CO_2 diffuse out of the slime into the flowing liquid. As oxygen diffuses into the biological film, it is consumed by microbial respiration, so that a defined depth of aerobic activity is developed. Slime growth below this depth is anaerobic as shown in figure 9.59. Conventional filters contain 2.5 to 4 inch rock packing in varying depths of 3 to 8 feet (0.9 to 2.5 m). Maximum hydraulic loadings to the filter of 0.5 gpm per ft^2 (0.02 m^3/m^2-min) will yield BOD removal efficiencies of up to 85 percent treating domestic wastewater. Recently, plastic packing (figure 9.60) has become more popular and are employed in filter depths of up to 40 feet (12 m) with hydraulic loadings as high as 4.0 gpm per ft^2 (0.16 m^3/m^2-min). Depending on the hydraulic loading and depth of filter, BOD removal efficiencies as high as 90 percent have been attained on some wastewaters. A typical trickling filter is shown in figure 9.61.

FIGURE 9.59
Biological Filter Schematic

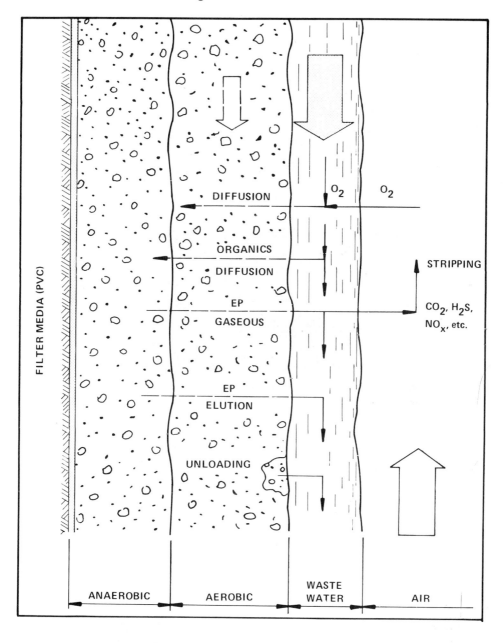

FIGURE 9.60
Plastic Trickling Filter Media
(Courtesy of Envirotech, Inc.)

FIGURE 9.61
Plastic Packed Trickling Filter
(Courtesy of Envirotech, Inc.)

PROBLEM 9–12

The following data was obtained on a trickling filter pilot plant study:

Depth ft	Hydraulic Loading, gpm/ft²		
	1.0	1.5	3.0
	% BOD Remaining		
0	100	100	100
3	84.5	85.0	92.0
6	70.0	75.5	83.5
9	62.0	63.0	77.0
12	51.5	56.0	73.5
15	45.0	48.5	68.0
18	37.5	43.0	63.5
21	31.5	38.5	59.5
S_o, mg/l	220	180	200

Determine the surface area for a trickling filter with and without recirculation for the following conditions:

S_{inf} = 350 mg/l
S_{eff} = 40 mg/l
Q = 2.5 MGD
N = 1.3
D = 20 ft

1. Based on organic loading the pilot plant data is correlated by

$$\frac{S}{S_o} = e^{-k_s D/Q^n S_o}$$

in which Q is the hydraulic loading rate.

(1) Plot in Figure I (S/S_o) vs. depth and the negative slopes of the resulting lines are:

Q, gpm/ft²	Slope
1.0	0.0549
1.5	0.0437
3.0	0.0260

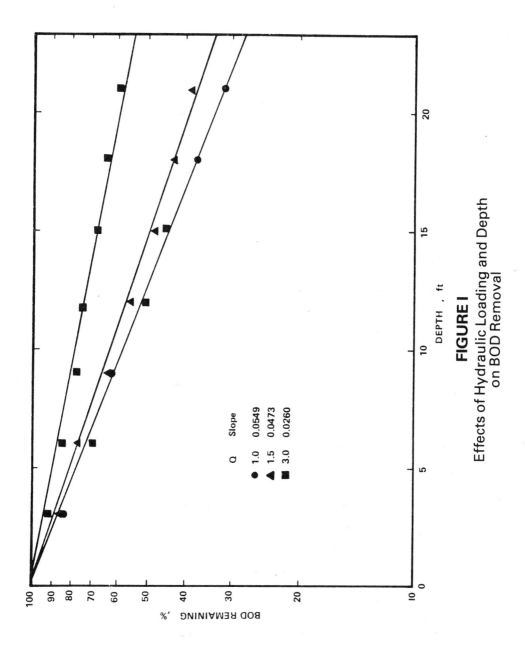

FIGURE I

Effects of Hydraulic Loading and Depth
on BOD Removal

(2) Plot above slopes and Q S_o on log-log paper and the negative slope of the resulting line is n (see Figure II).

$$n = 0.751$$

(3) Plot $\ln (S/S_o)$ vs. $D/(Q^nS_o)$ and the negative slope is K_s (see Figure III).

$$k_s = 11.7$$

Then, the design equation is

$$\frac{S}{S_o} = e^{-11.7D/Q^{0.751}S_o}$$

2. System without recirculation

$$\frac{40}{350} = e^{-11.7 \cdot 20/(Q^{0.751} \cdot 350)}$$

Solve for Q
$Q = 0.209$ gpm/ft^2
Required area

$$A = \frac{2.5 \times 10^6}{0.209 \cdot 1440} \text{ ft}^2$$

$$= 8{,}310 \text{ ft}^2$$

3. System with recirculation

$$S_o = \frac{S_a + NS_e}{1 + N}$$

$$= \frac{350 + 1.3 \cdot 40}{1 + 1.3}$$

$$= 175 \text{ mg/l}$$

$$\frac{40}{175} = e^{-11.7 \cdot 20/(Q^{0.751} \cdot 175)}$$

Solve for Q
$Q = 0.877$ gpm/ft^2
Required area

$$A = \frac{(1 + 1.3) \cdot 2.5 \times 10^6}{0.877 \cdot 1440}$$

$$= 4560 \text{ ft}^2$$

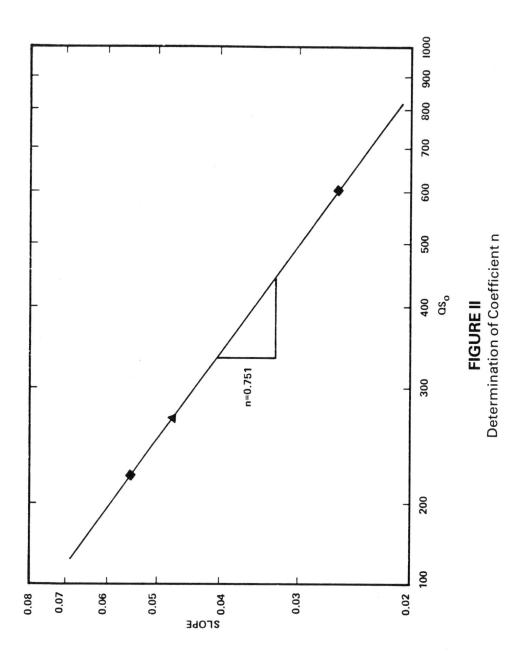

FIGURE II

Determination of Coefficient n

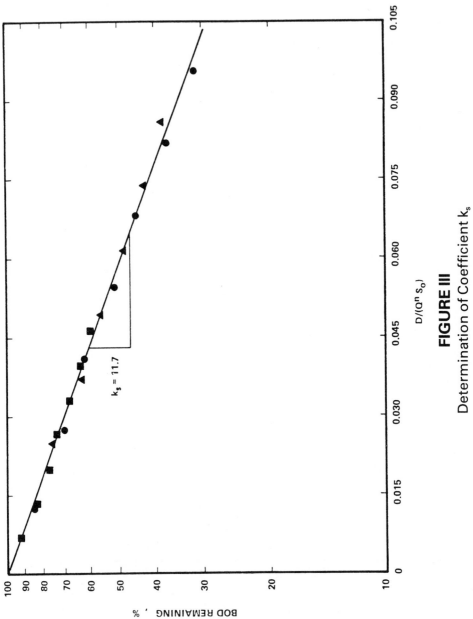

FIGURE III

Determination of Coefficient k_s

Design Considerations

Two mathematical models will be presented for the design of trickling filters. In a manner analogous to the activated sludge process, BOD removal through a trickling filter can be related to the available biological slime surface and to the time of contact of wastewater with that surface. Considering plug flow conditions, the general relationship for BOD removal in the activated sludge process [equation (9–16)] is:

$$\frac{S}{S_o} = e^{-KX_v t/S_o} \tag{9–46}$$

The mean time of contact of liquid with the filter surface is related to the filter depth, D, the hydraulic loading, Q (gpm/ft^2) and the characteristics of the filter packing, as shown below:

$$t = \frac{CD}{Q^n} \tag{9–47}$$

C and n are constants related to the configuration of the packing. The biological solids concentration in a filter is proportional to the specific surface, A_v, or X_v is proportional to A_v. Equation (9–46) can then be modified for a trickling filter as shown in equation (9–48):

$$\frac{S}{S_o} = e^{-K_s A_v D/Q^n S_o} \tag{9–48}$$

For a specific packing where A_v is constant, equation (9–48) can be expressed as equation (9–49):

$$\frac{S}{S_o} = e^{-k_s D/Q^n S_o} \tag{9–49}$$

In order to avoid filter plugging, a maximum specific surface, A_v, of about 30 ft^2/ft^3 (98 m^2/m^3) is recommended for the treatment of carbonaceous wastewaters. Specific surfaces in excess of 100 ft^2/ft^3 (328 m^2/m^3) can be employed for nitrification because of the low yield of biological cellular material.

In many cases, recirculation of filter effluent will improve the overall BOD removal through the filter. This is particularly true when the influent BOD is high and dilution is required to maintain aerobic conditions in the filter film. When recirculation is employed as a diluent to the influent wastewater, the applied BOD to the filter S_o becomes:

$$S_o = \frac{S_a + NS_e}{1 + N} \tag{9–50}$$

where S_a = influent wastewater BOD
          N = recirculation ratio R/Q

When the BOD of the wastewater exhibits a decreasing removal rate with decreasing concentration, the BOD in the recirculated flow would be removed at a lower rate than that prevailing in the influent. In this case, one must apply a coefficient of retardation to the recirculated flow. The treatment of dilute black liquor on plastic packing in accordance with equation (9–49) is shown in figure 9.62.

Recent studies [43] have related BOD removal through the filter to the applied organic loading expressed as lbs of BOD/1000 ft³/day as shown in equation (9–51):

$$\frac{S}{S_o} = e^{-kA_v/L} \qquad (9\text{--}51)$$

in which L is expressed as lbs of BOD/1000 ft³/day (kg of BOD/m³-day). Data correlated in accordance with equation (9–51) is shown in figure 9.63. It should be noted that when the exponent n in equation (9–49) is unity, equation (9–49) and equation (9–51) become mathematically the same. Filter performance characteristics for various wastewaters, in accordance with equation (9–51), are shown in table 9.20.

Effect of Temperature

The performance of trickling filters will be affected by changes in temperature of the filter films and the liquid passing over the films. It is usually assumed that these two temperatures will be essentially the same when only the aerobic portion of the film is considered. A decrease in temperature results in a decrease in respiration rate, a decrease in oxygen transfer rate, and an increase in oxygen saturation. The combined effect of these factors results in an increase in aerobic film depth at a lower activity level, yielding a somewhat reduced efficiency at lower temperatures. The relationship of efficiency and temperature through a trickling filter can be expressed by equation (9–52):

$$k_{s(T)} = k_{s(20°C)} \ 1.035^{(T-20)} \qquad (9\text{--}52)$$

Effect of Specific Surface

In general, the reaction rate coefficient k or k_s will be directly proportional to the specific surface of the filter packing. This is shown for BOD removal in figure 9.64 and for nitrification in figure 9.65. In some cases, however, with unequal sliming of the filter media, the specific surface in equation (9–48) may have an exponent of less than 1.

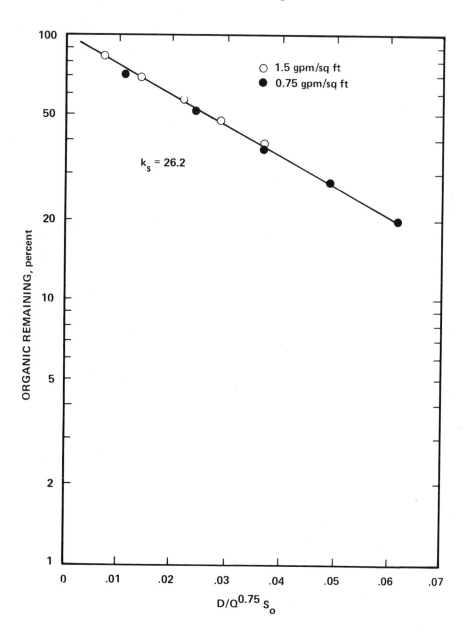

FIGURE 9.62

Treatment of Dilute Black Liquor (S_o = 400 mg/l) on
Plastic Packing

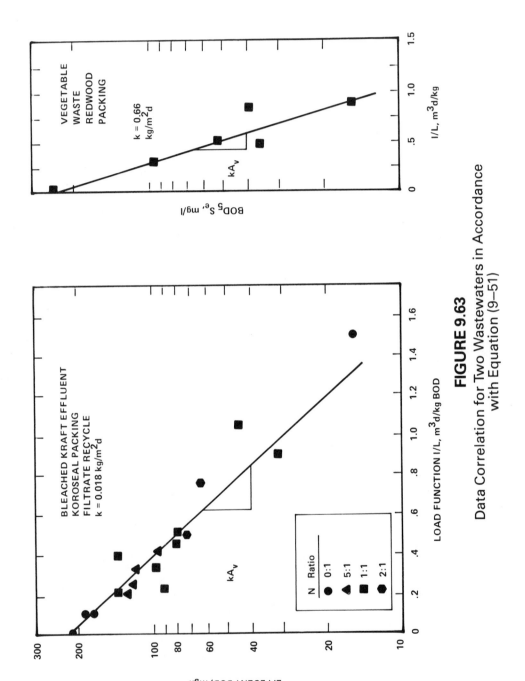

FIGURE 9.63

Data Correlation for Two Wastewaters in Accordance
with Equation (9–51)

TABLE 9.20
Trickling Filter Performance

Type of Waste	Type of Media	Mean Value S_o or S_a (mg/l)	Rate coefficient – k (kg/m²d)
1	2	3	4
Pharmaceutical	Vinyl Core	5248	0.2160
Phenolic	Vinyl Core	340	0.0210
Sewage	6 parallel: Basalt, slag, Surfpac I & II, Flocor, Cloisonyle	280	0.0104
Sewage	4 parallel: Slag 6″ and 4″, Flocor, Surfpac	332	0.0480
Sewage	4 parallel: Slag 6″ and 4″, Flocor, Surfpac	215	0.0500
Kraft mill sludge recycle	Vinyl Core	210	0.0160
Kraft mill filtrate recycle	Vinyl Core	220	0.0180
Vegetable	Del Pak	235	0.0660
Fruit canning	Surfpac	2200	0.0930
Kraft mill	Surfpac	130	0.0054
Fruit processing	Surfpac	3200	0.0017
Pulp & paper	Vinyl Core	280	0.0160

FIGURE 9.64

Effect of Specific Surface of Removal Rate Coefficient [49]

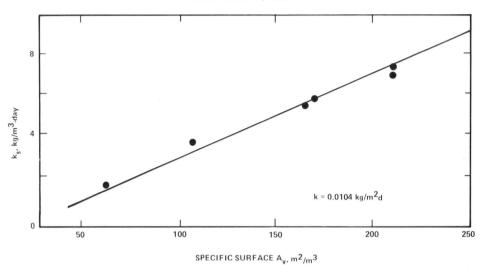

SPECIFIC SURFACE A_v, m^2/m^3

As previously indicated, an aerobic depth in filter film will be established as shown in figure 9.59. Increasing concentration of BOD applied to the filter will increase the respiration rate of the filter film and decrease the aerobic depth. There will, therefore, be a limiting BOD concentration to maintain aerobic conditions in the filter. Depending on the biodegradability of the wastewater, this may vary from a BOD of 500 mg/l to 1200 mg/l. Higher influent BOD concentrations require recirculation for dilution of the influent strength.

Applications of Trickling Filters to Wastewater Treatment

In the treatment of domestic sewage, biological flocculation and agglomeration occur through the filter in a manner analogous to the activated sludge process. BOD removals in the order of 85 percent are readily attained. The removal rate coefficient, K_s, for soluble industrial wastewaters, however, is relatively low, and, hence, filters are not economically attractive for high efficiency (greater than 85 percent BOD reduction) of such wastewaters. Plastic-packed filters, however, have been employed as a pretreatment for high BOD wastewaters in which BOD removals in the order of 50 percent have been achieved at hydraulic and organic loadings of greater than 4 gpm/ft^2 (0.16 m^3/m^2-min) and 500 lbs BOD/day per 1000 ft^3 (8 kg/m^3-day) of filter packing respectively at depths up to 40 feet (12 m).

FIGURE 9.65

Relationships Between Specific Surface and
Nitrification [50]

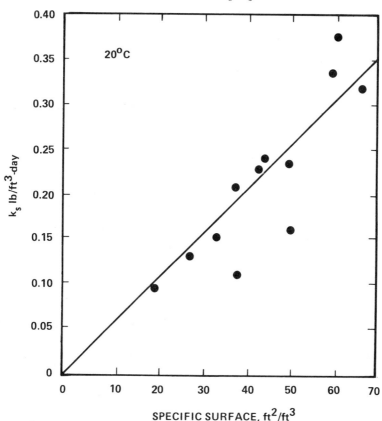

SPECIFIC SURFACE, ft^2/ft^3

Performance characteristics for the pretreatment of food packing waste-waters is shown in figure 9.66. As can be seen from figure 9.66 nutrient requirements for trickling filters are similar to those required for activated sludge and other aerobic processes.

ACTIVATED SLUDGE

The activated sludge process is a continuous system in which aerobic biological growths are mixed and aerated with wastewater and separated in a gravity clarifier. A portion of the concentrated sludge is recycled and mixed with additional wastewater. The process should provide an effluent with a soluble BOD of 10–30 mg/l, although the organic concentration of the effluent in terms of COD

FIGURE 9.66

Treatment of Food Processing Wastewater on High-Rate Trickling Filter

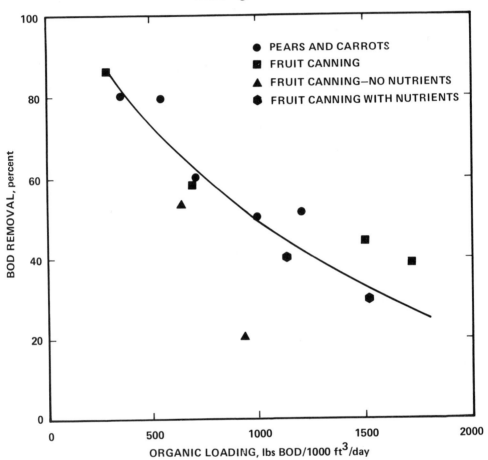

may be as high as 500 mg/l, depending on the concentration of bioresistant compounds originally in the wastewater. There are many impurities in industrial wastewaters such as oil and grease, which must be removed or altered by preliminary treatment before subsequent activated sludge treatment can be considered.

Many modifications of this process have been developed over the decades, the principal ones being shown in figure 9.67. The conventional process employs long rectangular aeration tanks, which approximate plug-flow with some longitudinal mixing. This process is primarily employed for the treatment of domestic wastewater. Returned sludge is mixed with the wastewater in a mixing box or chamber at the head end of the aeration tank. The mixed liquor then flows

FIGURE 9.67
Activated Sludge Systems

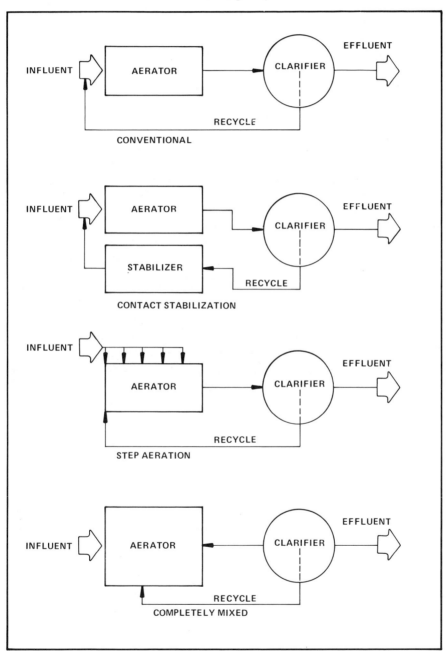

through the aeration tanks, during which progressive removal of organics occurs. The oxygen utilization rate is high at the beginning of the aeration tanks and decreases with aeration time. Where complete treatment is achieved, the oxygen utilization rate will approach the endogenous level toward the end of the aeration tanks. The principal disadvantages of this system for the treatment of industrial wastewaters are:

1. The oxygen utilization rate varies with tank length and requires irregular spacing of the aeration equipment or a modulated air supply.
2. Load variation may have a deleterious effect on the activated sludge when it is mixed at the head end of the aeration tanks.
3. The sludge is susceptible to slugs or spills of acidity, causticity or toxic materials.

In the completely mixed system the aeration tank serves as an equalization basin to smooth-out load variations and as a diluent for slugs and toxic materials. Since all portions of the tank are mixed, the oxygen utilization rate will not vary with time and the aeration equipment can be equally spaced.

Step aeration is a variant between the conventional process and the completely mixed process and has been successfully employed for the treatment of domestic wastewaters.

In the extended aeration process, sufficient aeration time is provided to oxidize virtually all of the biodegradable sludge synthesized from the BOD present in the wastewater.

The contact stabilization process is applicable to wastewaters containing a high proportion of the BOD in suspended or colloidal form. Since bio-adsorption and flocculation of colloids and agglomeration of suspended solids occur very rapidly, only short retention periods (15–30 minutes) are required to effect clarification as shown in figure 9.68. After the contact period the activated sludge is separated in a clarifier. A sludge reaeration or stabilization period is required to stabilize the organics removed in the contact tank. The retention period in the stabilization tank is dependent on the time required to assimilate the soluble and colloidal material removed from the wastewater in the contact tank.

Effective removal in the contact period requires sufficient activated sludge to remove the colloidal and suspended matter and a portion of the soluble organics. The retention time in the stabilization tank must be sufficient to stabilize these organics. If it is insufficient, unoxidized organics are carried back to the contact tank and the removal efficiency is decreased. If the stabilization period is too long, the sludge undergoes excessive auto-oxidation and looses some of its initial high removal capacity. Increasing retention period in the contact tanks increases the amount of soluble organics removed and decreases required stabilization time. A large increase in contact time negates the requirement for sludge stabilization. In this case, the process becomes the same as the conventional activated sludge process.

FIGURE 9.68

Schematic Representation of the Contact-Stabilization Process

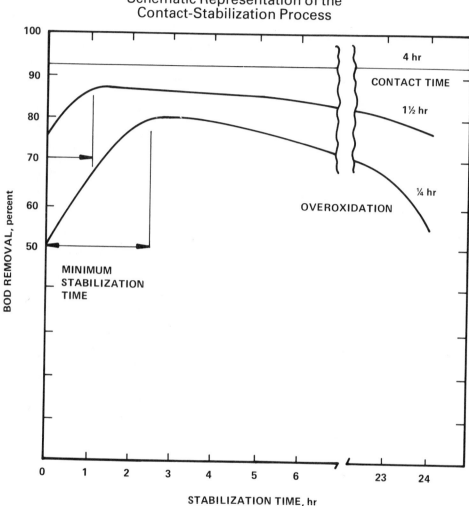

Increasing the biological solids level also decreases the stabilization time requirements since the organic loading for unit solids becomes less. The total oxygen requirements in the process are those required for synthesis of the organics removed and for endogenous respiration. The split of this oxygen between the contact tank and the stabilization tank depends on the solids level carried in both units and on the retention period in each tank. Increasing the contact tank solids level or retention period increases the percentage of total oxygen to that unit.

Domestic Sewage

A distinction must be made when considering the activated sludge process for the treatment of domestic sewage as compared to soluble industrial wastewaters, based upon the physical composition of the wastewaters and the resulting mechanisms of removal as discussed on page 389. The organic content of domestic sewage consists of three components, suspended organics resulting from ground garbage, paper, rubber, etc., colloidal matter, and soluble organics consisting mainly of carbohydrates and some nitrogenous material. Most of the organics are in the form of particulates. Investigators have shown 35, 40, and 25 percent of the total COD was present as suspended, colloidal, and soluble material, respectively in American sewage.

When sewage is mixed with activated sludge, several things immediately occur. Suspended solids are enmeshed in the biological floc, colloidal solids are adsorbed on the floc interface, and some soluble organics are adsorbed by enzymatic action. The net effect of these reactions is to attain near 90 percent BOD removal in less than 15 minutes of contact time of the wastewater with the sludge. BOD removal from domestic wastewaters is shown in figure 9.69. Note that while these organics have been removed from the wastewater by physical and biological mechanisms, achieving ultimate stabilization requires a considerably longer period of aeration.

The soluble organics that have been removed from solution are most readily assimilable by the microbial cell and are probably synthesized in less than one hour of aeration. The colloids must first be broken down by extracellular enzymes in order to be available to the cell so that complete stabilization of these organics requires a longer aeration period. The suspended solids must also undergo a slow hydrolytic breakdown so that it is probable that a portion of the organic suspended matter entering the aeration tanks is disposed of as excess sludge in the detention periods normally encountered in the activated sludge process. The one exception is the extended aeration process in which the biodegradable portion of the suspended solids are oxidized.

The various phases of the activated sludge process treating domestic sewage is shown in figure 9.69. When the return sludge is first contacted with the sewage, the suspended and colloidal, and a portion of the dissolved organics are removed as denoted by time t_1. As aeration continues additional soluble organics are removed. Maximum synthesis, and, hence, the highest sludge yield, occurs after exhaustion of the substrate (point A, figure 9.69b). Nitrogen in the form of ammonia will be utilized by the organisms during this period for synthesis of new cells. Extending the aeration time beyond A results in the autooxidation of synthesized sludge and the continuing oxidation of suspended solids present in the wastewater. A portion of the synthesized biological sludge is not biodegradable in the time periods under consideration. In addition, some of the volatile solids initially present in the sewage will not be biodegraded and will accumulate as a residue. The total nonbiodegraded solids are shown as point B in figure 9.69b. The

FIGURE 9.69
Schematic Representation of the Activated
Sludge Process

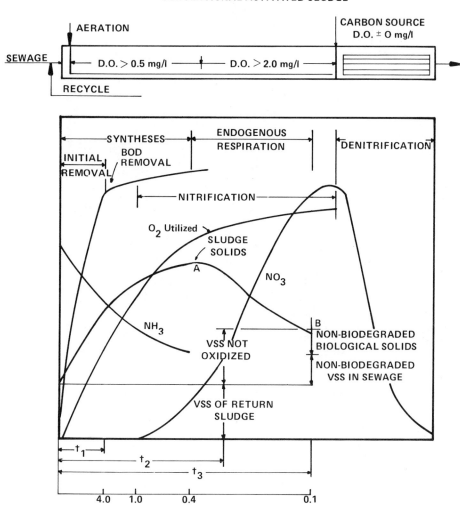

CONVENTIONAL ACTIVATED SLUDGE

rate of oxygen utilization is rapid during assimilation and decreases during endogenous respiration. A typical activated sludge plant is shown in figure 9.70.

Nitrification will occur during the activated sludge process when ammonia is oxidized to nitrite, which is then oxidized to nitrate. A sludge age greater than the reciprocal of the growth rate of the nitrifying organisms is necessary for nitrification.

FIGURE 9.70

Typical Activated Sludge Plant

(Courtesy Degremont, Paris, France.)

Due to the relatively high temperature coefficient, nitrification is more effected by winter operation than is carbonaceous BOD removal. Denitrification is the reduction of nitrate to gaseous end products $(-N_2, N_2O)$ under anaerobic conditions. Many of the facultative organisms present in activated sludge will use the nitrate as an oxygen source in the absence of dissolved oxygen. Denitrification may occur in final settling tanks resulting in rising sludge. The increasing emphasis on nutrient removal makes nitrification-denitrification one of the economical processes for nitrogen removal from domestic wastewater.

PROBLEM 9–13

An activated sludge plant is to be designed to meet the following requirements for a wastewater flow of 1 MGD and a BOD_5 of 720 mg/l. The summer effluent soluble BOD_5 (25°C) is to be 25 mg/l. The effluent suspended solids is estimated as 40 mg/l. The following data was generated from pilot plant studies (at 20°C).

BOD removal rate coefficient	16/day
Maximum F/M	0.7/day
a	0.58
a'	0.35
b	0.15/day
MLVSS	2800 mg/l
Θ	1.08

I. Compute a) the aeration basin volume
   b) the oxygen requirements
   c) the excess sludge, ΔX_v

II. Compute the soluble and total effluent BOD during winter operation with a temperature of 10°C

III. Size the secondary clarifier for the sludge with the following settling properties and the volatile fraction of 74%.

MLSS, mg/l	ZSV, ft/hr
1075	22.1
3950	6.95
8200	1.85
12400	0.67
16900	0.32
20500	0.16

The recycled sludge concentration is to be 20,000 mg/l as SS.

I. a) Aeration basin volume

$$\frac{S_o - S}{X_v t} = K \frac{S}{S_o}$$

$$K_{25°C} = K_{20°C} (1.08)^{(T-20)}$$

$$= (16)(1.08)^{(25-20)}$$

$$= 23.5/day$$

$$t = \frac{S_o (S_o - S)}{K X_v S}$$

$$= \frac{720 (720 - 25)}{(23.5)(2800)(25)}$$

$$= 7.3 \text{ hours } (0.304 \text{ days})$$

Check F/M

$$F/M = \frac{S_o}{X_v t} = \frac{720}{(2800)(0.304)} = 0.85/day > 0.7/day$$

Too high; use F/M = 0.7/day

$$t = \frac{720}{(2800)(0.7)} = 0.367 \text{ days}$$

$$V = 1 \text{ MGD } (0.367 \text{ days}) = 0.367 \text{ mg}$$

b) at F/M = 0.7, x = 0.64
    Oxygen requirements—summer conditions

$$b_{25°C} = 0.15 \cdot 1.04^{(25-20)}$$

$$= 0.18/day$$

lbs O_2/day = $0.35 \cdot 1.0 \cdot 8.34 (720 - 25) + 0.18 \cdot 1.4 \cdot 0.64 \cdot 2800 \cdot 0.367 \cdot 8.34$

$$= 3411 \text{ lbs/day}$$

c) Sludge Production—summer conditions

$$\Delta X_v = a (S_o - S) - b x X_v$$

$$= 0.58 \cdot (720 - 25) \cdot 1.0 \cdot 8.34 - 0.18 \cdot 0.64 \cdot 2800 \cdot 0.367 \cdot 8.34$$

$$= 2374 \text{ lbs/day}$$

$$N/day = 0.123 \frac{x}{0.77} \Delta X_v + 0.07 \frac{0.77 - x}{0.77} \Delta X_v$$

$$= 0.123 \frac{0.64}{0.77} (2374) + 0.07 \frac{0.13}{0.77} (2374)$$

$$= 270 \text{ lbs/day}$$

$$P/\text{day} = 0.026 \frac{x}{0.77} \Delta X_v + 0.01 \frac{0.77 - x}{0.77} \Delta X_v$$

$$= 55 \text{ lbs/day}$$

II. Winter effluent BOD

$$K_{10°C} = 16 \cdot 1.08^{(10-20)}$$

$$= 7.41/\text{day}$$

$$S = \frac{S_o{}^2}{KX_v t + S_o}$$

$$= \frac{720^2}{7.4 \cdot 2800 \cdot 0.367 + 720}$$

$$= 62 \text{ mg/l soluble BOD}$$

Total effluent $BOD_5 = 62 \text{ mg/l} + 40 \text{ mg/l SS} \cdot 0.3 \dfrac{BOD_5}{SS} = 74 \text{ mg/l}$

(Effluent suspended solids during the winter may be higher resulting in higher BOD levels.)
Winter sludge production

$$b = 0.15 \cdot 1.04^{(10-20)} = 0.10/\text{day}$$

$$\Delta X_v = 0.58 \cdot 1.0 \cdot (720 - 62) \cdot 8.34 - 0.1 \cdot 0.64 \cdot 2800 \cdot 0.367 \cdot 8.34$$

$$= 2634 \text{ lbs/day}$$

III. The example calculation of solids flux of SS = 1075 mg/l with ZSV = 22.10 ft/hr is

$$\text{Solids Flux} = 1075 \text{ mg/l} \cdot 22.10 \text{ ft/hr} \cdot \left(\frac{24 \text{ hr}}{\text{day}} \right) \left(\frac{28.321}{\text{ft}^3} \right) \left(\frac{\text{lb}}{453.6 \times 10^3 \text{ mg}} \right)$$

$$= 35.6 \text{ lb/ft}^2\text{-day}$$

The results are shown and plotted in Figures I and II.

MLSS mg/l	ZSV ft/hr	Solids Flux lb/ft²-day
1075	22.10	35.6
3950	6.95	41.1
8200	1.85	22.7
12400	0.67	12.5
16900	0.32	8.10
20500	0.16	4.91

The SS concentration to the secondary clarifier is

$$X_a = \frac{2800}{0.74} = 3800 \text{ mg/l}$$

The recycle flow rate can be calculated by material balance equation

$$(Q + R)X_a = RX_r$$

or

$$R = \frac{QX_a}{X_r - X_a} = \frac{1.0 \cdot 3800}{20000 - 3800} = 0.235 \text{ MGD}$$

1. Area required for clarification
    From zone settling velocity curve, pick ZSV = 7.5 ft/hr for SS concentration equal to 3800 mg/l. The overflow rate is calculated

    $$v = (7.5 \text{ ft/hr}) \left(\frac{7.48 \text{ gal}}{ft^3} \right) \left(\frac{24 \text{ hr}}{day} \right)$$

    $$= 1350 \text{ gpd/ft}^2$$

    $$A = \frac{Q + R}{v}$$

    $$= \frac{1.0 + 0.235}{1350 \times 10^{-6}} \text{ ft}^2$$

    $$= 915 \text{ ft}^2$$

2. Area required for thickening
    a) From the batch solids flux curve the design solids flux is obtained by drawing the tangent to the curve from the desired underflow SS concentration X_r = 20,000 mg/l.

FIGURE I
Zone Settling Velocity Curve

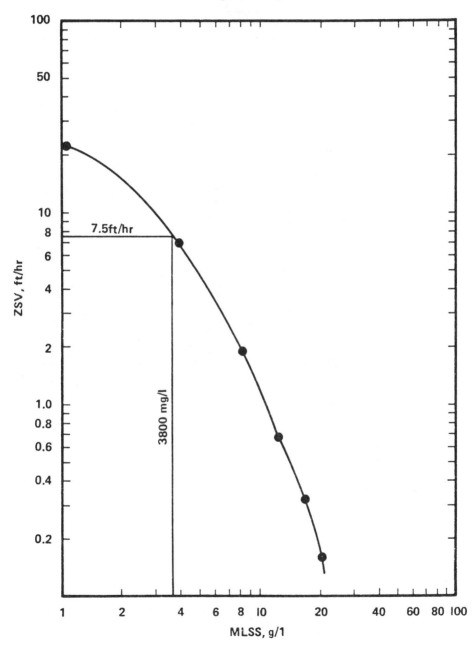

FIGURE II
Batch Solids Flux Curve

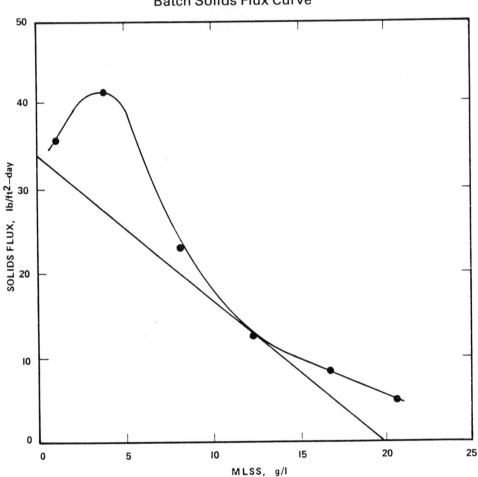

Design Solids Flux = 34 lb/ft²-day

$$A = \frac{\text{Total Solids Flux}}{\text{Design Solids Flux}}$$

$$= \frac{3800 \cdot (1.0 + 0.235) \cdot 8.34}{34} \text{ft}^2$$

$$= 1160 \text{ ft}^2$$

The required area for secondary clarifier is based on the largest of the two calculated area, thus

$$A = 1160 \text{ ft}^2$$

For operational purpose the relationship between X_a and X_r as functions of R/Q can be developed as follows:

$$R = \frac{QX_a}{X_r - X_a}$$

and solids flux

$$G = \frac{(Q + R) X_a}{A} \cdot 8.34$$

Combining the above two equations it becomes

$$X_a = \frac{GAX_r}{GA + 8.34 \, QX_r}$$

Select a value of X_r on which the G is estimated from Figure II by drawing a tangent to the batch solids flux curve and thus, the X_a is computed. Consequently the recycle flow rate can be calculated. The results are presented in Figure III.

Check F/M:

$$F/M = \frac{S_o}{X_v t} = \frac{720}{(2800)\,(0.304)} = 0.85/\text{day} > 0.7/\text{day}$$

Too high; use F/M = 0.7/day:

$$t = \frac{720}{(2800)\,(0.7)} = 0.367 \text{ days}$$

$$V = 1 \text{ MGD } (0.367 \text{ days}) = 0.367 \text{ mg}$$

At F/M = 0.7, $x = 0.64$,

b) Oxygen requirements—summer conditions:

$$b_{25°C} = 0.15 \cdot 1.04^{(25-20)}$$

$$= 0.18/\text{day}$$

$$\text{lbs } O_2/\text{day} = 0.35 \cdot 1.0 \cdot 8.34 \, (720 - 25) + 0.18 \cdot 1.4 \cdot 0.64 \cdot 2800 \cdot 0.367 \cdot 8.34$$

$$= 3411 \text{ lbs/day}$$

FIGURE III
Effect of R/Q on X_r and X_a

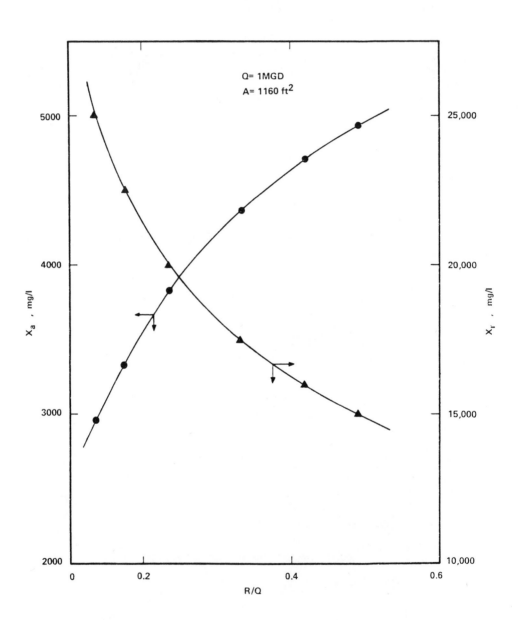

c) Sludge production—summer conditions:

$$\Delta X_v = a\,(S_o - S) - bxX_v$$

$$= 0.58 \cdot (720 - 25) \cdot 1.0 \cdot 8.34 - 0.18 \cdot 0.64 \cdot 2800 \cdot 0.367 \cdot 8.34$$

$$= 2374 \text{ lbs/day}$$

$$\text{N/day} = 0.123\,\frac{x}{0.77}\,\Delta X_v + 0.07\,\frac{0.77 - x}{0.77}\,\Delta X_v$$

$$= 0.123\,\frac{0.64}{0.77}\,(2374) + 0.07\,\frac{0.13}{0.77}\,(2374)$$

$$= 270 \text{ lbs/day}$$

$$\text{P/day} = 0.026\,\frac{x}{0.77}\Delta X_v + 0.01\,\frac{0.77 - x}{0.77}\,\Delta X_v$$

$$= 55 \text{ lbs/day}$$

Winter effluent BOD:

$$K_{10^\circ C} = 16 \cdot 1.08^{(10-20)}$$

$$= 7.41/\text{day}$$

$$S = \frac{S_o{}^2}{KX_v t + S_o}$$

$$= \frac{720^2}{7.4 \cdot 2800 \cdot 0.367 + 720}$$

$$= 62 \text{ mg/l soluble BOD}$$

Total effluent $BOD_5 = 62$ mg/l $+ 40$ mg/l SS $\cdot 0.3\ BOD_5/SS = 74$ mg/l (Effluent suspended solids during the winter may be higher resulting in higher BOD levels).
Winter sludge production:

$$b = 0.15 \cdot 1.04^{(10-20)} = 0.10/\text{day}$$

$$\Delta X_v = 0.58 \cdot 1.0 \cdot (720 - 62) \cdot 8.34 - 0.1 \cdot 0.64 \cdot 2800 \cdot 0.367 \cdot 8.34$$

$$= 2634 \text{ lbs/day}$$

Industrial Wastewaters

The majority of organic industrial wastewaters contain soluble BOD and are deficient in nitrogen and phosphorus. Depending on the biodegradability of the organics the F/M to the process must be controlled in order to avoid the generation of filamentous growths and bulking sludge. Since BOD removal from soluble wastewaters is totally dependent on the rate of the biological reaction, equation (9–18) defines the retention period necessary to achieve a specified degree of treatment.

High Purity Oxygen Systems

The high purity oxygen system is a series of well-mixed reactors employing concurrent gas-liquid contact in a covered aeration tank. Feed wastewater, recycle sludge, and oxygen gas are introduced into the first stage. The oxygen gas is fed at a pressure of only 1.4 inches (3.6 cm) of water above ambient, as shown in figure 9.71. Two gas-liquid contacting systems can be employed: submerged turbine aeration and surface aeration.

With turbine aeration, recirculating gas blowers pump the gas through a hollow shaft to a rotating sparger. The pumping action of the impeller on the same shaft as the sparger promotes adequate liquid mixing and yields relatively long residence times for the dispersed oxygen bubbles. Gas is recirculated within a stage at a rate that is usually higher than the rate of gas flow from one stage to another. A slight pressure drop occurs from stage-to-stage with no gas backmixing. Since the relative liquid mixing and oxygen transfer requirements vary from stage-to-stage, each stage is equipped with an independent mixer-compressor combination designed to provide the required level of mixing and oxygenation.

Utilizing surface aerators, the gas-liquid contact eliminates the need for gas recirculating compressors with associated piping. The required level of bulk fluid mixing to maintain the sludge in suspension and ensure a uniform liquid composition is provided in an efficient pumping, slow-speed low shear impeller. Oxygen gas is automatically fed to either system on a pressure demand basis with the entire unit operating, in effect, as a respirometer. As the organic loading increases the pressure decreases resulting in an automatic increase in feed-oxygen flow. Due to the high mixed liquor solids maintained in the oxygen system, the major portion of soluble BOD removal, and thus the highest oxygen demand, occur in the first stage requiring a highest mixer and compressor horsepower. The additional stages are then utilized to stabilize a sludge with an oxygen demand decreasing in the later stages due to increasing degrees of sludge stabilization. Effluent mixed liquor from the system is settled conventionally and the activated sludge is returned to the first stage for blending with the feed. A restricted exhaust line from the final stage vents the essentially odorless gas to the atmosphere. Normally the system will operate most economically with a vent-gas composition of about 50 percent oxygen. Due to the net transfer of gas to the liquid, the vent-gas flow rate will be a fraction (10 to 20 percent) of the gas feed rate. Based upon economic considerations about 90 percent of oxygen utilization with on-site oxygen generation is desired. Two basic oxygen generator designs are employed:

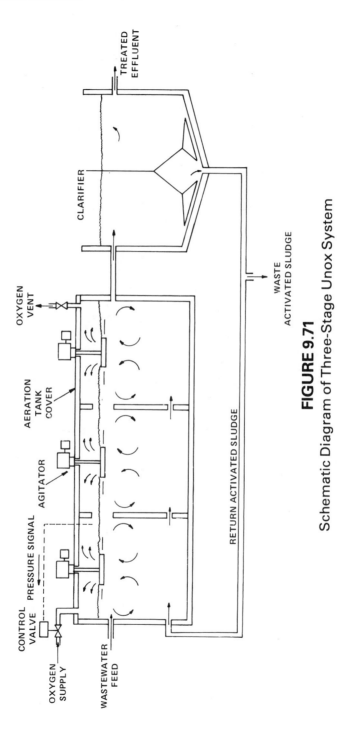

FIGURE 9.71

Schematic Diagram of Three-Stage Unox System

a traditional cryogenic air separation process for large installations (75 to 100 MGD) (3 to 4 × $10^5 m^3$/day) and a pressure swing adsorption (PSA) system for smaller installations. With the larger installations deep tank construction with submerged turbine aeration is normally preferable, while with the smaller plants a surface aerator-PSA combination is the most cost effective. The power requirements for the surface and turbine aeration equipment vary from 0.08 to 0.14 horsepower per thousand gallons (0.016–0.028 kw/m^3), depending on the waste strength, mixing requirement feed oxygen purity and the rate of capacity of the aeration equipment. At peak load conditions the oxygen systems are designed to maintain six mg/l DO in the mixed liquor. During unusually severe peak loads, additional oxygen can be transferred to the liquid if the DO level decreases to one mg/l. Liquid oxygen storage is designed for backup purposes with the same supply capacity as the installed plant. It is possible to double the feed oxygen flow to the aeration tank upon need. This results in an increased gas phase oxygen partial pressure and increased transfer, but reduced oxygen utilization. Although this is not an economic mode for operation over extended time periods, it is quite effective for short-term operation. A high purity oxygen plant is shown in figure 9.72.

The maintenance of a fully aerobic floc will maximize the endogenous rate coefficient *b* and thereby minimize the excess sludge under moderate to high loading F/M conditions. Sludge settling will also be at a maximum rate under fully aerobic conditions as shown in figure 9.73. The rationale for this is discussed earlier under Oxygen Penetration into Biological Flocs. The high purity oxygen process by virtue of maintaining high dissolved oxygen levels in the aeration basin meets these two conditions.

Extended Aeration Systems

The oxidation ditch and the Carousel systems both employ surface aeration devices and operate at extended aeration loading conditions. Nitrification and denitrification is possible under these conditions due to the establishment of anoxic conditions between the aerators. These systems are shown in figures 9.74 and 9.75.

For small installations of one or two shift operations such as a dairy, a batch activated sludge process may be employed. Operation at a milk processing plant is shown in figure 9.76. The process operates as an extended aeration plant with a mean design F/M of 0.1–0.2/day. Accumulated sludge is withdrawn on a biweekly or monthly basis for disposal. A major advantage of this process is its simplicity of operation, since each stage of the process operation is regulated by time controls. Because of the variable tank volume, the most effective aeration equipment is an educator-induced air system.

FIGURE 9.72
Oxygen Activated Sludge Plant
(Courtesy of Linde Division, Union Carbide Co.)

FIGURE 9.73
Effect of DO Level on Sludge Settling at
Various Locations

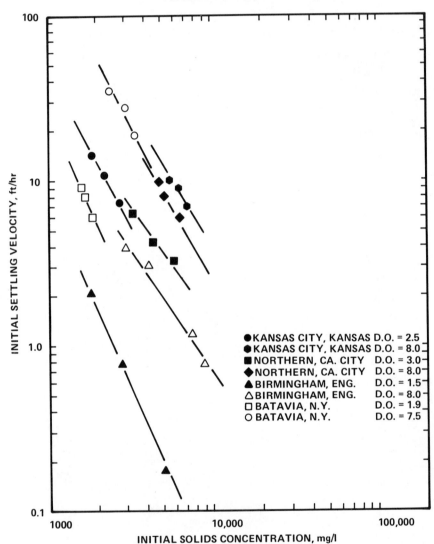

●KANSAS CITY, KANSAS D.O. = 2.5
●KANSAS CITY, KANSAS D.O. = 8.0
■NORTHERN, CA. CITY   D.O. = 3.0
◆NORTHERN, CA. CITY   D.O. = 8.0
▲BIRMINGHAM, ENG.   D.O. = 1.5
△BIRMINGHAM, ENG.   D.O. = 8.0
□BATAVIA, N.Y.   D.O. = 1.9
○BATAVIA, N.Y.   D.O. = 7.5

FIGURE 9.74
Oxidation Ditch

(Courtesy of Lakeside Equipment Co.)

FIGURE 9.75
Carousel Activated Sludge Plant
(Courtesy of Envirotech Process Equipment.)

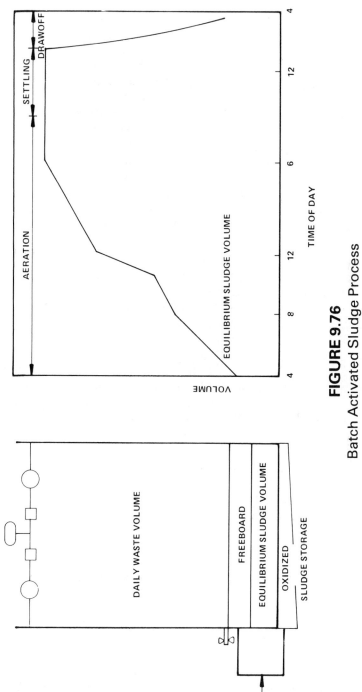

FIGURE 9.76
Batch Activated Sludge Process

Biological Nitrogen Removal

Nitrogen removal in the activated sludge process involves a series of reactions as discussed under nitrification. These reactions may proceed in separate reactors or in one reactor simultaneously. Total nitrogen removal may employ multistage sludge recycle systems with an external carbon source addition for denitrification or a single-stage sludge recycle system with internal carbon utilization for denitrification. The alternative systems are shown in figure 9.77.

The three-stage system is shown in figure 9.77a. BOD is removed in the first stage under relatively highly loaded conditions followed by nitrification in the second stage. The third stage (denitrification) can be either a mixed reactor or an upflow denitrifying filter to which an external carbon source (such as methanol) is added. A flash reactor should be added before the denitrified mixture enters the settling tank to avoid floating sludge.

The two-stage system (figure 9.77b) achieves BOD removal and nitrification in one stage under low loading conditions followed by denitrification as in figure 9.77a. In a single-stage system followed by a denitrification tank (figure 9.77c) the organic carbon and ammonia nitrogen are oxidized in an aerated basin while the denitrification occurs in a second tank. The carbon source necessary for denitrification is made available through endogenous metabolism. To avoid the occurrence of floating sludge in final clarifier a single-stage system proceeded by a denitrification tank (figure 9.77d) has been suggested with the advantage of a smaller volume for denitrification. In the single-stage system with simultaneous denitrification (figures 9.77e and 9.77f) the aeration system is controlled in such a way that oxic and anoxic zones occur along the path of flow of the mixed liquor. In oxic zones after an aerator NH_3-N is nitrified while in anoxic zones after oxygen depletion nitrate is reduced. This has been practiced by Pasveer [51] in the oxidation ditch and by Von der Emde [52] in Vienna-Blumental. The single-stage system preceded and followed by denitrification was developed by Barnard (figure 9.77g) [53]. A flash aerator should be provided before the mixed liquor enters the final clarifier to avoid floating sludge in the clarifier. This process modification also results in biological removal of phosphorus.

A single-stage system preceded by a separate and with simultaneous denitrification is shown in figure 9.77h. This system utilizes the high oxygen uptake rate for denitrification when blending incoming waste with mixed liquor. Chemical coagulation followed by nitrification and residual BOD removal and separate denitrification is a system similar to the three-stage system except the coagulant addition into the first aeration basin. The denitrification procedures would be identical to the options in figures 9.77a and 9.77b.

In cases where denitrification was not required the above design would be modified accordingly. The performance of the multistage system is shown in table 9.21. The performance of the single-stage system (Vienna-Blumental) is shown in table 9.22. System selection will depend on such factors as effluent quality re-

FIGURE 9.77
Biological Nitrogen Removal Systems

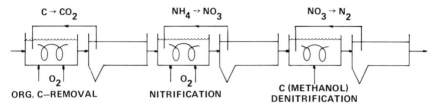

A. THREE-STAGE ACTIVATED SLUDGE PROCESS FOR DENITRIFICATION

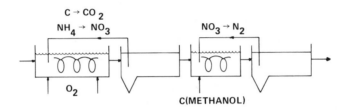

B. TWO-STAGE ACTIVATED SLUDGE PROCESS FOR DENITRIFICATION

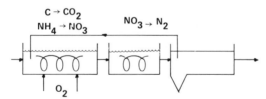

C. SINGLE-STAGE ACTIVATED SLUDGE PROCESS FOLLOWED
BY A DENITRIFICATION

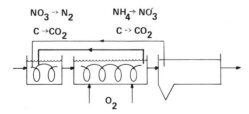

D. SINGLE-STAGE ACTIVATED SLUDGE PROCESS PRECEDED
BY A DENITRIFICATION TANK

FIGURE 9.77 (Continued)

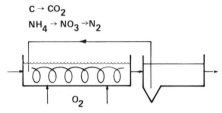

E. SINGLE-STAGE ACTIVATED SLUDGE
PROCESS WITH SIMULTANEOUS
DENITRIFICATION

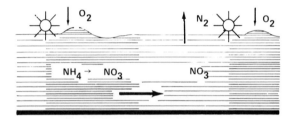

F. OXIC AND ANOXIC ZONES IN A SINGLE-
STAGE SYSTEM WITH SIMULTANEOUS
DENITRIFICATION

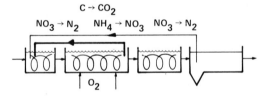

G. SINGLE-STAGE ACTIVATED SLUDGE
PROCESS PRECEDED AND FOLLOWED
BY A DENITRIFICATION TANK

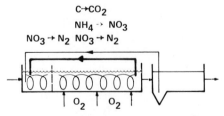

H. SINGLE-STAGE ACTIVATED SLUDGE
PROCESS PRECEDED BY A SEPARATE
AND WITH SIMULTANEOUS DENITRIFICATION

Process Stream	COD	SS	Total P	(mg/l)								Total N
				Org. N		NH₃ N		NO₂⁻ N		NO₃⁻ N		
Raw wastewater	320	157	12.6	10.3	+	11.3	+	—	—	—	=	21.6
Primary effluent	218	90	11.9	5.9	+	13.7	+	—	—	—	=	19.6
High-rate effluent	64	9	2.8	0.8	+	7.7	+	1.1	+	4.3	=	13.9
Nitrified effluent	43	7	2.6	0.4	+	0.6	+	0.3	+	11.5	=	12.8
Final (denitrified) effluent	44	7	1.5	0.4	+	0.3	+	0.3	+	0.9	=	1.9
Percentage removal	86	95	88.0									91.0

TABLE 9.21

Pilot Plant Results of Biological Nitrogen Removal[a]

[a]From [54].

TABLE 9.22

Operational Results From Treatment Plant

	Units	Sept. '71	Feb. '72	Aug. '75
Wastewater flow	m^3/d	36,190	52,860	55,550
Temperature Aeration Tank	°C	18	12	18.5
BOD$_5$-influent	mg/l	268	200	231
COD-influent	mg/l	475	384	421
NH$_4$-N-influent	mg/l	22	9	16
Total-N-influent	mg/l	36	24	24
BOD$_5$-effluent	mg/l	13	13	12
COD-effluent	mg/l	49	50	30
NH$_4$-N-effluent	mg/l	4	3	3
NO$_3$-N-effluent	mg/l	0	0	1
BOD$_5$-reduction	%	95	94	95
TKN-reduction	%	89	87	87
Total-N-reduction	%	89	87	83
Aeration time	h	8.0	5.4	5.2
BOD$_5$-loading	kg BOD$_5$/m^3/d	0.81	0.88	1.07
F/M ratio	kg/kg/d	0.12	0.14	0.17
MLSS	kg/m^3	6.7	6.2	6.4
Energy Consumption	kWh/kg BOD$_5$ removed	0.93	0.82	0.69
Number of rotors in operation		4 + 2	3 + 3	varying

[a] From Vienna-Blumental [52].

quirements, characteristics of the raw wastewater (BOD/N ratio, presence of toxic substances to nitrification, pH, etc.).

Engineering Considerations

The selection of aeration equipment should consider:

1.   Economics relative to power consumption—Oxygen transfer and mixing requirements depend on the basin geometry and wastewater characteristics. The presence of surfactants that lower the surface tension drastically affects the mixing pattern with surface aeration equipment.

2.   Temperature considerations—In cases where winter temperatures will control the effluent quality, aeration equipment that minimizes heat losses from the aeration basin should be considered. In some cases a trade-off between aeration volume requirements and aeration capital and operating cost will result.

3.   Large surface aerators in earthen basins require a concrete pad or rock bottom below the aerators to avoid erosion of the basin bottom.

There are a number of considerations in final clarifier design:

1.   A suction type sludge drawoff usually provides a maximum sludge concentration in the recycle over a wide range of recirculation ratios for large secondary clarifiers. A minimum bottom slope with this type mechanism may minimize excavation for the clarifier. Adequate removal of large solids is essential to avoid plugging of the sludge drawoff orifices as shown in figure 9.78.

2.   A minimum side water depth of 14 feet (4.3 m) is desirable to avoid solids carryover due to a density current striking the basin wall with resulting solids rising and flowing over the effluent weirs. The deeper depth permits the solids to resettle in the basin.

3.   To achieve a high clarity effluent the biological sludge should be well flocculated. A wide center well or a reflocculation section permits activated sludge reflocculation after it leaves the aeration basin. This is particularly important when high mixing intensities and shear are encountered in the aeration basin. Where the effluent contains dispersed suspended solids, provision should be made for the addition of polyelectrolytes prior to or in the final clarifier.

4.   Design sludge recirculation ratios are in the order of 25 to 35 percent. A maximum recirculation capacity of 100 percent should be provided in most cases. Note that for a given clarifier size and design condition increasing the recycle rate decreases the concentration of underflow suspended solids as shown in figure 9.79. This must be considered when determining the maximum capacity of the biological treatment plant (maximum MLSS).

Where nutrients (N and P) are to be fed in large plants, nitrogen will usually be fed either as aqua or anhydrous ammonia and phosphorus as phosphoric acid. The nutrients should be well mixed with the wastewater upstream of the aeration basin. In small plants diammonium phosphate may be batch fed to the aeration basin several times a day.

Process Optimization

Optimal performance from the activated sludge process related itself to maintaining a favorable sludge age (or F/M) and imposing constraints on the influent wastewater variability to avoid upsets to the process. It has been found that effluent quality from the activated sludge process can be related to sludge age or F/M (figure 9.80). When the sludge age is too low, filamentous and/or dispersed growth results yielding poor settling properties and a high suspended solids carryover from the final settling tank. This in turn increases the total BOD discharged from the plant. For example, in a kraft pulp and paper mill wastewater treatment

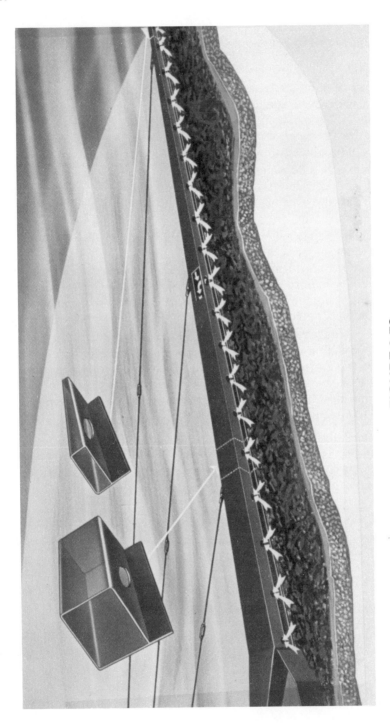

FIGURE 9.78
Vacuum Sludge Collector
(Courtesy of Envirex Inc.)

FIGURE 9.79
Effects of Recycle Ratio on Recycle SS
and MLSS Concentrations

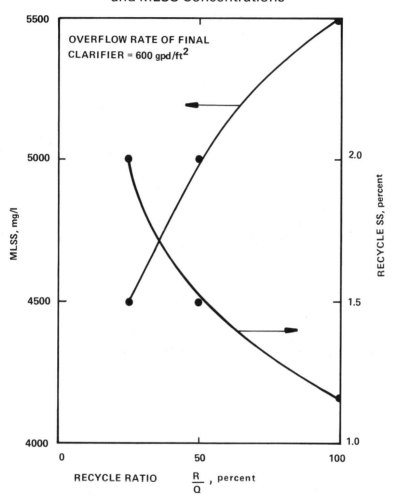

plant, the soluble and total BOD removal was 94 and 88 percent removal respectively at an F/M of 1.0 (based on active mass). The BOD contributed by the carryover suspended solids was 0.2 mg BOD/mg SS. Increasing the loading (F/M) to 2.0 reduced the soluble and total BOD removals to 90.5 and 76 percent, respectively; and increased the BOD contribution of the suspended solids to 0.5 mg BOD/mg SS. While there was only a small decrease in soluble BOD removal, the effluent deteriorated markedly due to increased carryover of suspended solids with a high active fraction.

FIGURE 9.80

Effluent Characteristics as Related to Sludge Age and F/M

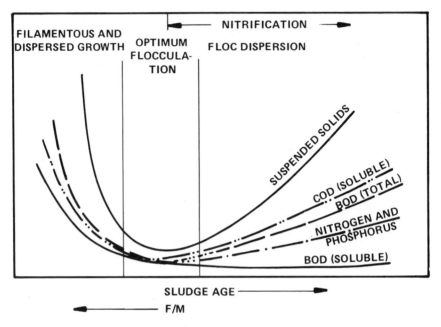

When the sludge age becomes too high (or the F/M too low) the biological floc is oxidized and dispersed.

Chudoba [55] has shown that the biological oxidation of degradable organic compounds yield refractory organics as a by-product equal to 0.5–1.2 percent of the original COD. By contrast, other investigators have reported residual COD values of 5 to 15 percent of the original COD. It would appear that the residual COD after treatment bears a relationship to the F/M employed in the process. As a result, while the effluent BOD remains relatively constant with increasing initial concentration of biodegradable substrate, the effluent COD increases due to the increased bio-resistant products of oxidation. This phenomena results in a changing BOD_5/COD ratio through treatment. Raw biodegradable wastewaters will have a BOD_5/COD ratio of 0.5–0.7. Activated sludge process effluents will decrease to 0.03–0.2. Chudoba showed that after very long periods of aeration, the ratio further reduced to 0.007–0.04.

The COD of the effluent will therefore be composed of bioresistant materials present in the wastewater, refractory metabolic by-products and residual compounds resulting from cell lysis and autooxidation. Chudoba [56] showed that the residual COD of an activated sludge effluent increased with sludge age as a result of the release of refractory organic compounds to solution. While most of the residual COD and BOD is in suspension, the soluble COD increased from about 20 mg/l at an F/M of 0.2 and a sludge age of 10 days to 40 mg/l at a sludge age of 50 days and an F/M of 0.082. The soluble BOD remained constant over this period. Filtration of the biological effluent effects a substantial reduction in COD and BOD due to removal of the finely suspended organics. Coagulation preceding filtration is necessary to remove the finely dispersed organics as expressed by the difference between the course filtered and soluble COD. At high sludge ages, the autooxidation of the cell mass releases nitrogen and phosphorus back to solution. The nitrogen and phosphorus discharged in the effluent therefore depend on the BOD/N and the BOD/P ratio in the wastewater and the sludge age in the process. (See table 9.23.)

Instrumentation and Automatic Control

A minimum amount of instrumentation is essential to monitor the performance of the activated sludge plant. In the case of large treatment plants, automatic control of some parts of the process is desirable. Figure 9.81 shows control elements presently available. The use of automatic controls should consider economics (the cost of control is the savings achieved) as well as the ability of the plant personnel to operate and maintain the equipment.

ROTATING BIOLOGICAL CONTACTORS

The rotating biological contactor consists of large diameter plastic media mounted on a horizontal shaft in a tank as shown in figures 9.82 and 9.83. The contactor is slowly rotated with approximately 40 percent of the surface area submerged. A 1–4 mm layer of slime biomass is developed on the media. (This would be equivalent to 2,500–10,000 mg/l in a mixed system.) As the contactor rotates it carries a film of wastewater through the air resulting in oxygen and nutrient transfer. Additional removal occurs as the contactor rotates through the liquid in the tank. Shearing forces cause excess biomass to be stripped from the media in a manner similar to a trickling filter. This biomass is removed in a clarifier. The attached biomass is shaggy with small filaments resulting in a high surface area for organic removal to occur as shown in figure 9.84. Present media consists of high density polyethylene with a specific surface of 37 ft^2/ft^3 (121 m^2/m^3). Single units are up to 12 ft (3.7 m) diameter and 25 ft (7.6 m) long containing up to 100,000 ft^2 (9290 m^2) of surface in one section.

The primary variables affecting treatment performance are:
1. Rotational speed
2. Wastewater retention time
3. Staging
4. Temperature

TABLE 9.23
Effluent Quality Attainable from the Activated Sludge Process

Parameter	Concentration, mg/l
Soluble BOD[a]	$< 10\ (1)$
Suspended Solids[b]	$< 20\ (2)$
Total BOD[c]	$< 30\ (3)$
COD[d]	$-\ (4)$
Nitrogen[e]	$-\ (5)$
Phosphorus[f]	$-\ (6)$
TDS	No significant change

[a]Economics depend on biodegradability of waste; low biodegradability might justify advanced wastewater treatment technology for removal of soluble BOD from 50–100 mg/l.

[b]30 mg/l settleable solids from normal clarifier operation; increase in TDS will increase nonsettleable suspended solids to levels as high as 80 mg/l in the presence of 7000 mg/l TDS. Effluent suspended solids for industrial wastewaters will also be a function of temperature. For example, in an organic chemicals activated sludge plant, summer effluent suspended solids averaged 42 mg/l and winter effluent suspended solids 107 mg/l.

[c]Based on 0.3 mg BOD/mg SS from conventional activated sludge operation. Total BOD increases with an increase in effluent SS.

[d]$$COD_{effluent} = COD_{influent} - \frac{BOD_{ult}}{0.92} + a + b$$

where $\dfrac{BOD_{ult}}{0.92}$ = the degradable COD removed in the process

a = refractory COD as a by-product of the oxidation of degradable organics (This has been reported to vary from 1 to 5 percent of the initial COD.)

b = refractory soluble organics produced from the endogenous oxidation of the cell mass

[e]Effluent nitrogen will be residual ammonia nitrogen plus organic nitrogen not hydrolyzed to ammonia in the biological process. Effluent total nitrogen can be computed:

$$N_{effluent} = N_{influent} - (0.123)\frac{x}{0.77}\Delta X_v - (0.07)\frac{0.77 - x}{0.77}\Delta X_v$$

where ΔX_v = the excess volatile suspended solids (assumed to be excess biomass and must be adjusted for other influent volatile suspended solids)

x = the biodegradable fraction of the volatile suspended solids

[f]Effluent phosphorus will be nonsoluble phosphorus plus residual soluble phosphorus:

$$P_{effluent} = P_{influent} - (0.026)\frac{x}{0.77}\Delta X_v - (0.01)\frac{0.77 - x}{0.77}\Delta X_v$$

In the treatment of domestic wastewater (BOD to 300 mg/l) performance increased with rotational speed to 60 ft/min (18 m/min) with no improvement noted at higher speeds. Increasing rotational speed increases contact, aeration, and mixing, and would therefore improve efficiency for high BOD wastewaters. However, increasing rotational speed rapidly increases power consumption, so that an economic evaluation should be made between increased power and increased area.

In the treatment of domestic wastewater performance increases with liquid volume to surface area up to 0.12 gal/ft^2 (0.0049 m^3/m^2). No improvement was noted above this value.

In treating domestic wastewater significant improvement was observed by increasing from two to four stages (figure 9.85) with no significant improvement with greater than four stages. Several factors could account for this phenomena. The reaction kinetics would favor plug flow or multistage operation. With a variety of wastewater constituents acclimated biomass for specific constituents may develop in different stages. Nitrification will be favored in the later stages where low BOD levels permit a higher growth of nitrifying organisms on the media. In the treatment of industrial wastewaters with high BOD levels or low reactivity more than four stages may be desirable. For high strength wastewaters an enlarged first stage may be employed to maintain aerobic conditions. An intermediate clarifier may be employed where high solids are generated to avoid anaerobic conditions in the contactor basins.

Wastewater temperature had no effect on contactor performance between 55 and 85° F (13 and 29° C) for domestic wastewater. Treatment efficiency decreased at temperatures below 55° F (13° C). One reason for this is the nature of domestic wastewater BOD and the equilibrium between aerobic depth and oxygen diffusion.

Domestic Wastewater Design

Antonie [57] has developed design relationships for BOD removal and nitrification of domestic wastewater. The contactor is designed to operate at 60 ft/min (18 m/min) rotational speed with 0.12 gal of basin volume/ft^2 (0.0049 m^3/m^2) of contactor surface. Antonie's performance relationship for BOD removal from municipal wastewater is shown in figure 9.86. The contact or surface area required is computed from figure 9.87a for BOD removal and figure 9.87b for nitrification. Where necessary a temperature correction is applied. It has been shown that a loading rate of 1.0 gpd/ft^2 (0.041 m^3/m^2-day) will yield an effluent BOD as low as 8 mg/l and an effluent suspended solids as low as 10 mg/l with greater than 90 percent nitrification.

FIGURE 9.81
Monitoring and Control Diagram For Wastewater
Treatment Facility

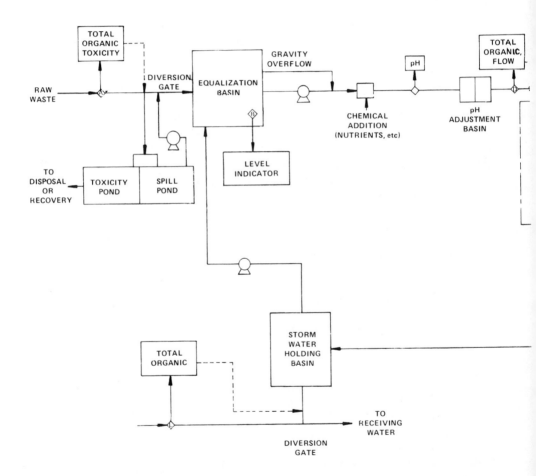

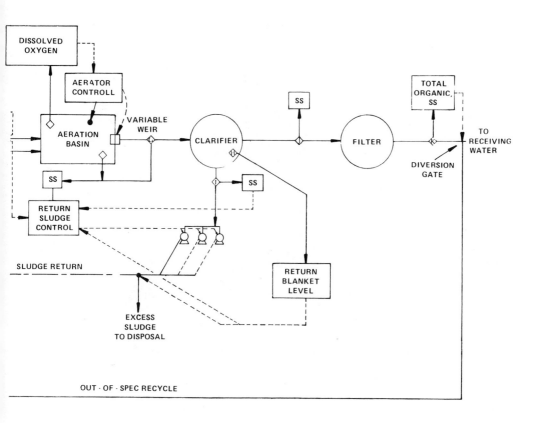

FIGURE 9.82
Rotating Biological Contactors
(Courtesy of Autotrol Co.)

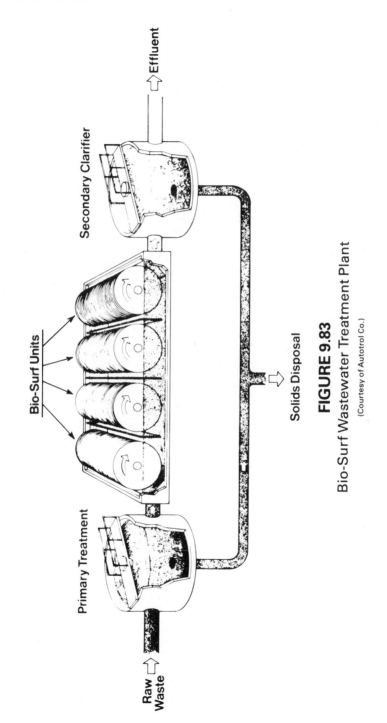

FIGURE 9.83
Bio-Surf Wastewater Treatment Plant
(Courtesy of Autotrol Co.)

FIGURE 9.84
Schematic Diagram of Wastewater Distribution and Flow in Corrugated Media

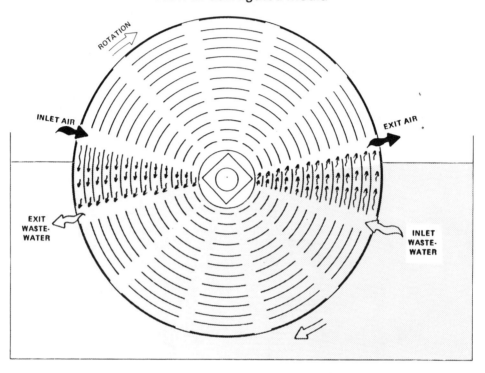

Industrial Wastewater Design

Design for industrial wastewaters will usually require a pilot plant study. A kinetic model similar to that developed for activated sludge may be employed:

$$\frac{Q}{A}\,(S_o - S) = kS \tag{9-53}$$

or for wastewaters with a highly variable influent strength:

$$\frac{Q}{A}\,(S_o - S) = K\frac{S}{S_o} \tag{9-54}$$

Equation (9–53) for several industrial wastewaters is shown in figure 9.88. For high BOD wastewaters performance can be improved by increasing the rotational speed or by using enriched oxygen surrounding the media to enhance oxygen transfer and BOD removal as shown in figure 9.89.

FIGURE 9.85
Staging Effects on Rotating Biological Contactor Performance

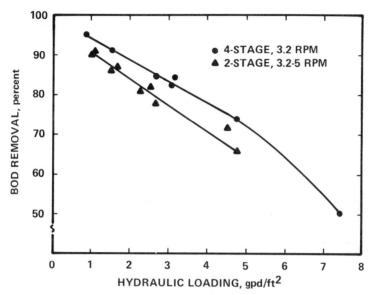

(a) COMPARISON OF BOD REMOVALS FOR TWO-STAGE AND FOUR STAGE OPERATION

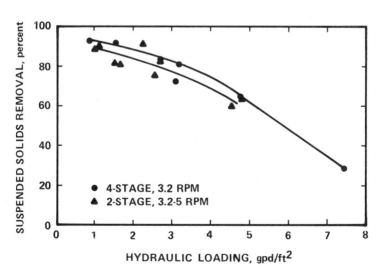

(b) COMPARISON OF SUSPENDED SOLIDS REMOVAL FOR TWO-STAGE AND FOUR-STAGE OPERATION

FIGURE 9.86
Performance Data for BOD Removal— Municipal Wastewater

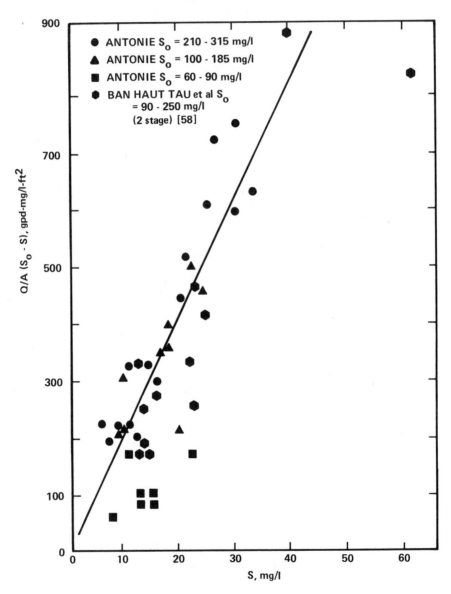

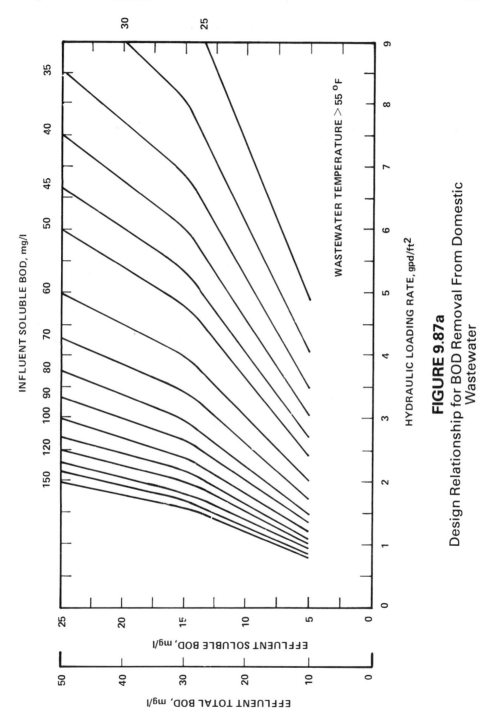

FIGURE 9.87a

Design Relationship for BOD Removal From Domestic
Wastewater

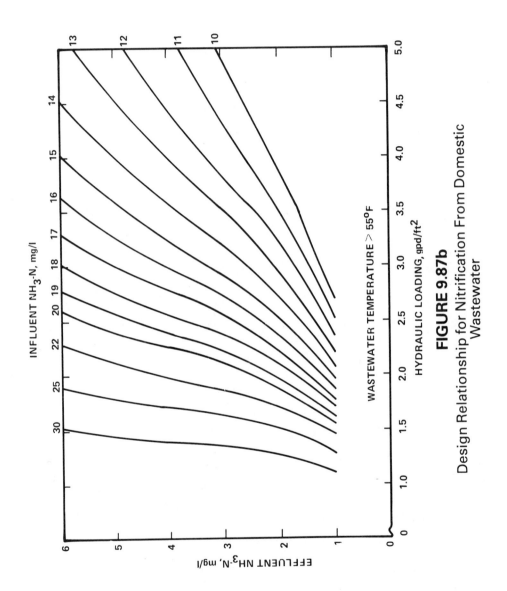

FIGURE 9.87b

Design Relationship for Nitrification From Domestic
Wastewater

FIGURE 9.88

BOD Removal Characteristics of a Rotating Biological Contactor Treating Industrial Wastewater

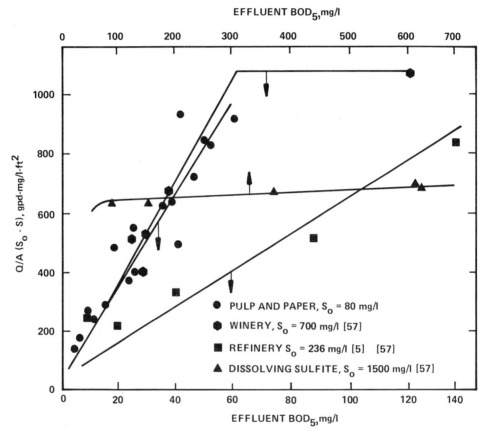

Several factors become apparent from figure 9.88. The maximum BOD removal rate Q/A (S_o-S) for a given operating condition (rotational speed, gas oxygen content, etc.) will relate to both the concentration of influent BOD and the biodegradability of the wastewater. The performance of multiple contactors in series can be defined by the relationship:

$$\frac{S}{S_o} = \left(\frac{1}{1 + kA/Q}\right)^n$$

in which n is the number of stages. Since at some loading oxygen will be limiting, it is important that the loading for each stage be checked for a multistage system. As previously noted the economics of the alternatives should be evaluated for each application.

FIGURE 9.89

Oxygen Limitation Effects on Performance of Rotating
Biological Contactor

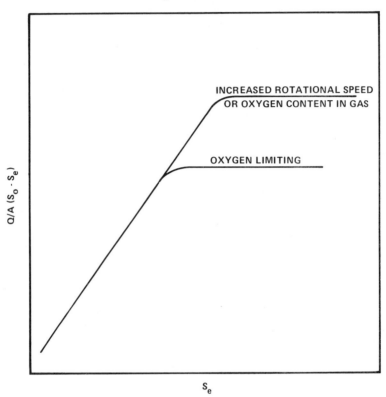

Sludge Production and
Nutrient Requirements

The excess sludge production from rotating biological contactors is similar to the activated sludge process or trickling filters and generally follows the relationship discussed in the beginning of the chapter under Cell Yield. Nutrient requirements are also computed as for activated sludge or trickling filters or on a basis of a BOD:N:P ratio of 100:5:1.

AEROBIC DIGESTION

Aerobic digestion when applied to excess biological sludges involves the oxidation of cellular organic matter through endogenous metabolism. When the process is applied to primary sludges or mixtures of activated and primary sludges both synthesis and endogenous metabolism are occurring. The oxidation of cellular organics have been found to follow first order kinetics when applied to the degradable volatile suspended solids as shown in figure 9.90. Under batch or plug flow conditions:

$$\frac{(X_d)_e}{(X_d)_o} = e^{-k_dt} \tag{9-55}$$

where $(X_d)_e$ = degradable volatile solids after time t
$(X_d)_o$ = initial degradable volatile solids
k_d = reaction rate coefficient, day^{-1}
t = time of aeration, days

If the total volatile suspended solids are considered equation (9–55) becomes:

$$\frac{X_e - X_n}{X_o - X_n} = e^{-k_dt} \tag{9-56}$$

where X_o = initial VSS
X_e = effluent VSS
X_n = nondegradable VSS

When considering a completely mixed reactor [59] the relationship is modified to:

$$\frac{X_e - X_n}{X_o - X_n} = \frac{1}{1 + k_dt} \tag{9-57}$$

and the required retention time is:

$$t = \frac{X_o - X_e}{k_d (X_e - X_n)} \tag{9-58}$$

For multiple mixed reactors in series:

$$\frac{X_e - X_n}{X_o - X_n} = \frac{1}{(1 + k_dt_n)^n} \tag{9-59}$$

In accordance with the kinetic relationships, multiple reactors in series are more efficient than one mixed reactor.

The oxygen requirements for aerobic digestion can be estimated from the relationships given earlier in the chapter under Oxygen Utilization in Aerobic Systems. 1.4 lbs of oxygen will be consumed for each lb of VSS destroyed. Nitrogen and phosphorus will usually be present in sufficient quantity released by the oxidation process. Under mesophilic digestion conditions nitrification will

FIGURE 9.90
Kinetics of Aerobic Sludge Digestion

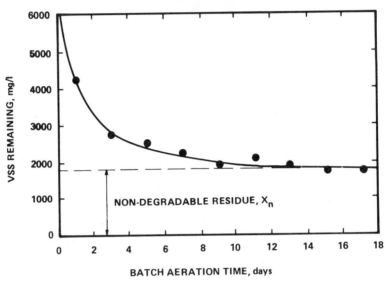

(a) CHRONOLOGICAL DESTRUCTION OF VSS IN BATCH REACTOR

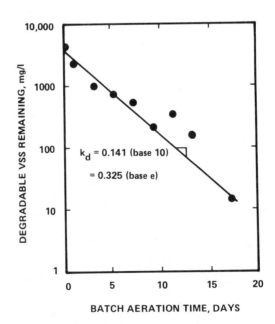

(b) CORRELATION OF DEGRADABLE VSS WITH DETENTION TIME

usually occur because of the long sludge ages in the reactor. Under thermophilic operation nitrification will be inhibited.

Temperature will affect the rate coefficient k_d. Reported data for both the mesophilic and thermophilic range by Matsch and Drnevich [60] is shown in figure 9.91. Conventional aerobic digestion design employs secondary clarifier under-flow (0.5–1.5 percent solids) in one or more completely mixed aeration basins. Power levels of 15–20 scfm/1000 ft^3 using diffused air or 100 hp/MG (0.02 kW/m^3) using surface mechanical aerators are usually adequate for providing both mixing and oxygen requirements. Prethickening the sludge offers a number of advan-tages, namely reducing the basin volume requirements and increasing the temper-ature due to the exothermic heat of reaction. Andrews [61] has estimated the heat of combustion as 9,000 BTU/lb (16.2 kcal/g) VSS destroyed. Difficulties are en-countered with conventional aeration systems to provide adequate mixing and aeration at sludge solids levels above 2 percent.

Studies have shown [60] that thermophilic digestion at high solids levels could be attained using high purity oxygen in covered basins in which the released heat of combustion is retained. Reported data is summarized in table 9.24.

ANAEROBIC TREATMENT

The anaerobic conversion of organic materials to inoffensive end-products is very complex, and is the result of many reactions as shown in figure 9.43. In the conventional high rate anaerobic system, all the reactions shown in figure 9.43 occur simultaneously in the same tank. Under equilibrium operating conditions (steady state) all the reactions must be occurring at the same rate because there is no build-up of intermediate by-products. Although many factors such as sludge organic composition and concentration, pH, temperature and mixing influence the reaction rates, it is generally assumed that the overall rate is controlled by the rate of conversion of volatile acids to methane gas and carbon dioxide (figure 9.44). Digester upset and failure occurs when there is an imbalance in the rate mechanism resulting in a build-up of intermediate volatile acids.

Anaerobic processes have been employed for the treatment of high strength organic industrial wastewaters. Two process schemes have been used: the anaerobic contact process, and the anaerobic filter.

The anaerobic contact process (figure 9.92) operates similarly to the acti-vated sludge process. The effluent from the anaerobic reactor passes through a degasifier and a settling tank from which the settled solids are recycled to the reactor. The performance of the anaerobic contact process treating several indus-trial wastewaters is shown in table 9.25.

The anaerobic filter (figure 9.93) employs a packing on which the anaerobic organisms adhere and treatment occurs as the wastewater passes up-ward through the filter. The advantage of this approach is that the major portion of

FIGURE 9.91
Effect of Temperature on the Decay Constant, K_d

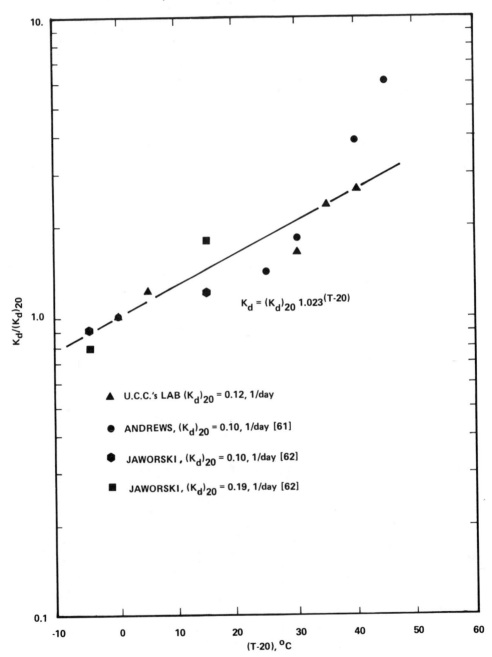

$$K_d = (K_d)_{20} \, 1.023^{(T-20)}$$

▲ U.C.C.'s LAB $(K_d)_{20}$ = 0.12, 1/day

● ANDREWS, $(K_d)_{20}$ = 0.10, 1/day [61]

⬣ JAWORSKI, $(K_d)_{20}$ = 0.10, 1/day [62]

■ JAWORSKI, $(K_d)_{20}$ = 0.19, 1/day [62]

$K_d/(K_d)_{20}$

(T-20), °C

Sludge	Food Solids Concentration (%)	Retention (days)	Digester Temp. (°C)	Volatile Solids Loading (lb VSS/ft³/day)	Percent VSS Reduction (%)
Activated	0.9	16.3	32.8	0.0629	43.9
Activated/ primary	1.3	11.6	31.6	0.106	43.0
Activated	2.4	4.2	47.3	0.36	37.4
Activated	2.5	4.2	46.4	0.38	29.7
Activated	2.5	4.2	50.4	0.37	40.1
Activated/ primary	2.9	4.0	50.2	0.45	30.0
Activated[a]	2.4	4.6	50	0.32	41.9
Activated[a]	2.7	3.7	55	0.46	29.1
Activated[a]	3.2	5.0	54	0.40	36.0

[a]Two-stage system.

TABLE 9.24

High Temperature Aerobic Digester Performance

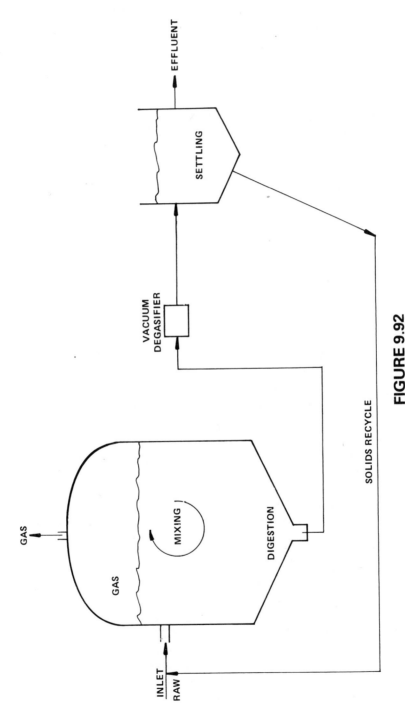

FIGURE 9.92
Anaerobic Contact Process

TABLE 9.25

Treatment Performance by the Anaerobic
Contact Process

Waste	Hydraulic Detention Time, days	Digestion Temperature, °C	Raw Waste, mg/l	Removal, %
Maize starch	3.3	23	6,280	88
Whisky distillery	6.2	33	25,000	95
Cotton kiering	1.3	30	1,600	67
Citrus	1.3	33	4,600	87
Brewery	2.3	–	3,900	96
Starch-gluten	3.8	35	14,000[a]	80[a]
Wine	2.0	33	23,400[a]	85[a]
Yeast	2.0	33	11,99[a]	65[a]
Molasses	3.8	33	32,800[a]	69[a]
Meat packing	1.3	33	2,000	95
Meat packing	0.5	33	1,380	91
Meat packing	0.5	35	1,450	95
Meat packing	0.5	29	1,310	94
Meat packing	0.5	24	1,110	91

[a] Volatile suspended solids, rather than BOD_5.

the gas is generated in the lower portion of the filter and as it rises upward rapidly, turbulence is produced which helps to keep the passageways for the fluid open and also helps to achieve some mixing in the filter, which is necessary at times to prevent low pH conditions at the entrance to the filter. The performance of an anaerobic filter treating food processing wastewater is shown in figure 9.94.

Anaerobic treatment is also employed for the stabilization of primary and secondary sludges from municipal and industrial wastewater treatment. As shown in figure 9.45 the solid organic materials undergo hydrolysis, which are in turn broken down to volatile fatty acids that are subsequently converted to methane and carbon dioxide by methane forming bacteria. As the loading rate to the digester is increased, the last step, the methane fermentation, becomes the rate limiting step. Unless the volatile acids are converted at the same rate they are being created, there is an increase in volatile acid concentration which ultimately leads to digester failure when the buffer capacity of the system is exceeded and the pH drops excessively. The two sludge digestion systems are shown in figure 9.95. Traditionally the hydraulic detention time for anaerobic sludge digestions has been in the range of 30 to 60 days under unmixed conditions. A high rate mixed

FIGURE 9.93
The Anaerobic Filter

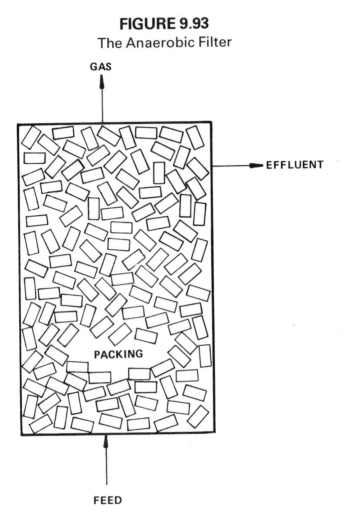

digester, stabilization occurs in a period of 10 to 20 days, as shown in figure 9.96. The digester contents are heated to about 95° F (35° C) to promote optimum microbial activity. At colder temperatures the generation time of the methane bacteria becomes exceptionally long, thus requiring the loading rate on the digester to be decreased or the detention time of the sludge in the digester to be increased. Since solids retention time (sludge age) is the determining factor in the destruction of volatile matter, improvements in digester operation can be obtained by separating and concentrating the solids from the digester so they can be recycled back to the digester. New York City's experience demonstrated that by incorporating a sludge thickener, which received all the raw primary and excess biological sludge, plus a fraction of the digested sludge from the mixed digester,

FIGURE 9.94
Treatment of Food Processing Wastewater
on the Anaerobic Filter

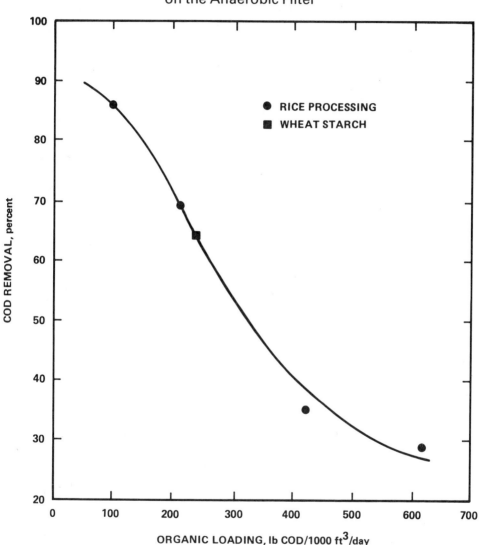

the concentration pumped to the digester is about double that existing under comparable operation, which does not utilize a thickener for solids recovery and recycle. Since the cost for disposal of digested sludge is proportional to the volume, and not to the concentration of solids, they have found that the disposal cost for final digester sludge has been reduced about 50 percent, because the solids

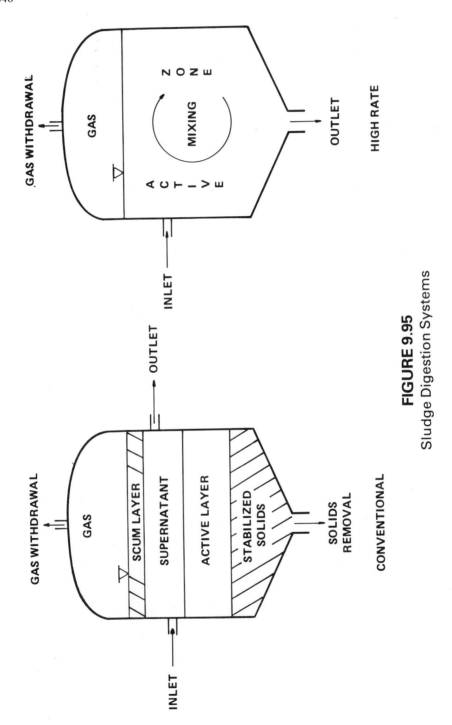

FIGURE 9.95
Sludge Digestion Systems

FIGURE 9.96
High-Rate Anaerobic Digester Performance

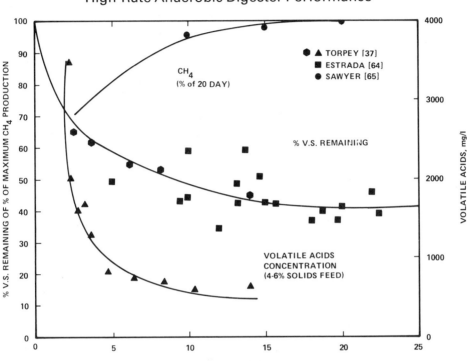

concentration of sludge to be disposed of has been doubled. Thus, they have essentially accomplished the goal of maintaining the required solids retention time and yet have been able to reduce the hydraulic retention time in the digester considerably.

The flow sheets for two alternate digester design approaches are shown in figure 9.97. The reported gas production for volatile solids reduction in a well-operating anaerobic digestion tank is 17–20 ft^3 per pound (1.06–1.25 m^3/kg) VS destroyed, with a methane content of about 65 percent.

It has been found advantageous in some cases to heat only the raw sludge going into the digester to the operating temperature, plus a temperature increment that offsets heat losses, which have been approximated at about 1° F per day of detention time (this depends on the geographical location). This type of operation allows a much smaller quantity of sludge to be processed through the heat exchanger than the method of heating, which continuously recycles digester contents through the exchanger.

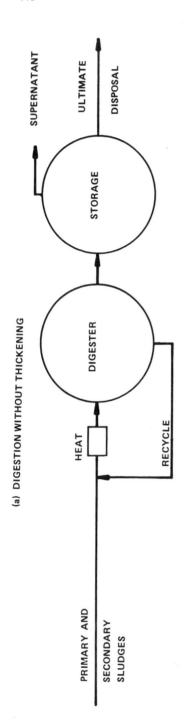

FIGURE 9.97

Anaerobic Sludge Digestion Processes

Mixing is important in the digestion tank to maintain optimal conversion to methane. In practice, a mixing intensity of about 30–50 hp per million gallons (0.0059–0.0099 kw/m³) is employed. This can be provided by mechanical agitation or by gas mixing.

Anaerobic System Design

In designing the anaerobic contact process, the required reactor volume is computed from the kinetic relationship $t = S_r/X_v kS_e$. X_v, the volatile suspended solids, is generally selected as 3000 to 5000 mg/l. It is important to check the solids retention time (sludge age) in the digester to insure the growth of the methane fermentation organisms. The sludge age should be in excess of 20 days for normal operation. The sludge yield is computed from the relationship $\Delta X_v = aS_r - bX_v$. The gas (CH_4) production is computed G (ft³/day) $= 5.62$ $(S_r - 1.42\Delta X_v)$. The heating requirements for the digester can then be estimated based on one degree Fahrenheit heat loss per day of detention time and raising the temperature of the incoming wastewater to 95° F. A sludge digester should be designed based on a solids retention time of 10 days (figure 9.96) and for an organic loading of 0.4 lbs VSS per ft³/day. The larger volume of the two should be used for design. The anaerobic filter can be designed using a relationship developed by Young and McCarty [63]:

$$t = \frac{S_o - S_e}{kS_e}$$

Young and McCarty found a value of $k = 0.45$/hr for high strength soluble organic wastewaters.

PROBLEM 9–14

Design an anaerobic contact process to achieve 90 percent removal of COD from a wastewater flow of 100,000 gpd.

Total influent COD	=	10,300 mg/l
Non-removal COD	=	2,200 mg/l
Removable COD	=	8,100 mg/l
COD to be removed	=	90%

Process parameters:
SRT	= 20 days minimum
Temp	= 35° C
a	= 0.136
b	= 0.021/day

$$k = 0.0004 \text{ l/mg-day}$$
$$X_v = 5000 \text{ mg/l}$$

1. Digester Volume

   From the kinetic relationship:

   $$t = \frac{S_r}{X_v k S_e}$$

   $$= \frac{8100 \cdot 0.9}{5000 \cdot 0.0004 \cdot 810}$$

   $$= 4.5 \text{ days}$$

   The digester volume is therefore:

   $$4.5 \text{ days} \cdot 0.1 \text{ mgd} = 0.45 \text{ mg}$$

   Check SRT (sludge age):

   $$SRT = \frac{X_v}{\Delta X_v} = \frac{X_v t}{a S_r - b X_v t}$$

   $$= \frac{5000 \cdot 4.5}{0.136 \cdot 7290 - 0.021 \cdot 5000 \cdot 4.5}$$

   $$= 43.4 \text{ days}$$

   This is in excess of a recommended SRT of 20 days to insure methane growth.

2. Sludge Yield

   The sludge yield from the process is:

   $$\Delta X_v = a S_r - b X_v$$

   $$= 0.136 \, (7290 \cdot 0.1 \cdot 8.34) - 0.021 \, (5000 \cdot 0.45 \cdot 8.34)$$

   $$= 433 \text{ lb/day}$$

3. Gas Production

   $$G = 5.62 \, (S_r - 1.42 \, \Delta X_v)$$

   $$= 5.62 \, (7290 \cdot 0.1 \cdot 8.34 - 1.42 \cdot 433)$$

   $$= 30{,}700 \text{ ft}^3/\text{day } CH_4$$

4. Heat Requirements

   Average wastewater temperature $= 75°$ F

   Heat transfer efficiency $= 50\%$

$$BTU = \frac{W(T_i - T_e)}{E}$$

$$BTU_{required} = \frac{0.1 \text{ mgd} \cdot \left(8340000 \ \frac{lb}{MG}\right)(95° + 4.5° - 75°)}{0.5}$$

$$= 40,900,000 \ BTU/day$$

The heat available from gas production is:

$$BTU_{available} = 30700 \times 960$$

$$= 29,500,000 \ BTU/day$$

The external heat of $40,900,000 - 29,500,000 = 11,400,000$ BTU/day should be supplied to maintain the reactor at 95° F.

5. The clarifier should be designed as discussed under thickener design. A vacuum degasifier or flash aerator should be provided between the digester and the clarifier to purge gas from the sludge. A flash aerator inhibits further methane production in the clarifier with resulting floating sludge.

6. Nutrient Requirements
    The nitrogen requirement is:

$$N = 0.11 \cdot 433 = 47.6 \ lbs/day$$

The phosphorus requirement is:

$$P = \frac{47.6}{7} = 6.8 \ lbs/day$$

PROBLEM 9–15

A sludge digester (at 95° F) is to be designed for 13,730 lbs/day of volatile suspended solids. The volume of sludge at 8 percent solids is 25,400 gal/day at a temperature of 40° F.

1. Digester volume
    The digester volume at a loading of 0.3 lb of volatile solids/ft³/day is:

$$V = \frac{13,730}{0.3}$$

$$= 45,800 \text{ ft}^3 \text{ or } 343,000 \text{ gal}$$

The solids retention time is:

$$\frac{343,000 \text{ gal}}{25,400 \text{ gal/day}} = 13.5 \text{ days} > 10 \text{ days, O.K.}$$

2. The volatile solids reduction (figure 9.96) is 56%. The gas production at 18 ft³/lb VSS destroyed is:

$$13,730 \cdot 0.56 \cdot 18 = 138,400 \text{ ft}^3/\text{day}$$

The methane produced is:

$$138,400 \cdot 0.65 = 90,000 \text{ ft}^3/\text{day } CH_4$$

3. Heat the raw sludge to the digester temperature plus a 1° F/day to offset digester heat losses. Assuming a 50 percent heat transfer efficiency, the energy required is:

$$BTU = \frac{(25,400 \text{ gal/day}) (9 \text{ lb/gal}) (95° + 13.5° - 40°)}{0.5}$$

$$BTU = 31,300,000 \text{ BTU/day}$$

The digester gas required to supply this energy is calculated as:

$$G = \frac{31,300,000 \text{ BTU/day}}{960 \text{ BTU/ft}^3 \cdot 0.65} = 50,200 \text{ ft}^3/\text{day}$$

This is less than the estimated gas production of 138,400 ft³/day.

REFERENCES

1. McCarty, P. L. 1966. Thermodynamics of biological synthesis and growth. In *Advances in water pollution research* vol. 2. Oxford: Pergamon Press.
2. Servizi, J. A. and Bogan, R. H. 1963. Free energy as a parameter in biological treatment. *J. San. Eng. Div., ASCE* 89, SA3, 17.
3. Eckenfelder, W. W. and O'Connor, D. J. 1961. *Biological waste treatment.* Oxford: Pergamon Press.
4. Andrews, J. F. *et al.* 1962. *Kinetics and characteristics of multi-stage methane fermentations.* SERL Rep. 64-11. Berkeley: University of California.
5. Porges, N. *et al.* 1956. Principles of biological oxidation. In *Biological treatment of sewage and industrial wastes* vol. 1, B. J. McCabe and W. W. Eckenfelder, Eds. New York: Reinhold Publ. Co.
6. Wuhrmann, K. Hauptwirkungne and Wechsel Wirk Kugen einiger Betriebsparameter im Belebschlammsystem Ergebnissemehrijahirger, Grossversuche Verlag, Zurich (1964).
7. Grau, P. *et al.* 1975. Kinetics of multi-component substrate removal by activated sludge. *Water Research* 9: 637.
8. Wuhrmann, K. 1956. Factors affecting efficiency and solids production in the activated sludge process. In *Biological Treatment of Sewage and Industrial Wastes* vol. 1, B. J. McCabe and W. W. Eckenfelder, Eds. New York: Reinhold Publ. Co.
9. Tischler, L. F. and Eckenfelder, W. W. 1969. Linear substrate removal in the activated sludge process. In *Advances in Water Pollution Research.* Oxford: Pergamon Press.
10. Gaudy, A. F. *et al.* 1963. Sequential substrate removal in heterogeneous populations. *J. Water Pollution Control Federation* 35: 903.
11. Tischler, L. F. 1969. "A mathematical study of the kinetics of biological oxidation," M.S. Thesis, University of Texas.
12. Adams, C. E. *et al.* 1975. A kinetic model for design of completely mixed activated sludge treating variable-strength industrial wastewater. *Water Research* 9: 37.
13. Pitter, P. 1976. Determination of biological degradability of organic substances. *Water Research* 10: 231.
14. McCauley, D. C. Union Carbide Research Report (October 1974).
15. Engelbrecht, R. S. and McKinney, R. E. 1957. Activated sludge cultures developed on pure organic compounds. *Sew. Ind. Wastes* 29: 1350.
16. Jasewicz, L. and Porges, N. 1956. Biochemical oxidation of dairy wastes, VI. Isolation and study of sludge microorganisms. *Sew. Ind. Wastes* 28: 1130.
17. Porges, N. 1953. Yeast, a valuable product from wastes. *J. Chemical Education.* 30: 562.
18. Hoover, S. R. and Porges, N. 1952. Assimilation of dairy wastes by activated sludge, II. The equation of synthesis and rate of oxygen utilization. *Sew. Ind. Wastes* 24: 306.

19. Rickard, M. D. and Gaudy, A. F. 1968. Effect of mixing energy on sludge yield and cell composition. *J. Water Pollution Control Federation* 40: R129.

20. Busch, A. W. 1971. *Aerobic biological treatment of waste waters: principles and practice.* Houston:1 Oligodynamics Press.

21. Pasveer, A. 1954. Research on activated sludge, III. Distribution of oxygen in activated sludge floc. *Sew. Ind. Wastes* 26: 28.

22. Mueller, J. A. *et al.* 1966. Nominal diameter of floc related to oxygen transfer. *J. San. Eng. Div., ASCE* 92, SA2: 9.

23. Zahradka, V. 1967. The role of aeration in the activated sludge process. In *Advances in Water Pollution Research,* vol. 2. Washington, D.C.: Water Pollution Control Federation.

24. Okun, D. A. 1963. Discussion. In *Advances in Biological Waste Treatment.* W. W. Eckenfelder, Jr. and B. J. McCabe, eds. Oxford: Pergamon Press.

25. Keller, P. J. and Ochs, L. D. 1973. Control of sludge bulking with hydrogen peroxide. *Proc. 46th Annual Conf. Water Pollution Control Federation.* Cleveland, Ohio.

26. Ford, D. L. 1966. "The effect of process variables on sludge floc formation and settling characteristics." Ph.D. Thesis, University of Texas, Austin.

27. von der Emde, W. 1963. Aspects of the high rate activated sludge process. In *Advances in Biological Waste Treatment.* W. W. Eckenfelder, Jr. and B. J. McCabe, Eds. Oxford: Pergamon Press.

28. Forster, C. F. 1968. The surface of activated sludge particles in relation to their settling characteristics. *Water Research* 2: 767.

29. Chudoba, J. *et al.* 1973. Control of activated-sludge filamentous bulking, II. Selection of microorganisms by means of a selector. *Water Research* 7: 1389.

30. Downing, A. L. 1968. Factors to be considered in the design of activated sludge plants. In *Advances in Water Quality Improvement,* vol. 1. E. F. Gloyna and W. W. Eckenfelder, Jr., Eds. Austin: University of Texas Press, Texas.

31. Wuhrmann, K. 1964. Nitrogen removal in sewage treatment processes. Verhandl. Intern. Verein. Limnol. XV, 580.

32. Wong-Chong, G. M. and Loehr, R. C. 1975. The kinetics of microbial nitrification. *Water Research* 9: 1099.

33. McCarty, P. L. *et al.* 1969. Biological denitrification of wastewaters by addition of organic materials *Proc. 24th Industrial Waste Conference.* West Lafayette, Indiana: Purdue Univ.

34. Hockenbury, M. R. and Grady, C. P. L. 1977. Inhibition of nitrification-effects of selected organic compounds. *J. Water Pollution Control Federation* 49: 768.

35. Anthonisen, A. C. *et al.* 1976. Inhibition of nitrification by ammonia and nitrous acid. *J. Water Pollution Control Federation* 48: 835.

36. Speece, R. E. and McCarty, P. L. 1966. Nutrient requirements and biological solids accumulation in anaerobic digestion. In *Advances in Water Pollution Research,* vol. 2. Oxford: Pergamon Press.

37. Torpey, W. N. 1955. Loading to failure of a pilot high-rate digester. *Sew. Ind. Wastes* 27: 121.

38. Lawrence, A. W. and McCarty, P. L. 1969. Kinetics of methane fermentation in anaerobic treatment. *J. Water Pollution Control Federation* 41: R1.

39. Dirasian, H. A. *et al.* 1963. Electrode potentials developed during sludge digestion. *J Water Pollution Control Federation* 35: 424.

40. Gloyna, E. F. 1968. Basis for waste stabilization pond designs. In *Advances in Water Quality Improvement,* E. F. Gloyna and W. W. Eckenfelder, Jr., Eds. Austin: University of Texas Press.

41. Oswald, W. J. 1968. Advances in anaerobic pond systems design. In *Advances in Water Quality Improvement,* E. F. Gloyna and W. W. Eckenfelder, Jr., Eds. Austin: University of Texas Press.

42. Gellman, I. and Berger, H. F. 1968. Waste stabilization pond practices in the pulp and paper industry. In *Advances in Water Quality Improvement,* E. F. Gloyna and W. W. Eckenfelder, Jr., Eds. Austin: University of Texas Press.

43. Hermann, E. R. and Gloyna, E. F. 1958. Waste Stabilization Ponds, III. Formulation of design equations. *Sew. Ind. Wastes* 30: 963.

44. Marais, G. V. R. and Shaw, V. A. 1961. *Trans. South African Inst. Civil Engineers* 3: 205.

45. Fisher, C. P. *et al.* 1968. Waste stabilization pond practices in Canada. In *Advances in Water Quality Improvement,* E. F. Gloyna and W. W. Eckenfelder, Jr., Eds. Austin: University of Texas Press.

46. Argaman, Y. and Adams, C. E. Jr. 1976. Comprehensive temperature model for aerated biological systems. *Proc. 8th Int. Conf. on Water Pollution Research,* Sydney.

47. Eckenfelder, W. W., Jr., *et al.* 1972. A rational design procedure for aerated lagoons treating municipal and industrial wastewaters. In *Advances in Water Pollution Research.* Oxford: Pergamon Press.

48. Oleszkiewicz, J. A. and Eckenfelder, W. W., Jr. 1973. *The mechanism of substrate removal in high rate plastic media trickling filters.* Technical Report no. 33. Dept. of Environmental and Water Resources Engineering, Vanderbilt University Nashville, Tenn.

49. Bruce, A. M. and Merkens, J. C. 1973. Further studies of partial treatment of sewage by high-rate biological filtration *Water Pollution Control* 72: 499.

50. Eckenfelder, W. W., Jr. and Barnhart, E. L. 1963. Performance of a high rate trickling filter using selected media *J. Water Pollution Control Federation* 35: 1535.

51. A. Pasveer, "Contribution on nitrogen removal from sewage," Muncher Beitrage zur Abwasser-, Fischerei- und Flussbiologie, Bd 12, 197 (1965).

52. W. von der Emde, Die Klaranlage Wien-Blumental-Betriebsergebrisse einer Belebungsanlage ohne Vorklarung gur weitgehenden Entfernung von Kohlenstoff-und Stickstoffverbin-dungen, Osterr. Abwasser-Rundschau, Jubilaumsausgabe, 73-82 (1975).

53. Barnard, J. L. 1973. Biological denitrification. *Water Pollution Control* 72: 705.

54. Barth, E. F. 1972. Design of treatment facilities for the control of ni-
     trogenous materials *Water Research* 6:1 481.
55. Chudoba, J. 1967. Scientific papers of the institute of chemical technol-
     ogy, Prague, Technology of Water, F 12.
56. Chudoba, J. 1971. Scientific papers of the institute of chemical technol-
     ogy, Prague, Technology of Water, F 16.
57. Antonie, R. L. 1976. *Fixed biological surfaces-wastewater treatment.*
     West Palm Beach, Florida: CRC Press.
58. Ban Haut Tau, *et al.* Epuration biologique des eaux usees urbaines au
     moyen de disques biologique tournants, techniques et science (1976).
59. Adams, C. E. *et al.* 1974. Modification to aerobic digester design. *Water
     Research* 8: 213.
60. Matsch, L. C. and Drnevich, R. F. 1977. Autothermal aerobic digestion. *J.
     Water Pollution Control Federation* 49: 296.
61. Andrews, J. F. and Kambhu, K. 1970. Thermophilic aerobic digestion of
     organic solid wastes. Final Progress Report. Clemson, S.C.: Clemson
     University.
62. Jaworski, N. *et al.* 1963. Aerobic sludge digestion. In *Advances in Biolog-
     ical Waste Treatment.* W. W. Eckenfelder, Jr. and B. J. McCabe, Eds.
     Oxford: Pergamon Press.
63. Young, J. C. and McCarty, P. L. 1967. The anaerobic filter for waste
     treatment. *Proc. 22nd Industrial Waste Conference,* West Lafayette, In-
     diana: Purdue University.
64. Estrada, A. Cost and performance of sludge digestion systems. *J. San.
     Eng. Div., ASCE* 86, SA3 111 (1960).
65. Sawyer, C. N. 1958. An evaluation of high rate digestion. In *Biological
     Treatment of Sewage and Industrial Wastes,* vol. 2. B. J. McCabe and
     W. W. Eckenfelder, Jr., Eds. New York: Reinhold Publ. Co.

10

Physical Chemical Treatment

Physical chemical treatment may be categorically defined as treatment for the removal of pollutants not removed by conventional biological treatment processes (activated sludge, trickling filters, aerated lagoons, etc.). These pollutants include suspended solids, BOD (usually less than 10 to 15 mg/l), refractory organics (usually reported as COD or TOC), nutrients (nitrogen and phosphorus), heavy metals, and inorganic salts. The treatment processes to remove these pollutants are frequently referred to as tertiary or advanced wastewater treatment processes.

The characteristics of secondary effluents vary widely, particularly in the case of industrial wastes, and are, to a great degree, a function of the characteristics of the untreated waste. The general characteristics of secondary effluents from sewage and industrial waste secondary treatment plants are summarized in table 10.1.

Prior to considering tertiary treatment needs, it is first necessary to establish water quality requirements for specific water uses. In the United States today, increasing emphasis is being placed on the removal of phosphorus and unoxidized nitrogen that exhibit a long-term oxygen demand in the receiving waters. For example, a State of Michigan requirement for a major municipality is removal of 80 percent of the total phosphate and an average and maximum daily 20-day BOD in the effluent of 8 mg/l and 15 mg/l, respectively, during the summer months. The average suspended solids must not exceed 10 mg/l.

It is probable that requirements of this type will become prevalent in many urbanized parts of the United States and in other parts of the world in the near future. More stringent requirements relating to refractory organics and total inorganic solids will become necessary in more arid areas, where extensive water reuse is necessary for industrial expansion.

Constituent, mg/l	Sewage, Stevenage, G.B.	Sewage, Amarillo, Tex.	Refinery[a]	Refinery[b]	Petrochemical[b]
Total Solids	728	557	2,900	3,000	–
Suspended solids	15	11	14	17	–
Volatile suspended solids	–	–	10	10	11
BOD	9	10	2	4	11
COD	63	–	99	112	132
Phosphate, as P	9.6	9.0	–	–	–
Nitrogen, as N	43.9	22.3	–	–	–
Chlorides	69	83	–	1,640	–
pH	7.6	7.7	6.8	6.6	7.9

[a]Activated sludge.
[b]Extended aeration.

TABLE 10.1

Characteristics of Secondary Effluents

TABLE 10.2
Tertiary Treatment Processes
for the Removal of Specific Pollutants

Unit Process	Major Elements Removed	Additional Features
Filtration—sand, diatomite, or mixed media	Suspended solids	Removal of BOD, COD, PO_4 in suspended form
Filtration plus coagulation—mixed media	Suspended solids and phosphate, color and turbidity	As above plus colloidal solids
Coagulation	Color, turbidity, and PO_4	Some COD and BOD removed
Air stripping	NH_3	High pH required
Nitrification and denitrification	Nitrogen	Ultimate BOD reduced
Carbon adsorption	COD or TOC	Reduction in color and residual suspended solids
Ion exchange	PO_4, nitrogen, total dissolved solids	Resins selected for specific purposes
Reverse osmosis	Organics and inorganics	Pretreatment required to avoid membrane fouling; treatment or disposal of residue
Electrodialysis	Inorganic salts	Pretreatment required to avoid membrane fouling; treatment or disposal of residue

In recent years, laboratory and pilot-plant research (and in some cases, large demonstration plants) has established various combinations of tertiary treatment to meet varying effluent quality requirements. The unit processes that have been investigated for specific treatment requirements are tabulated in table 10.2. An example of tertiary treatment performance at Lake Tahoe is shown in figure 10.1.

To place tertiary treatment in perspective, a more detailed discussion of the principal processes now available for the removal of specific pollutants is presented next.

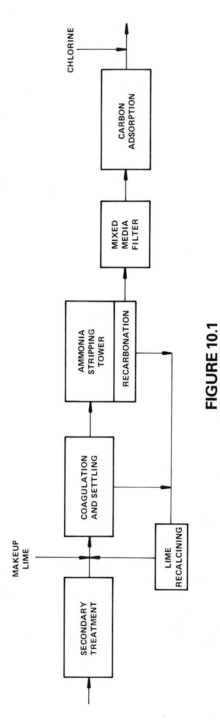

FIGURE 10.1

Tertiary Treatment Processes, Lake Tahoe [1]

COAGULATION AND PRECIPITATION

Inorganic and organic colloidal suspensions in wastewater can be removed by chemical coagulation. Coagulation has been defined as the addition of a chemical to a colloidal dispersion that results in particle destabilization by the reduction in forces that tend to keep the particles apart. Coagulation involves the reduction of surface charges and the formation of complex hydrous oxides that form flocculent suspensions. Precipitation involves formation of insoluble precipitates of the pollutants themselves. Examples of the former include colloidal dispersions of turbidity and color, and examples of the latter include precipitates of phosphorus and heavy metals.

COAGULATION

The colloids usually found in wastewater vary in size from approximately 10 nm to 10 μm and are characterized by a zeta potential of -15 to -20 mv. Colloids are stable due to repulsive forces induced by a high zeta potential or due to adsorption of a relatively small lyophilic protective colloid on a larger hydrophilic colloid or adsorption of a nonionic polymer. Most microscopic and colloidal particles are stabilized by the formation of layers of ions, which tend to collect around the particle and form a protective barrier for stabilization. These ionic layers act as a part of the particle and travel with it through solution inhibiting the close approach of respective particles to each other. Both the thickness of the ionic layers and the surface charge density are sensitive to the concentration and valence of ions in solution. Therefore, the stability of a suspension can be markedly affected or altered by adding suitable ions to the solution.

The zeta potential is a measure of the stability of a particle and indicates the potential that would be required to penetrate the layer of ions surrounding the particle for destabilization as shown in figure 10.2. The higher the zeta potential the more stable the particle. The purpose of coagulation is to reduce the zeta potential by adding specific ions and then induce motion for the destabilized particles to agglomerate [2].

Coagulation is accomplished by:

1. Lowering the zeta potential by the addition of a strong cationic electrolyte such as $Al_2(SO_4)_3$. This reduces the repulsive forces, permitting the van der Waals attractive forces to become effective, resulting in agglomeration. The dosage of cationic electrolyte is dependent of the concentration of colloid.
2. The addition of a cationic electrolyte and an alkali resulting in the formation of charged hydrous oxides, $Me_x(OH)_{y2+}$. The particles become adsorbed on the colloid, resulting in a coating or sheath.
3. Agglomeration by the addition of sufficient cationic polyelectrolyte to lower the zeta potential to zero. Attractive forces are then

FIGURE 10.2
Concept of Zeta Potential

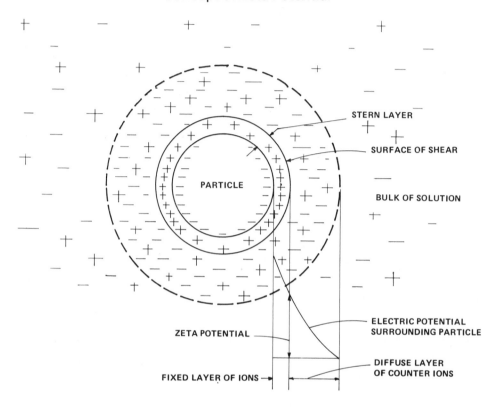

operative and mechanical bridging is achieved by the polymer.
The larger the chain length, the more effective the bridging.

4.   Mutual coagulation of anionic and cationic polyelectrolytes in a
     system.

5.   Agglomeration of a negative colloid with an anionic or nonionic
     polyelectrolyte.

6.   Entrapment in flocs of hydrous oxide. Mixing preformed floc
     with the wastewater prior to coagulation produces a more granu-
     lar concentrated floc. Most precipitated hydrous oxide flocs
     settle and concentrate very poorly.

The most common coagulants in use today are lime, alum, and ferric salts.
The use of coagulants is summarized in table 10.3.

It is believed that the addition of a cationic electrolyte and an alkali (2) is
the principal mechanism in the coagulation of colloids in wastewater treated with
aluminum or iron salts with some further removal by the entrapment in flocs of
hydrous oxide (6). To achieve optimum floc formation, the pH must be at or near

TABLE 10.3

Chemical Coagulant Applications

Chemical Process	Dosage Range mg/l	pH	Comments
Lime	150–500	9.0–11.0	For colloid coagulation and P removal.
			Wastewater with low alkalinity, and high and variable P.
			Basic Reactions:
			$Ca(OH)_2 + Ca(HCO_3) \rightarrow 2CaCO_3 + H_2O$
			$MgCO_3 + Ca(OH)_2 \rightarrow Mg(OH)_2 + CaCO_3$
Alum	75–250	4.5–7.0	For colloid coagulation and P removal.
			Wastewater with high alkalinity and low and stable P.
			Basic Reactions:
			$Al_2(SO_4)_3 + 6H_2O \rightarrow 2Al(OH)_3 + 3H_2SO_4$
$FeCl_3$, $FeCl_2$	35–150	4.0–7.0	For colloid coagulation and P removal.
$FeSO_4 \cdot 7H_2O$	70–200	4.0–7.0	Wastewater with high alkalinity and low and stable P.
			Where leaching of iron in the effluent is allowable or can be controlled
			Where economical source of waste iron is available (steel mills, etc.)
			Basic Reactions:
			$FeCl_3 + 3H_2O \rightarrow Fe(OH)_3 + 3HCl$
Cationic Polymers	2–5	no change	For colloid coagulation or to aid coagulation with a metal.
			Where the buildup of an inert chemical is to be avoided.
Anionic and some nonionic polymers	0.25–1.0	no change	Use as a flocculation aid to speed flocculation and settling and to toughen floc for filtration.
Weighting aids and clays	3–20	no change	Used for very dilute colloidal suspensions for weighting.

the isoelectric point, which for alum is in the range of 5.0 to 7.0. Sufficient alkali must be present for reaction with the aluminum (stochiometrically) and have a resulting pH at the isoelectric point. (Alkalinity in natural waters is usually in the form of HCO_3^-. When insufficient alkali is present in the wastewater, it can be added as Na_2CO_3, $NaOH$, or lime.)

The most effective destabilization results from contact of the colloidal particles with small positively charged microflocs of hydrous oxide. The hydrous oxide microflocs are generated in less than 0.1 sec., so high-intensity mixing for a short period is desirable. Following destabilization, flocculation is employed to permit the flocs to grow in size for subsequent removal from the treated wastewater. Alum and iron flocs tend to be rather fragile and easily dispersed by mixing. Activated silica at dosages of 2 to 5 mg/l is added to toughen the floc. Long-chain anionic or nonionic polymers at dosages of 0.2 to 1.0 mg/l can be added to gather and enlarge the flocs toward the end of the flocculation period. The presence of salts such as NaCl tends to lower the zeta potential and increase the coagulant dosage. The dosage is also increased when anionic surfactants are present in the wastewater that tend to stabilize the colloids.

Destabilization can also be accomplished by the addition of cationic polymers, which can bring the system to the isoelectric point without a change in pH. Although cationic polymers are 10 to 15 times as effective as alum as a coagulant, they are considerably more expensive. The coagulation process is schematically shown in figure 10.3.

Coagulation with Aluminum Compounds

Aluminum Sulfate (alum), $Al_2(SO_4)_3 \cdot 14H_2O$, reacts with alkalinity to form the insoluble hydroxide:

$$Al_2(SO_4)_3 + 6OH^- \longrightarrow 2Al(OH)_3 + 3SO_4^{--}$$

or

$$Al_2(SO_4)_3 + 6HCO_3^- \longrightarrow 2Al(OH)_3 + 3SO_4^{--} + 6CO_2$$

Actually the aluminum ions enter into a series of hydrolytic reactions with water to form a series of multivalent charged hydrous oxide species. These species may range from positive compounds at low pH to negative at high pH.

Therefore aluminum hydroxide is actually $Al_2O_3 \cdot x\ H_2O$ and is amphoteric (acts as either acid or base) depending on the pH. Under acidic conditions

$$[Al^{+++}]\ [OH^-]^3 = 1.9 \times 10^{-33}$$

At pH 4.0, 51.3 mg/l of aluminum is in solution. Under alkaline conditions, the hydrous oxide dissociates:

$$Al_2O_3 + 2OH^- \longrightarrow 2AlO_2^- + H_2O$$

$$[AlO_2^-]\ [H^+] = 4 \times 10^{-13}$$

At pH 9.0, 10.8 mg/l of aluminum is in solution. The alum floc is least soluble at pH near 7.0 and the floc charge positive below pH 7.6 and negative above pH 8.2.

If insufficient alkalinity is present in the water for the reaction to occur, alkalinity must be added (usually as lime, soda ash, etc.). The quantity required for the reaction is:

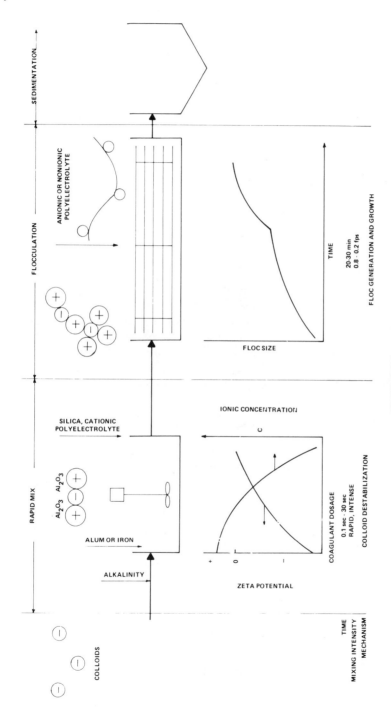

FIGURE 10.3
Mechanism of Coagulation Process

1 mg/l alum reacts with:

> 0.5 mg/l natural alkalinity, expressed as $CaCO_3$
> 0.33 mg/l 85% quicklime as CaO
> 0.39 mg/l 95% hydrated lime as $Ca(OH)_2$
> 0.53 mg/l soda ash as Na_2CO_3

High alum dosages may result in post precipitation depending on the pH of flocculation. This is particularly prevalent when a treated wastewater at low or high pH is discharged to a stream at neutral pH.

Coagulation with Iron Compounds

Ferric salts, $FeCl_3$, $Fe_2(SO_4)_3$, react to produce the insoluble ferric hydroxide:

$$Fe^{+3} + 3OH^- \longrightarrow Fe(OH)_3$$

$$[Fe^{+3}]\ [OH^-]^3 = 10^{-36}$$

The insoluble oxide is produced over a pH range of 3.0–13.0. The floc charge is positive in the acid range, negative in the alkaline range and mixed over a pH range of 6.5–8.0. Ferrous sulfate ($FeSO_4 \cdot 7H_2O$) is also used as a coagulant.

For any wastewater treated with a coagulant, such as Al^{+++} or Fe^{+++}, there is an optimum pH as shown in figure 10.4. This pH is usually at or near the pH value for minimum solubility. Exceptions have been found in the treatment of some textile and paint wastewaters. There will also be an optimum coagulant dosage corresponding to a zeta potential at or near zero as shown in figure 10.5.

Coagulation with Lime

Lime ($Ca(OH)_2$) functions differently as a coagulant than aluminum or iron salts. Lime reacts with bicarbonate alkalinity to precipitate calcium carbonate and with orthophosphate to precipitate calcium hydroxyapatite. Magnesium hydroxide will precipitate at high pH levels. It should be noted, however, that the magnesium hydroxide formed is gelatinous in nature and difficult to dewater and will affect the sludge handling procedure. The lime requirements to raise the pH to 11.0 are shown in figure 10.6.

$$Ca^{++} + HCO_3^- + OH^- \longrightarrow CaCO_3 + H_2O$$

$$5Ca^{++} + 4OH^- + 3HPO_4^{--} \longrightarrow Ca_5(OH)(PO_4)_3 + 3H_2O$$

$$Mg^{++} + 2OH^- \longrightarrow Mg(OH)_2$$

The lime sludge can be thickened, dewatered and calcined to convert calcium carbonate to lime for reuse. In most cases this is only economical for municipal plants in excess of 10 MGD ($3.8 \times 10^4 m^3$/day) capacity. Coagulation of raw sewage involves the problem of separating inerts from the recovered lime before reuse.

FIGURE 10.4
Optimum pH Determination

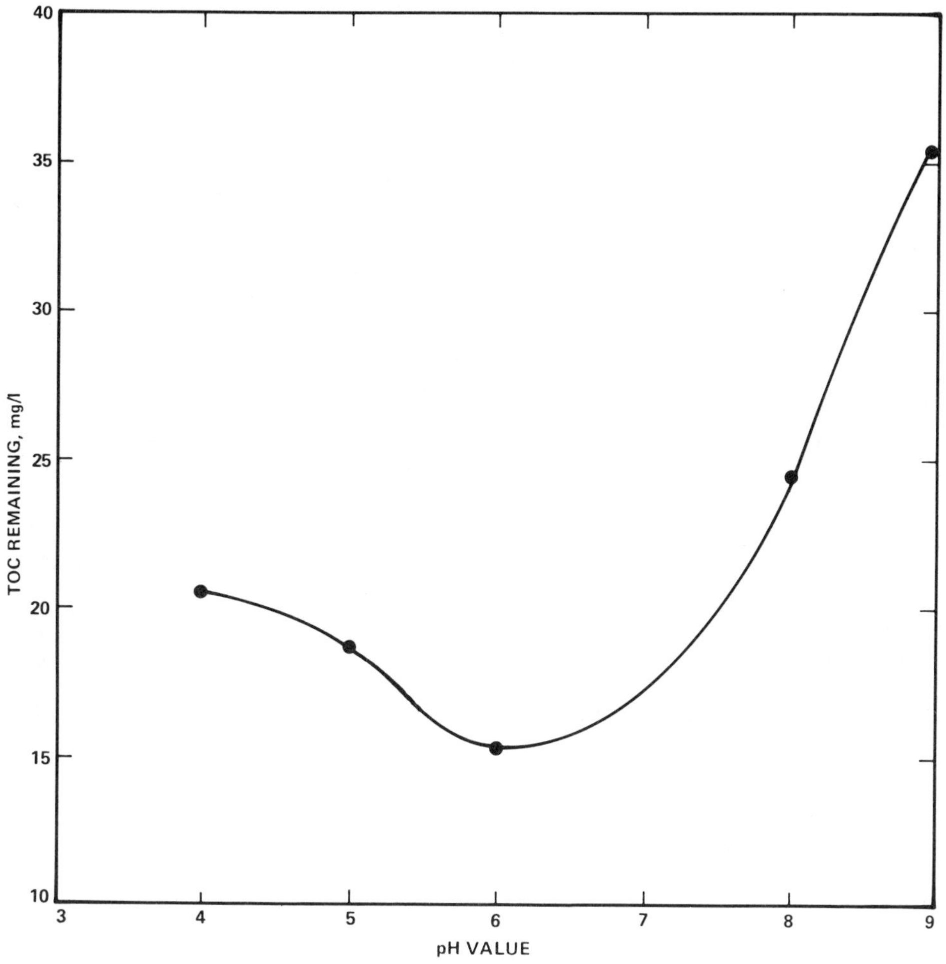

Role of Polyelectrolytes in Coagulation

As previously noted the coagulation process can be enhanced by the addition of coagulant aids which strengthen the floc, agglomerate the floc to larger size, and gather dispersed particles for improved effluent clarity. Activated silica is a short-chain polymer that promotes the growth of large, rapid-settling flocs and binds together the particles of microfine aluminum hydrate. At high dosages inhibition of floc formation results due to the electronegative properties of the silica. For most applications the usual dosage is 5–10 mg/l.

FIGURE 10.5
Coagulation of Raw Sewage with Alum

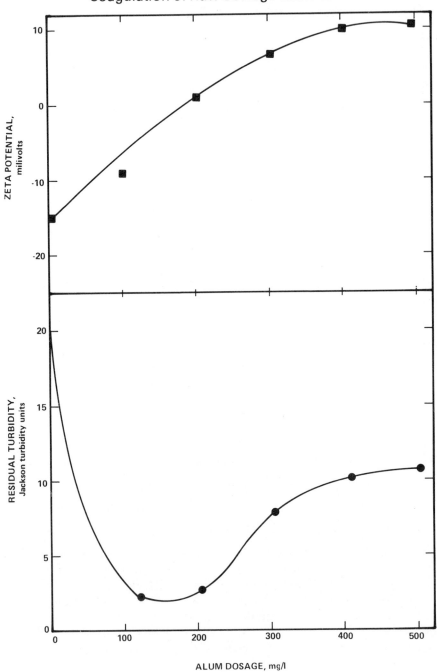

ALUM DOSAGE, mg/l

FIGURE 10.6

Lime Required to Raise the pH to 11 as a Function of the
Raw Wastewater Alkalinity [27]

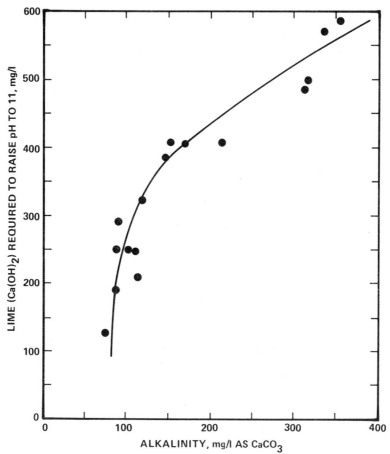

Polyelectrolytes may be synthetic or natural and are high-molecular-weight polymers that contain adsorbable groups, which form bridges between particles or charged flocs. Cationic polymers adsorb on a negatively charged colloid or floc particle. Anionic polymers replace anionic groups on a colloidal particle and permits hydrogen bonding between the colloid and the polymer. Anionic polymers are a function of pH, the alkalinity of the water, the hardness of the water and turbidity. Nonionic polymers adsorb and flocculate by hydrogen bonding between solid surfaces and polar groups in the polymer.

The optimum polymer dosage increases linearly with the coagulant, such as alum dosage. The usual polymer dosage is 1–5 mg/l in conjunction with alum or ferric salts. The polyelectrolytes can serve as a coagulant themselves although in most cases the dosages necessary are uneconomical.

Processes and Equipment

Conventional coagulation process employs several steps:

1.   A rapid high intensity mixing of the coagulant with the wastewater. Alkalinity, if required, can be added to the pump or piping manifold prior to the addition of the coagulant. Rapid, high intensity mixing is essential since the most effective destabilization occurs with the initial formation of the hydrous oxide microfloc in a matter of microseconds. As the floc increases in size, its effectiveness for colloidal destabilization decreases.

2.   Activated silica or cationic polymer, if used, should be added at the end of the rapid mix.

3.   Flocculation for a period of 20–30 minutes to develop large flocs. The objective is to obtain contact and promote agglomeration but not to shear the flocs. Riddick [3] has suggested the use of four flocculation bays in series with peripheral speeds tapering from 0.8 to 0.2 fps (24.4 to 6.10 cm/sec). The velocity gradient in the flocculation basin is defined by

$$G = \sqrt{W/\mu} \tag{10-1}$$

where G = Mean velocity gradient, fps/ft or sec^{-1}
        W = Horsepower dissipated per unit volume
        μ = Absolute viscosity of the water

G should taper from 50 to 15 sec^{-1} through the flocculation basin.

4.   An anionic or nonionic polymer is added toward the end of the flocculation basin for floc gathering and agglomeration. This sequence is shown in figure 10.7.

With lime and a few other coagulants the time required to form a settleable floc is a function of the time necessary for calcium carbonate or other calcium precipitates to form a nuclei on which other calcium materials can deposit and grow large enough to settle. It is possible to reduce both the coagulant dosage and the time of floc formation by seeding the influent wastewater with previously formed nuclei or by recycling a portion of the precipitated sludge. The effect of solids recycle is shown in figure 10.8.

Sludge blanket units or solids-contact units can replace the conventional mixing, flocculation, and settling basins. The sludge blanket units combine mixing, flocculation, and settling in a single unit. The advantages of recirculating preformed floc are that chemical dosages are frequently reduced, the blanket serves as a filter for effluent clarity, and denser sludge are frequently attainable. A sludge blanket unit is shown in figure 10.9.

Application Results

Coagulation has been applied to the treatment of domestic wastewater as a physical chemical treatment (PCT) for removal of colloidal and suspended or-

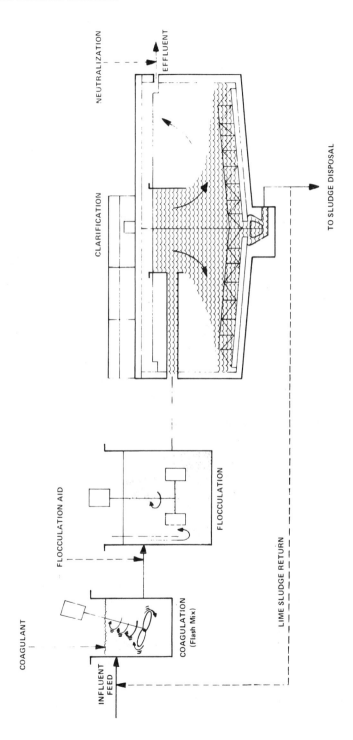

FIGURE 10.7
Three-Stage Chemical Treatment System

FIGURE 10.8
Determination of Optimum Coagulant Dosage with Solids Recycle

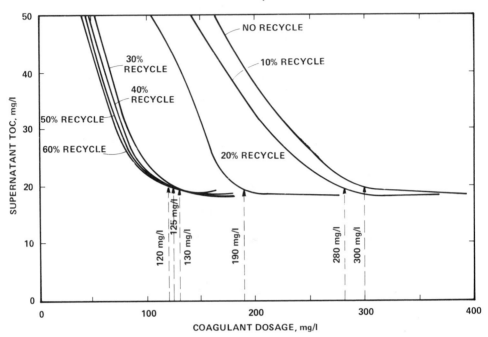

ganics prior to removal of soluble organics in granular activated carbon. A typical flowsheet for this process is shown in figure 10.10. Results from the coagulation of raw domestic sewage is shown in table 10.4. Note that the presence of soluble industrial wastewaters in the sewage increases the effluent BOD after coagulation, since these materials are not removed by coagulation. Coagulation has been employed for textile mill wastewaters for the removal of color. Some results are summarized in table 10.5 and figure 10.11. Note that the selection of a coagulant should consider sludge handling and disposal, since each coagulant produces a different quantity of sludge with different thickening and dewatering properties. This is discussed in further detail under sludge handling. The removal of color from pulp and paper mill effluents is summarized in table 10.6. The clarification of board and tissue mill effluents is shown in table 10.7.

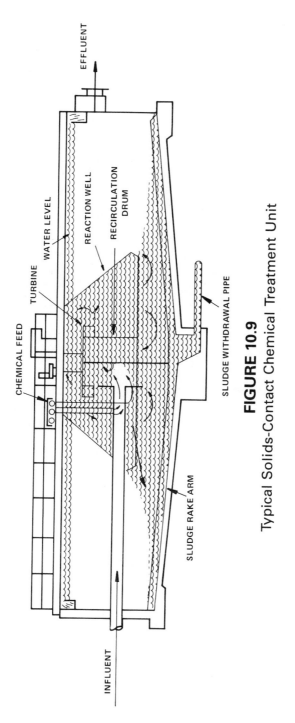

FIGURE 10.9

Typical Solids-Contact Chemical Treatment Unit

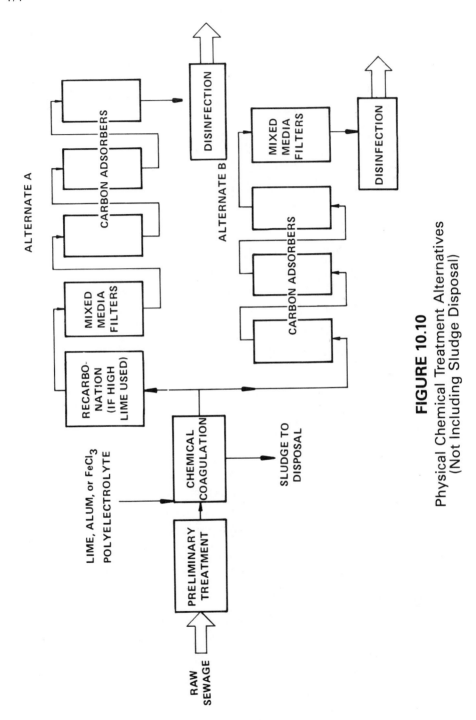

FIGURE 10.10

Physical Chemical Treatment Alternatives
(Not Including Sludge Disposal)

Constituent	Ca(OH)$_2$			Al$_2$(SO$_4$)$_3$ · 14H$_2$O			FeCl$_3$ · 6H$_2$O		
	Feed mg/l	Effluent mg/l	Removed %	Feed mg/l	Effluent mg/l	Removed %	Feed mg/l	Effluent mg/l	Removed %
SS	138	22	84	143	40	72	113	10	91
Turbidity, JTU	(410)	(24)	—	(430)	(67)	—	(425)	(22)	—
BOD	135	24	82	172	67	62	113	37	67
COD	327	98	70	345	136	61	285	99	65
Total-P	14.8	1.1	93	13.6	2.3	83	12.8	4.3	66
Soluble-P	11.9	0.6	95	8.3	0.6	93	10.1	1.6	84
NH$_3$-N	27.4	28.9	0	33.7	33.4	0	—	—	—
Organic-N	14.2	3.2	77	14.1	3.8	73	—	—	—
pH	(7.4)	(11.2)		(7.2)	(6.7)		(7.5)	(6.6)	
Chemical Dose mg/l	360			250			200		

TABLE 10.4

Chemical Treatment of Raw Sewage—Pilot Plant Studies

TABLE 10.5
Coagulation of Textile Wastewaters[a]

Plant	Coagulant	Dosage mg/l	pH	Color[b] Influent	Color[b] % Removal	COD Influent mg/l	COD % Removal
1	$Fe_2(SO_4)_3$	250	7.5–11.0	0.25	90	584	33
	Alum	300	5–9		86		39
	Lime	1200			68		30
2	$Fe_2(SO_4)_3$	500	3–4/9–11	0.74	89	840	49
	Alum	500	8.5–10		89		40
	Lime	2000			65		40
3	$Fe_2(SO_4)_3$	250	9.5–11	1.84	95	825	38
	Alum	250	6–9		95		31
	Lime	600			78		50
4	$Fe_2(SO_4)_3$	1000	9–11	4.60	87	1570	31
	Alum	750	5–6		89		44
	Lime	2500			87		44

[a]From [4].
[b]Color sum of absorbances at wavelengths of 450, 550, and 650 nm.

TABLE 10.6
Color Removal from Pulp and Paper Mill Effluents[a]

Plant	Coagulant	Dosage mg/l	pH	COLOR Influent	COLOR % Removal	COD Influent	COD % Removal
1	$Fe_2(SO_4)_3$	500	3.5–4.5	2250	92	776	60
	Alum	400	4.0–5.0		92		53
	Lime	1500	–		92		38
2	$Fe_2(SO_4)_3$	275	3.5–4.5	1470	91	480	53
	Alum	250	4.0–5.5		93		48
	Lime	1000	–		85		45
3	$Fe_2(SO_4)_3$	250	4.5–5.5	940	85	468	53
	Alum	250	5.0–6.5		91		44
	Lime	1000	–		85		40

[a]From [5].

Precipitation

Precipitation is employed for the removal of phosphorus and heavy metals from wastewaters. Phosphorus removal was discussed in an earlier chapter. Heavy metals are generally precipitated as the hydroxide through the addition of lime or caustic to a pH of minimum solubility. However several of these compounds are amphoteric and exhibit a point of minimum solubility. The pH of minimum solubility varies with the metal in question as shown in figure 10.12.

FIGURE 10.11
Removal of Color From Textile Dye Wastes
Using Alum [4]

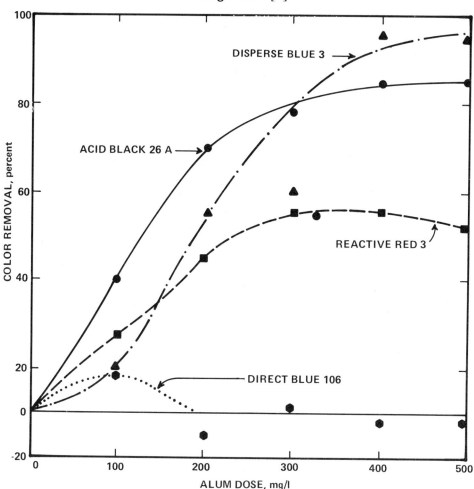

The solubilities for chromium and zinc are minimum at pH 7.5 and 10.2, respectively and show a significant increase in concentration above these pH values. When treating industrial wastewaters containing metals, it is frequently necessary to pretreat the wastewaters to remove substances which will interfere with the precipitation of the metals. Cyanide and ammonia form complexes with many metals that limit the removal that can be achieved by precipitation. Cyanide can be removed by alkaline chlorination or other processes such as catalytic oxidation on carbon. Cyanide wastewaters containing nickel or silver are difficult to treat by alkaline chlorination because of the slow reaction rate of these metal

TABLE 10.7

Chemical Treatment of Paper and Paperboard Wastes

| | Influent | | Effluent | | | Coagulants | | | Sludge | |
Waste	BOD mg/l	SS mg/l	BOD mg/l	SS mg/l	pH	Alum mg/l	Silica mg/l	Other mg/l	% Solids	Remarks
Board	–	350–450	–	15–60	–	3	5	–	2–4	–
Board	–	140–420	–	10–40	–	1	–	10[a]	2	Flotation
Board	–	240–600	–	35–85	–	–	–	–	2–5	–
Board[b]	127	593	68	44	6.7	10–12	10	–	1.76	–
Tissue	140	720	36	10–15	–	2	4	–	–	–
Tissue	208	–	22	–	6.6	4	–	–	–	–

[a] Glue.
[b] 15,000 gal/ton waste paper.

complexes. Ferrocyanide $[Fe(CN)_6^{-4}]$ is oxidized to ferricyanide $[Fe(CN)_6^{-3}]$ which results further oxidation. Ammonia can be removed by stripping, break point chlorination or other suitable methods prior to the removal of metals. The solubility of metals in a wastewater with and without ammonia removal is shown in figures 10.13(a) and 10.13(b).

The removal of heavy metals from domestic sewage by lime precipitation as reported by Argaman and Weddle [6] is shown in table 10.8. Heavy metals may also be precipitated as the sulfide and in some cases as the carbonate as in the case of lead. In order to meet low effluent requirements, it may be necessary in some cases to provide filtration to remove floc carried over from the precipitation process. Using precipitation and clarification alone effluent metals concentrations may be as high as 1 to 2 mg/l. Filtration should reduce these concentrations to 0.5 mg/l or less. For chromium wastes treatment hexavalent chromium must first be reduced to the trivalent state (Cr^{+3}) and then precipitated with lime. This is referred to as the process of reduction and precipitation.

The reducing agents commonly used for chromium wastes are ferrous sulfate, sodium metabisulfite, or sulfur dioxide. Ferrous sulfate and sodium metabisulfite may be dry- or solution-fed; SO_2 is diffused into the system directly from gas cylinders. Since the reduction of chromium is most effective at acidic pH values, a reducing agent with acidic properties is desirable. When ferrous sulfate is used as the reducing agent, the Fe^{++} is oxidized to Fe^{+3}; if metabisulfite or sulfur dioxide is used, the negative radical $SO_3^=$ is converted to $SO_4^=$.

The general reactions are:

$$Cr^{+6} + \left(\frac{Fe^{++}, SO_2,}{\text{or } Na_2S_2O_5} \right) + H^+ \longrightarrow Cr^{+3} + \left(\frac{Fe^{+3},}{SO_4^=} \right)$$

$$Cr^{+3} + OH^- \longrightarrow Cr(OH)_3$$

Ferrous ion reacts with hexavalent chromium in an oxidation-reduction reaction, reducing the chromium to a trivalent state and oxidizing the ferrous ion to the ferric state. This reaction occurs rapidly at pH levels below 3.0. The acidic properties of ferrous sulfate are low at high dilution; acid must therefore be added for pH adjustment. The use of ferrous sulfate as a reducing agent has the disadvantage that a contaminating sludge of $Fe(OH)_3$ is formed when an alkali is added. In order to obtain a complete reaction, an excess dosage of 2 and ½ times the theoretical addition of ferrous sulfate must be made.

Reduction of chromium can also be accomplished by using either sodium metabisulfite or SO_2. In either case reduction occurs by reaction with the H_2SO_3 produced in the reaction. The H_2SO_3 ionizes according to the mass action:

$$\frac{[H^+][HSO_3^-]}{[H_2SO_3]} = 1.72 \times 10^{-2}$$

Above pH 4.0, only 1 percent of sulfite is present as H_2SO_3, and the reaction is very slow.

FIGURE 10.12
Solubility of Metals Versus pH

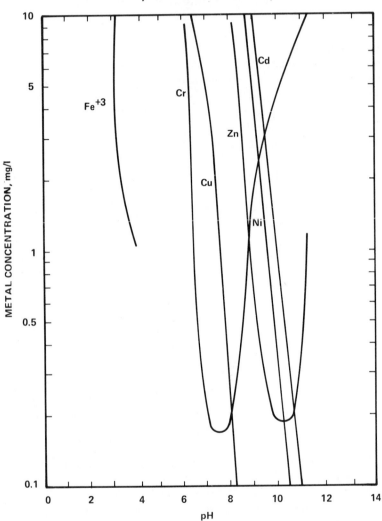

When metabisulfite is used, the salt hydrolyzes to sodium bisulfite:

$$Na_2S_2O_5 + H_2O \rightleftharpoons 2NaHSO_3$$

The sodium bisulfate in turn dissociates:

$$Na^+ + HSO_3^- + H_2O \rightleftharpoons H_2SO_3 + NaOH$$

FIGURE 10.13(a)
Optimum pH Values for Metals Removals in the Presence of Ammonia

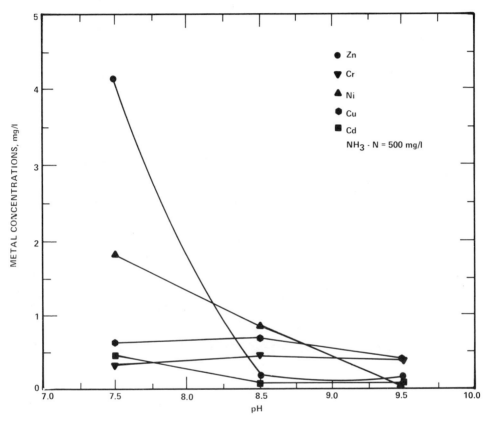

During this reaction, acid is required to neutralize the NaOH formed. The reaction is highly dependent on both pH and temperature. At pH levels below 2.0 the reaction is practically instantaneous at close to the theoretical requirements.

Many small plating plants have a total daily volume of waste of less than 30,000 gpd (114 m³/day). The most economical system for such plants is a batch treatment in which two tanks are provided, each with a capacity of one day's flow. One tank is undergoing treatment while the other is filling. Accumulated sludge is either drawn off and hauled to disposal or dewatered on sand drying beds. A spadable dry cake can be obtained after 48 hours on the sand bed. A typical batch treatment system is schematically shown in figure 10.14.

When the daily volume of waste exceeds 30,000 to 40,000 gal (114–151 m³), batch treatment is usually not feasible because of the large tankage required.

FIGURE 10.13(b)
Optimum pH Values for Metal Removal

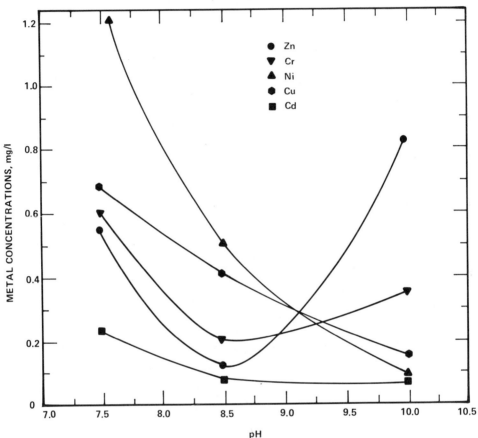

Continuous treatment requires a tank for acidification and reduction, then a mixing tank, for lime addition, and a settling tank. The retention time in the reduction tank is dependent on the pH employed but should be at least four times the theoretical time for complete reduction. Twenty minutes will usually be adequate for flocculation. Final settling should not be designed for an overflow rate in excess of 500 gal/ft²-day (20 m³/m²-day).

In cases where the chrome content of the rinse water varies markedly, equalization should be provided before the reduction tank to minimize fluctuations in the chemical feed system. The fluctuation in chrome content can be minimized by provision of a drain station before the rinse tanks.

Successful operation of a continuous chrome reduction process requires instrumentation and automatic control. pH and redox control are provided for the

TABLE 10.8

Removal of Heavy Metals by Lime,
Coagulation, Settling, and Recarbonation[a]

| Metal | Concentration Range (mg/l) | | Removed (%) |
	Influent	Effluent	
Ag	0.24–1.51	0.01–0.02	96–99
As	7.00–8.40	0.20–0.30	96–97
Ba	0.36–1.08	0.04–0.14	87–89
Cd	0.54–5.78	0.01–0.19	95–99
Co	0.42–1.29	0.04–0.09	90–96
Cr^{+6}	0.45–1.40	0.30–1.25	11–33
Cu	0.60–1.47	0.04–0.23	84–93
Hg	3.26–4.45	0.29–0.61	86–91
Mn	1.37–2.26	0.01–0.02	99
Ni	0.75–1.36	0.11–0.20	85
Pb	0.41–1.21	0.04–0.05	90–96
Zn	7.34–9.61	0.12–0.18	97–99

[a]From [6].

FIGURE 10.14

Batch Treatment of Chromium Wastes

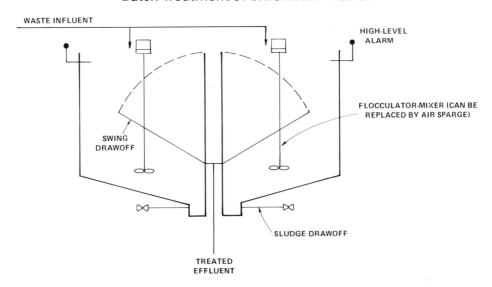

reduction tank. The addition of lime should be modulated by a second pH control system. A continuous treatment system is shown in figure 10.15.

In some cases regulations require use of a lagoon or holding pond following the treatment so that any wastes receiving incomplete chrome removal can be recycled for additional treatment.

An ingenious system has been developed by Lancy [7] in which the reduction and precipitation process is integrated into the plating line. In this process the first tank at the end of the plating line continuously receives SO_2, so that an excess is maintained at all times. The contents of the unit are continuously recirculated through a storage reservoir. The second tank recirculates through a lime-mixing tank in which the reduced chrome is precipitated. The precipitate is removed through a diatomic filter, and the clarified liquid recirculated through the second treatment tank. A third tank containing a clear water rinse is provided to remove any residues from the plated parts. This process is shown schematically in figure 10.16.

PROBLEM 10-1

Consider that 30,000 gpd of a waste containing 49 mg/l Cr^{+6}, 11 mg/l Cu, and 12 mg/l Zn are to be treated daily by using SO_2. Compute the theoretical requirements of chemicals and the daily sludge production. (Assume the waste contains 5 mg/l O_2.)

1. SO_2 requirements are as follows: for Cr^{+6},

$$2Cr^{+6} + 3SO_2 + 6H_2O \longrightarrow 2Cr^{+3} + 3SO_4^= + 12H^+$$

where one part of Cr^{+6} requires 1.85 parts of SO_2,

$$1.85 \times 49 \times 8.34 \times 0.03 = 22.7 \text{ lb/day}$$

and for O_2,

$$O_2 + 2SO_2 + 2H_2O \longrightarrow 2SO_4^= + 4H^+$$

where one part of O_2 requires 4 parts of SO_2,

$$4 \times 5 \times 8.34 \times 0.03 = 5.0 \text{ lb/day}$$

$$\text{Total} = 27.7 \text{ lb/day of } SO_2$$

2. Lime requirements are as follows: for Cr^{+3},

$$2Cr^{+3} + 3Ca(OH)_2 \longrightarrow 2Cr(OH)_3 + 3Ca^{++}$$

where one part of Cr^{+3} requires 2.37 parts of 90% lime,

$$2.37 \times 49 \times 8.34 \times 0.03 = 29.1 \text{ lb/day}$$

for Cu (one part of Cu requiring 1.29 parts of 90% lime),

$$1.29 \times 11 \times 8.34 \times 0.03 = 3.6 \text{ lb/day}$$

and for Zn (one part of Zn requiring 1.26 parts of 90% lime),

$$1.26 \times 12 \times 8.34 \times 0.03 = 3.8 \text{ lb/day}$$

Total = 36.5 lb/day of 90% lime

3. Sludge production is:

$$1.98 \times 49 \times 8.34 \times 0.03 = 24.3 \text{ lb/day Cr(OH)}_3$$

$$1.54 \times 11 \times 8.34 \times 0.03 = 4.2 \text{ lb/day Cu(OH)}_2$$

$$1.52 \times 12 \times 8.34 \times 0.03 = 4.6 \text{ lb/day Zn(OH)}_2$$

Total = 33.1 lb/day dry sludge

If sludge concentrates to 1.5 percent by weight, a volume of 265 gal will require disposal each day. It should be noted that some of the copper and zinc will be soluble unless the final pH after lime addition exceeds pH 9.0.

Applications [8]

Arsenic Arsenic and arsenical compounds are present in wastewaters from the metallurgical industry, glassware and ceramic production, tannery operation, dye stuff, pesticide manufacture, some organic and inorganic chemicals manufacture, petroleum refining, and the rare-earth industry. Arsenic is removed from wastewater by chemical precipitation. Effluent arsenic levels of 0.05 mg/l are obtainable by precipitation of the arsenic as the sulfide by the addition of sodium or hydrogen sulfide at pH of 6 to 7. In order to meet reported effluent levels, polishing of the effluent by filtration would usually be required.

Arsenic present in low concentrations can also be reduced by filtration through activated carbon. Effluent concentrations of 0.06 mg/l of arsenic have been reported from an initial concentration of 0.2 mg/l. Arsenic is removed by coagulation producing a ferric hydroxide floc that ties up the arsenic and removes it from solution. Effluent concentrations of less than 0.05 mg/l have been reported from this process.

Barium Barium is present in wastewaters from the paint and pigment industry, the metallurgical industry, glass, ceramics and dye manufacturers and in the vulcaniz-

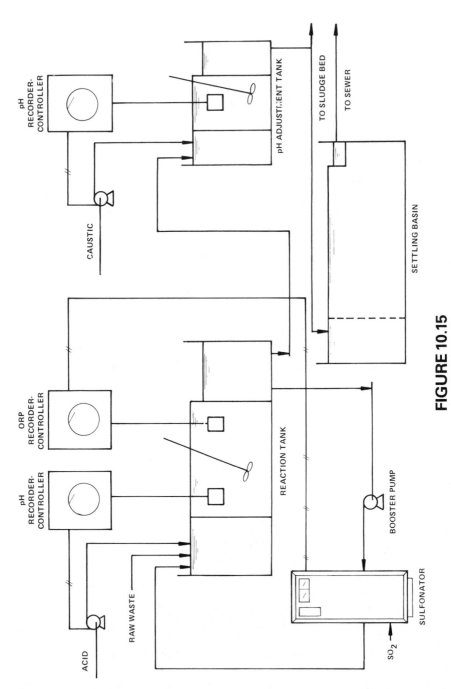

FIGURE 10.15

Continuous Chrome-Wastes Treatment System

(Courtesy of Fischer-Porter, Inc.)

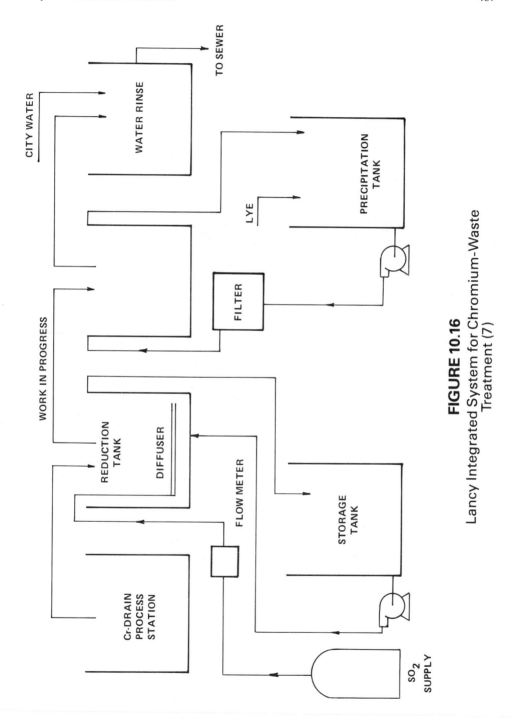

FIGURE 10.16

Lancy Integrated System for Chromium-Waste Treatment (7)

ing of rubber. Barium has also been reported in explosives manufacturing wastewater. Barium is removed from solution by precipitation as barium sulfate.

Barium sulfate is extremely insoluble having a maximum theoretical solubility of approximately 1.4 mg/l as barium at stochiometric concentrations of barium and sulfate. The solubility level of barium can be reduced in the presence of excess sulfate. Coagulation of barium slats as the sulfate would be capable of reducing barium to effluent levels of 0.5 mg/l. Barium can also be removed from solution by ion exchange and electrodialysis, although these processes would be more expensive than chemical precipitation.

Cadmium Cadmium is present in wastewaters from metallurgical alloying, ceramics, electro-plating, photography, pigment works, textile printing, chemical industries and lead mine drainage. Cadmium is removed from wastewaters by precipitation or ion exchange. In some cases, electrolytic and evaporative recovery processes can be employed providing the wastewater is in a concentrated form. Cadmium forms an insoluble and highly stable hydroxide at an alkaline pH. Cadmium in solution is approximately 1 mg/l at pH 8 and 0.1 mg/l at pH 10. Coprecipitation with iron hydroxide at pH 8.5 effects complete removal of cadmium. Cadmium is not precipitated in the presence of complexing ions, such as cyanide. In these cases, it is necessary to pretreat the wastewater to destroy the complexing agent. In the case of cyanide, cyanide destruction is necessary prior to cadmium precipitation. A hydrogen peroxide oxidation precipitation system has been developed that simultaneously oxidizes cyanides and forms the oxide of cadmium resulting in cadmium oxide precipitation. Ion exchange can be employed for the removal of cadmium where recovery of the cadmium is feasible.

Copper The primary sources of copper in industrial wastewaters are metal process pickling baths and plating baths. Copper may also be present in wastewaters from a variety of chemical manufacturing processes employing copper salts or a copper catalyst. Copper is removed from wastewaters by precipitation or recovery processes, which include ion exchange, evaporation, and electrodialysis. The value of recovered copper metal will frequently make recovery processes attractive. Ion exchange or activated carbon are feasible treatment methods for wastewaters containing copper at concentrations of less than 200 mg/l. Chemical precipitation is applicable for copper levels of 1.0 to 1000 mg/l. Electrolytic recovery is advantageous for removal of copper at concentrations above 10,000 mg/l. Copper is precipitated as a relatively insoluble metal hydroxide at alkaline pH. In the presence of high sulfates, the calcium sulfate will also be precipitated that will interfere with the recovery value of the copper sludge. This may dictate the use of a more expensive alkali such as NaOH to obtain a pure sludge. Cupric oxide has a minimum solubility of between pH 9.0 and 10.3 with a reported solubility of 0.01 mg/l. Field practice has indicated that the maximum technically feasible treatment level for copper by chemical precipitation is 0.2 mg/l for soluble copper and 1.0 mg/l for total copper. Low residual concentrations of copper are difficult to achieve in the presence of complexing agents such as cyanide and ammonia. Removal of the complexing agent by pretreatment is essential for high copper removal. Copper cyanide is effectively removed on activated carbon.

Fluorides Fluorides are present in wastewaters from glass manufacturing, electroplating, steel and aluminum, and pesticide and fertilizer manufacture. Fluoride is removed by precipitation with lime as calcium fluoride. Effluent concentrations in the order of 10 to 20 mg/l are readily obtainable. Enhanced removal of fluoride has been reported in the presence of magnesium. The increased removal is attributed to adsorption of the fluoride ion onto the magnesium hydroxide floc resulting in effluent fluoride concentrations of less than 1.0 mg/l. Low concentrations of fluoride can be removed by ion exchange. Fluoride removal through ion exchange pretreated and regenerated with aluminum salts is attributable to aluminum hydroxide precipitated in a column bed. Fluoride is removed through contact beds of activated aluminum, which may be employed as a polishing unit to follow lime precipitation. Fluoride concentrations of 30 mg/l from the lime precipitation process have been reduced to approximately 2 mg/l upon passage through an activated alumina contact bed.

Iron Iron is present in a wide variety of industrial wastewaters including mining operations, ore milling, chemical industrial wastewater, dye manufacture, metal processing, textile mills, petroleum refining, and others. Iron exists in the ferric or ferrous form, depending upon pH and dissolved oxygen concentration. At neutral pH and the presence of oxygen, soluble ferrous iron is oxidized to ferric iron, which readily hydrolizes to form the insoluble ferric hydroxide precipitate. At high pH values ferric hydroxide will solubilize through the formation of the $Fe(OH)_4^-$ complex. Ferric and ferrous iron may also be solubilized in the presence of cyanide due to the formation of ferro and ferri cyanide complexes. The primary removal process for iron is conversion of ferrous to the ferric state and precipitation of ferric hydroxide at a pH of near 7 corresponding to minimum solubility. Conversion of ferrous to ferric iron occurs rapidly upon aeration at pH 7.5. In the presence of dissolved organic matter, iron oxidation rate is reduced.

Lead Lead is present in wastewaters from storage battery manufacture. Lead is generally removed from wastewaters by precipitation. Lead is precipitated as the carbonate, $PbCO_3$ or the hydroxide $Pb(OH)_2$. Minimum solubility of lead at pH of 10 is below 0.5 mg/l. Lead is effectively precipitated as the carbonate by the addition of soda ash, resulting in effluent dissolved lead concentrations of 0.001 mg/l at a pH of 8.0 to 9.0. Precipitation as the sulfide can be accomplished with sodium sulfide at a pH of 7.5 to 8.5.

Manganese Manganese and its salts are found in wastewaters from steel alloy, dry-cell battery manufacture, glass and ceramics, paint and varnish, ink and dye works. Among the many forms and compounds of manganese only the manganous salts and the highly oxidized permanganate anion are appreciably soluble. The latter is a strong oxidant that is reduced under normal circumstances to insoluble manganese dioxide. Treatment technology for the removal of manganese involves conversion of the soluble manganous ion to an insoluble precipitate. Removal is effected by oxidation of the manganous ion and separation of the resulting insoluble oxides and hydroxides. Manganous ion has a low reactivity with oxygen and simple aeration is not an effective technique below pH 9. It has been reported that at even high pH levels, organic matter in solution can combine with manganese and prevent its oxidation by simple aeration. A reaction pH above 9.4 is required to achieve significant manganese reduction by precipitation. The use of chemical oxidants to convert manganous ion to insoluble manganese dioxide in conjunction with coagulation and filtration has been employed. The presence of copper ion enhances air oxidation of manganese and chlorine dioxide rapidly oxidizes manganese to the insoluble form. Permanganate has successfully been employed in the oxidation of manganese. Ozone has been employed in conjunction with lime for the oxidation and removal of manganese removal. The drawback in the application of ion exchange is the nonselective removal of other ions that increases operating costs.

Mercury The major consumptive user of mercury in the United States is the chlor-alkali industry. Mercury is also used in the electrical and electronics industry, explosives manufacturing, photographic industry and the pesticide and preservative industry. Mercury is used as a catalyst in the chemical and petrochemical industry. Mercury is also found in most laboratory wastewaters. Power generation is a large source of mercury release into the environment through the combustion of fossil fuel. When scrubber devices are installed on thermal power plant stacks for sulfur dioxide removal, accumulation of mercury is possible if extensive recycle is practiced. Mercury can be removed from wastewaters by precipitation, ion exchange and adsorption. Mercury ions can be reduced upon contact with other metals such as copper, zinc, or aluminum. In most cases mercury recovery can be achieved by distillation. The precipitation of mercury compounds must be oxidized to the mercuric ion. Studies using sodium sulfide plus lime employing iron or aluminum sulfate as coagulant aids resulted in a supernatant mercury residual of 0.1 mg/l. Ion exchange has been successfully employed for mercury removal resulting in effluent concentrations of 0.1 to 0.15 mg/l. A second stage resin that acts as an adsorbant for the mercury reduces the final effluent concentration to less than 5 micrograms per liter.

Nickel Wastewaters containing nickel originate from the metal processing industries, steel foundries, motor vehicle, and aircraft industries, printing and in some cases the chemicals industry. In the presence of complexing agents such as cyanide, nickel may exist in a soluble complex form. The presence of nickel cyanide complexes interferes with both cyanide and nickel treatment. Nickel forms insoluble nickel hydroxide upon the addition of lime, resulting in a minimum theoretical solubility of 0.01 mg/l at pH 10. Nickel can also be precipitated as the carbonate or the sulfate associated with recovery systems. In practice, lime addition (pH 11.5) may be expected to yield residual nickel concentrations in the order of 0.15 mg/l after sedimentation and filtration. Recovery of nickel can be accomplished by ion exchange, or evaporative recovery providing the nickel concentrations in the wastewaters are at a sufficiently high level.

Selenium Selenium may be present in various types of paper, fly-ash, and in metalic sulfide ores. The selenous ion appears to be the most common form of selenium in wastewater except for pigment and dye waste, which contain the selenide (yellow cadmium selenide). Selenium can be removed from wastewaters by precipitation as the sulfide at a pH of 6.5. Effluent levels of 0.05 mg/l are reported.

Silver Soluble silver, usually in the form of silver nitrate, is found in wastewaters from the porcelain, photographic, electroplating and ink manufacturing industries. Treatment technology for the removal of silver usually considers recovery because of the high value of the metal. Basic treatment methods include precipitation, ion exchange, reductive exchange and electrolytic recovery. Silver is removed from wastewater by precipitation as silver chloride which is an extremely insoluble precipitate resulting in the maximum silver concentration of approximately 1.4 mg/l. An excess of chloride ion will reduce this value, but greater excess concentrations will increase the solubility of silver through the formation of soluble silver chloride complexes. Silver can be selectively precipitated as silver chloride from a mixed metal wastestream without initial wastewater segregation or concurrent precipitation of other metals. If the treatment conditions are alkaline, resulting in precipitation of hydroxides of other metals along with the silver chloride, acid washing of the precipitated sludge will remove contaminated metal ions leaving the insoluble silver chloride. Plating wastes contains silver in the form of silver cyanide, which interferes with the precipitation of silver chloride. Therefore, cyanide removal is necessary prior to precipitation of silver as the chloride salt. Oxidation of the cyanide with chlorine releases chloride ions into solution, which in turn react to form silver chloride directly. Sulfide will precipitate of silver from photographic solutions as the extremely insoluble silver sulfide. Ion exchange has been employed for the removal of soluble silver from wastewaters. Activated carbon will remove low concentrations of silver. The mechanism reported is one of reductive recovery by formation of elemental silver at the carbon surface. Reported results indicate that the carbon is capable of retaining silver to 9 percent of its weight at a pH of 2.1 and 12 percent of its weight at a pH of 5.4.

Zinc Zinc is present in wastewater streams from steel works, rayon yarn and fiber manufacture, ground wood pulp production, and recirculating cooling water systems employing cathodic treatment. Zinc is also present in wastewaters from the plating and metal processing industry. Zinc can be removed by precipitation as zinc hydroxide with either lime or caustic. The disadvantage of lime addition is the concurrent precipitation of calcium sulfate in the presence of high sulfate levels in the wastewater.

FILTRATION

Filtration is employed for the removal of suspended solids as a pretreatment for low suspended solids wastewaters, following coagulation in physical chemical treatment or as a tertiary treatment following a biological wastewater treatment process.

Suspended solids are removed on the surface of a filter by straining and through the depth of a filter by both straining and adsorption. Adsorption is related to the zeta potential on the suspended solids and the filter media. Particles normally encountered in a wastewater vary in size and particle charge and some will pass the filter continuously. The efficiency of the filtration process is therefore a function of:

1. The concentration and characteristics of the solids in suspension
2. The characteristics of the filter media and other filtration aids
3. The method of filter operation

Types of Filters

Granular Media Filter

Dual or multimedia filters are preferred to sand filters for secondary effluents because the gelatinous nature of the suspended solids leads to rapid surface blinding on a sand bed. In order to achieve maximum utilization of the filter media, coarse to fine gradation of the media is placed on fine media of a higher specific gravity. The filters are usually designed for a 6–12-hour run. Runs of 20 hours or more may result in mud balling in the filter bed. Granular media filters may be either gravity or pressure. Gravity filters may be operated at a constant rate with influent flow control and flow splitting or at a declining rate with four or more units fed through a common header. To achieve constant flow an artificial head loss (flow regulator) is used. As suspended solids are removed and the head loss increases the artificial head loss is reduced so the total head loss remains constant. In a declining rate filter design the decrease in flow rate through one filter as the head loss increases raises the head and rate through the other filters. A maximum filtration rate of 6 gpm/ft^2 (0.24 m^3/m^2-min) is used when one unit is out of service. The filter run terminates when the total head loss reaches the available driving force or when excess suspended solids or turbidity appear in the effluent. A typical filter is shown in figure 10.17.

The suspended solids in a secondary effluent tend to blind a filter. For this reason, both an air and water backwash is commonly used. The media is air scoured for a period of 1–2 minutes followed by an air and water wash. This is followed by a water wash. The water wash rate and duration will depend on media type, nature of the suspended solids and temperature. The combination of air scour, air and water backwash appears to be a most effective way for removing accumulated suspended solids and biological slimes from the filter media.

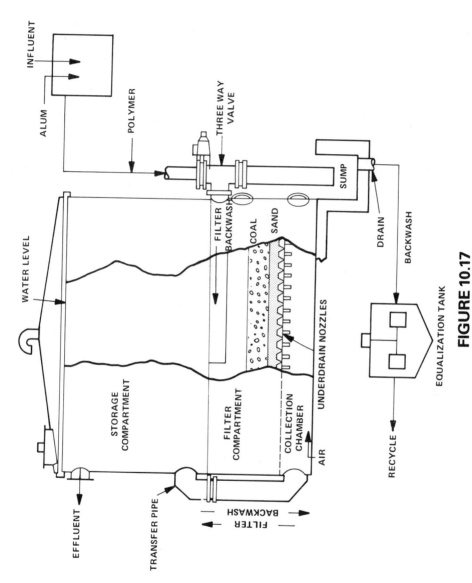

FIGURE 10.17

Typical Automatic Granular Media Filter

Media size is an important consideration in filter design. The sand size is chosen on the basis that it provides slightly better removal than is required. The coal size is selected to provide 75–90 percent suspended solids removal across 1.5–2.0 feet (0.46–0.6/m) of media. For example, if 90 percent suspended solids removal is desired across a filter bed, 68–80 percent should be removed through the coal layer and the remaining 10–25 percent through the sand layer. If the feed suspended solids particle size is larger than 5 percent of the granular media particles, mechanical straining will occur.

A 25 μm particle will be mechanically strained by a 0.5 mm filter media. If the feed solids particles have a density of 2–3 times that of the suspending medium then particles as small as 0.5 percent of the filter media particle size can be effectively removed by in-depth granular media filtration.

The effect of media size on suspended solids removal from a municipal secondary effluent is shown in figure 10.18.

Filtration rate will affect the build up of head loss and the effluent quality attainable. The optimum filtration rate is defined as the filtration rate, which results in the maximum volume of filtrate per unit filter area while achieving an acceptable effluent quality.

FIGURE 10.18
Effect of Filtration Rate and Sand Size
on Removal Efficiency [9]

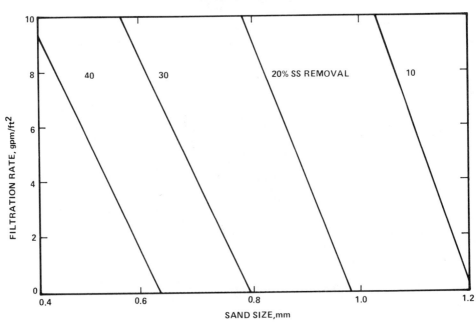

The effect of filtration rate on head loss is shown in figure 10.19. The filtration rate of 6 gpm/ft^2 (0.24 m^3/m^2-min) was too great since solids penetrated the coarse media and accumulated on the fine media. A filtration rate of 2 gpm/ft^2 (0.08 m^3/m^2-min) was insufficient to achieve good solids penetration of the coarse media resulting in head loss build up at the top of the coarse media. Filtration rate will also influence effluent quality depending on the nature of the particles to be removed as shown in figure 10.20.

The head loss through the filter is related to the solids loading as shown in figure 10.21.

$$H = aS^n$$

(10–2)

where H = head loss, ft
      S = solids captured lb/ft^2
      a,n = constants

For a given head loss, the filtration cycle depends upon the influent suspended solids and the hydraulic flow rate. The type of coagulant used may also influence the head loss as shown in figure 10.22.

FIGURE 10.19
Effect of Filtration Rate on Head Loss Buildup

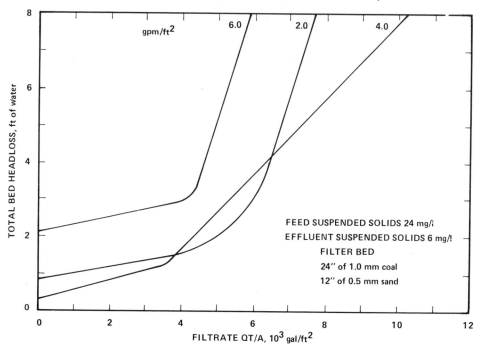

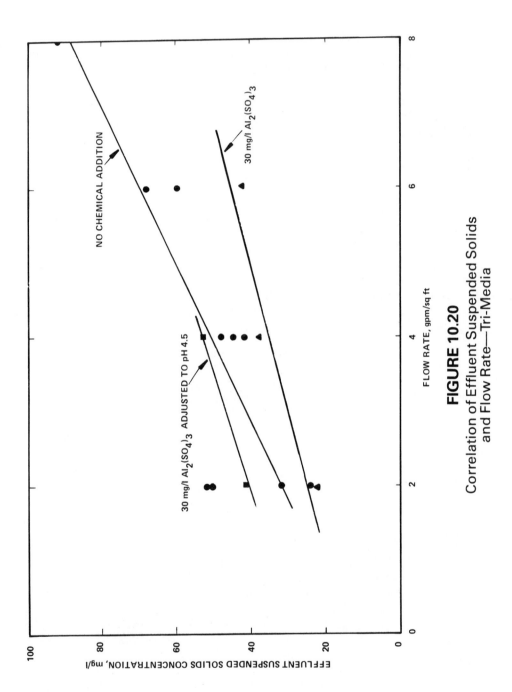

FIGURE 10.20

Correlation of Effluent Suspended Solids
and Flow Rate—Tri-Media

FIGURE 10.21
Head Loss Versus Solids Retention In Filter

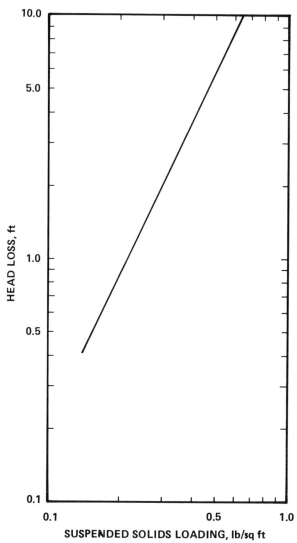

Improved suspended solids removals can be achieved by the addition of coagulants to the wastewater prior to filtration. The use of alum also results in the precipitation and removal of phosphorus through the filter. Flocculation is not needed since the filter serves as a flocculator. Effective mixing is required to disperse the chemicals and initiate the reaction. Since the suspended solids are removed by filtration rather than by sedimentation, 25–50 percent less chemicals

FIGURE 10.22
Correlation of Head Loss with Deposit

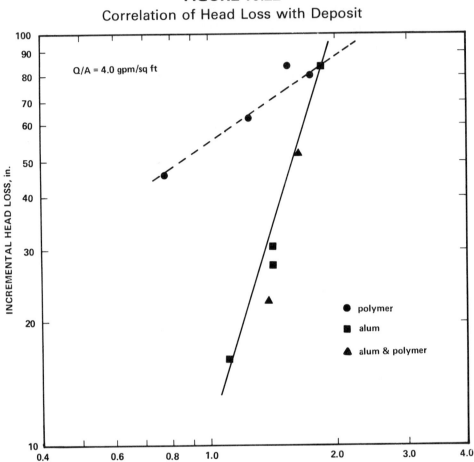

are required in many cases. For most applications a maximum of 100 mg/l suspended solids removal is used in order to avoid excessive backwash volumes.

Data showing the effect of polyelectrolyte dosage are shown in figure 10.23. The effluent quality deteriorated once breakthrough had occurred for a polyelectrolyte dosage of 0.9 mg/l. At a polyelectrolyte dosage of 1.4 mg/l the effluent quality was maintained to the terminal head loss.

Typical operating conditions for the filtration of domestic secondary effluent is shown in figure 10.24. Performance data is summarized in table 10.9. Powdered activated carbon can be applied to granular media filtration at dosages up to 100 mg/l during emergency operation to maintain effluent quality. Continual addition at lower dosages serves to polish the effluent. Maximum utilization of the

FIGURE 10.23
Effect of Polymer Dose of Effluent Turbidity and Head Loss

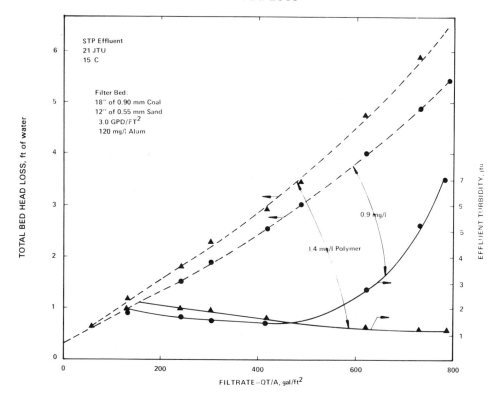

powdered activated carbon would occur when the filter backwash is recycled back to the aeration basin of an activated sludge plant.

Granular media filters can be fully automated by the use of pressure sensors that monitor the head loss across the filter. When the terminal head loss is achieved the filter would automatically backwash. Turbidimeters serve as a secondary control such that the backwash sequence would be initiated when the turbidity reached the allowable level in the effluent.

While considerable data is available for the design of multimedia filters treating domestic secondary effluents, industrial wastewaters require pilot plant studies to define the type of media, filter flow rate, coagulant requirements, head loss relationships and backwash requirements.

FIGURE 10.24

Granular Media Filtration of Biological STP Effluent

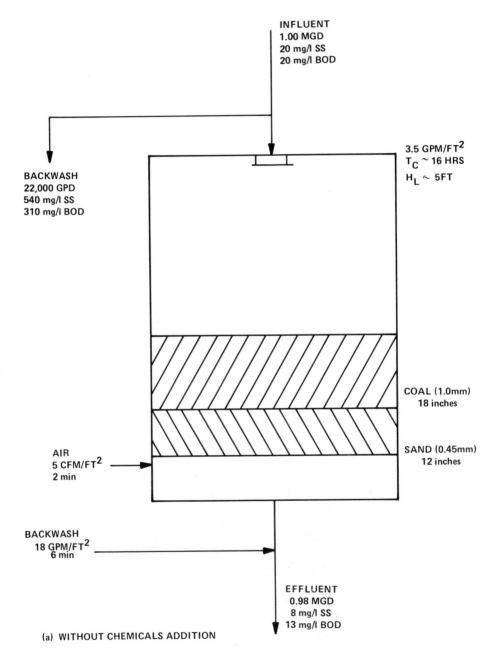

(a) WITHOUT CHEMICALS ADDITION

Filter Type	Feed Type[a]	Media Size mm	Filter Depth ft.	Hydraulic Loading gpm/sq. ft.	SS Removal %	BOD Removal %	Eff. SS mg/l	Eff. BOD mg/l
Deep-bed								
Gravity downflow	TF Eff.	1.0–2.0	–	6	70	55	5–7	–
Gravity downflow	TF Eff.	0.9–1.7	2–3	3	67	58	–	2.5
Pressure upflow	TF Eff.	0.9–1.7	5	3	85	74	5.0	2.5
Pressure upflow	AS Eff.	0.9–1.7	5	3	77	–	–	–
Pressure upflow	AS Eff.	1.0–2.0	5	2.2	50	62	7.0	6.4
Pressure upflow	AS Eff.	1.0–2.0	5	4.0	67	73	4.9	6.4
Pressure upflow	AS Eff.	1.0–2.0	5	4.9	56	65	5.7	7.1
Mixed media	EA Eff.	0.25–2.0	2.5	5.0	74	88	4.6	2.5
Mixed media	AS Eff.			2.0	73	74	3.8	6.0
Mixed media	AS Eff.			4.0	73	85	4.3	3.9
Mixed media	AS Eff.	0.25–2.0	2.5	2.5				
Surface filters								
Moving bed	TF Eff.[b]	0.6–0.8	4.2	2	47	71	–	–
Moving bed	TF Eff.[c]	0.6–0.8	4.2	2	67	80	–	–
Gravity downflow	AS Eff.	–	–	2.2	55	64	7.2	7.3
Gravity downflow	AS Eff.	–	–	4.0	69	70	4.8	7.4
Gravity downflow	AS Eff.	–	–	8.0	48	64	6.1	6.7
Gravity downflow	AS Eff.	0.9–1.7	2.0	1.6–4.0	72–91	52–70	–	–
Gravity downflow	AS Eff.	0.95	1.0	2.0	46	57	–	–
Gravity downflow	AS Eff.	0.58	–	2.0–6.0	70	80	–	–
Gravity downflow	CS Eff.	0.45	1.0	5.3	62	78	5	4

TABLE 10.9
Filtration Performance

[a] TF—Trickling Filter, EA—Extended Aeration, AS—Activated Sludge, CS—Contact Stabilization.
[b] 100 mg/l alum & 0.2–0.75 mg/l anionic polymer added.
[c] 200 mg/l alum & 0.2–0.75 mg/l anionic polymer added.

PROBLEM 10–2

A 1 MGD secondary effluent treating with 30 mg/l $Al_2(SO_4)_3$ is to be polished by a granular media filter for 70% SS removal. If 8 ft head loss is applied, calculate the filter run time for secondary effluent with SS of 70 mg/l.

$$\text{Total influent SS to filter} = 70 + 30 \cdot \frac{2 \cdot 78}{342}$$

$$= 84 \text{ mg/l}$$

$$\text{Effluent SS from filter} = 84 \cdot (1 - 0.70)$$

$$= 25 \text{ mg/l}$$

From figure 10.20, for effluent SS of 25 mg/l, the flow rate is estimated as 2.0 gpm/ft². Thus, the required surface area is:

$$A = \frac{1 \times 10^6}{2.0 \cdot 1440}$$

$$= 347 \text{ ft}^2$$

For head loss of 8 ft the suspended solids loading is to be 0.57 lb/ft² (see figure 10.21). The expected filter run is then computed:

$$t = \frac{0.57 \cdot 347}{1 \cdot (84 - 25) \cdot 8.34}$$

$$= 0.402 \text{ day or 9.6 hours}$$

Microscreen

A microscreen is a rotary drum revolving on a horizontal axis covered with stainless steel fabric (figure 10.25). The water enters the open end of the drum and is filtered through the fabric with solids being retained on the inside surface of the fabric. As the drum rotates, the solids are transported and continuously removed at the top of the drum by pumping effluent under pressure through a series of spray nozzles that extend the length of the drum. The head loss is less than 12–18 inches (30–46 cm) in water. The backwash water is 4–6 percent of the total throughput water. Peripheral drum speeds vary up to 100 ft/min (30.5 m/min) with hydraulic loadings of 2.5–10 gpm/ft² (0.1–0.4 m³/m²-min). Periodic cleaning of the drum is required for slime control.

For filtration of secondary effluent a maximum solids loading of 0.88 lb/ft²-day (4.3 kg/m²-day) at a hydraulic loading of 6.6 gpm/ft² (0.27 m³/m²-min) has

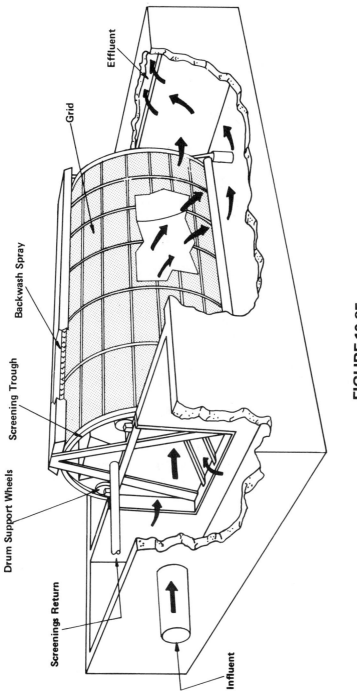

FIGURE 10.25
Microscreen
(Courtesy of Envirex Inc.)

been reported. Lynam *et al* [10] reported effluent suspended solids and BOD of 6–8 mg/l and 3.5–5 mg/l, respectively, with a 20–20 activated sludge effluent on a 23 μm microscreen at 3.5 gpm/ft^2 (0.14 m^3/m^2-min). For design purposes, the removal efficiencies treating a secondary effluent are [11]:

Aperture Screen	Flow Rate	Percent Removal	
μm	gpm/ft^2 (submerged)	SS	BOD
35	10.0	50–60	40–50
23	6.7	70–80	60–70

Operating data from several installations is summarized in table 10.10. The efficiency of the unit is suspended solids sensitive as indicated by a decrease in throughput rate from 60 to 13 gpm with an increase in influent suspended solids from 25 to 200 mg/l [12].

Other Filters

An upflow sand filter (figure 10.26) consists of filter media graded from bottom to top of 12.5–15 mm gravel, 3.75–6.25 mm gravel, 2–3 mm sand and 0.5–1.0 mm sand. The maximum upflow filtration rate is limited by the media size and specific gravity. Surface loadings range from 4–8 gpm/ft^2 (0.16–0.33 m^3/m^2-min). An air and water backwash is used similar to deep bed filtration. An advantage of the upflow design is the use of raw water for backwashing. Suspended solids removals are comparable to downflow filter designs. Typical performance for filtration of secondary effluent from a petroleum refinery is shown in figure 10.27.

Surface filters of several types are available. In general, these are better for fragile chemical flocs, have shorter filter runs and use less backwash water. The moving bed filter (figure 10.28) mechanically moves the most heavily clogged portion of the medium out of the zone of filtration. Wastewater (A) flows through the inlet pipe when chemicals, if required, are added at (B). The wastewater enters the head tank (C) and then passes through the sand bed (D). The filtered water leaves through the exit screens. When excessive head loss develops, the bed is pushed toward the head tank by pressurizing a chamber separated from the bed by a flexible diaphragm. A mechanical cutter (F) sweeps down over the face of the bed cutting off the top layers. These then fall into hopper (G) of the head tank. The sludge and sand are removed from the head tank with the aid of an ejector using feed water. The solids are hydraulically conveyed to the sand washer (H) where filtered water and air are used to backwash the sand. Clean sand moves by gravity back to the base of the filter. The spent backwash water is sent to a sedimentation tank for removal of the wastewater solids. The operation of the system is automated. Flow rates up to 7.0 gpm/ft^2 (0.29 m^3/m^2-min) are used.

Location	Plant Size	Feed Type	Fabric Opening	SS Removal	Eff. SS	BOD Removal	Eff. BOD	Backwash
	mgd		μm	%	mg/l	%	mg/l	% of flow
Brampton, Ontario	0.1	AS[a] Effluent	23	57	–	54	–	–
Lebanon, Ohio	Pilot	AS Effluent	23	89	1.9	81	–	5.3
	Pilot	AS Effluent	35	73	7.3	61	–	5.0
Chicago, Illinois	3.0	AS Effluent	23	71	3.0	74	3.0	3.0
Lutton, England	3.6	Effluent from AS and TF[b]	35	55	7.3	30	–	3.0
Bracknell, England	7.2	TF Effluent	35	66	5.7	32	8.4	–

TABLE 10.10

Microstrainer Performance

[a] AS—Activated sludge.
[b] TF—Trickling filter.

FIGURE 10.26

Cutaway View of an Upflow Pressure Filter

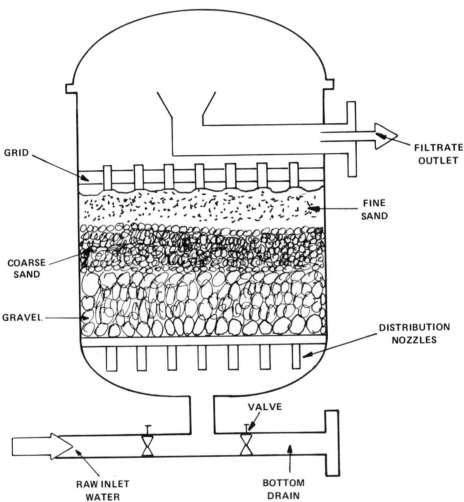

The efficiency of removal of suspended solids by filtration is primarily related to the nature of the solids. For example, a trickling filter effluent may show only 50 percent removal of suspended solids due to poor flocculation while an extended aeration effluent may show up to 98 percent removal. Mixed media filtration of an activated sludge effluent with up to 120 mg/l of suspended solids yielded a filter run of 15–24 hours at 5 gpm/ft² (0.2 m³/m²-min) with a 15 ft terminal head loss.

FIGURE 10.27

Typical Results for Upflow Sand Filter

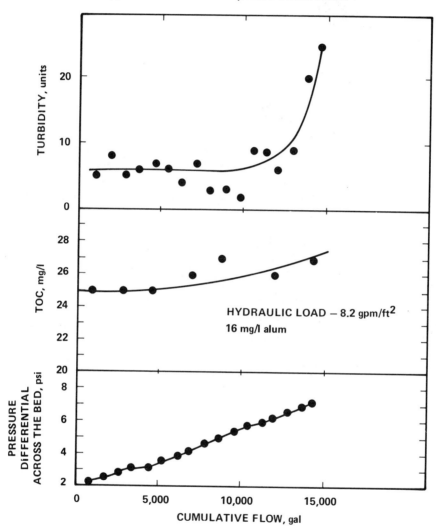

HYDRAULIC LOAD – 8.2 gpm/ft^2

16 mg/l alum

CUMULATIVE FLOW, gal

Process	Effluent Suspended Solids mg/l
High rate trickling filter	10–20
Contact Stabilization	6–15
Activated Sludge	3–10
Extended Aeration	1–5

The effluent quality can be improved by the addition of a coagulant immediately prior to the filter to coagulate colloidal solids and to coat the filter media. Cationic polyelectrolytes are commonly used for this purpose.

Filtration can be employed following biological treatment for removal of residual soluble BOD as well as suspended solids with the development of a biological growth on the filter media. Studies in Japan using an upflow sand filter at a hydraulic rate of 2.4–8.6 gpm/ft^2 (0.1–0.35 m^3/m^2-min) reduced the secondary

FIGURE 10.28
Schematic Drawing of the Johns-Manville Moving Bed Filter

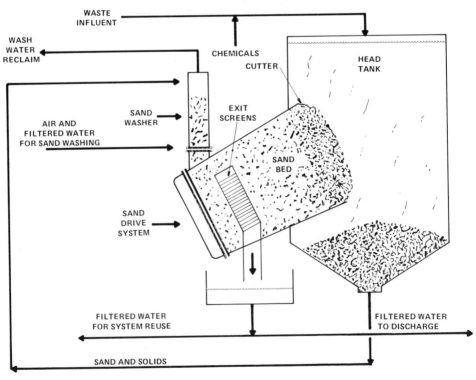

effluent soluble BOD from 12 mg/l to 5 mg/l. Faup *et al.* [13] using an expanded clay media with a high specific surface achieved a reduction in soluble COD from 51 to 35 mg/l and a suspended solids reduction from 34 to 5 mg/l at a hydraulic loading of 4.3 gpm/ft^2 (0.18 m^3/m^2-min). The conclusions from these studies indicated that enhanced biological removal through post filtration should include aeration prior to filtration to insure adequate dissolved oxygen in the filter and to aid coagulation of residual suspended solids and operation at a filtration rate in the order of 3.2 gpm/ft^2 (0.13 m^3/m^2-min). Lower rates did not improve performance and higher rates reduced performance. Chlorination prior to filtration should be avoided if biological oxidation is desired.

ADSORPTION

Adsorption is employed for the removal of refractory organics (COD) in tertiary treatment and for the removal of biodegradable and nonbiodegradable organics in physical-chemical treatment processes.

In the adsorption process molecules attach themselves to the solid surface through attractive forces (van der Waals) resulting from unbalanced forces between the adsorbent and the molecules in solution. Adsorption continues until equilibrium is established with the concentration in solution (driving force). The weight of material adsorbed is therefore related to the concentration remaining in solution.

The rate of adsorption is a function of:
1. Diffusion to the adsorbent surface through a liquid film resistance
2. Diffusion into the pores of the adsorbent and adsorption at the interior surface sites

In most applications, the first two mechanisms are controlling.

The rate of adsorption increases with:
1. An increase in adsorbate concentration
2. A decrease in adsorbent particle size
3. A smaller adsorbate molecule and an increase in surface area of adsorbent

Adsorption to attain equilibrium is proportional to the square root of the time of contact. For granular activated carbon, the adsorbate must penetrate the channels in the carbon and a long time is required to attain equilibrium.

Adsorption capacity may increase with:
1. An increase in adsorbate concentration
2. An increase in surface area of the adsorbent
3. An increase in the molecular weight of the adsorbate
4. A decrease in pH of the solution that changes the organic molecules into a less soluble form

Most wastewaters are highly complex and vary widely in the adsorbability of the compounds present. Molecular structure, solubility, etc., all affect the adsorbability. These effects are shown in table 10.11 and the relative adsorption of several compounds in wastewaters is shown in table 10.12.

The applicability of adsorption as a wastewater treatment process can be defined on a batch basis that will approximate carbon effectiveness and predict organic residual levels.

TABLE 10.11
Influence of Molecular Structure and Other Factors on Adsorbability[a]

1. An increasing solubility of the solute in the liquid carrier decreases its adsorbability.
2. Branched chains are usually more adsorbable than straight chains. An increasing length of the chain decreases solubility.
3. Substituent groups affect adsorbability:

Substituent Group	Nature of Influence
Hydroxyl	Generally reduces adsorbability; extent of decrease depends on structure of host molecule.
Amino	Effect similar to that of hydroxyl but somewhat greater. Many amino acids are not adsorbed to any appreciable extent.
Carbonyl	Effect varies according to host molecule; glyoxylicare more adsorbable than acetic but similar increase does not occur when introduced into higher fatty acids.
Double bonds	Variable effect as with carbonyl.
Halogens	Variable effect.
Sulfonic	Usually decreases adsorbability.
Nitro	Often increases adsorbability.

4. Generally, strong ionized solutions are not as adsorbable as weakly ionized ones; i.e., undissociated molecules are in general preferentially adsorbed.
5. The amount of hydrolytic adsorption depends on the ability of the hydrolysis to form an adsorbable acid or base.
6. Unless the screening action of the carbon pores intervene, large molecules are more sorbable than small molecules of similar chemical nature. This is attributed to more solute carbon chemical bonds being formed, making desorption more difficult.
7. Molecules with low polarity are more sorbable than highly polar ones.

[a] After [14].

TABLE 10.12

Relative Amenability to Adsorption of
Typical Petrochemical Wastewater Constituents[a]

Compound at 1000 mg/l Initial Concentration[b]	Percentage Removal of Compound at 5 gm/l powdered carbon dosage
Ethanol	10
Isopropanol	13
Acetaldehyde	12
Butyraldehyde	53
Di-N-propylamine	80
Monoethanolamine	7
Pyridine	47
2-Methyl 5-ethyl pyridine	89
Benzene	95
Phenol	81
Nitrobenzene	96
Ethyl acetate	50
Vinyl acetate	64
Ethyl acrylate	78
Ethylene glycol	7
Propylene glycol	12
Propylene oxide	26
Acetone	22
Methyl ethyl ketone	47
Methyl isobutyl ketone	85
Acetic acid	24
Propionic acid	33
Benzoic acid	91

[a] After [15].
[b] Benzene test at near saturation level, 420 mg/l.

Adsorption data is usually correlated according to the Freundlich Isotherm (empirical) developed from batch tests:

$$\frac{X}{M} = KC^{1/n} \tag{10-3}$$

where X/M = amount of solute adsorbed per unit weight of adsorbent
C = concentration of solute remaining in solution at equilibrium
K, n = constants

The constants K and n define both the nature of the carbon and the adsorbate. Generally K and n decrease with increasing wastewater complexity. High K and n values indicate high adsorption throughout the concentration range studies; conversely, low values indicate a low adsorption at dilute concentrations. A low value of n (steep slope) indicates high adsorption at strong solute concentrations and poor adsorption at dilute concentrations. In the case of complex wastewaters a portion of the organics present in the wastewater may not be adsorbed yielding a residual regardless of the carbon dosage as shown in figure 10.29.

Properties of Activated Carbon

Activated carbons are made from a variety of materials including wood, lignin, bituminous coal, lignite, and petroleum residues. Granular carbons produced from medium volatile bituminous coal or lignite have been most widely applied to the treatment of wastewater. Activated carbons have specific properties depending on the material source and the mode of activation. Property standards are helpful in specifying carbons for a specific application. In general granular carbons from bituminous coal have a small pore size, the largest surface area and the highest bulk density. Lignite carbon has the largest pore size, least surface area and the lowest bulk density. Adsorptive capacity is the effectiveness of the carbon in removing desired constituents such as COD, color, phenol, etc., from the wastewater. Several tests have been employed to characterize adsorptive capacity. The phenol number is used as an index of a carbons ability to remove taste and odor compounds. The iodine number relates to the ability of activated carbon to adsorb low molecular weight substances (micropores having an effective radius of less than 2 nm) while the molasses number relates to the carbons ability to adsorb high molecular weight substances (pores ranging from 1–50 nm). In general, high iodine numbers will be most effective on wastewaters with predominately low molecular weight organics, while high molasses numbers will be most effective for wastewaters with a dominance of high molecular weight organics. Properties of commercial carbons are shown in table 10.13.

Depending on the characteristics of the wastewater, one type of carbon may be superior to another (figure 10.30) since the capacity is greater at equilibrium effluent concentrations. Carbon A is better for carbon column operation since the capacity is greater in equilibrium with the influent (C_o) at column exhaustion, while carbon B would be better for batch treatment. Figure 10.31 shows the increase in adsorption capacity with reduction of pH due to a change in solubility at the organics.

The complex nature of most wastewaters makes it necessary to empirically generate design data. Batch isotherm tests are useful to define the degree of treatment attainable by adsorption. Either a jar test assembly or a shake flask assembly can be employed. It has been shown that batch isotherm results cannot

FIGURE 10.29
Freundlich Isotherm Application

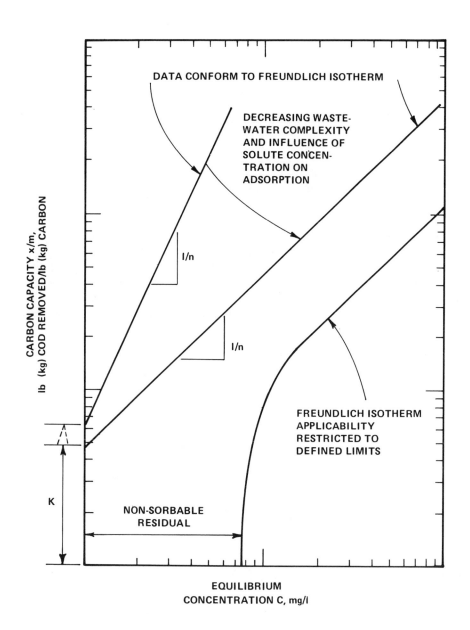

EQUILIBRIUM
CONCENTRATION C, mg/l

	ICI America Hydrodarco 3000 (Lignite)	Calgon Filtrasorb 300 (8×30) (Bituminous)	Westvaco Nuchar WV-L (8×30) (Bituminous)	Witco 517 (12×30) (Bituminous)
Physical Properties				
Surface area, m²/gm (BET)	600–650	950–1050	1000	1050
Apparent density, gm/cc	0.43	0.48	0.48	0.48
Density, backwashed and drained, lb/cu. ft.	22	26	26	30
Real density, gm/cc	2.0	2.1	2.1	2.1
Particle density, gm/cc	1.4–1.5	1.3–1.4	1.4	0.92
Effective size, mm	0.8–0.9	0.8–0.9	0.85–1.05	0.89
Uniformity coefficient	1.7	1.9 or less	1.8 or less	1.44
Pore volume, cc/gm	0.95	0.85	0.85	0.60
Mean particle diameter, mm	1.6	1.5–1.7	1.5–1.7	1.2
Specifications				
Sieve size (U.S. std. series)				
Larger than No. 8 (max. %)	8	8	8	c
Larger than No. 12 (max. %)	c	c	c	5
Smaller than No. 30 (max. %)	5	5	5	5
Smaller than No. 40 (max. %)	c	c	c	c
Iodine No.	650	900	950	1000
Abrasion No., minimum	b	70	70	85
Ash (%)	b	8	7.5	0.5
Moisture as packed (max. %)	b	2	2	1

[a] After [16].
[b] No available data from the manufacturer.
[c] Not applicable to this size carbon.

TABLE 10.13

Properties of Several Commercially Available Carbons[a]

FIGURE 10.30
Adsorption Isotherms for Carbon A and Carbon B

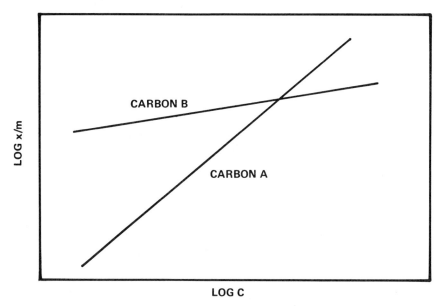

be effectively scaled up to plant design and columnar studies should be conducted. The carbon columns are usually 3–8 inches (7.6–20 cm) in diameter and 5–8 ft (1.5–2.5 m) in height, piped and valved for upflow or downflow operation as shown in figure 10.32.

Carbon Regeneration

It is generally feasible to regenerate spent carbon for economic reasons. In the regeneration process the object is to remove from the carbon porous structure the previously adsorbed materials. The modes of regeneration are thermal, steam, solvent extraction, acid or base treatment and chemical oxidation. The latter methods (excluding thermal) are usually to be preferred when applicable since they can be accomplished *in situ*. The difficulty arises that adsorption from multicomponent wastewaters usually does not lend itself to high efficiency regeneration by these methods. Exceptions are phenol, which can be regenerated with caustic in which the phenol is converted to the more soluble phenate and a single chlorinated hydrocarbon which can be removed with steam. In most wastewater cases, however, thermal regeneration is required. Thermal regeneration is the process of drying, thermal desorption and high temperature heat treatment (1200–1800° F) in the presence of limited quantities of water vapor, flue gas and

FIGURE 10.31

Adsorption Isotherm Relationships

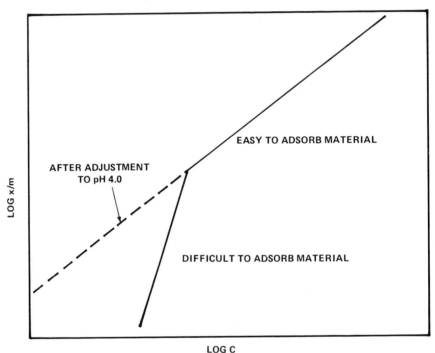

oxygen. Multiple hearth furnaces are most commonly used as shown in figure 10.33.

The carbon initially contains about 50 percent water by weight. The first step in the regeneration process is the evaporation of water and volatile constituents at 700–900° F. This occurs in the first few hearths. Carbonization then occurs in which the organics are burned to elemental carbon at 1500–1700° F (pyrolysis). The last hearths regenerate the pore structure by oxidation of carbon in a steam atmosphere at 1700–1800° F. The effect of steam is to reduce the apparent density and decrease the iodine number of the regenerated carbon. Normally about one pound of steam per pound of carbon is used. Natural gas or fuel oil is usually added to supply auxiliary heat. Generally about 3000 BTU/lb (5400 cal/g) carbon and 1300 BTU/lb (2340 cal/g) of steam is required. Spent carbon is removed from the bottom of the adsorption column through an educator with 3 lbs carbon/gallon water (360 kg/m³). The carbon is transferred by a screw to the regeneration furnace. The regenerated carbon is stored in a tank with 1 and ½ times the capacity of the adsorber being regenerated.

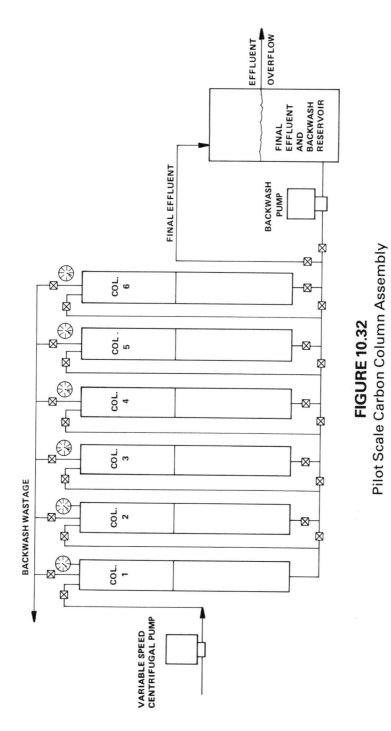

FIGURE 10.32

Pilot Scale Carbon Column Assembly

FIGURE 10.33
Cross-Sectional View of Multiple-Hearth Furnace

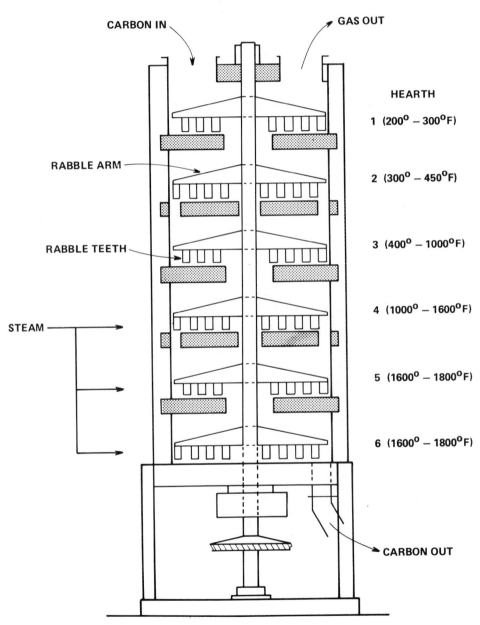

CARBON IN

GAS OUT

HEARTH

1 (200° – 300°F)

RABBLE ARM

2 (300° – 450°F)

RABBLE TEETH

3 (400° – 1000°F)

4 (1000° – 1600°F)

STEAM

5 (1600° – 1800°F)

6 (1600° – 1800°F)

CARBON OUT

Weight losses of carbon result from attrition and carbon oxidation. Depending on the type of carbon and furnace operation this usually amounts to 5–10 percent by weight of the carbon regenerated. There is also a change in carbon capacity through regeneration that may be caused by a change in pores size (usually an increase resulting in a decrease in iodine number) and a loss of pores by deposition of residual materials. Some typical data is shown in table 10.14. In the evaluation of carbons for a wastewater treatment application the change in capacity through successive regeneration cycles should be evaluated.

Powdered carbon can be regenerated in a fluidized bed system. In this process a bed of inert granular material such as sand is fluidized by the upward flow of hot gases and the wet spent carbon is injected directly into the bed. The inert bed particles provide a reservoir of heat that is rapidly transferred to the spent carbon particles.

Adsorption System Design

Granular carbon columns are most practical for the treatment of wastewaters containing high concentrations of organics to be removed because:

1. Separation of the carbon from the wastewater after contact is not required.

TABLE 10.14
Effect of Regeneration on Bituminous Coal Carbon's Physical Properties and Performance

	Regeneration Number	Iodine Number	Molasses Number	Ash, %	Capacity, lb Soluble COD/ lb Carbon	
Pittsburgh Activated Carbon Company	0	942	–	5.3	0.17[a]	
	10	588	–	11.9	0.13[a]	
MSA Research Corp.	0	1,090	250	5.7	0.17[b]	
	3	940	355	9.5	0.15[b]	
MSA Research Corp.	0	1,090	190	–	0.53	
	10	630	250	–	0.39	
Pomona	0	1,100	–	–	0.46	
	7	700	–	–	0.29	
Pomona	0	1,028	–	–	0.22	
	10	686	–	–	0.17	
Tahoe	0	935	–	5.0	0.36	
	4	820	–	7.1	0.38	
Pomona	0	1,100	–	–	0.59	
	4	690	–	–	0.49	

[a]ABS.
[b]Actual weight increase during use.

2. The concentration of the adsorbed solute is in equilibrium with
    the influent concentrations rather than the effluent concentration
    and a greater flexibility of operation is attained.

Adsorption in a column is a non-steady process in which the adsorption
zone moves down the bed until breakthrough occurs as shown in figure 10.34. The
time of breakthrough decreases with:

1. Decrease in depth of the carbon bed
2. An increase in the effective particle size of the carbon
3. An increase in the rate of flow through the bed and an increase in
    the solute concentration in the influent

The carbon capacity at breakthrough as related to exhaustion is a function of
waste complexity as shown in figure 10.35. A single organic such as dichloro-
ethane will yield a sharp breakthrough curve such that the column is greater than
90 percent exhausted when breakthrough occurs. By contrast, a multicomponent
petrochemical wastewater shows a drawnout breakthrough curve. This is caused
by the varying rates of sorption and desorption in the mixed wastewater.

There are several modes of carbon column operation depending on the
results desired. These include:

1. Downflow—fixed beds in series. When breakthrough occurs in the
last column the first column is in equilibrium with the influent concentration (C_o)
in order to achieve a maximum carbon capacity. After carbon replacement in the
first column it becomes the last column in a series, etc. (figure 10.36a).

2. Multiple units—operated in parallel with the effluent blended to
achieve the final desired quality. The effluent from a column ready for regeneration or replacement, which is high in COD is blended with the other effluents from
fresh carbon columns to achieve the desired quality (figure 10.36b). This mode of
operation is most adaptable to waters in which the capacity at breakthrough/
capacity at exhaustion is great (near 1.0).

3. Upflow—expanded beds are used when suspended solids are present
in the influent or when biological action occurs in the bed (figure 10.36c).

4. Continuous counterflow—column or pulsed beds with the spent carbon from the bottom (in equilibrium with influent solute concentration) sent to
regeneration. Regenerated and makeup carbon is fed to top of the reactor (figure
10.36d).

As mentioned maximum economy requires that spent carbon be in
equilibrium with the influent wastewater. The depth of carbon removed for regeneration (and hence the depth of the total carbon system) will therefore depend on
the depth of the adsorption zone as shown in figure 10.37. A carbon adsorption
system including regeneration is shown in figure 10.38.

The most reliable approach to design a carbon adsorption system is to
conduct pilot column tests under conditions similar to those expected in the

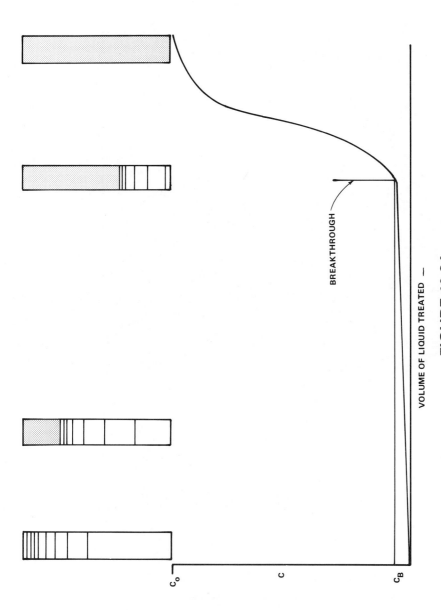

FIGURE 10.34

Mechanism of Adsorption in Continuous Columns

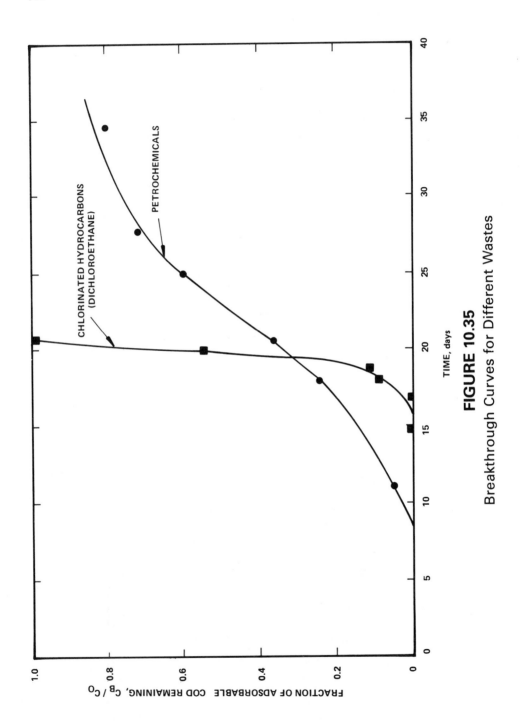

FIGURE 10.35

Breakthrough Curves for Different Wastes

FIGURE 10.36
Carbon Column Configurations

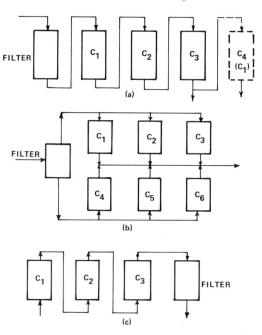

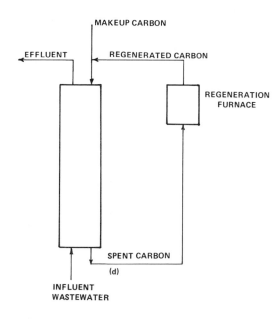

FIGURE 10.37
Advantage of Multistage Columns in Systems with Deep Adsorption Zones

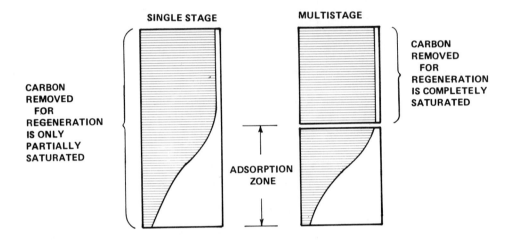

a. DEEP ADSORPTION ZONE

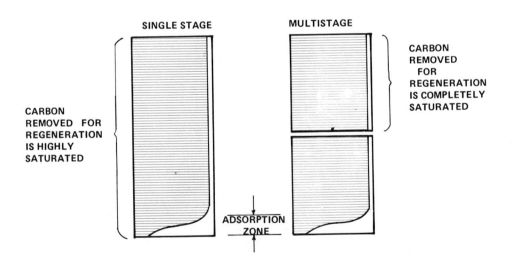

b. SHALLOW ADSORPTION ZONE

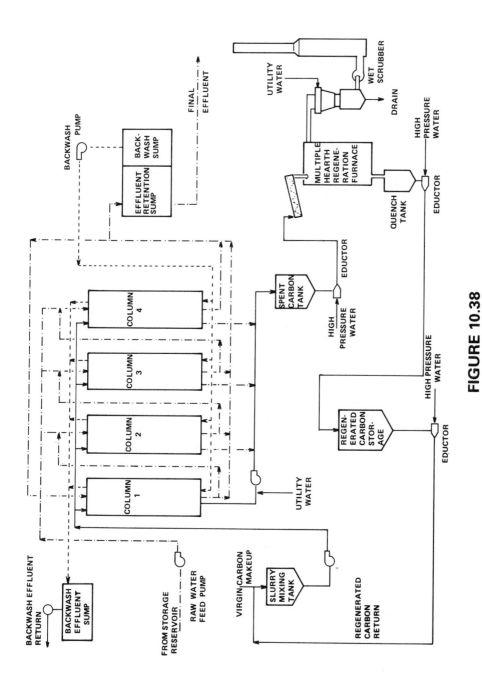

FIGURE 10.38

Flow Diagram of Carbon Adsorption Process

prototype. Since these conditions are generally not known at the time of testing, a method for extrapolating the data to various conditions is needed in order to reduce the number of experiments. Hutchins [17] has employed the Bohart-Adams equation in the form of Bed Depth Service Time (BDST) for interpretation of column data and process design, including extrapolation to conditions other than tested.

In order to develop a BDST correlation a number of pilot columns of equal depth are operated in series and breakthrough curves are plotted for each as shown in figure 10.39. These data are then used for producing a BDST correlation by recording the operating time required to reach a certain removal at each depth. Plots of BDST correlations for 10 and 90 percent breakthrough are shown in figure 10.40. These correlation lines correspond to 90 and 10 percent removals, respectively. The Bohart-Adams equation for the BDST is:

$$t = \frac{N_o}{C_o v} \, X - \frac{1}{KC_o} \, \ln \left(\frac{C_o}{C_B} - 1 \right) \qquad (10\text{-}4)$$

where t = service time, hr
X = bed depth, ft
v = hydraulic loading, or linear velocity of fluid, ft/hr
C_o = concentration of impurity in influent, lb/cu ft
C_B = concentration of impurity in effluent, lb/cu ft
N_o = adsorption efficiency, lb/cu ft
K = adsorption rate constant, cu ft/lb-hr

The slope of the BDST line is equal to the reciprocal velocity of the adsorption zone, and the x-intercept is the critical depth defined as the minimum bed depth required for obtaining the desired effluent quality at time zero. Hutchins [17] has shown that N_o is dependent on C_B as can also be seen from the variation in slope for different breakthrough values in figure 10.40.

If the adsorption zone is arbitrarily defined as the carbon layer through which the liquid concentration varies from 90 to 10 percent of the feed concentration, the depth of this zone is given by the horizontal distance between these two lines in the BDST plot. As can be seen from figure 10.40 the depth of the adsorption zone increases with time or with depth of bed.

Multistage carbon columns are used in an attempt to maximize carbon utilization thereby reducing operating costs. This is accomplished by regenerating only the carbon that has been in contact with the feed solution at its highest concentration, thus approaching the maximum level of saturation for the given feed. The advantage of a multistage system over a single column is more pronounced in the case of a deep adsorption zone as illustrated in figure 10.37. The figure shows the level of saturation along a single and a multistage system when the carbon is removed for regeneration. Whereas in the case of a shallow adsorption zone the carbon from both systems is highly saturated, in the case of a deep

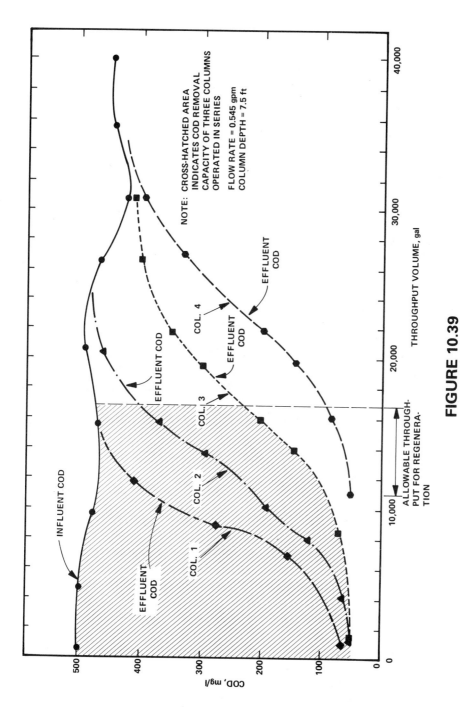

FIGURE 10.39

Breakthrough Curves from Continuous-Flow Studies

FIGURE 10.40
Bed Depth-Service Time Curves for a Single Stage System

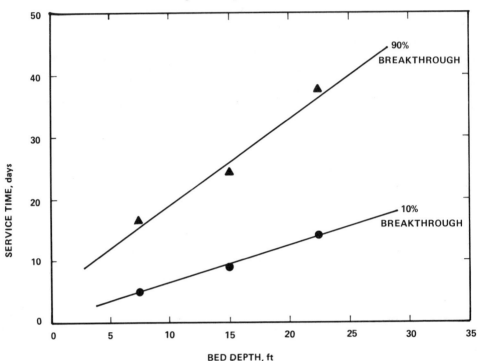

adsorption zone the carbon removed for regeneration in the multistage system is considerably more saturated compared to the single stage.

It should be emphasized that the BDST equation described above is only valid for a single stage column in which the feed is always applied to the head end of the bed, and the entire bed is in service throughout the run. These conditions differ from the continuous operation of a multistage system since in the latter case the head section of the column is removed and a fresh section added to the tail end when the effluent quality is no longer acceptable. Hutchins [18] suggested to use a modified BDST equation to calculate the forward velocity of the adsorption zone in the column of freshly added carbon. By assuming an exponential variation of concentration in the effluent from the last column in a series, and using the relationship of equation (10–4), he calculated the adsorption zone velocity in the new column, from which he obtained the rate of carbon utilization in a multistage system.

The main disadvantage of this method is that it is based on experimental data obtained during the early stage of operation that do not simulate the conditions that prevail in a continuous operation of a multistage system. A more reliable testing approach would be to simulate a multistage operation by repeatedly removing the head column when the concentration C_B in the last column exceeds the acceptable level and adding a new column containing fresh or regenerated carbon at the tail end. Under these conditions a steady-state operation will be reached on which a reliable design can be based. The results of such an experiment can be presented in a BDST type plot as shown in figure 10.41. In this hypothetical plot the lines start at different slopes indicating a different velocity of the adsorption zone for different breakthrough level. However, after several cycles a steady-state is approached and the forward velocity of the adsorption zone is the same for all levels of breakthrough. Deviation from this behavior can be expected if the carbon adsorption capacity changes upon successive regenerations.

In order to design an adsorption system with maximum carbon utilization the carbon removed for regeneration should be in a high level of saturation. For practical design purposes a 90 percent breakthrough level may be selected as the minimum effluent concentration at which the carbon is regenerated. If the

FIGURE 10.41

Hypothetical BDST Curves for a Multistage Operation

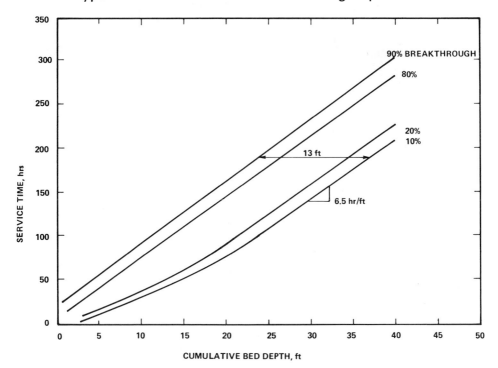

effluent quality requirements call for a 90 percent removal, the addition of a new column to the tail end of the series should coincide with a 10 percent breakthrough from the last column. Thus, the horizontal distance between 90 and 10 percent breakthrough lines in the BDST plot is taken as the depth of the adsorption zone. This is the minimum bed depth in a moving bed system producing the desired effluent with maximum carbon utilization. For a multistage system the number of stages and the bed depth in each stage are related to the depth of the adsorption zone as follows:

$$n = \frac{D}{d} + 1 \tag{10-5}$$

where n = number of stages in series (not including one stage in regeneration or standby)

D = depth of adsorption zone, ft

d = depth of single stage, ft

Selection of d should be based on practical considerations and should be an integer fraction of D. Selecting a small d will result in small size equipment with lower carbon inventory but a high number of stages and consequently more costly equipment. A very low d approaches a moving bed operation. A high value of d will result in simpler operation but higher carbon inventory and larger plant size.

In the example shown in figure 10.41 the depth of the adsorption zone appears to be 13 ft (4 m) as indicated by the horizontal distance between the 10 and 90 percent breakthrough lines. Assuming a total of 3 columns in the series the depth of a single column should be 6.5 ft (2 m) by virtue of equation (10–5).

The rate of carbon utilization can be calculated from the reciprocal of the slope of the BDST line. This value multiplied by the column's cross-sectional area and the apparent density of the carbon is the rate of carbon utilization that is also equal to the required regeneration capacity. In an example shown in figure 10.41, the forward velocity of the adsorption zone is 0.15 ft/hr (4.6 cm/hr).

The preceding procedure may be applied directly for the design of systems operating under conditions similar to those in the experiment. If other operating conditions are contemplated extrapolation can be done although the accuracy of such extrapolation is still not certain. The two most likely parameters to be extrapolated are the feed concentration C_o and the hydraulic loading v. Each of these parameters affects both the depth and the forward velocity of the adsorption zone.

An increase in feed concentration will result in an increase in the forward velocity of the adsorption zone although not proportionately. As can be seen from equation (10–4) the velocity of the adsorption zone is directly proportional to C_o and inversely proportional to N_o. Since N_o is related to C_o the net effect of C_o on the forward velocity of the adsorption zone depends on the C_o to N_o correlation. In systems where no biological activity is present and the contact time is sufficiently long, this correlation can be derived from isotherm data. The effect of C_o on the depth of the adsorption zone cannot be readily calculated. However, for equal percent removal it is reasonable to assume that the effect is insignificant. Obviously, if equal effluent concentrations are required, the system operated with higher C_o will require higher percentage of removal, which in turn will lead to a deeper adsorption zone.

The hydraulic loading also affects both the depth and the forward velocity of the adsorption zone. The forward velocity is directly proportional to the hydraulic loading as can be seen from equation (10–4). It is assumed that the contact time is long enough and N_o does not change with hydraulic loading. Due to increased dispersion the depth of the adsorption zone will generally increase with hydraulic loading. The exact relationship between the two is yet unknown.

PROBLEM 10–3

From a pilot column test to treat a refinery wastewater at hydraulic loading of 5 gpm/ft² the steady-state results are shown in table I.

TABLE I
Results from Activated Carbon Column Test

Bed Depth,	Service Time, hr	
ft	$C_B/C_o = 0.9$	$C_B/C_o = 0.1$
4	42	6.6
8	66	31
12	90	55

$C_o = 480$ mg/l COD

    (a) Using the Bohart-Adams approach, calculate the constants, N_o and K.

    (b) Determine the depth of adsorption zone.

    (c) For effluent COD of 50 mg/l, determine the number of stages in series required and the predicted service time.

(a) Develop the BDST curve by plotting service time against bed depth as shown in figure I.

$$\text{Slope} = \frac{N_o}{C_o v} = 6 \text{ hr/ft}$$

$$v = \frac{5 \text{ gpm}}{\text{ft}^2} \cdot \frac{0.1337 \text{ ft}^3}{\text{gal}} \cdot \frac{60 \text{ min}}{\text{hr}}$$

$$= 40.1 \text{ ft/hr}$$

$$C_o = \frac{480 \text{ mg}}{1} \cdot \frac{2.205 \times 10^{-6} \text{ lb}}{\text{mg}} \cdot \frac{28.32 \text{ l}}{\text{ft}^3}$$

$$= 0.03 \text{ lb/ft}^3$$

Then:

$$N_o = 6 \cdot 0.03 \cdot 40.1 \text{ lb/ft}^3$$

$$= 7.22 \text{ lb/ft}^3$$

$$\text{Intercept} = -\frac{1}{KC_o} \ln\left(\frac{C_o}{C_B} - 1\right) = 18.0 \text{ (based on 90 percent breakthrough curve)}$$

$$K = -\frac{1}{18.0 \cdot 0.03} \ln\left(\frac{100}{90} - 1\right)$$

$$= 4.07 \text{ ft}^3/\text{lb-hr}$$

$K = 4.31 \text{ ft}^3/\text{lb-hr}$ if based upon 10 percent breakthrough curve.

FIGURE I
BDST Curves for Continuous Column Tests

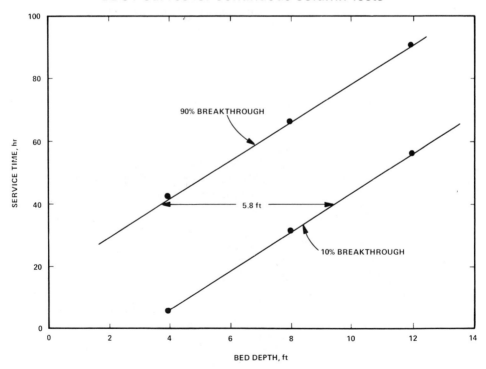

Hence, averaged value of 4.19 ft³/lb-hr is to be used for K in the following calculation.

(b) The horizontal distance between 90 percent and 10 percent breakthrough lines in figure I is taken as the depth of the adsorption zone, 5.8 ft.

(c) For effluent COD of 50 mg/l the critical bed depth of activated carbon column can be computed by:

$$X_o = \frac{v}{N_o K} \ln\left(\frac{C_o}{C_B} - 1\right)$$

$$= \frac{40.1}{7.22 \cdot 4.19} \ln\left(\frac{480}{50} - 1\right)$$

$$= 2.85 \text{ ft}$$

If the column depth of 3 ft is selected, the number of stages in series is:

$$n = \frac{D}{d} + 1$$

$$= \frac{5.8}{3.0} + 1$$

$$= 2.93, \text{ say 3 columns in series}$$

The predicted service time is:

$$t = \frac{N_o}{C_o v} X - \frac{1}{KC_o} \ln\left(\frac{C_o}{C_B} - 1\right)$$

$$= \frac{7.22}{0.03 \cdot 40.1} \cdot 3 \cdot 3 - \frac{1}{4.19 \cdot 0.03} \ln\left(\frac{480}{50} - 1\right)$$

$$= 36.9 \text{ hr}$$

Performance of Activated Carbon Systems

Activated carbon columns have been applied for tertiary treatment following biological treatment and for the treatment of sewage and industrial wastewaters following a chemical precipitation step (PCT). In the former case the biodegradable organics (BOD) have been substantially removed and little or no biological activity occurs in the carbon column. In the latter case, soluble BOD passes to the columns and biological activity (aerobic or anaerobic) occurs in the columns. The performance of tertiary carbon systems is shown in table 10.15. The performance of PCT systems is shown in table 10.16.

TABLE 10.15
Characteristics of Activated Carbon Tertiary Treatment

Wastewater	Treatment Sequence[a]	Average Effluent Characteristics (mg/l)				
		BOD	COD	TOC	Oil	Phenol
Petroleum	AS – F – CA	–	27	17	–	–
Petroleum	AS – F – CA	–	–	6	1.7	0.013
Petroleum	AS – F – CA	3	26	7	7.0	0.001
Organic Chemicals	AS – CA	38	118	49	–	–
Domestic	AS – C – F – CA	1	1–25	1–6	–	–

[a]AS = Activated sludge; F = Filtration; C = Coagulation; CA = Carbon adsorption.

Biological action in the carbon columns provides biological regeneration of the carbon, thus increasing the apparent capacity of the carbon. Weber *et al.* [20] showed that the capacity in columnar operation with facultative biological activity was in excess of 1.0 lbs COD/lb C as compared to 0.5 lbs COD/lb C determined from an adsorption isotherm. Biological activity may be an asset or a liability. When the applied BOD is in excess of 50 mg/l anaerobic activity in the columns may cause serious odor problems while aerobic activity may cause serious common plugging problems due to the large amount of biomass generated by aerobic biological action. The presence of biological activity resulting in the production of biomass increases the head loss in conventional packed bed adsorbers thus making upflow, expanded bed adsorbers more feasible. A high quality effluent requires filtration following the adsorbers.

Ford [14] found that anaerobic activity could be inhibited and some non-sorbable compounds converted to sorbable compounds by ozonation prior to the carbon columns. The ozone also increase the dissolved oxygen level in the carbon column influent thereby reducing the possibility of sulfide production and anaerobic biological activity. Ozone dosages at the Cleveland Westerly Plant ranged from 4–9 mg/l. In the PCT pilot study at Pomona an average dosage of 514 mg/l of sodium nitrate (as N) was employed to retard sulfide generation in the column. Studies by the EPA [21] on a petroleum refinery wastewater found high removal of heavy metals through the carbon columns. This data is summarized in table 10.17. It was noted, however, that aluminum salts present in the influent can remain on the surface of the carbon during regeneration, reducing the effective surface area of the carbon and its adsorption capacity. Iron salts can catalyze oxidation reactions of the carbon and the gases during regeneration thus permanently damaging the carbon structure.

A PCT system treating a wastewater from the production of sulfuric acid, formaldehyde, pentaerythritol, sodium sulfate and sulfite, orthophenylphenol, and a number of synthetic resins reduce the BOD and COD from 390 mg/l and 650

	Effluent COD			Effluent TOC			Effluent BOD		
	Raw COD mg/l	mg/l	Removal, %	Raw TOC mg/l	mg/l	Removal, %	Raw BOD mg/l	mg/l	Removal, %
Blue Plains Pilot Plant	320	16	95	100	8	92	150	6	96
Owosso, Michigan	250–350	24–30	≈91	–	–	–	140	8	84
Pomona, California	321	19	94	–	–	–	120[b]	7.8	78.5[c]
Rosemount, Minnesota (first year)	–	–	–	–	–	–	230	23	90
Rosemount, Minnesota (last 3 to 4 months)									
Battelle Pilot Plant at Westerly	527	42	92	–	–	–	240	26	89
CRSD Pilot Plant at Westerly	437	56	87	90	21	77	206	32	84

[a] From [19].
[b] Estimated based on BOD similar to COD removals across clarifier.
[c] Just around carbon columns.

TABLE 10.16
Summary of PCT Pilot-Plant and Full-Scale
Plant Performances[a]

TABLE 10.17

Refinery Wastewater Treatment Results[a]

Parameter	API Separator, mg/l	Biotreated, mg/l	Carbon Treated, mg/l	Biocarbon Treated, mg/l
Chromium	2.2	0.9	0.20	0.20
Copper	0.5	0.1	0.03	0.05
Iron	2.2	3.0	0.30	0.90
Lead	0.2	0.2	0.20	0.20
Zinc	0.7	0.4	0.08	0.15

[a]From [21].

mg/l to 35 mg/l and 70 mg/l, respectively. Effluent variability is of major concern in all wastewater treatment processes. The long term average COD concentrations in effluents from carbon adsorbers treating refinery wastewaters is shown in figure 10.42. The carbon adsorption capacity for various plants summarized by Ford [14] is shown in figure 10.43. The results of adsorption tests on various industrial wastewaters is shown in table 10.18.

Powdered Activated Carbon (PAC)

Powdered activated carbon can be added directly to the aeration basin, to a post filter and backwashed to the aeration basin or in two or more carbon contactors in series for PCT. Carbon dosages of 25–50 mg/l are employed to improve effluent quality by adsorption of color, surfactant, and degradable and nondegradable organics. A buildup of PAC in the mixed liquor suspended solids occurs with the steady state value which is a function of the sludge age of the system. At steady state, the fraction of carbon in the MLSS is related to the ratio of biosolids produced to carbon added:

$$\% \, C = \frac{M}{\Delta X_B + M} \cdot 100 \qquad (10\text{--}6)$$

where ΔX_B = biosolids produced
       M = the carbon added

The fresh carbon continually added to the aeration tank provides a degree of tertiary treatment. At a municipal plant a 20–25 mg/l PAC dosage reduced effluent suspended solids, improved settling and compaction properties and decreased color, odor, and aeration tank foaming. PAC addition to a contact stabili-

FIGURE 10.42

Effluent COD Attainable from Activated Carbon Systems [8]

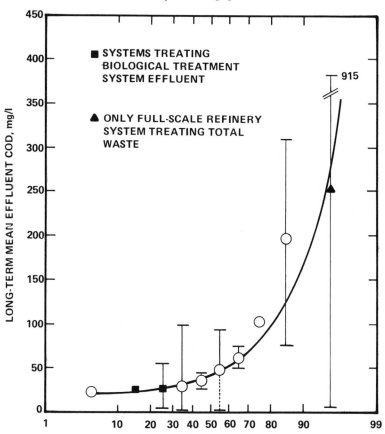

PERCENT OF PILOT-PLANT OR FULL-SCALE SYSTEMS
TREATING REFINERY OR RELATED WASTEWATERS
WITH LONG TERM MEAN COD LESS THAN
STATED VALUE

zation plant provided resistance to shock loading by trichlorophenol. PAC addition to sludges have been shown to increase anaerobic digester performance. These are indications that PAC acts as a settling aid and a catalyst for the digestion reaction.

PAC systems have been evaluated for the treatment of municipal wastewater. Since carbon capacity relates to the effluent concentration, multiple contactor in counterflow series should be employed. Pilot plant studies reported a carbon dosage of 450–500 mg/l with a polyelectrolyte dosage of 1.3 mg/l state [23].

FIGURE 10.43
Carbon Adsorption Capacity for Various Plants

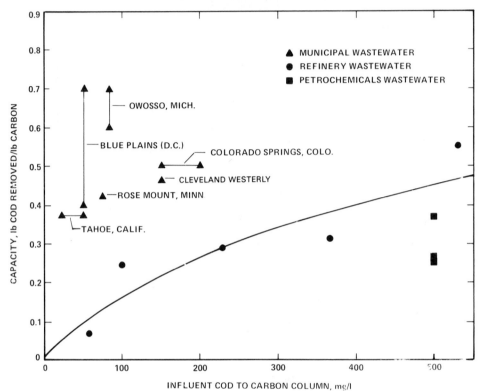

ION EXCHANGE header image:

ION EXCHANGE

Ion exchange has been applied to the treatment of plating rinse waters for the recovery of chromic acid and deionized water. A typical flowsheet is shown in figure 10.44. Ion exchange has been used in the fertilizer industry for the removal and recovery of ammonium nitrate. Ion exchange can be considered if the total dissolved solids is less than 1000 mg/l for the partial demineralization of wastewaters.

Ion exchange is a process in which ions held by electrostatic forces to functional groups on the surface of a solid are exchanged for ions of a different species in solution. Ion exchange is employed for the removal or exchange of dissolved inorganic salts in waters or wastewaters such as hardness (Ca^{++} and Mg^{++}) or metal ions (Cr^{++6}, Zn^{++}) etc.

TABLE 10.18

Results of Adsorption Isotherm
Tests on Different Industrial Wastes

Type of Industry	Initial TOC (or phenol), mg/l	Initial color O.D.	Average % reduction	Carbon exhaustion rate, lb/1000 gal
Food and kindred products	25–5,300	–	90	0.8–345
Tobacco manufacturers	1,030	–	97	58
Textile mill products	9–4,670	–	93	1–246
	–	0.1–5.4	97	0.1–83
Apparels and allied products	390–875	–	75	12–43
Paper and allied products	100–3,500	–	90	3.2–156
	–	1.4	94	3.7
Printing, publishing and allied industries	34–170	–	98	4.3–4.6
Chemicals and allied products	19–75,500	–	85	0.7–2,905
	(0.1–5,325)	–	99	1.7–185
	–	0.7–275	98	1.2–1,328
Petroleum refining and related industries	36–4,400	–	92	1.1–141
	(7–270)	–	99	6–24
Rubber and miscellaneous plastic products	120–8,375	–	95	5.2–164
Leather and leather products	115–9,000	–	95	3–315
Stone, clay and glass products	12–8,300	–	87	2.8–300
Primary metal industries	11–23,000	–	90	0.5–1,857
Fabricated metal products	73,000	–	25	606

Most ion exchange resins in use today are synthetic materials consisting of a network of hydrocarbon radicals to which are attached soluble ionic functional groups. The total number of functional groups per unit weight of resin determines the exchange capacity and the group type determines the exchange equilibrium and the ion selectivity.

Ion exchange resins may be strongly or weakly acidic cation exchangers and strongly or weakly basic anion exchangers. The cation exchangeable ion may be hydrogen or some other monovalent cation such as sodium. In the anion exchanger the exchangeable ion is hydroxyl or some other monovalent anion.

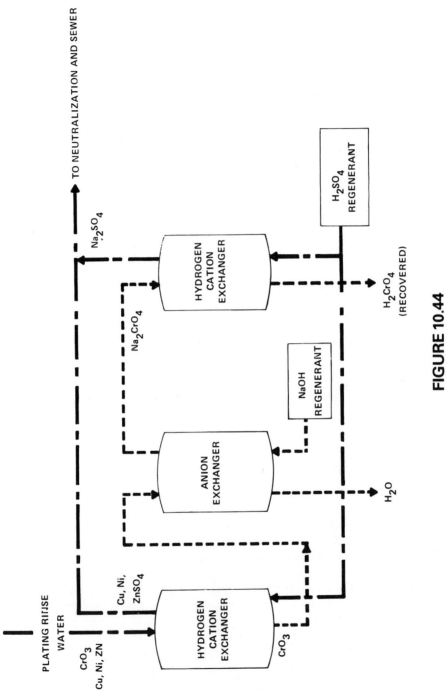

FIGURE 10.44
Ion Exchange Flow Diagram

When the resin is exhausted, it is regenerated with a concentrated solution containing the original ions which reverse the equilibrium. The reactions occurring in a cation exchanger are:

$$2R \cdot H + Ca^{++} \rightleftharpoons R_2 \cdot Ca + 2H^+$$

The regeneration reaction is:

$$R_2 \cdot Ca + 2H^+ \rightleftharpoons 2R \cdot H + Ca^{++}$$

An anion exchanger similarly replaces anions with hydroxyl ions:

$$R \cdot (OH)_2 + SO_4^= \rightleftharpoons R \cdot SO_4 + 2OH^-$$

and regeneration yields

$$R \cdot SO_4 + 2OH^- \rightleftharpoons R \cdot (OH)_2 + SO_4^=$$

Ion exchangers may be operated as batch or column operations. The most common is column operation in which the liquid is applied through a down flow packed bed. In order to minimize head loss through the ion exchange bed, prefiltration of the water is usually employed.

Nearly all of the cation exchangers employed in the water pollution field are strongly acidic and remove all exchangeable cations. Anion exchangers, however, may be strongly or weakly basic depending on the application.

In exchanger operation, water or wastewater is passed through the column until breakthrough occurs (the concentration of ions in the effluent from the column exceeds the desired level). The exchanger is then removed from service, backwashed to remove accumulated dirt and solids and regenerated by passing a concentrated solution containing the exchangeable cation or anion through the bed. The bed is then rinsed to remove residual regenerant and placed back in service.

In the design and selection of an exchange resin the capacity of the resin and the efficiency of the process must be established. The theoretical capacity of the resin is the equivalent number of exchangeable ions per unit weight or unit volume of resin and is expressed as equivalents/gm of resin, kg $CaCO_3/ft^3$ of resin, lbs $CaCO_3/ft^3$ of resin, etc. The degree of theoretical capacity achieved depends on the quantity of regenerant used and an economic balance must be derived between the degree of theoretical capacity attained (degree of column utilization) and the quantity of regenerant employed (regenerant efficiency). For example, in water softening applications, levels of regeneration of 3 to 10 pounds of NaCl per cubic foot of resin are used yielding 45 to 70 percent regenerant efficiency and 30 to 65 percent column utilization. Typical data are shown in figure 10.45.

The operating characteristics of an exchanger system are shown below:

Applied feed rate 2–8 gpm/ft³ resin (0.3–1.1 m³/min/m³ resin)
Exchanger depth 2–6 feet (0.6–1.8 m)

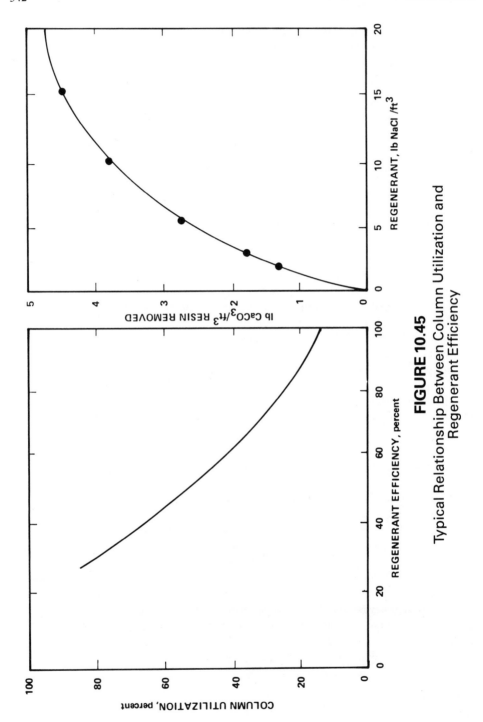

FIGURE 10.45

Typical Relationship Between Column Utilization and
Regenerant Efficiency

Backwash expansion 50–100 percent
Backwash flow rate 5–10 gpm/ft^2 (0.2–0.4 m^3/min-m^2)
Regenerant flow rate 1–2 gpm/ft^2 (0.04–0.08 m^3/min-m^2)
Rinse water volume 30–100 gal/ft^3 (4–13 m^3/m^3)
Rinse water flow rate 1–1.5 gpm/ft^2 (0.04–0.06 m^3/min-m^2)

Table 10.19 may serve as a guide.

TABLE 10.19
Suggested Regeneration Levels

Ion Exchange Resin	Ionic Form	Regenerant	Requirement $\left(\dfrac{\text{Meq Regen.}}{\text{Meq Resin}} \right)$
Strong Acid	H$^+$	HCl	3–5
Cation	Na$^+$	H$_2$SO$_4$	3–5
		NaCl	3–5
Weak Acid	H$^+$	HCl	1.5–2
Cation	Na$^+$	H$_2$SO$_4$ / NaOH	1.5–2
Strong Base	OH$^-$	NaOH	4–5
Anion I	Cl$^-$	NaCl / HCl	4–5
	SO$_4^=$	Na$_2$SO$_4$ / H$_2$SO$_4$	4–5
Strong Base Anion II	OH$^-$	NaOH	3–4
Weak Base Anion	Free Base	NaOH / NH$_4$OH / Na$_2$CO$_3$	1.5–2
	Cl$^-$	HCl	1.5–2
	SO$_4^=$	H$_2$SO$_4$	1.5–2

PROBLEM 10–4

Design an ion exchange system to treat 100,000 gallons of wastewater per day. The principal metal components of the waste water are chromium with a concentration of 65 mg/l as CrO$_3$. The wastewater also contains 20 mg/l Cu, 10 mg/l Zn and 15 mg/l Ni. The flow sheet for the process is shown in figure 10.44.

(a) The number of equivalents of metal ions to be removed in the cation exchanger is:

$$20 \text{ mg/l Cu} = 0.63 \text{ meq/l}$$
$$10 \text{ mg/l Zn} = 0.31 \text{ meq/l}$$
$$\underline{15 \text{ mg/l Ni} = 0.51 \text{ meq/l}}$$
$$\text{Total} = 1.45 \text{ meq/l}$$

The total equivalents/day to be removed are:

$$= (1.45 \text{ meq/l}) \ (3.78 \text{ l/gal}) \ (100{,}000 \text{ gal/day}) \left(\frac{1 \text{ eq}}{1000 \text{ meq}} \right)$$

$$= 549 \text{ eq/day}$$

Using a resin operating capacity of 80 eq/ft^3 resin and a 5-day operation period between regenerations the total resin requirement is:

$$\text{Resin requirement} = \frac{549 \text{ eq/day} \times 5}{80 \text{ eq/ft}^3}$$

$$= 35 \text{ ft}^3 \text{ of resin}$$

If a column diameter of 2 feet is used, the total resin depth required is 11.1 feet. Use two 2-ft-diameter columns each 8.5 feet deep. This will permit sufficient depth for 50 percent expansion of the bed for backwashing and cleaning. If a 5 percent H_2SO_4 solution is used for regeneration and the regeneration required is 12 lbs/ft^3, the required H_2SO_4 is:

$$\text{lbs } H_2SO_4 = 12 \text{ lbs/ft}^3 \cdot 35 \text{ ft}^3 \cdot 1/0.05 = 8400 \text{ lbs}$$

Approximately 150 gallons of water will be required for rinsing each cubic foot of resin. The required rinse water is:

$$\text{Rinse water} = 150 \text{ gal/ft}^3 \cdot 35 \text{ ft}^3 = 5250 \text{ gal}$$

(b) The chromic acid passing through the cation unit will be removed in the anion exchange unit. The total chrome removed each day is:

$$65 \text{ mg/l} \cdot 0.1 \text{ mgd} \cdot 8.34 = 54 \text{ lb/day}$$

If the anion unit has a capacity of 3.8 lb CrO_3 per cubic foot, then the resin requirement for a five day regeneration is:

$$\text{Resin requirement} = \frac{54 \cdot 5}{3.8} = 71 \text{ ft}^3$$

For a 2 ft column, the required depth is 22.6 ft. Use four columns each 8.5 ft deep to permit 50% expansion during backwash. The required regenerant at 4.8 lb NaOH per cubic foot of resin at 10% solution is:

$$\text{Regenerant} = 4.8 \cdot 71 \cdot 1/0.10 = 3410 \text{ lbs}$$

The rinse water requirement at 100 gal per cubic foot of resin is:

$$\text{Rinse water} = 100 \text{ gal/ft}^3 \cdot 71 \text{ ft}^3 = 7100 \text{ gal}$$

(c)   Chromic acid can be recovered from the spent regenerant by passing the sodium chromate from the anion unit through a cation exchanger. The total sodium in the spent regenerate is 341 lbs and the total equivalents of sodium is:

$$\frac{341 \text{ lbs} \cdot 454 \text{ g/lb}}{40 \text{ g}} = 3870 \text{ equivalents}$$

DISINFECTION

Disinfection is the destruction of disease producing organisms by the use of a chemical or other agent. Disinfection is practiced in most municipal wastewater treatment plants. Industrial wastewater treatment plants may or may not require disinfection depending on the presence of pathogenic organisms in the wastewater. Disinfection can be accomplished by physical action such as direct heat or ultraviolet light, by a reaction with the cell to form toxic substances and a kill by either chemical poisoning or disruptive structural changes within the cell. Halogens will kill bacteria by interference with the glycogen cycle or by oxidation of the -SH endings of protein enzyme combinations. Oxidizing agents such as $KMnO_4$, H_2O_2, O_3 and superchlorination literally "burn up" the organism. An effective disinfectant should have a high germicidal power and rapid speed of action but low toxicity to higher life and a low affinity for extraneous matter at a low cost. Of the disinfectants available today, chlorine and ozone are in most common use. An objection to chlorine in many applications is the formation of chlorinated hydrocarbons, which in themselves may be toxic or carcenogenic. There are a number of factors that will affect disinfection efficiency. Among these are the specific nature of the organism, the concentration of disinfectant, the time of contact, temperature and the presence of extraneous materials that will react with the disinfectant.

Under ideal conditions all cells would be equally susceptible to a given concentration of a disinfectant and the rate of disinfection can be defined by Chick's Law:

$$\frac{dN}{dt} = -kN \qquad\qquad (10\text{--}7)$$

which integrates to:

$$N/N_o = e^{-kt}$$

where k = decay rate coefficient of organisms
       N = number of organisms remaining at any time
       N_o = initial number of organisms
       t = time of contact

Because of a variability in organism susceptibility deviation frequently occurs and the fraction remaining is proportional to t^2 instead of t as shown in figure 10.46.

The effect of the concentration of disinfectant can be experienced by the relationship

$$C^n t_r = constant \tag{10-8}$$

where C = concentration of disinfectant
       t_r = time required for a constant percent kill
       n = coefficient of dilution

when $n > 1$ the time of contact is more important than the dosage. The efficiency of disinfection will decrease with a decrease in temperature.

CHLORINATION

Chlorine is the most common disinfectant at present and may be employed as Cl_2 (gas), hypochlorite, chlorine dioxide, or as a chloramine. Chlorine will immediately hydrolyze in water to hypochlorous acid:

$$Cl_2 + H_2O \rightleftharpoons HOCl + H^+ + Cl^-$$

The hypochlorous acid will ionize:

$$HOCl \rightleftharpoons H^+ + OCl^-$$

HOCl is the principal disinfectant and therefore the efficiency of disinfection will be a function of pH as shown in figure 10.47.
Hypochlorite will ionize:

$$Ca(OCl)_2 + H_2O \rightleftharpoons Ca^{++} + H_2O + 2OCl^-$$

and:

$$OCl^- + H^+ \rightleftharpoons HOCl$$

Chloramines resulting from the reaction of chlorine with ammonia have a disinfecting power and while less than chlorine and have the advantage of longer persistence in water than chlorine. The reactions are:

$$HOCl + NH_3 \rightleftharpoons NH_2Cl + H_2O \text{ (monochloramine)}$$

$$NH_2Cl + HOCl \rightleftharpoons NHCl_2 + H_2O \text{ (dichloramine)}$$

$$NHCl_2 + HOCl \rightleftharpoons NCl_3 + H_2O \text{ (nitrogen trichloride)}$$

The percentage of each form is highly dependent on pH. The percentage of mono and dichloramine as a function of pH is shown in figure 10.48. Nitrogen trichloride will exist only below pH 4.4. Under normal conditions chloramine may take as much as 100 times longer than chlorine to effect equivalent disinfection. The chloramines are more effective at high pH.

Usual requirements for chlorination vary from 0.2 mg/l after 15 minutes contact to 1.0 mg/l after 30 minutes contact with the most common being 0.5 mg/l residual after 15 minutes contact. In general the dosages required to yield 0.2 mg/l residual after 10–15 minutes contact are shown in table 10.20.

Viruses

Viruses are present in municipal wastewaters and present problems in potable water supplies and water reuse systems. The density of viruses in wastewaters will vary with the season of the year, the percapita water supply usage, the socio-economic level of the community, etc. It has also been shown that under certain conditions enteric viruses may survive in water for a significant period of time, e.g., up to 200 days. Engelbrecht [24] has summarized the removal of viruses by conventional wastewater treatment processes as shown in table 10.21. Under favorable conditions the activated sludge process can effect > 96 percent removal of enteric viruses. Chlorination is less effective for virus inactivation than for pathogens. It was further shown that little virus inactivation is obtained in wastewater after the chlorine has reacted with ammonia. In general high chlorine residuals are required to inactivate enteric viruses. It has been *tentatively* concluded by some investigators that enteric viruses can be effectively destroyed by a free chlorine residual of 1.0 mg/l for 30 minutes. Ozone has been shown to be more effective than chlorine for virus inactivation. Katzenelson *et al.* [25] showed a 99 percent inactivation of Poliovirus 1 in less than 8 seconds with a 0.3 mg/l residual ozone. Inactivation occurred in two steps, an initial inactivation in less than 8 seconds of 99 to 99.5 percent depending on the ozone residual followed by a second stage of 1 to 5 minutes, which gave complete inactivation of the virus.

NITROGEN REMOVAL

Nitrogenous matter introduced into waterways from wastewater discharges may lead to accelerated eutrophication and to depleted oxygen concentration in the receiving waters. If water is to be reused for municipal supplies, ammonia will adversely affect the efficiency of chlorination of the water as well as

FIGURE 10.46

Length of Survival of *E. Coli*

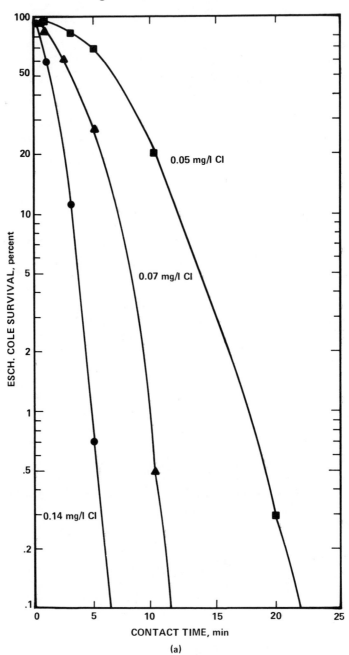

(a)

FIGURE 10.46 (Continued)

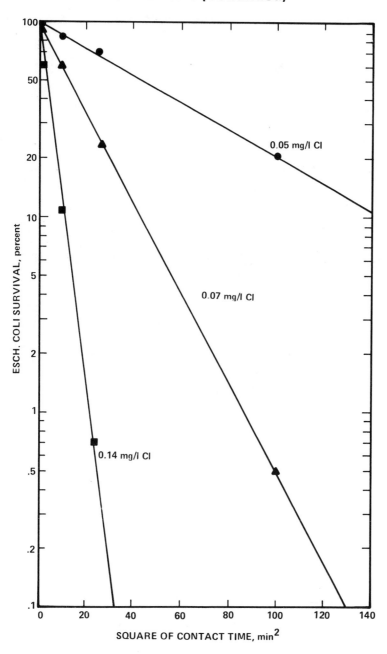

(b)

FIGURE 10.47
Relative Amounts of HOCl and OCl⁻ at Various pH Levels

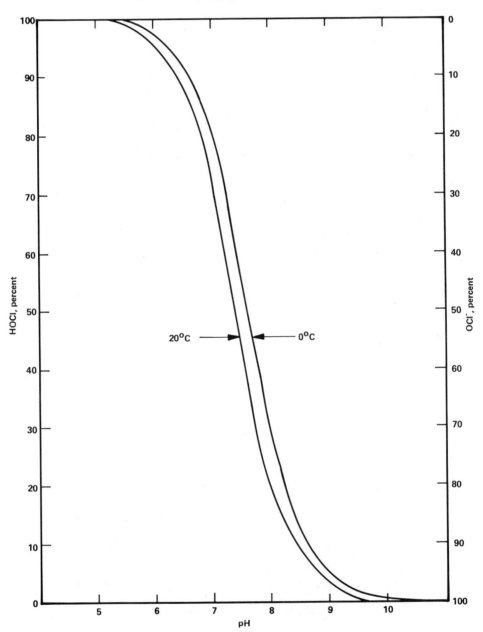

FIGURE 10.48
Relative Amounts of Chloramines Formed
at Various pH Levels

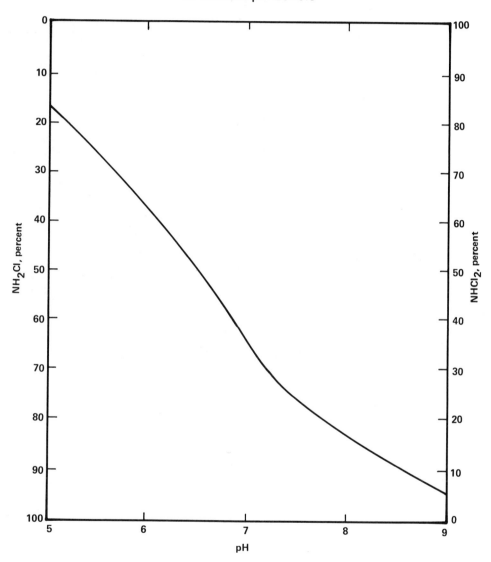

TABLE 10.20

Chlorine Dosages Required to Yield 0.2 mg/l Residual after 10–15 Minutes Contact Time

Sewage type	Dosage, mg/l
Raw:	
Fresh to stale	6–12
Septic	12–25
Settled:	
Fresh to stale	5–10
Septic	12–40
Effluent chemical precipitation	3–6
Trickling Filter	
Normal	3–5
Poor	5–10
Activated Sludge	
Normal	2–4
Poor	3–8
Intermittent Sand Filter:	
Normal	1–3
Poor	3–5

TABLE 10.21

Removal of Viruses by Wastewater Treatment Processes

Processes	Removal or Inactivation, %
Primary treatment	
Grit chamber—comminutor	0–50
Plain sedimentation	0–?
Secondary treatment	
Activated sludge	75–99
Trickling filter	0–85
Stabilization pond	0–96
Chemical coagulation-alum, iron salts	20–60
Advanced treatment	
Chemical coagulation-alum, iron salts	90–99
Phosphate precipitation	90–98
Activated carbon adsorption	10–99

NOTE: Approx. virus input concentration: 1×10^4–1×10^5 PFU/mℓ

increase chlorine consumption. Recently nitrogen removal has assumed greater significance as a means of protecting and preserving the environment.

The most feasible methods for nitrogen removal from wastewater may be grouped as either biological or chemical-physical processes. The biological processes include microbial nitrification-denitrification and algal harvesting. Nitrification-denitrification systems have been discussed in detail in chapter 9. The assimilation of nitrogen into algal cells is a potential method of nitrogen removal provided that the algae can be separated from the wastewater effluent. Removals by the use of algal stripping can be expected to range from 60 to 80 percent under favorable conditions of sunlight and temperature.

Chemical and physical processes for nitrogen removal have the advantage of being more amenable to control than biological processes and more adaptable to the fluctuating flows and concentrations inherent to most of industrial wastewater systems. The commonly used processes are air stripping, ion exchange, and breakpoint chlorination. Other methods such as reverse osmosis and electrodialysis might also be used. Electrodialysis can be expected to remove about 40 percent of ammonium or nitrate; reverse osmosis, 80 percent.

Air Stripping

In a wastewater stream ammonium ions exist in equilibrium with ammonia:

$$NH_3 + H_2O \rightleftharpoons NH_4^+ + OH^-$$

At pH 7 only ammonium ions (NH_4^+) exist in solution while at pH 12 the solution contains NH_3 as a dissolved gas. The relative percentages of ammonium ions and ammonia at different pH levels and temperatures are shown in figure 10.49. Air stripping of ammonia consists of raising the pH of the wastewater to pH 10.5–11.5 and providing sufficient air-water contact to strip the ammonia gas from solution. Conventional cooling towers have generally been employed for the stripping process. pH adjustment of the wastewater may employ caustic or lime. If lime is used with municipal wastewater the values shown in figure 10.6 should generally approximate the lime requirements. Under turbulent conditions in a stripping tower the theoretical air requirements per unit of water can be calculated from Henry's Law assuming that the water leaving and air entering the bottom of the tower are free of ammonia. The equilibrium relationship is shown in figure 10.50. At 20° C the theoretical gas/liquid ratio can be calculated from figure 10.50 as 1.33 moles air/mole H_2O or 221 ft^3 (STP)/gallon.

In practice, Tchobanoglous [27] estimated the air requirements as shown in figure 10.51. Conventional cooling towers that have been employed for ammonia stripping are shown in figure 10.52. A design procedure for stripping towers has been developed by Roesler et al. [28]. Ammonia removal from domestic wastewa-

FIGURE 10.49

Distribution of Ammonia and Ammonium Ion
with pH and Temperature

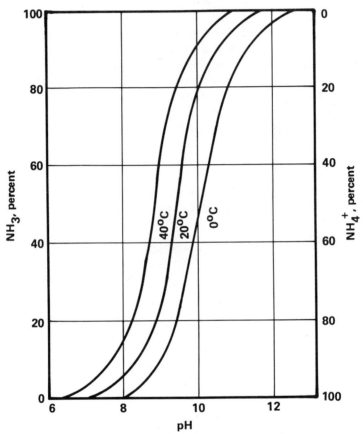

ter in excess of 90 percent was found with 480 ft³/gallon above pH 9.0. Hydraulic loading to stripping towers is recommended over the range 1–3 gpm/ft² (0.04–0.12 m³/min-m²). Tower depth and packing configuration will also affect performance as shown in figures 10.53 and 10.54.

Problems associated with ammonia stripping are reduced efficiency and ice formation in colder climates, deposition of calcium carbonate on the media when lime is used for pH adjustment, possible air pollution problems and deterioration of wood packing. A process alternative has been developed in which the exhaust air from the stripper is passed through H_2SO_4 and recycled. In this way air pollution problems are eliminated, ammonium sulfate is recovered and air temperatures are maintained high [29].

A flowsheet for the removal of nitrogen and phosphorus from domestic wastewater employing ammonia stripping is shown in figure 10.55.

FIGURE 10.50
Equilibrium Curves for Ammonia in Water [27]

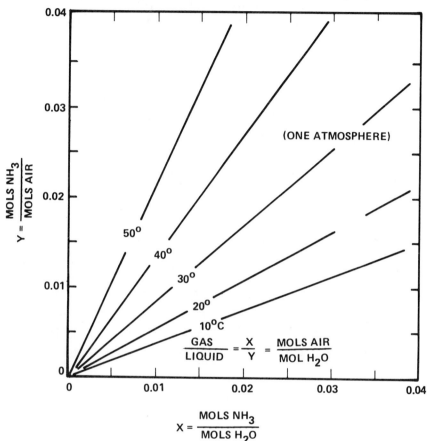

Considerable work has been done employing ammonia stripping ponds in Israel [30]. In unaerated ponds, ammonia was reduced 50 percent at a pH of 10.5 over a period of 130 hours. When aerated the retention period necessary to reduce the ammonia 50 percent was reduced to 9–16 hours. Post ponding resulted in a pH reduction through natural recarbonation.

Selective Ion Exchange

The natural zeolite, clinoptilolite is effective for ammonia removal from wastewaters since it selectively removes ammonia over calcium, magnesium and sodium ions. The capacity of the zeolite will decrease however with increased mineral composition of the wastewater [31]. The resin is regenerated with sodium or calcium hydroxide at high pH. The spent regenerant can then be air stripped for

FIGURE 10.51

Temperature Effects on Air/Liquid Requirements (G/L)
for Ammonia Stripping [27]

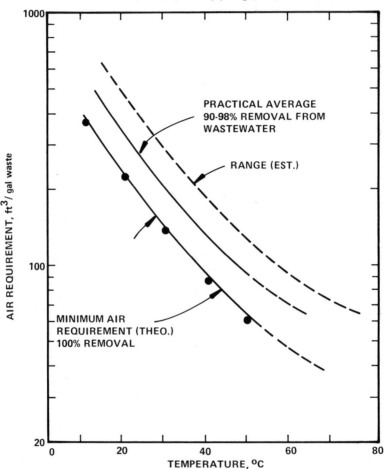

ammonia recovery and regenerant reuse. A typical flow sheet is shown in figure 10.56. Properties and operating conditions for clinoptilolite are shown in table 10.22.

Breakpoint Chlorination

Ammonia nitrogen may be removed from a wastewater by breakpoint chlorination in which the ammonia is chemically oxidized to nitrogen gas. Results on sewage effluents have shown 95–99 percent oxidation to nitrogen gas with the remainder converted to nitrate and nitrogen trichloride. The reactions are:

FIGURE 10.52
Conventional Cooling Towers Commonly Used for Ammonia Stripping

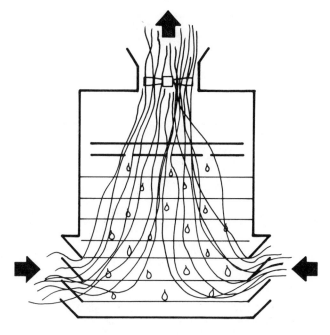

COUNTERFLOW COOLING TOWER

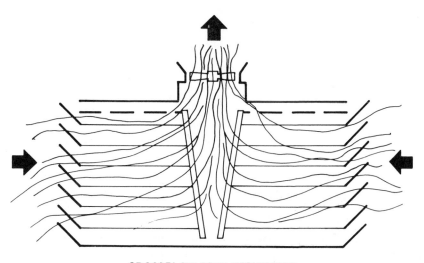

CROSSFLOW COOLING TOWER

FIGURE 10.53

Effect of Tower Depth on Ammonia Removal [1]

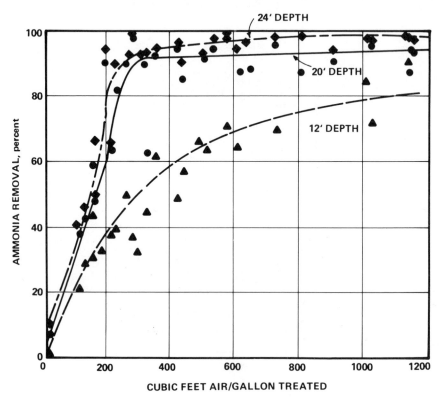

$$NH_4^+ + HOCl \longrightarrow NH_2Cl + H_2O + H^+$$

and:

$$NH_2Cl + 0.5\ HOCl \longrightarrow 0.5N_2 + 0.5H_2O + 1.5H^+ + 1.5Cl^-$$

The breakpoint curve for these reactions is shown in figure 10.57. In Zone 1 the major reaction is the conversion of ammonium ions to monochloramine. The peak of the breakpoint curve theoretically occurs at a 5:1 ratio of Cl_2 to NH_3-N (a molar ratio of 1:1). In Zone 2 the oxidation results in the formation of dichloramine and oxidation of ammonia. At the breakpoint where the theoretical ratio of chlorine to ammonia nitrogen is 7.6:1. The ammonia concentration is at a minimum. After the breakpoint (in Zone 3) free chlorine residual as well as small quantities of dichloramine, nitrogen trichloride and nitrate accumulate. While the theoretical ratio is 7.6:1 (figure 10.57), in practice this has been found to vary from 8-10:1. The

FIGURE 10.54

Effects of Packing Spacing on Ammonia Removal [1]

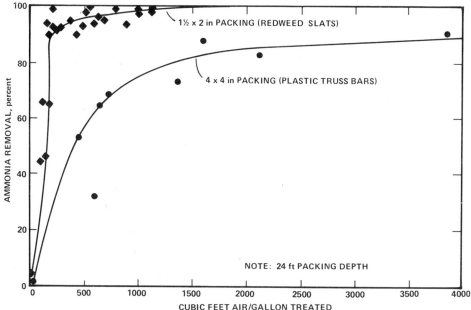

optimum pH is near pH 7.0 to minimize formation of nitrogen trichloride and nitrate. Stoichiometrically 14.3 mg/l of alkalinity is required for each 1.0 mg/l NH_3-N might be expected to be consumed. There will be an increase in TDS in the effluent due to the chloride ions and neutralization. The chlorination will result in 6.2 mg/l TDS/mg/l NH_6-N oxidized. Neutralization with lime (CaO) results in a total of 12.2 mg/l TDS/mg/l NH_3-N oxidized. For example, if a wastewater contained 20 mg/l ammonia nitrogen, chlorine in the gaseous form would result in a 124 mg/l increase in TDS. Neutralization with lime (CaO) would result in an increase of 244 mg/l TDS.

At a pH of 6 to 7 the breakpoint reaction is completed in less than 15 sec. Dechlorination will usually be required, the most common techniques being sulfur dioxide and activated carbon. In practice 0.9–1.0 parts of SO_2 are required to dechlorinate 1.0 part of Cl_2.

$$SO_2 + HOCl + H_2O \longrightarrow Cl^- + SO_4^= + 3H^+$$

The resulting acidity is seldom a problem in practice due to the low concentrations involved. About 2 mg/l alkalinity are consumed for each mg/l SO_2 applied.

When using activated carbon the reaction is:

FIGURE 10.55

Flow Sheet for the Removal of Nitrogen and Phosphorus from Raw Wastewater

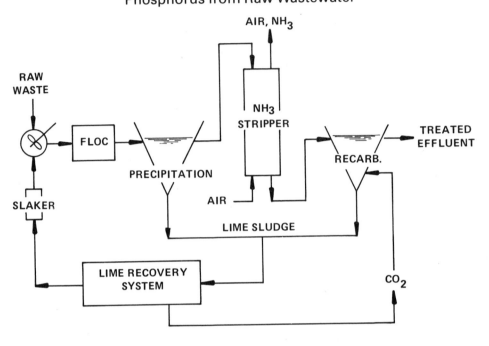

$$C + 2HOCl \longrightarrow CO_2 + 2H^+ + 2Cl^-$$

Activated carbon is expensive and should only be considered in those cases where functions other than chlorine residual control is important.

PHOSPHORUS REMOVAL

Phosphorus in its various forms presents problems in receiving waters due to the stimulation of green plant growth. Biological and physical-chemical processes have been successfully employed to remove the different phosphorous constituents from wastewaters.

Domestic wastewaters contain about 10 mg/l of P of which synthetic detergents account for about one-half. Typical concentrations of the various forms of phosphorus in domestic wastewaters is shown in table 10.23.

Phosphorous may be removed from wastewaters by chemical precipitation with aluminum, iron, or calcium salts. Phosphorus may also be removed by a modification of the biological oxidation process usually referred to as the luxury uptake of phosphorus.

FIGURE 10.56
Flow Sheet for the Removal of Ammonia Nitrogen by Zeolite Exchange

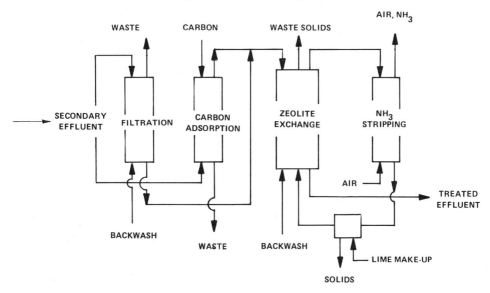

Aluminum

Theoretical phosphorus residuals from precipitation with aluminum are a function of pH and the Al/P ratio. The precipitate is usually a mixture of $Al(OH)_3$ and $AlPO_4$ although the $AlPO_4$ precipitation is favored over $Al(OH)_3$. The precipitate tends to be amorphous rather than crystalline. The relationship between removal and pH and Al/P ratio is shown in figure 10.58. Aluminum may be used as a coagulant in pretreatment, or posttreatment of directly added to the biological treatment process.

In order to approach the predicted solubility, a dosage of 1.5–3.0 moles of aluminum per mole of phosphorus as P is required over a pH range of 6.0–6.5. If the water is alkaline the pH should be lowered prior to alum addition to minimize $Al(OH)_3$ precipitation. Some turbidity may result from the alum addition.

When the alum is added to the activated sludge process, addition should be immediately prior to the final clarifier in the case of completely mixed systems or at the end of plug flow aeration basins. This is to avoid phosphorus precipitation in the biological process before microbial utilization and to minimize shear of the chemical flocs in the aeration basin.

TABLE 10.22

Properties and Operating Conditions for Clinoptilolite

Item	Value
Total exchange capacity[a]	
milliequivalents/gram	1.9[b]
equivalents/liter	1.4
kilograins/cu. ft.	30.6
Physical properties	
Gross particle specific gravity	1.6
Net particle specific gravity	2.4
Bulk density, lb/cu. ft.	47[a]
Particle size, U.S. mesh	20×50
Proposed operating conditions	
Exhaustion flow rate, BV/hr	15
Bed depth, in.	36–72[c]
Regenerant flow rate, BV/hr	15
Rinse volume, gal/cu. ft.	80
Backwash rinse rate, 50% bed	
expansion, gal/sq. ft.-min	11
Head loss, ft/ft at 5.6 gal/sq. ft.-min	0.7

[a] Ammonia exchange capacity will usually be less than this value depending on water composition.
[b] Dry weight.
[c] Greater depths might be used, but allowance must be made for greater head loss.

Iron

Iron precipitation of phosphorus is complicated by the existence of iron in the ferrous (Fe^{++}) and the ferric state (Fe^{+++}), which is dependent on the dissolved oxygen level, the pH, biological catalysis, and the presence of sulfur and carbonates. The solubility of P with respect to $FePO_4$ is shown in figure 10.59. Iron has been employed for phosphorus precipitation in biological treatment processes but has the disadvantage of some iron in the treated effluent.

The iron dosage will range from 2.0–5.0 moles of iron (Fe^{++}) per mole of phosphorus (as P). The optimum pH is 5.0 which is too low for conventional biological treatment. Precipitation at neutral pH values may produce a colloidal precipitate requiring a polymer to obtain a minimum total phosphorus residual.

Thomas [33] has obtained 92 percent removal of total phosphate on plant scale by the addition of 10 mg/l $FeCl_3$ to activated sludge aeration tanks in Mannedorf, Switzerland.

Studies by Barth and Ettinger [34] showed phosphorus removal in excess of 90 percent by the addition of coagulants to the aeration tank. Some of their results and others are summarized in table 10.24.

FIGURE 10.57
Theoretical Breakpoint Chlorination Curve [32]

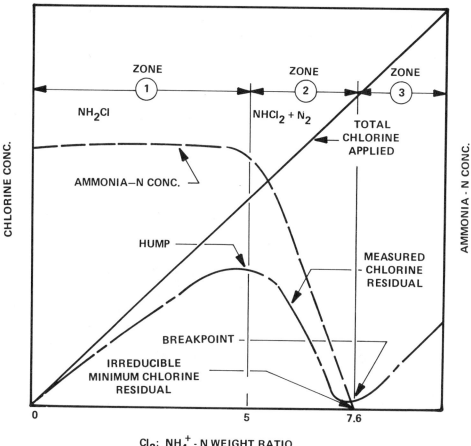

A chemical requirement of approximately 1.1 to 2.0 times the theoretical phosphorus requirement is needed to attain effluent qualities with a phosphorus concentration less than 0.5 mg/l.

Calcium

Phosphorus is precipitated with calcium salts to low residuals depending on the pH. The precipitate is a hydroxyapatite, $Ca_5OH(PO_4)_3$. Between pH of 9.0 and pH 10.5 precipitation of calcium carbonate competes with calcium phosphate. Unlike aluminum and iron, calcium phosphate solids nucleate and grow very slow especially at neutral pH. The addition of seed enhances the reaction indicating the

TABLE 10.23
Approximate Concentrations of Phosphate Forms in a Typical Raw Domestic Sewage

| | Concentration | |
Phosphate Form	mg/l	mole/l
Total	10	3.2×10^{-4}
Ortho	5	1.6×10^{-4}
Tripoly	3	3.2×10^{-4}
Pyro	1	1.6×10^{-5}
Organic	≈ 1	$\approx 3.2 \times 10^{-5}$

FIGURE 10.58
Effluent Phosphorus as a Function of Molar Al/P Ratio and pH

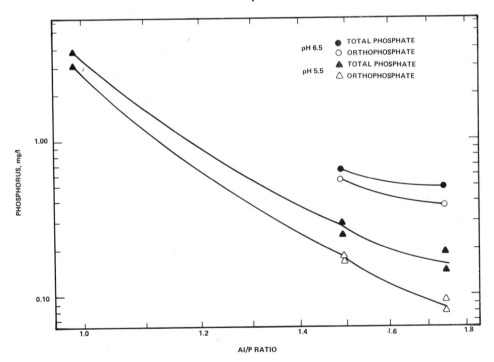

FIGURE 10.59
Equilibrium Solubility Diagram for Fe, Al, and Phosphates

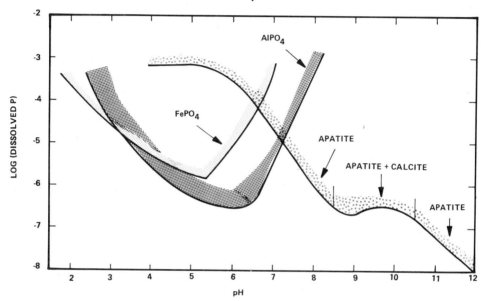

TABLE 10.24
Removal of Phosphorus by Coagulant Addition to the Aeration Tank

Coagulant Added	Initial Phosphorus(P), mg/l	Percent P removal	Reference
Calcium, 150 mg/l	4–12	64	[5]
Magnesium, 20 mg/l	4–12	50	[5]
[1] Fe^{+++}, 15 mg/l[a]	4–12	75	[5]
[1] Al^{+++}, 20 mg/l[a]	4–12	70	[5]
Al^{+++}, 30 mg/l + 20 mg/l of Ca	4–12	90	[5]
$NaAl(OH)_4$, 10 mg/l	8–15	94	[5]
$FeCl_3$, 15 mg/l	8.5	82	[4]
$FeCl_3$, 10 mg/l	7.9	92	[4]

[a] Turbid effluent due to reduction in pH.

advantage of solids recycle. Phosphorus solubility with respect to calcium phosphate is shown in figure 10.60. The calcium phosphate precipitate is finely divided so that the presence of $Mg(OH)_2$ floc aids in the removal of the calcium phosphate precipitate.

The lime requirements will be dictated by the hardness and the alkalinity (figure 10.6). At high pH levels, low soluble phosphorus levels are achieved but residual particulates may require post filtration. Recarbonation for pH adjustment after precipitation may redissolve particulate phosphorus if incomplete removal exists prior to pH adjustment.

Biological Removal of Phosphorus

It has been found that biological sludge when subjected to anaerobic conditions will release phosphorus to solution. When the sludge is subsequently

FIGURE 10.60
Solubility Diagrams for Calcium Phosphates

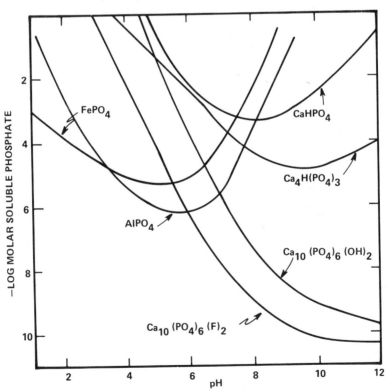

mixed with wastewater and aerated, an immediate high phosphorus uptake occurs. This phenomena has been employed for the removal of phosphorus in the PhoStrip process as shown in figure 10.61.

The high phosphorus decant from the anaerobic basin is lime treated for phosphorus removal and the treated liquor recycled back to the process. Overall phosphorus removals in excess of 90 percent have been reported for this process.

ELECTRODIALYSIS

Electrodialysis involves removing the inorganic ions from water by impressing an electrical potential across the water, resulting in the migration of cations and anions to the cathode and anode, respectively. By alternately placing anionic and cationic permeable membranes a series of concentrating and diluting compartments result. The amount of electric current required for the demineralization is proportional to the number of ions removed from the diluting compartments. An electrodialysis cell is shown in figure 10.62.

To minimize membrane fouling and maintain long runs, it is essential that turbidity suspended solids, colloids, and trace organics be removed prior to the electrodialysis cell.

Large organic ions and colloids are attracted to the membranes by electrical forces and result in fouling of the membranes. This results in an increase in electrical resistance that at constant stack voltage decreases the current and hence the demineralizing capacity of the equipment.

Turbidity in the feed has a major effect on membrane fouling.

Conventional methods of coagulation, filtration, and adsorption would be employed. Brunner [35] obtained 40 percent total dissolved solids removal with a concentrate of 10 percent (by volume) of the influent water.

The estimated operating costs are 12 to 16 cents/1000 gal (3.2 to 4.2 cents/1000 l) for a 10-MGD (37,850 m^3/day) plant, not including pretreatment or concentrate disposal.

Electrolytic treatment of wastewater has been employed in some cases. Performance data has been reported by Poon and Brueckner [36] as shown in table 10.25.

REVERSE OSMOSIS

Membrane filtration includes a broad range of separation processes from filtration and ultra-filtration to reverse osmosis. Generally, those processes defined as filtration refer to systems in which discrete holes or pores exist in the filter media generally, in the order of 10^2 to 10^4 nm or larger. The efficiency of this type of filtration depends entirely on the difference in size between the pore and the particle to be removed.

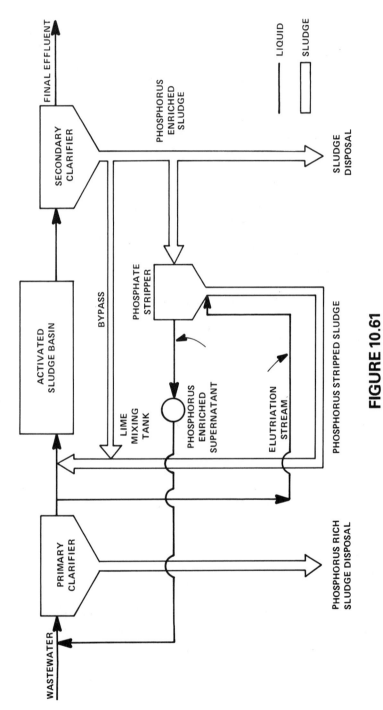

FIGURE 10.61
PhoStrip Process for Phosphorus Removal

(Courtesy of Union Carbide Co.)

FIGURE 10.62
Electrodialysis Cell

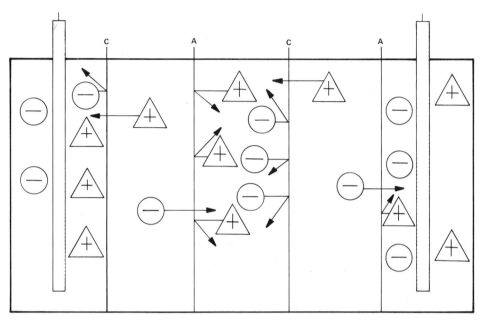

△⊕ –CATION; ⊖ –ANION; A–ANION
PERMEABLE MEMBRANE; C–CATION-PERMEABLE MEMBRANE

TABLE 10.25
Performance of a Flow-Through Model in Electrolytic Treatment of Wastewater*

Parameter	Treatment Efficiency (%)
SS removal	66.7– 85.0
BOD removal	59.8– 73.4
NH_3-N removal	67.5–100
Total N removal	59.0– 94.5
Ortho PO_4 removal	74.0– 85.6
Total PO_4 removal	78.6– 90.2

*Experimental conditions are: wastewater:seawater ratio, 2:1 to 3:1; wastewater detention time, 40 to 60 min; power consumption, 6 to 8 kwh/3,785 l; and the mixing time of treated wastewater and spent seawater, 40 to 50 min.

The various filtration processes relative to molecular size are shown in table 10.26.

This section will discuss those processes applicable to the removal of ions and organics in true solution. The reverse osmosis process employes a semi-permeable membrane and a pressure differential to drive fresh water to one side of the cell, concentrating salts on the input or rejection side of the cell. In this process, fresh water is literally squeezed out of the feed water solution.

The reverse osmosis process can be described by considering the normal osmosis process. In osmosis, a salt solution is separated from a pure solvent or solution of less concentration by a semi-permeable membrane. The semi-permeable membrane is permeable to the solvent and impermeable to the solution. Such an arrangement is shown in figure 10.63. The chemical potential of the pure solvent is greater than that of the solvent in solution and, therefore, pure solvent will pass through the membrane to dilute the solution. This flow of solvent increases the chemical potential of the solvent in solution, and therefore, drives the system to equilibrium. If an imaginary piston applies an increasing pressure on the solution compartment, the solvent flow through the membrane will continue to decrease. When sufficient pressure has been applied to bring about thermodynamic equilibrium, the solvent flow will stop. The pressure developed in achieving equilibrium is the osmotic pressure of the solution, or the difference in the osmotic pressure between the two solutions if a less concentrated salt solution is used instead of pure solvent in the right chamber of the cell.

Now, if a pressure in excess of the osmotic pressure is applied to the more concentrated solution chamber, pure solvent is caused to flow from this chamber to the pure solvent side of the membrane, leaving a more concentrated solution behind. This phenomena is the basis of the reverse osmosis process.

TABLE 10.26

Waste Management Application and Required Process

Material to be Removed	Approximate Size nm	Process
Ion removal	1–20	Diffusion or reverse osmosis
Removal of organics in true solution	5–200	Diffusion
Removal of organics: subcolloidal—not in true molecular dispersion	200–10,000	Pore flow
Removal of colloidal and particulate matter	75,000	Pore flow

FIGURE 10.63
Osmosis and Reverse Osmosis [37]

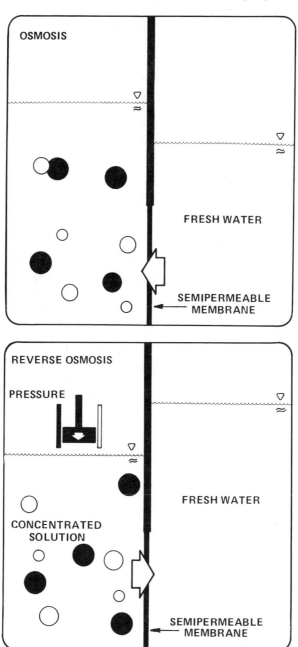

The criteria of membrane performance are the degree of impermeability or how well the membrane rejects the flow of the solute and the degree of permeability, or how easily the solvent is allowed to flow through the membrane. Cellulose acetate membranes provide an attractive combination of these criteria. Several membrane types are shown in figure 10.64. Each of these design configurations

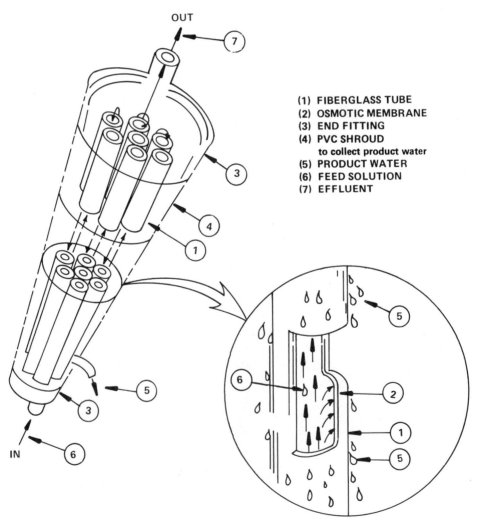

FIGURE 10.64
Membrane Filtration Systems
(a) Tubular Reverse Osmosis Unit

OUT

(1) FIBERGLASS TUBE
(2) OSMOTIC MEMBRANE
(3) END FITTING
(4) PVC SHROUD
    to collect product water
(5) PRODUCT WATER
(6) FEED SOLUTION
(7) EFFLUENT

IN

FIGURE 10.64
Membrane Filtration Systems
(b) Spiral Wound Reverse Osmosis Module

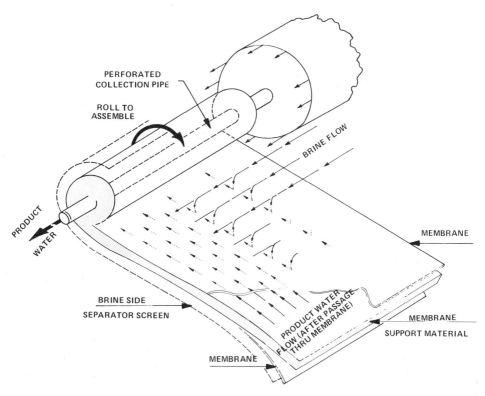

uses cellulose acetate membranes, except certain of the hollow fine fiber systems that employ a nylon polymer.

In order to minimize membrane fouling, pretreatment is required for the removal of suspended matter, bacteria and precipitatable ions. A typical reverse osmosis process schematic is shown in figure 10.65. The design and operating parameters for a reverse osmosis system is summarized from Agardy [38].

Pressure

The water flux is a function of the pressure differential between the applied pressure and the osmotic pressure across the membrane. The higher the applied pressure, the greater the flux. However, the pressure capability of the membrane is limited, so the maximum pressure is generally taken to be 1000 psig. Operating experience dictates in the 400 to 600 psig range, with 600 psig normally being the design pressure.

FIGURE 10.64
Membrane Filtration Systems
(c) Dupont Hollow Fiber Reverse Osmosis Unit

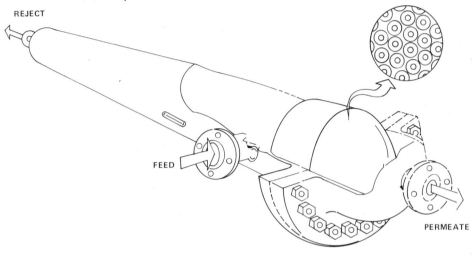

FIGURE 10.65
Basic Reverse Osmosis Process Schematic [38]

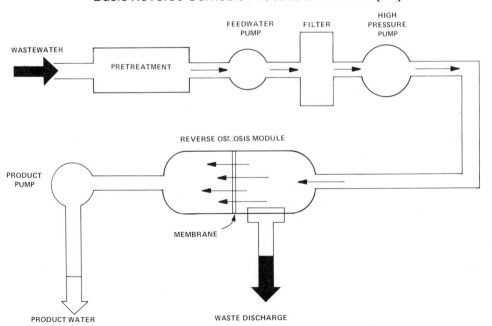

Temperature

The water flux increases with increasing feed water temperature. A standard of 70° F is generally assumed and temperatures of up to 85° F are acceptable. However, temperatures in excess of 85° F and up to 100° F will accelerate membrane deterioration and cannot be tolerated for long operating periods.

Membrane Packing Density

This is an expression of the unit area of the membrane, which can be placed per unit volume of pressure vessel. The greater this factor the greater will be the overall flow through the system. Typical values range from 50 to 500 square feet per cubic foot (160–1640 m^2/m^3) of pressure vessel.

Flux

Assuming a typical pressure of 600 psig, flux values range from 10–80 gpd per square foot (0.4–3.3 m^3/m^2-day) with 12 to 35 gpd per square foot (0.5–1.4 m^3/m^2-day) being common. This flux tends to decrease with length of run and over a period of one to two years of operation might be reduced by 10 to 50 percent.

Recovery Factor

This consideration actually represents plant capacity and is generally in the range of 75 to 95 percent with 80 percent being the practical maximum. At high recovery factors there is a greater salt concentration in the process water as well as in the brine. At higher concentrations salt precipitation on the membrane increases causing a reduction in operation efficiency.

Salt Rejection

Salt rejection depends on the type and character of the selected membrane and the salt concentration gradient. Generally, rejection values of 85 to 99.5 percent are obtainable with 95 percent being commonly used.

Membrane Life

Membrane life can be drastically shortened by undesired constituents in the feed water, such as phenols, bacteria and fungi as well as high temperatures and high or low pH values. Generally, membranes will last up to two years with some loss in flux efficiency.

pH

Membranes consisting of cellulose acetate are subject to hydrolysis at high and low pH values. The optimum pH is approximately 4.7 with operating ranges between 4.5 to 5.5.

Turbidity

Reverse osmosis units can be used to remove turbidity from feed waters. They operate best if little or no turbidity is applied to the membrane. Generally it is considered that the turbidity not exceed one Jackson turbidity unit (JTU) and that the feed water should not contain particles larger than 25 μm.

Feed Water Stream Velocity

The hydraulics of reverse osmosis systems are such that velocities range of 0.04 to 2.5 fps (1.2 to 76.2 cm/sec) are common. Plate and frame systems operate at the higher velocity, while the hollow fine fiber units operate at the lower velocities. High velocities and turbulent flow are necessary to minimize concentration polarization at the membrane surface.

Power Utilization

Power requirements are generally associated with the system pumping capacity and operational pressures. Values range from 9 to 17 kwh per 1000 gallons (2.4 to 4.5 kwh/m^3) with the lower figure taking into account some power recovery from the brine stream.

Pretreatment

The present development of membranes limits their direct application to feed waters having a TDS not exceeding 10,000 mg/l. Further the presence of scale-forming constituents, such as calcium carbonate, calcium sulfate, oxidies and hydroxides of iron, manganese and silicon, barium and strontium sulfates, zinc sulfide and calcium phosphate, must be controlled by pretreatment or they will require subsequent removal from the membrane. These constituents can be controlled by pH adjustment, chemical removal, precipitation, inhibition, and filtration. Organic debries and bacteria can be controlled by filtration, carbon, pretreatment, and chlorination. Oil and grease must also be removed to prevent coating and fouling of the membranes.

Cleaning

Recognizing that under continuous use membranes will foul, provision must be made for mechanical and/or chemical cleaning. Methods include periodic depressurizations, high velocity water flushing, flushing with air-water mixtures, backwashing, cleaning with enzyme detergents, ethylene diamine, tetra-acetic acid, and sodium perborate. The control of pH during cleaning operations must be maintained to prevent membrane hydrolysis. Approximately 1 to 1.5 percent of the process water goes to waste as a part of the cleaning operation, with the cleaning cycle being every 24 to 48 hours.

A summary of operational parameters is shown in table 10.27.

TABLE 10.27
Summary of System Operational Parameters[a]

Parameter	Range	Typical
Pressure	400–1,000 psig	600 psig
Temperature	60° F–100° F	70° F
Packing Density	50–500 sq ft/cu ft	–
Flux	10–80 gpd/sq ft	12–35 gpd/sq ft
Recovery Factor	75–95%	80%
Rejection Factor	85–99.5%	95%
Membrane Life	–	2 years
pH	3–8	4.5–4.5
Turbidity	–	1 JTU
Feedwater Velocity	0.04–2.5 fps	–
Power Utilization	9–17 kwhr/1,000 gal	–

[a] After [38].

Results

In the treatment of municipal wastewaters, removal of total dissolved solids range from 78 to 92 percent, total hardness, 94 to 99 percent, turbidity, 62 to 100 percent, COD from 89 to 98 percent, total nitrogen from 67 to 72 percent, ammonia from 65 to 89 percent, nitrate from 44 to 81 percent, phosphate from 93 to 100 percent, sulfate from 94 to 100 percent.

Reverse osmosis has been applied to the treatment of plating wastewaters for the removal of cadmium, copper, nickle, and chromium at pressures of 200 to 300 psi. The concentrated stream is returned to the plating bath and the treated water to the next to the last rinse tank.

Pulp mill effluents have been treated by reverse osmosis at a pressure of 600 psi. Waste streams were concentrated up to 100,000 mg/l total solids. The flux was found to be a function of total solids level and varied from 2 gpd per square foot to 15 gpd per square foot ($0.08–0.61$ m^3/m^2-day) [39].

CHEMICAL OXIDATION

Chemical oxidation of a wastewater may be employed to oxidize pollutants to terminal end products or to intermediate products that are more readily biodegradable or more readily removable by adsorption. Common oxidants are chlorine, ozone, air and potassium permanganate. Chemical oxidation is frequently markedly dependent on pH and the presence of catalysts.

Ozone

Ozone is a gas at normal temperature and pressure. As with oxygen the solubility of ozone in water depends on temperature and the partial pressure of ozone in the gas phase and has recently been thought to also be a function of pH. Ozone is unstable and the rate of self-decomposition increases with temperature and pH. The decomposition is catalyzed by hydroxide ion (OH^-), the radical decomposition products of O_3, the organic solute decomposition products, and by a variety of other substances such as solid alkalis, transition metals, metal oxides and carbon. Under practical conditions complete degradation of fairly unreactive compounds such as saturated hydrocarbons and halogenated aliphatic compounds does not occur with O_3 alone, but current research has shown that O_3 with an additional energy source, i.e., sonification or ultra violet readily decomposes these refractory compounds.

The mechanisms of ozone oxidation of organics are:

1. Oxidation of alcohols to aldehydes and then to organic acids

$$RCH_2OH \xrightarrow{O_3} RCHO \xrightarrow{O_3} RCOOH$$

2. Substitution of an oxygen atom into an aromatic ring
3. Cleavage of carbon double bonds

Ozone is generated from dry air or oxygen by a high voltage electric discharge with oxygen yielding twice the O_3 concentration (0.5–10 wt. %) as air. Theoretically, 1058 g of ozone can be produced per kilowatt hour of electrical energy and in practical application only a production of 150 g/kwh can be expected.

Ozonation can be employed for the removal of color and residual refractory organics in effluents. Ozonation of a secondary dye waste effluent for color removal is shown in figure 10.66. While there was a decrease in TOC in the final filtered effluent the soluble BOD increased from 10 mg/l to 40 mg/l due to the conversion from long chain biologically refractory organics to biodegradable compounds. Similar results were obtained from the ozonation of a secondary effluent from low and high rate activated sludge units treating a tobacco processing wastewater as shown in table 10.28. TOC will not be reduced until the organic carbon has been oxidized to CO_2 while COD will generally be reduced with any oxidation.

Ozonation of unsaturated aliphatic or aromatic compounds react with water and oxygen to form acids, ketones and alcohols. At a pH greater than 9.0 in the presence of redox salts such as Fe, Mn and Cu aromatics may form some hydroxyaromatic structures (phenolic) which may be toxic. Many of the by-products of ozonation are readily biodegradable.

Organic removal is improved with ultraviolet radiation. It is postulated that the UV activates the O_3 molecule and may also activate the substrate.

FIGURE 10.66

Correlation of Color Removal with Ozone
Utilized Dye Wastewater

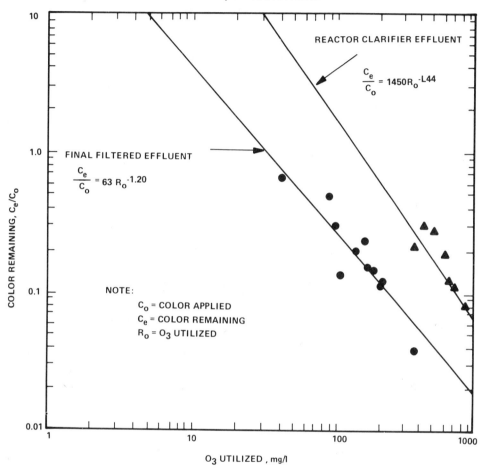

O_3 UTILIZED , mg/l

Ozone-UV is effective for the oxidative destruction of pesticides to terminal end products of CO_2 and H_2O.

For many of the priority pollutants there is a rapid, first order reduction in COD to some level followed by a slow or even zero removal indication the non-reactive nature of the by-products.

COD reduction by ozonation of an aerated lagoon effluent treating an organic chemicals wastewater is shown in figure 10.67.

Phenol can be oxidized with ozone producing as many as 22 intermediate products between phenol and CO_2 and H_2O. The reaction is first order with

TABLE 10.28

Ozonation Test Results on Secondary Clarifier Effluent

Parameter	Low F/M (0.15) Time (min)[a]		High F/M (0.60) Time (min)[a]	
	0	60	0	60
BOD (mg/l)	27	22	97	212
COD (mg/l)	600	154	1100	802
pH	7.1	8.3	7.1	7.6
Org-N (mg/l)	25.2	18.9	40	33
NH_3-N (mg/l)	3.0	5.8	23	25
Color (Pt-Co)	3790	30.0	5000	330

[a] Loading of 155 mg O_3/min.

respect to phenol and proceeds optimally over a pH of 8 to 11. The ozone consumption is 4–6 moles O_3 consumed/mole phenol oxidized. This requires in the order of 25 moles O_3/mole phenol to be generated in the gas phase.

Ammonia is oxidized by ozone to nitrate in the NH_3 form. Since the percentage of ammonia in the NH_3 form is a function of pH the oxidation occurs under alkaline conditions.

$$NH_3 + 4O_3 \longrightarrow NO_3^- + 4O_2 + H_2O + H^+$$

The optimum pH is pH 9–10. At pH levels less than 9.0 most of the ammonia is in the NH_4^+ form and the oxidation rate is very slow. At pH levels greater than 9.0 ozone decomposition is very rapid and occurs faster than the oxidation. The presence of carbonate and other organic solutes will consume ozone in parallel reactions. The effect of pH on the rate of ammonia oxidation is shown in figure 10.68.

Chlorine

Chlorine may be used as a chemical oxidant. In reactions with inorganic materials terminal end products usually result while organic oxidations usually produce chlorinated hydrocarbons. Chlorination of a secondary effluent for color removal is shown in figure 10.69.

The alkaline chlorination process oxidizes cyanide by the addition of chlorine. The initial reaction forms cyanogen chloride (CNCl); this reaction is instantaneous at all pH levels:

$$NaCN + Cl_2 \rightleftharpoons CNCl + NaCl$$

FIGURE 10.67

Effect of pH on COD Removal by Ozonation-Aerated Lagoon Effluent

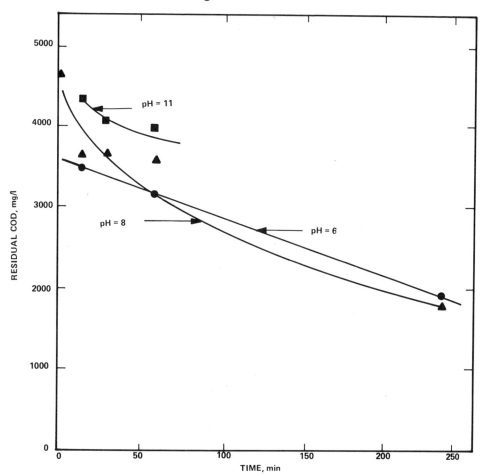

One part of cyanide requires 2.73 parts of chlorine. The cyanogen chloride formed in this reaction is volatile and odorous. In the presence of caustic, sodium cyanate is formed from the cyanogen chloride:

$$CNCl + 2NaOH \rightleftharpoons NaCNO + 2H_2O + NaCl$$

The rate of this reaction is pH-dependent and is slow below pH 8.0. Above pH 8.5 the reaction goes to completion in 30 min. The reaction is temperature-dependent and can be increased by use of an excess of chlorine. The caustic requirement is 1.13 parts caustic/part Cl_2 applied.

FIGURE 10.68
Effect of pH on Rate of Ammonia Oxidation

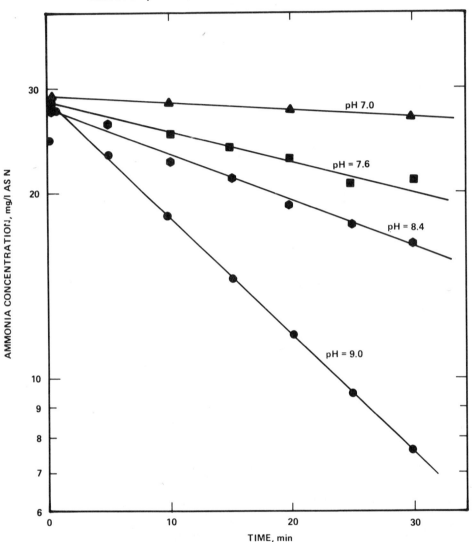

Interferences to this reaction result from the presence of organic compounds and oxidizable metals and complexes. Ferro- and ferri-cyanide are oxidized at a slow rate.

The cyanate produced by this reaction is only one one-thousandth as toxic as cyanide and will hydrolyze under acidic conditions in lakes or streams to

FIGURE 10.69

Color Removal by Chlorination

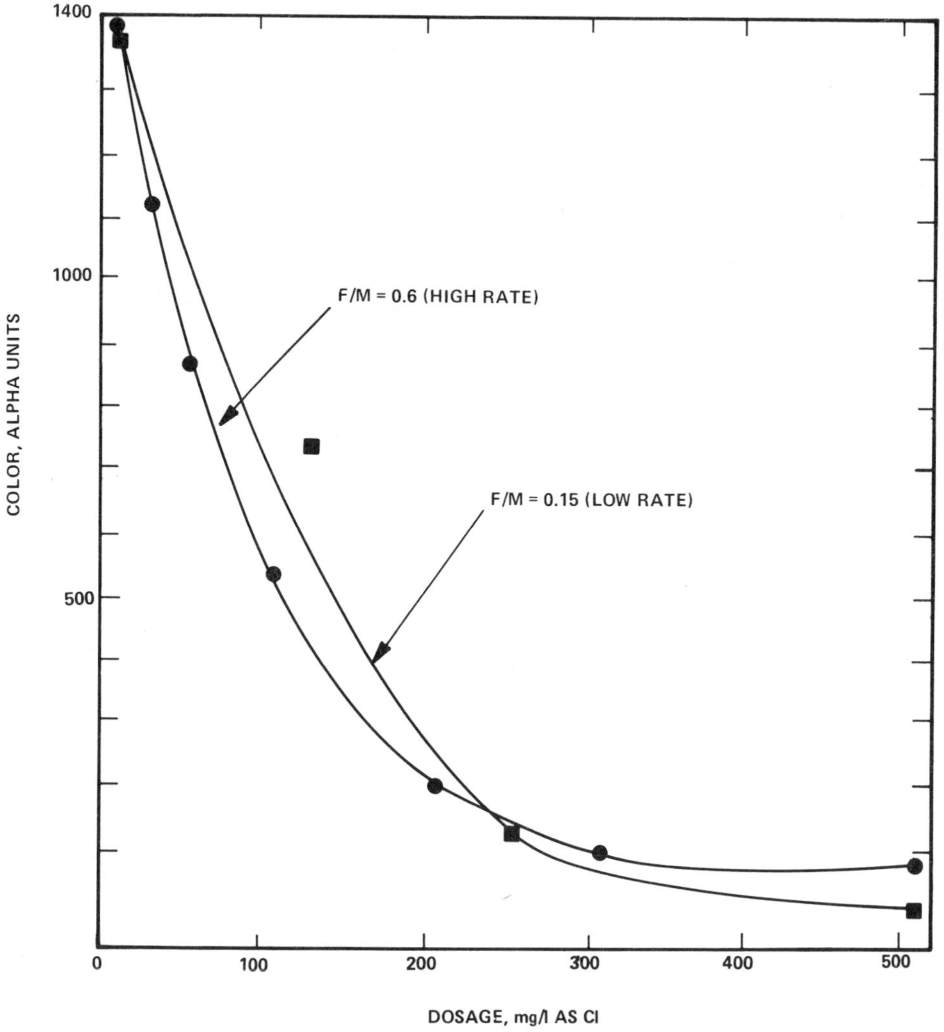

ammonia and carbon dioxide. The cyanate will be further oxidized by the addition of chlorine to carbon dioxide and nitrogen:

$$2NaCNO + 4NaOH + 3Cl_2 \rightleftharpoons 2CO_2 + 6NaCl + N_2 + 2H_2O$$

This reaction requires 4.09 parts Cl_2 and 3.08 parts $NaOH$/part of CN^-. Although the reaction is most rapid at pH 6.5 to 6.8, a pH of 8.5 is usually used to

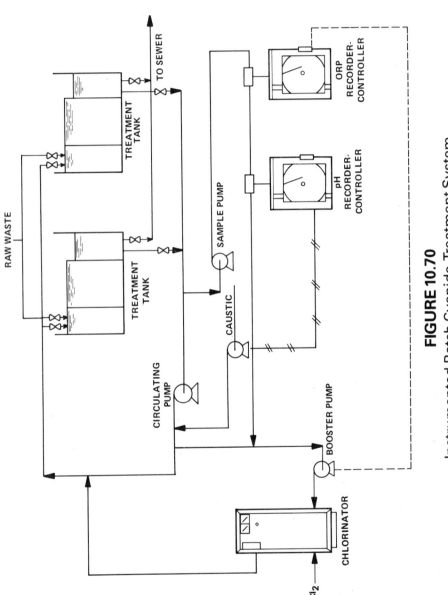

FIGURE 10.70

Instrumented Batch Cyanide-Treatment System

(Courtesy of Fischer-Porter, Inc.)

FIGURE 10.71

Oxidation-Reduction Potential Relationships in the Alkaline Chlorination of Cyanide Waste

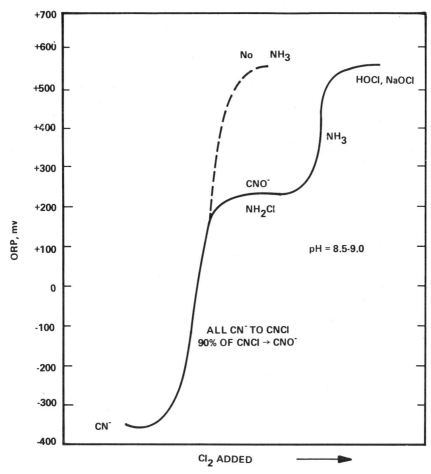

simplify operation, to eliminate the formation of nitrogen trichloride, and to permit the last two reactions to occur sequentially in a single treatment unit. At pH 8.4 the reaction is complete in 10 minutes; at pH 9.6 it is complete in 40 minutes.

Excess chlorine is used to complete the reaction. Although the theoretical chlorine requirement for the complete oxidation of cyanide is 6.82 parts Cl_2/part CN^-, the reaction is completed with 7.35 parts Cl_2/part CN^-. Oxides of nitrogen that increase the chlorine requirement are formed in the presence of free chlorine. It is essential to avoid an excess temperature buildup, which leads to the formation of chlorates.

Cyanide waste treatment can be accomplished in either batch or continuous treatment systems. The batch treatment system is similar to that described for chromium reduction and precipitation except that only a small volume need be provided for the precipitation of heavy metal sludges. Usually it is necessary to remove sludge only at infrequent intervals. An instrumented batch treatment system is shown in figure 10.70.

In order to avoid the formation of toxic HCN strict segregation of cyanide wastes from acidic waste streams is essential.

The caustic requirements must usually be determined experimentally because of the presence of other contaminants in the waste stream.

In continuous treatment systems oxidation-reduction potential can be used to control the addition of chlorine, as shown in figure 10.71. In the presence of excess chlorine, complete oxidation is assured if the pH and retention time are maintained at their proper levels.

Chlorine or hydrogen peroxide may be employed for the oxidation of H_2S primarily for the purpose of odor control. The resulting reactions are:

$$H_2S \rightleftharpoons HS^- + H^+$$

$$HS^- + 4Cl_2 + 4H_2O \longrightarrow SO_4^= + 9H^+ + 8Cl^-$$

$$HS^- + 4H_2O_2 \longrightarrow SO_4^= + H^+ + 4H_2O$$

REFERENCES

1. Slechta, A. F. and Culp, G. L. 1967. Water reclamation studies at the South Tahoe Public Utility District. *J. Water Pollution Control Federation* 39: 787.
2. Stumm, W. and O'Melia, C. R. 1968. Stoichiometry of coagulation. *J. American Water Works Association* 60: 514.
3. Riddick, T. M. 1968. *Control of colloid stability through zeta potential.* Wynnewood, Penn.: Livingston Publishing.
4. Olthof, M. G. and Eckenfelder, W. W. 1976. Coagulation of textile wastewaters. *Textile Chemist and Colorist* vol. 8, 7, 18.
5. Olthof, M. G. and Eckenfelder, W. W. 1975. Color removal from pulp and paper wastewaters by coagulation. *Water Research* 9: 853.

6. Argaman, Y. and Weddle, C. L. 1974. Fate of heavy metals in physical-chemical treatment processes *Water-1973*. AIChE Symp. Series, 70, 136: 400.

7. Lancy, L. E. 1954. A new process for plating waste treatment. *Sew. Ind. Wastes* 26: 1117.

8. Patterson, J. W. 1976. *Technology and Economics of Industrial Pollution Abatement*. Chicago: Illinois Institute for Environmental Quality.

9. Tchobanoglous, G. 1968. Filtration techniques in tertiary treatment. *Proc. 40th Annual Conf. Calif. Water Pollution Control Assoc.* Santa Rosa, Calif.

10. Lynam, B. *et al.* 1969. Tertiary treatment at Metro Chicago by means of rapid sand filtration and microstrainers. *J. Water Pollution Control Federation* 41: 247.

11. Diaper, E. W. J. Microstraining and ozonation of sewage effluents. *Proc. 41st Annual Conf. Water Pollution Control Federation* Chicago, Illinois (September 1968).

12. Convery, J. J. 1968. Solids removal processes. *FWPCA Symposium on Nutrient Removal and Advanced Waste Treatment*, Tampa, Florida, Washington, D.C.: EPA.

13. Faup, G. M. *et al.* 1976. Improvement of tertiary filtration efficiency by upgrading biological activity. *Proc. 8th Int. Conf. on Water Pollution Research*, Sydney, Australia: Pergamon Press (Oxford).

14. Ford, D. L. "Current state of the art of activated carbon treatment," Open Forum on Management of Petroleum Refinery Wastewaters, Tulsa, Oklahoma, Washington, D.C.: EPA.

15. Lawson, C. T. 1975. *Activated carbon adsorption for tertiary treatment of activated sludge effluents from organic chemicals and plastics manufacturing plants—application studies and concepts*. South Charleston, West Virginia: Union Carbide Co.

16. U.S. Environmental Protection Agency. 1973. *Process Design Manual for Carbon Adsorption, Technology Transfer*.

17. Hutchins, R. A. 1973. New method simplifies design of activated carbon systems. *Chemical Engineering*, Vol. 80, 19, 133.

18. Hutchins, R. A. "Method for sizing granular activated carbon systems," Proc. 47th Annual Conf. Water Pollution Control Federation, Denver, Colorado, Unpublished paper.

19. Engineering-Science, Inc. and Cleveland Regional Sewer District. 1975. *Report on Evaluation of Continuing Westerly Pilot-Plant Studies*. Prepared for CRSD.

20. Weber, W. J., Jr. *et al.* 1970. Physical-chemical treatment of municipal wastewater. *J. Water Pollution Control Federation* 42: 83.

21. Short, T. E. and Myers, L. A. 1975. *Pilot Plant Activated Carbon Treatment of Petroleum Refinery Wastewaters*. Ada, Oklahoma: Robert S. Kerr Environmental Research Laboratory.

22. Engineering-Science, Inc., Petroleum Refinery Industry—Technology and Cost of Wastewater Control, report to the National Commission on Water Quality (June 1975).

23. Shell, G. L. and Burns, D. E. 1972. Powdered activated carbon application, regeneration and reuse in wastewater treatment systems. *Proc. 6th Int. Conf. on Water Pollution Research*. Jerusalem, Israel.

24. Engelbrecht, R. S. Removal and inactivation of enteric viruses by wastewater and water treatment processes. *Advanced Wastewater Treatment, JRGWP Seminar*, Tokyo, Japan (1976).

25. Katzenelson, E. *et al*. 1974. Inactivation kinetics of viruses and bacteria in water by use of ozone. *J. Amer. Water Works Assoc*. 66: 725.

26. Culp, R. L. and Culp, G. L. 1971. *Advanced Wastewater Treatment*, New York: Van Nostrand Reinhold.

27. Tshobanoglous, G. 1970. Physical and chemical processes for nitrogen removal: theory and applications. *Proc. 12th Sanitary Engineering Conf*. Urbana: University of Illinois.

28. Roesler, J. F. *et al*. 1971. Simulation of ammonia stripping from wastewater. *J. Sanit. Eng. Div., ASCE* 97:1 269.

29. Kepple, L. G. Ammonia removal and recovery becomes feasible. *Water Sew. Works* 121, 4, vol. 42 (1974).

30. Folkman, Y. and Wachs, A. M. 1972. Nitrogen removal from ammonia release from ponds. *Proc. 6th Int. Conf. on Water Pollution Research*, Jerusalem, Israel.

31. Koon, J. H. and Kaufman, W. J. Optimization of ammonia removal by ion exchange using clinoptilolite, *U.S. EPA Water Pollution Control Research Series* 17080 DAR 09/71 (1971).

32. U.S., Environmental Protection Agency. 1975. *Process Design Manual for Nitrogen Control, Technology Transfer*.

33. Thomas, E. A. Phosphate precipitation in the Uster treatment plant and removal of iron phosphate sludge. Viertel. Naturforsch. Ges. Zurich, 111, 309 (1966).

34. Barth, E. F. and Ettinger, M. B. 1967. Mineral controlled phosphorus removal in the activated sludge process. *J. Water Pollution Control Federation* 39: 1361.

35. Brunner, C. A. 1967. Pilot-plant experiences in demineralization of secondary effluent using electrodialysis. *J. Water Pollution Control Federation* 39: R1.

36. Poon, C. P. and Brueckner, T. G. 1975. Physicochemical treatment of wastewater-seawater mixture by electrolysis. *J. Water Pollution Control Federation* 47: 66.

37. Kremen, S. S. 1975. Reverse osmosis makes high quality water now. *Environ. Sci. Technol*. 9: 314.

38. Agardy, F. J. 1972. Membrane Processes. *Process design in water quality engineering*. E. L. Thackston and W. W. Eckenfelder, Jr., Eds. Austin, Texas: Jenkins Publishing.

39. Okey, R. W. 1970. The application of membranes to sewage and waste treatment. *Water Quality Improvement by Physical and Chemical Processes*. E. F. Gloyna and W. W. Eckenfelder, Jr., Eds. Austin:1 University of Texas Press.

Sludge Handling and Disposal

Most of the treatment processes normally employed in water pollution control yield a sludge from a solids-liquid separation process (sedimentation, flotation, etc.) or produce a sludge as a result of a chemical (coagulation) or a biological reaction. These solids undergo a series of treatment steps involving thickening, dewatering, and final disposal. Organic sludges may also undergo treatment for reduction of the organic or volatile content prior to final disposal.

The increase in solids content that might be expected through the treatment sequence is shown in figure 11.1. In general, gelatinous-type sludges such as alum or activated sludge yield the lower concentrations, whereas primary and inorganic sludges yield the higher concentrations in each process sequence.

FIGURE 11.1

Solids Concentration Through
Sludge-Dewatering Sequence

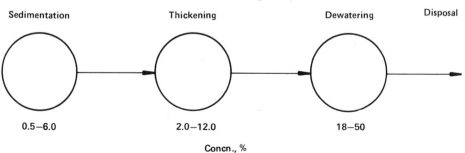

Sedimentation	Thickening	Dewatering	Disposal
0.5–6.0	2.0–12.0	18–50	

Concn., %

589

Figure 11.2 shows a substitution flow sheet for the various alternative processes available for sludge dewatering and disposal. The processes selected depend primarily on the nature and characteristics of the sludge and on the final disposal method employed. For example, activated sludge is considerably more effectively concentrated by flotation than by gravity thickening. Final disposal by incineration desires a solids content that supports its own combustion. In some cases the process sequence is apparent from experience with similar sludges or by geographical or economic constraints. In other cases an experimental program must be developed to determine the most economical solution to a particular problem.

CHARACTERISTICS OF SLUDGES
FOR DISPOSAL

The physical and chemical characteristics of sludges dictate the most technically and economically effective means of disposal. For thickening, the concentration ratio C_u/C_o (the concentration of the underflow divided by the concentration of the influent) is related to the mass loading (lbs solids/ft²/day), which indicates the feasibility of gravity thickening.

The dewaterability of a sludge by filtration is related to the specific resistance. While the specific resistance of a sludge can be reduced by the addition of coagulants, economic considerations may dictate alternative dewatering methods. When selecting a processing sequence for physical chemical treatment of domestic wastewater both the quantity of sludge produced and its dewatering characteristics need to be considered. Data reported by Krissel and Westrick [1] as shown in table 11.1 illustrate this point. While the quantities of lime sludge produced are considerably higher than alum or iron sludge, the thickening and dewatering characteristics are considerably better. An economic analysis of the total sludge handling operation becomes necessary to ascertain the most cost effective methodology. In some cases chemical recovery and reuse should be considered.

Ultimate disposal usually considers land disposal or incineration. When considering incineration the heat value of the sludge and the concentration attainable by dewatering dictates the economics of the operation. The heat content of various sludges is summarized in table 11.2. Land disposal may use the sludge as a fertilizer or soil conditioner as in the case of treated domestic sewage sludges or in a confined land fill for industrial sludges. The chemical composition of some sewage sludges is shown in table 11.3. Minor chemical elements in activated and digested sludge is shown in table 11.4. The concentrations vary widely due to industrial waste discharges. It is important that if a sludge is to be used for land disposal, heavy metals be removed by pretreatment.

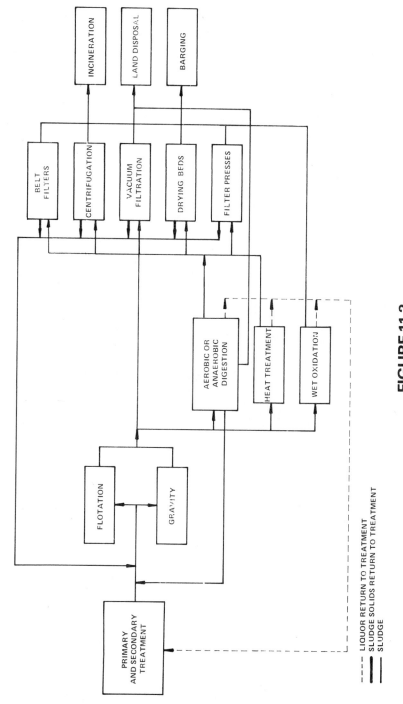

FIGURE 11.2
Process Alternatives

| | SLUDGE | | GRAVITY THICKENING | | | VACUUM FILTRATION | |
	lb/MG	gal/MG	% SOLIDS IN	% SOLIDS OUT	lb/day·ft²	lb/ft²·hr	% CAKE MOISTURE
ACTIVATED SLUDGE	2,000-2,500	20,000-25,000	⋯	⋯	⋯	⋯	⋯
HIGH LIME pH >11.5	3,100-12,000	3,000-28,000	4.8	9.4	18	4.5	72
LOW LIME pH <11.0	⋯	⋯	11.5	17.9	46	12.8	67
IRON	1,280-2,500	9,000-12,000	1.5	4.5	12	1.0	80
ALUM	930-1,780	⋯	.36	1.5	6	0.4	85

TABLE 11.1

Biological-Physical-Chemical Sludge Characteristics,
Municipal Wastewater

TABLE 11.2
Heat Content of Sludges

Material	Combustibles %	Ash %	BTU/lb
Grease and scum	88.5	11.5	16,750
Raw sewage solids	74.0	26.0	10,285
Fine screenings	86.4	13.6	8,990
Ground garbage	84.8	15.2	8,245
Digested sewage solids and ground garbage	49.6	50.4	8,020
Digested sludge	59.6	40.4	5,290
Grit	33.2	69.8	4,000

SCREENING, DEGRITTING, AND SKIMMING

Although not normally considered part of the sludge disposal sequence, grit, screenings, and skimmings from domestic sewage treatment require disposal. Screenings are materials present in the raw sewage that are collected on .5- to 2-in. (1- to 5-cm) screens. These materials are usually buried, but in some cases are incinerated or ground into small particles to be disposed of with the primary sludge. The characteristics and disposal of sewage screenings are summarized in table 11.5.

Grit is usually removed from sewage as a preliminary step prior to primary and secondary treatment. The heavier grit particles are removed by selective deposition through velocity-control chambers (either hydraulic or mechanical). Grit is disposed of by burial, or if washed, by land fill.

Skimmings are scum, grease, and other floatables removed from primary sedimentation tanks. Mechanical or manual skimming facilities may be provided. Depending on the size of the installation, one of several methods of disposal can be employed. These are summarized in table 11.5.

THICKENING

The first step in most sludge-disposal processes is thickening, either by gravity or by dissolved air flotation. Thickening is advantageous because it:

1. Improves digester operation and reduces capital cost where this process is to be employed.
2. Reduces sludge volume for land or sea disposal.
3. Increases economy of sludge-dewatering systems (centrifuges, vacuum filters, etc.).

TABLE 11.3
Chemical Composition of Sewage Sludges
(Dry Weight Basis)

Sewage Treatment Plant	Nitrogen %	Carbon %	Carbon-Nitrogen Ratio	Phosphoric Oxide %	Ash %
Washington, D.C. (Primary Treatment)					
Influent solids:					
Spring	2.42	43.46	18.0	1.14	32.35
Summer	2.39	43.69	18.3	1.09	37.59
Digested sludge	2.06	28.59	13.9	1.44	52.83
Baltimore, Md.					
Influent solids	2.23	47.09	21.1	1.29	24.16
Activated sludge	2.36	30.37	12.9	11.01	29.70
Humus tank sludge	5.34	37.90	7.1	3.96	32.30
Heat-dried digested sludge	3.05	36.53	12.0	2.97	39.73
Jasper, Indiana					
Influent solids	2.90	42.31	14.6	1.62	32.29
Activated sludge	3.51	23.01	6.6	2.81	52.43
Digested sludge	5.89	22.95	3.9	3.49	36.96
Richmond, Indiana					
Influent solids	3.80	28.21	7.4	5.19	40.94
Activated sludge	3.02	44.04	14.6	3.64	31.37
Digested sludge	2.24	26.36	11.8	4.34	50.09
Chicago, Illinois (Southwest plant)					
Raw sludge	2.70	46.62	17.3	2.71	28.24
Activated sludge	4.98	28.62	5.7	5.58	34.82
Heat-dried sludge	5.56	29.41	5.3	6.56	37.42
Milwaukee, Wisconsin					
Heat-dried sludge	5.96	20.88	3.5	3.96	27.73

Gravity Thickening

Gravity thickening is accomplished in a tank equipped with a slowly rotating rake mechanism that breaks the bridge between sludge particles, thereby increasing settling and compaction. A typical gravity thickener is shown in figure 11.3.

The primary objective of a thickener is to provide a concentrated sludge underflow. The area of thickener for a specified underflow concentration is related to the mass loading (lb/ft²al/day, kg/m²/day) or to the unit area (ft²/lb/day, m²/kg/day).

TABLE 11.4
Minor Chemical Elements in Activated Sludge and Digested Sludge

Sludge	Elements, mg/l				
	Copper	Zinc	Boron	Manganese	Molybdenum
Activated sludge					
Minimum	385	950	6	65	6
Average	916	2,500	33	134	16
Maximum	1,500	3,650	74	190	45
Digested sludge					
Minimum	315	1,350	4	30	2
Average	643	2,459	9	262	6
Maximum	1,980	3,700	15	790	12

TABLE 11.5
Characteristics and Disposal of Sewage Screenings, Grit, and Skimmings[a]

Material	Quantity ft³/MG (m³/m³)	Moisture Content, %	Organic Content, %	Thermal Value, BTU/lb (kcal/kg)	Disposal
Screenings	0.5–6.0[b] (0.004–0.045)	85–95	50–80	1400–3500 (775–1940)	Burial incineration grinding, and return to sewage
Grit	1–12 (0.00075–0.090)	14–34	50[d]	–	Grit chambers, hydro-clones on primary sludge; burial, drying beds, incineration
Skimmings	0.1–7.0[e] (0.008–0.053)	60–90	90–95	8000–18,000 (4440–9960)	Burial, pumped to digesters, dewatering, incineration[f]

[a] From Burd [2].
[b] Screen openings of .5 to 2 in. (1.27 to 5.08 cm).
[c] Separate unit, skimmings, refuse, or dewatered sludge incinerator; reduce moisture content to 60 to 65%.
[d] Grit washing can reduce organics to 15%.
[e] Depends on industrial waste discharges to sewerage system.
[f] Separate incineration or combined with sludge, provision for high-temperature burning.

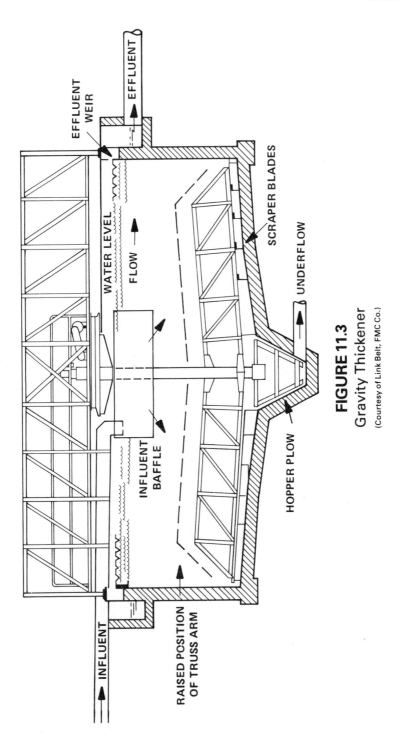

FIGURE 11.3
Gravity Thickener
(Courtesy of Link Belt, FMC Co.)

     The mass loading can be computed from a stirred laboratory cylinder test. For municipal sewage, the mass loading might be expected to vary from 4 lb/ft²/day (19.5 kg/m²/day) for waste-activated sludge to 22 lb/ft²/day (107 kg/m²/day) for primary sludge [2]. Average reported data is shown in table 11.6.

     A procedure for the design of gravity thickeners has been developed by Dick [3]. The most important criteria in thickener design and operation is the mass loading or solids flux expressed as lbs of solids fed/ft²/day. The limiting flux that produces the desired underflow for a given area can be defined:

$$G_L = \frac{C_o Q_o}{A} = \frac{M}{A} \qquad (11\text{--}1)$$

where Q_o = influent flow, ft³/day
          C_o = influent solids, lb/ft³
          M = solids loading, lb/day
          G_L = limiting solids flux, lbs/ft²/day

The limiting flux can be obtained from the following rationale.
     The capacity of a thickener for removing solids under batch conditions is:

$$G_B = C_i V_i \qquad (11\text{--}2)$$

where G_B = batch flux, lbs/ft²/day
          C_i = solids concentration, lbs/ft³
          V_i = settling velocity at C_i, ft/day

A relationship can be developed between C_i and V_i that is usually linear over a wide range of concentrations as shown for an activated sludge in figure 11.4.

     In a continuous thickener the solids are removed both by gravity and by the velocity resulting from the removal of sludge from the tank bottom:

$$G = C_i V_i + C_i U \qquad (11\text{--}3)$$

where G = continuous solids flux, lb/ft²/day
          U = downward sludge velocity due to sludge removal, ft/day

     G can be varied by controlling U since this is determined by the underflow pumping rate. Assuming total solids removal from the bottom:

$$U = \frac{Q_u}{A} = \frac{C_u Q_u}{C_u A} = \frac{M}{C_u A} = \frac{G_L}{C_u} \qquad (11\text{--}4)$$

where Q_u = underflow, ft³/day
          C_u = underflow concentration, lb/ft³

It is important to note from equation (11–4) that increasing U, the withdrawal rate decreases the underflow concentration C_u. A batch flux curve as shown in figure

TYPE OF SLUDGE	SOLIDS—SURFACE LOADING ($lb/day\text{-}ft^2$)	THICKENED SLUDGE SOLIDS CONCENTRATION (%)
SEPARATE SLUDGES		
PRIMARY	20-30	8-10
MODIFIED ACTIVATED	15-25	7-8.5
ACTIVATED	5-6	2.5-3
TRICKLING FILTER	8-10	7-9
COMBINED SLUDGES		
PRIMARY AND MODIFIED ACTIVATED	20-25	8-12
PRIMARY AND ACTIVATED	6-10	5-8
PRIMARY AND TRICKLING FILTER	10-12	7-9

TABLE 11.6
Gravity Thickener Surface Loadings
and Operational Results

FIGURE 11.4
Settling Characteristics of Waste Sludge

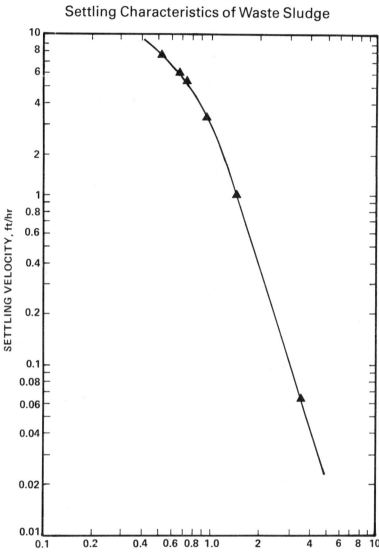

11.5 can be employed to determine the limiting flux G_L for a given underflow concentration C_u, since the slope of any line connecting G_L on the y-axis with C_u on the x-axis on the batch flux curve is the corresponding required underflow velocity U. The batch flux curve as shown in figure 11.5 is obtained by plotting G_B as computed from equation (11–2) against its corresponding concentration C_i (see figure 11.4).

FIGURE 11.5

Batch Flux Plot Illustrating How to Determine Limiting Flux for a Continuous Thickener

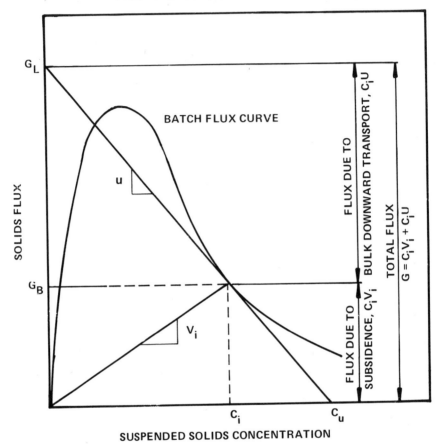

SUSPENDED SOLIDS CONCENTRATION

The required thickener area, A, is then computed from equation (11-1). It should be noted that the selected underflow concentration C_u must be less than the ultimate concentration attainable, C_∞. C_∞ is determined from thickening studies

$$C_\infty = \frac{C_o H_o}{H_\infty}$$

where C_o = initial solids concentration
  H_o = initial height
  C_∞ = final or ultimate concentration
  H_∞ = final height

In some cases thickener performance can be improved by the addition of coagulants. Data for municipal sludge is shown in figure 11.6. Performance data for gravity thickening is shown in table 11.7.

Flotation Thickening

Thickening through dissolved air flotation is becoming increasingly popular and is particularly applicable to gelatinous sludges such as activated sludge. In flotation thickening, small air bubbles released from solution attach themselves to

FIGURE 11.6

Thickening Performance as Affected by Mass Loading at
Constant Chemical Dosage

(Courtesy of Rexnord Inc.)

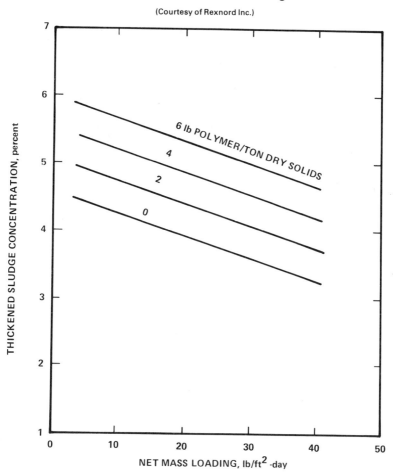

TABLE 11.7

Performance Data from Gravity Thickening

DESCRIPTION	LOADING (lb/sq ft-day)	INFLUENT SS (%)	THICKENED SS (%)
DOMESTIC			
TRICKLING FILTER			8.0
ACTIVATED SLUDGE		0.8	3.5
PRIMARY SLUDGE		1.8	9.0
	24.2	0.4	4.5
	31.2	0.6	4.6
	24.9	0.4	4.9
	17.2	.0.3	4.1
	13.2	0.2	3.8
	16.3	0.2	4.2
PRIMARY AND SECONDARY	3.1	0.2	4.5
	7.8	0.6	6.3
	17.9	1.2	8.1
	37.5	0.7	4.0
	28.5	0.5	4.5
TRICKLING FILTER	8-10		7.9
PULP AND PAPER			
(53% PRIMARY, 47% ACTIVATED)	25.0	2.0	4.0
(67% PRIMARY, 33% ACTIVATED)	25.0	2.0	6.0
(100% PRIMARY)	25.0	2.0	9.0
BREWERY-DOMESTIC		2.0	8.0
CARBON PROCESSING	18-76	0.2-0.5	5.0
LIME SOFTENING		4.5	18.7

and become enmeshed in the sludge flocs. The air-solid mixture rises to the surface of the basin, where it concentrates and is removed as shown in figure 11.7. The primary variables are recycle ratio, feed solids concentration, air-to-solids (A/S) ratio, and solids and hydraulic loading rates. Pressures between 40 and 60 psi are commonly employed. Recycle ratio is related to the air-to-solids ratio and the feed solids concentration. The float solids are related to the A/S ratio as shown in figure 11.8.

Experience has shown that in some cases dilution of the feed sludge to a lower concentration increases the concentration of the floated solids. Performance data for the thickening of excess activated sludge is shown in table 11.8. The use of polyelectrolytes will usually increase the solids capture and the thickened sludge concentration.

CENTRIFUGATION

Centrifugation is employed both for the thickening and the dewatering of sludges. The process of centrifugation is an acceleration of the process of sedimentation by the application of centrifugal forces. There are three types of

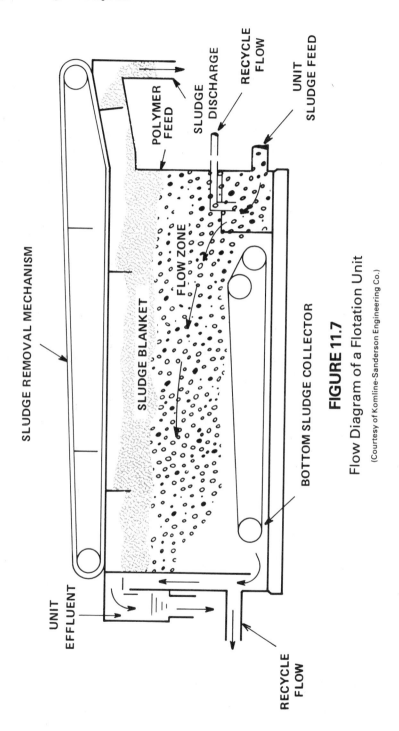

FIGURE 11.7

Flow Diagram of a Flotation Unit

(Courtesy of Komline-Sanderson Engineering Co.)

FIGURE 11.8

Influence of Air-to-Solids Ratio on Float Solids Content

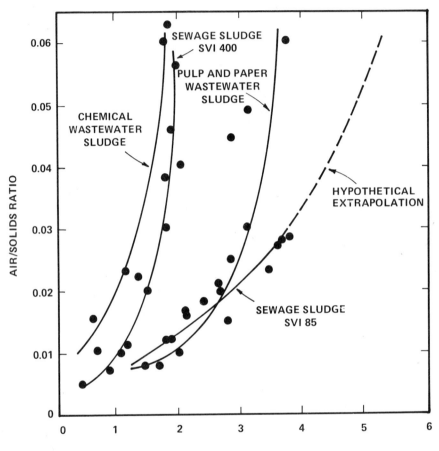

SOLIDS IN FLOAT, percent

centrifuges available, the solid bowl decanter, the basket type and the disc-nozzle separator. These units are shown in figures 11.9, 11.10, and 11.11. The basic difference between the types of centrifuges is the method in which solids are collected in and discharged from the bowl.

Centrifuge performance is affected by both machine and process variables as shown in table 11.9. The significant machine variables for the solid bowl decanter are bowl speed, pool volume, and conveyor speed. Disc-nozzle separator variables include bowl speed, recycle mode of operation, disc spacing and nozzle configuration. Basket centrifuge machine variables are bowl speed, cycle feed time, skimmer nozzle travel rate, and skimmer nozzle dwell time.

LOCATION	POLYMERS USED	FEED SUSPENDED SOLIDS CONCENTRATION, %	THICKENED SLUDGE SUSPENDED SOLIDS CONCENTRATION, %	SOLIDS LOADING lb/sq ft·hr	OVERFLOW RATE gpm/sq ft
NASSAU COUNTY, N.Y.	NO	0.81	4.9	--	--
WAYNE COUNTY, MICH.	NO	0.45	4.6	--	--
COORS BREWERY, GOLDEN, COLO.	NO	0.77	4.1	--	--
BERNARDSVILLE, N.J.	YES	1.7	4.3	4.3	0.5
HATBORO, PA	YES	0.7	4.0	2.9	0.8
BELLEVILLE, ILL	YES	1.8	5.7	3.8	0.4
COLUMBUS, OHIO	YES	0.7	5.0	3.3	1.0
FORT WORTH, TEXAS	YES	0.9	4.5	4.5	1.0

TABLE 11.8

Some Results Obtained with Activated Sludge
in Dissolved Air Flotation Units

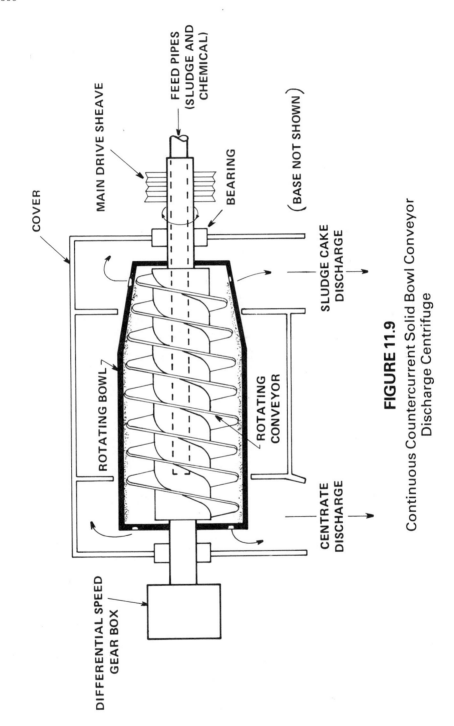

FIGURE 11.9

Continuous Countercurrent Solid Bowl Conveyor
Discharge Centrifuge

FIGURE 11.10
Schematic Diagram of a Basket Centrifuge

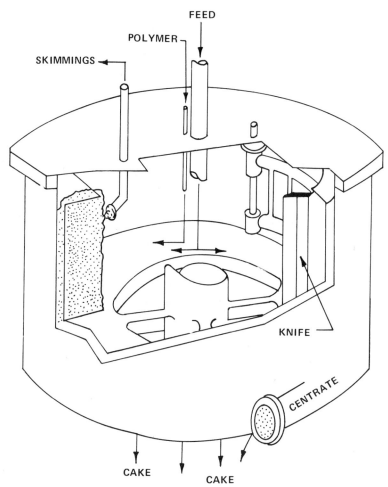

The solid bowl decanter consists of an imperforate cylindrical-conical bowl with an internal helical conveyor as shown in figure 11.9. The feed sludge enters the cylindrical bowl through the conveyor discharge nozzles. Centrifugal force compacts the sludge against the bowl wall, and the interval scroll or conveyor, which rotates slightly slower than the bowl, conveys the compacted sludge along the bowl wall toward the conical section (beach area) and out.

In the basket type centrifuge, feed is introduced in the bottom of the basket. At equilibrium, solids settle out of the annular moving liquid layer to the cake layer, which builds up on the bowl wall while the centrate overflows the lip

FIGURE 11.11
Disc-Type Centrifuge

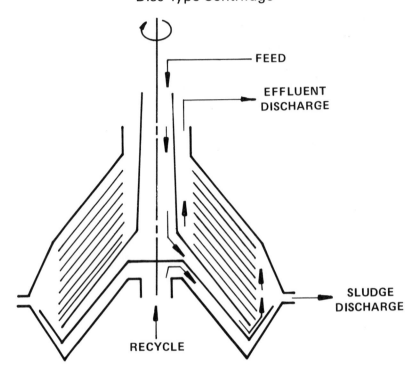

TABLE 11.9
Change in Centrifuge Variables for Improved Performance

Change Desired	Machine Variables			Process Variables			
	Bowl speed	Pool volume	Conveyor speed	Feed rate	Feed consistency	Temp.	Floccu-lents
To improve recovery	+	+	−	−	+	+	+
To improve cake solids	+	−	−	+	−	+	−

ring at the top. When solids have filled the basket, feed is stopped, the basket speed is reduced and a knife moves into the cake, discharging it from the bottom of the casing. Cycles are automated and cake unloading requires less than 10 percent of the cycle time. Chemical addition is generally not required for high solids recovery. However, the unit operates at low centrifugal forces, has a discontinuous cake discharge and a fairly low solids handling capacity.

In the disc-nozzle separator, the feed enters at the top and is distributed between a multitude of channels, or spaces between the stacked conical discs. Solid particles settle through the liquid layer, which is flowing in these channels to the underside of the disc and then slide down to a sludge compaction zone. The thickened sludge is flushed out of the bowl with a portion of the wastewater thus limiting the solids concentration to 10–20 times the feed rate. The disc-nozzle separator finds its major application in the thickening of activated and similar sludges. They are very efficient in thickening waste activated sludge at high feed rates without the addition of polymers. In an industrial waste treatment plant in Germany excess activated sludge has been thickened from 1 percent to 8–10 percent solids. Centrifugal thickening performance is shown in table 11.10.

Sludge dewatering usually uses a horizontal, cylindrical-conical, solid bowl machine. The centrifuge consists of three major components, a rotating solid bowl that takes the shape of a cylinder and truncated cone section, an inner rotating screw conveyor, and a planetary gear system. Machine variables include bowl speed, pool volume, and conveyor speed. Process variables include the feed rate of solids to the machine, solids characteristics, chemical addition, and temperature.

Centrifuges separate solids from liquids by sedimentation and centrifugal force. Sludge solids settle through the liquid pool and are compacted by centrifugal force against the walls of the bowl and are then conveyed by the screw conveyor to the drying or beach end of the bowl. The beach area is an inclined section of the bowl where further dewatering occurs before the solids are discharged over adjustable weirs at the opposite end of the bowl.

Bowl speed affects clarification. The clarification capacity depends upon the gravitational force and the detention time of the feed mixture in the pool. Centrifuges operate in excess of 3500 times the force of gravity.

The pool volume regulates the liquid retention time and the drying beach length in the bowl. As a result, increasing the pool volume improves solids recovery, but decreases the cake dryness. Since the pool volume is controlled by an adjustable weir, field adjustments can be made for optimum operation. The conveyor speed can be adjusted to yield maximum cake dryness and solids recovery for a specific application.

When the feed rate to a centrifuge is increased, the retention time in the unit is decreased and the recovery decreases. Flow rates are usually limited to 0.5–2.0 gpm/hp (3.65–14.6 m³/day–kw) to obtain satisfactory solids recovery. Since the lower recovery results in only the removal of larger particles, a drier

Type of Sludge	Centrifuge Type	Capacity gpm	Feed Solids, %	Underflow Solids, %	Solids Recovery, %	Polymer Requirement, lb/ton
AS[a]	Disc	150	0.75–1.0	5–5.5	90+	None
AS	Disc	400	–	4.0	80	None
AS (after Roughing Filter)	Disc	50–80	0.7	5–7	93–87	None
AS (after Roughing Filter)	Disc	60–270	0.7	6.1	97–80	None
AS	Basket	33–70	0.7	9–10	90–70	None
AS	Solid-Bowl	10–12	1.5	9–13	90	–
AS	Solid-Bowl	75–100	0.44–0.78	5–7	90–80	None
AS	Solid-Bowl	110–160	0.5–0.7	5–8	65	None
					85	<5
					90	5–10
					95	10–15

TABLE 11.10

Centrifugal Thickening Performance Data

[a]AS = Activated Sludge.

cake is produced as shown in figure 11.12. Increasing the feed solids concentration reduces the liquid overflow from the machine, resulting in an increased recovery of solids.

Chemical flocculents (polyelectrolytes) are used to increase recovery. The flocculents both increase the structural strength of the solids and flocculate fine particles. Because of the increased removal of the fine particles, chemical addition usually lowers the cake dryness.

Barnard and Englande [4] correlated centrifuge performance data in accordance with the relationship as shown in figures 11.13 and 11.14.

$$R = \frac{C_1 (C_2 + P)^m}{Q^n}$$ (11–6)

where R = percent recovery
P = polymer dosage, lbs/tons dry solids feed
Q = feed rate, gpm/ft^2
C_1, C_2 = constants
m, n = exponents

FIGURE 11.12
Cake Dryness Versus Recovery

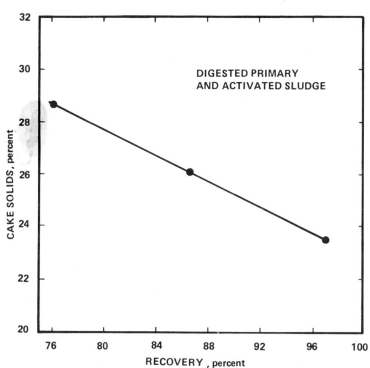

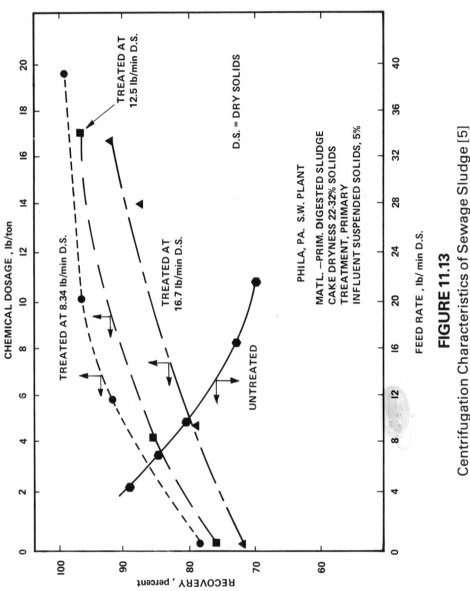

FIGURE 11.13

Centrifugation Characteristics of Sewage Sludge [5]

FIGURE 11.14

Solids Recovery for Digested Activated Sludge with Cationic Polymer

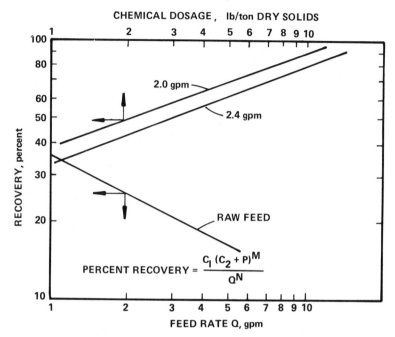

Values for the exponents m and n are shown in table 11.11.

Performance data for the centrifugal dewatering of several sludges are summarized in table 11.12.

TABLE 11.11

Values of Exponential Constants in Centrifuge Model

Waste	Type Sludge	m	n
Municipal	Primary digested	0.090	0.150
Municipal	Waste activated	0.370	0.520
Municipal	Raw primary	0.068	0.089
Municipal	Digested primary	0.045	0.037
Tannery	Raw sludge	0.170	0.400

VACUUM FILTRATION

Vacuum filtration is one of the most common methods for dewatering wastewater sludges. Vacuum filtration dewaters a slurry under applied vacuum by means of a porous media, which retains the solids but allows the liquid to pass through. Media used include cloth, steel mesh, or tightly wound coil springs.

In vacuum-filter operation (figures 11.15, 11.16, and 11.17) a rotary drum passes through a slurry tank in which solids are retained on the drum surface under applied vacuum. The drum submergence can vary from 12 to 60 percent. As the drum passes through the slurry, a cake is built up and water removed by filtration through the deposited solids and the filter media. The time the drum remains submerged in the slurry is the form time (t_f). As the drum emerges from the slurry tank, the deposited cake is further dried by liquid transfer to air drawn through the cake by the applied vacuum. This period of the drum's cycle is called the dry time (t_d). At the end of the cycle, a knife edge scrapes the filter cake from the drum to a conveyor. The filter media is usually washed with water sprays prior to again being immersed in the slurry tank.

FIGURE 11.15
Vacuum Filter Schematic

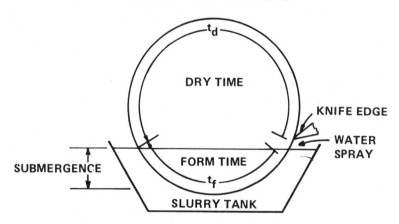

Solids Concentration %	Settling Velocity ft/hr	Solids Flux lb/ft²−day
0.4	7.0	41.9
0.8	1.9	22.8
1.2	0.7	12.6
1.6	0.28	6.7
2.0	0.13	3.9

SLUDGE	CONCENTRATION percent	CAKE percent	RECOVERY percent	CHEMICALS lb/ton[a] (kg/ton)[b]
BOARD MILL	2-5	22-30	85-95	--
KRAFT MILL	1-5	22-34	82-95	--
PULP	6.1	29.8	87	--
WHITE WATER	3.1	42.7	99	--
RAW SLUDGE (DOMESTIC)	--	30-40	70-90	--
DIGESTED SLUDGE	--	30-40	70-90	--
PRIMARY AND SECONDARY DOMESTIC	--	15-20	85-100	10-15 (5-7.5)
SECONDARY DOMESTIC	--	5-15	90-100	5-10 (2.5-5)
WASTE ACTIVATED	1.5	5-15	90-100	5-10 (2.5-5)

TABLE 11.12
Centrifugation of Waste Sludge

[a]SHORT TON
[b]METRIC TON

PROBLEM 11–1

A 0.1 MGD chemical sludge at a 1.5 percent solids concentration is to be concentrated in a gravity thickener to obtain a solids in the underflow at 2.5 percent SS concentration. The settling data of the chemical sludge are as follows:

Solids Concentration %	Settling Velocity ft/hr
0.4	7.0
0.8	1.9
1.2	0.7
1.6	0.28
2.0	0.13

Develop the solids flux curve by computing the solids flux at each solids concentration as shown below:

for SS = 0.4%

$$\text{solids flux} = SS\ (\%)\ \frac{62.4}{100}\ (ZSV)\ (24)$$

$$= 0.4 \cdot \frac{62.4}{100} \cdot 7.0 \cdot 24$$

$$= 41.9\ \text{lb/ft}^2\text{–day}$$

The results are plotted in figure I.

The design solids flux or limiting solids flux is obtained by drawing the tangent to the solids flux curve from the desired underflow SS concentration of 2.5 percent.

$$G_L = 18\ \text{lb/ft}^2\text{–day}$$

The required surface area of the gravity thickener is:

$$A = \frac{0.1 \cdot \dfrac{1.5}{100} \cdot 10^6 \cdot 8.34}{18}$$

$$= 695\ \text{ft}^2$$

FIGURE I
Batch Solids Flux Curve

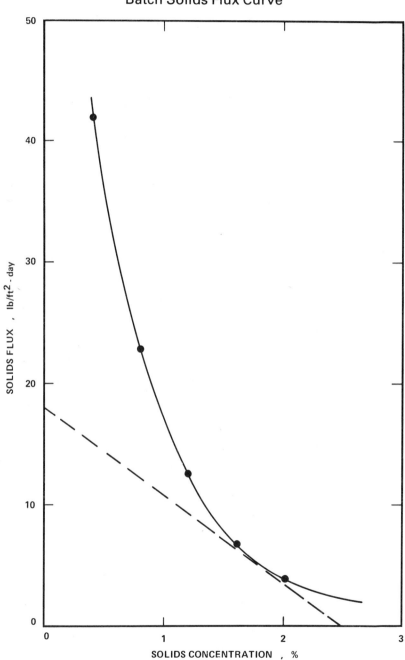

SOLIDS FLUX , lb/ft^2 - day

SOLIDS CONCENTRATION , %

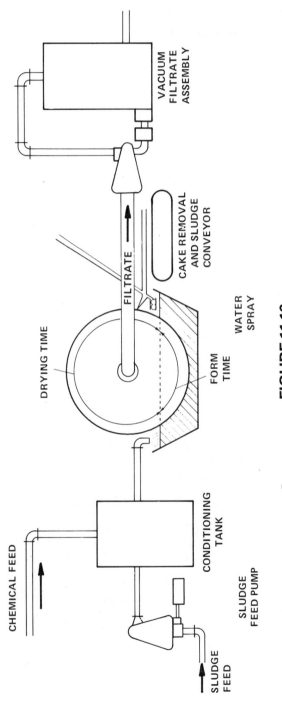

FIGURE 11.16

Schematic of Vacuum Filter Process

FIGURE 11.17
Vacuum-filter Installation
(Courtesy of Eimco BSP Div., Envirotech Co.)

PROBLEM 11–2

Determine the polymer required and size the centrifuge with respect to surface area for dewatering 10,000 lb/day of sludge previously thickened to 4 percent. The centrifuge will operate over 8 hours with a 95 percent solids recovery. The pilot plant data yielded the following relationship:

$$R(\%) = \frac{48\ (0.47 + P)^{0.37}}{Q^{0.52}}$$

where R = solid recovery efficiency, %
       P = polymer dosage, lb/ton of sludge
       Q = hydraulic loading, gpm/ft^2

For 4% solids concentration, the sludge flow is:

$$Q' = 10000 \cdot \frac{1}{0.04} \cdot \frac{1}{8.34}\ \text{gal/day}$$

$$= 30{,}000\ \text{gal/day}$$

If the centrifuge will operate for 8 hours per day, the total sludge flow to the centrifuge is:

$$Q' = 30{,}000 \cdot \frac{24}{8}\ \text{gal/day}$$

$$= 90{,}000\ \text{gal/day or 62.5 gpm}$$

At 95% solids recovery, the relationship between polymer requirement and hydraulic loading can be calculated by:

$$95 = \frac{48\ (0.47 + P)^{0.37}}{Q^{0.52}}$$

The centrifuge size is computed by:

$$A = \frac{62.5}{Q}\text{ft}^2$$

The polymer dosage, therefore, is computed by:

$$P' = P \cdot \frac{10000}{2000}\ \text{lb/day}$$

The results are plotted in figure I. Then, the capital and operating costs for each combination of required area and polymer dosage should be found and the least cost situation determined.

FIGURE I
Required Centrifuge Area and Polymer Dosage for Various Hydraulic Loadings

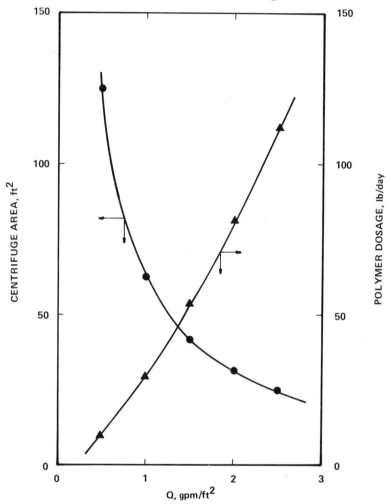

The variables that influence the dewatering process are solids concentration, sludge and filtrate viscosity, sludge compressibility, chemical composition, and the nature of the sludge particles (size, shape, water content, etc.).

The filter operating variables are vacuum, drum submergence and speed, sludge conditioning, and the type and porosity of the filter media.

The rate of filtration of sludges has been formulated and according to Poiseuille's and Darcy's laws by Carman and Coackley [6]:

$$\frac{dV}{dt} = \frac{PA^2}{\mu \, (rcV + R_m A)} \tag{11-7}$$

where V = volume of filtrate
t = cycle time (approximate form time in continuous drum filters)
P = vacuum
A = filtration area
μ = filtrate viscosity
r = specific resistance
c = weight of solids/unit volume of filtrate

$$= \frac{1}{\left(\dfrac{C_i}{100 - C_i} - \dfrac{C_f}{100 - C_f}\right)}$$

where c_i = initial % moisture content
c_f = final % moisture content

R_m, the initial resistance of the filter media, can usually be neglected as small compared to the resistance developed by the filter cake. The specific resistance r is a measure of the filterability of the sludge and is numerically equal to the pressure difference required to produce a unit rate of filtrate flow of unit viscosity through a unit weight of cake.

Integration and rearrangement of equation (11–7) yields

$$\frac{t}{V} = \frac{\mu r c}{2PA^2} V + \frac{\mu R_m}{PA} \tag{11-7a}$$

From equation (11–7a) a linear relationship will result from a plot of t/V against V. The specific resistance can be computed from the slope of this plot:

$$r = \frac{2bPA^2}{\mu c} \tag{11-8}$$

where b = slope of the t/V against V plot

Although specific resistance has limited value for the calculation of filtra-tion rates on drum filters, it provides a valuable tool for the evaluation of vacuum-filtration variables and the relative filterability of sludges. Typical values are given in table 11.13.

Most wastewater sludges form compressible cakes in which the filtration rate and the specific resistance is a function of the pressure difference across the cake:

$$r = r_o P^s \tag{11-9}$$

where s = coefficient of compressibility

TABLE 11.13

Specific Resistance of Sludges[a]

Description		Specific Resistance (sec²/gram) × 10⁷	Coefficient of Compressibility
Domestic activated sludge		2,800	
Activated (digested)		800	
Primary (raw)		1,310–2,110	
Primary (digested)		380–2,170	
Primary (digested)		1,350	25.00
Primary (digested)			
Detention time	Stage		
7.5 days	1	1,590	
10.0 days	1	1,540	
15.0 days	1	1,230	
20.0 days	1	530	
30.0 days	1	760	
15.0 days	2	400	
20.0 days	2	400	
30.0 days	2	480	
Activated sludge + 13.5% $FeCl_3$		45	
Activated sludge + 10.0% $FeCl_3$		75	
Activated sludge + 125% (by weight) newsprint		15	
Activated digested sludge + 6% $FeCl_3$ + 10% CaO		5	
Activated digested sludge + 125% newsprint + 5% CaO		4.5	
Vegetable processing sludge		46	7.00
Vegetable tanning		15	20.00
Lime neutralization acid mine drainage		30	10.50
Alum sludge (water works)		530	14.50
Neutralization of sulfuric acid with lime slurry		1–2	
Neutralization of sulfuric acid with dolomitic lime slurry		3	0.77
Aluminum processing		3	0.44
Paper industry		6	
Coal (froth flotation)		80	1.60
Distillery		200	1.30
Mixed chrome and vegetable tannery		300	
Chemical wastes (biological treatment)		300	
Petroleum industry (from gravity separators)			
Refinery A		10–100	0.50
Refinery B		100	0.70

[a]All values were recorded at 500 g/sq cm pressure.

The greater the value of s, the more compressible is the sludge. When $s = O$, the specific resistance is independent of pressure and the sludge is incompressible.

Some generalizations on filtration characteristics can be made. Raw sewage sludge is easier to filter than digested sewage sludges, and primary sewage sludges are easier to filter than secondary sludges. Filterability is influenced by particle size, shape, and density and by the electrical charge on the particle. Smaller particles exert a greater chemical demand than larger particles. The larger the particle size, the higher is the filter rate and the lower the cake moisture. Municipal and industrial sludges filter very poorly and coagulants must be added to enhance filterability and reduce the specific resistance. Combinations of lime and/or ferric salts have been the most common coagulating agents used in the past. Recently, however, polyelectrolytes have proved effective coagulants in many applications. Frequently, the dual use of anionic and cationic polymers is the most economic and effective procedure. The cationic polymer effects charge neutralization and the anionic effects polymer particle bridging and agglomeration of the particles. Note that excessive coagulant dosages result in a charge reversal and an increase in specific resistance. Typical data is shown in figure 11.18.

PROBLEM 11–3

The sludge dewatering data were collected from a Buchner funnel test as shown in table I. The specific conditions of the test were:

TABLE I
Results of Buchner Funnel Test

Time, Sec	Volume, ml	t/V, sec/ml
14.5	66	0.22
29.5	92	0.31
45.0	112	0.40
59.0	129	0.46
70.0	134	0.52
89.0	156	0.57
105.0	167	0.63
120.0	174	0.69

$A = 104.6 \text{ cm}^2$
$P = 15 \text{ in Hg} = 526 \text{ g/cm}^2$
$\mu = 0.00895 \text{ g/cm-sec}$
$C_i = 0.044 \text{ g/ml (solids) or 95.6% moisture}$
$C_f = 0.20 \text{ g/ml (solids) or 80% moisture}$

Determine the specific resistance.

By plotting t/V against V as shown in figure I the slope of the line is:

$$b = 0.004 \text{ sec/ml}^2$$

The solids deposited per unit volume of filtrate is calculated by:

$$c = \cfrac{1}{\left(\cfrac{C_i}{100 - C_i} - \cfrac{C_f}{100 - C_f}\right)}$$

$$= \frac{1}{(95.6/4.4) - (80/20)} \text{ g/ml}$$

$$= 0.056 \text{ g/ml}$$

FIGURE I
Correlation of Buchner Funnel Test Results

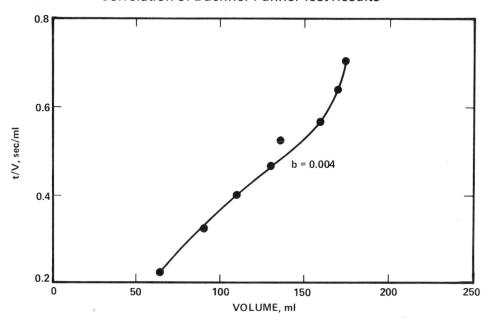

The specific resistance is:

$$r = \frac{2bPA^2}{\mu c}$$

$$= \frac{2 \cdot 0.004 \cdot 526 \cdot (104.6)^2}{0.00895 \cdot 0.056} \text{ sec}^2/\text{g}$$

$$= 9.19 \times 10^7 \text{ sec}^2/\text{g}$$

Vacuum Filtration Design

Equation (11–7) can be modified to express filter loading (neglecting the initial resistance of the filter media):

$$L_f = 35.7 \left(\frac{cP^{1-s}}{\mu R_o t_f} \right)^{1/2} \tag{11–10}$$

where $R_o = r_o \times 10^{-7}$, sec^2/g
   P = vacuum, psi
   c = solids deposited per unit volume filtrate, g/ml
   μ = filtrate viscosity, contipoises
   t_f = form cycle time, min
   L_f = filter loading, lb/ft^2-hr

For routine calculations c_i is usually used in equation (11–8) as c, the initial concentration.

Equation (11–10) should be modified for the prediction of filtration rates for various types of sludges:

$$L_f = 35.7 \left(\frac{P^{1-s}}{\mu R_o} \right)^{1/2} \frac{c^m}{t_f^n} \tag{11–11}$$

Filter operation has shown the exponent n to vary from 0.4 to 1.0. Schepman and Cornell [7] have attributed this variation to variations in cake permeability as additional cake is formed. The effect of variation on solids content fed to the filter will likewise vary m from 0.25 to 1.0. The exponents m and n must be determined experimentally for any specific application.

Final cake moisture is related to the cake thickness, the drying time, the pressure drop across the cake, the liquid viscosity, and the air rate through the cake. The drying time increases to a maximum beyond which very little increase will occur. In a very porous cake, the change in cake moisture with increased drying time or increased vacuum is small, because the high air rates through the cake cause a rapid initial drying to equilibrium. Nonporous cakes require longer drying times and high vacuum to attain a maximum cake solids content. Typical plots of experimental vacuum filtration analysis are shown in figure 11.19. The

FIGURE 11.18

Effect of Coagulant Dosage on Sludge Specific Resistance and Capillary Suction Time

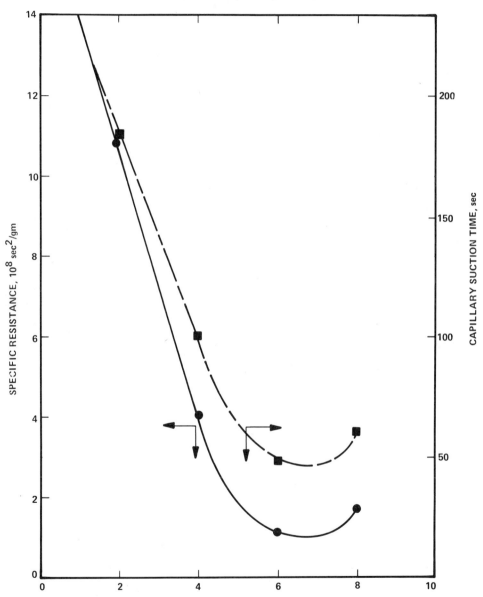

COAGULANT DOSAGE, lb /ton

FIGURE 11.19
Experimental Plots for Vacuum-Filtration Design

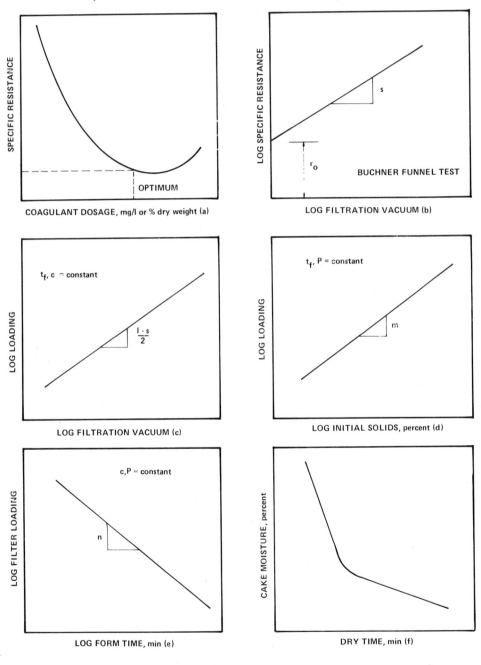

filter loading from equation (11–11) is in terms of form time. This is conventionally converted to cycle time for final specification and design.

$$L_c = L_f \left(\frac{\% \text{ submergence}}{100} \right) \cdot 0.8 \qquad (11\text{–}12)$$

The factor 0.8 compensates for the area of the filter drum where the cake is removed and the media washed. The total cycle time on a filter may vary from 1 to 6 minutes. Submergence of the drum may vary from 10 to 60 percent resulting in a maximum spread of form time of 0.1 to 3.5 minutes. This also yields a maximum spread of dry time of 2.5 to 4.5 minutes. In general, the filter yield from highly compressible cakes is relatively unaffected by increases in form vacuum varying from 12 to 17 in. (30 to 43 cm) of Hg. The vacuum-filtration characteristics of several sludges are summarized in table 11.14. The specific resistance can be estimated from a capillary suction apparatus as shown in figure 11.20.

PRESSURE FILTRATION

Pressure filtration is applicable to almost all water and wastewater sludges. As shown in figures 11.21 and 11.22, the sludge is pumped between plates that are covered with a filter cloth. The liquid seeps through the cloth leaving the solids behind between the plates. The filter media may or may not be precoated. When the spaces between the plates are filled, the plates are separated and the solids removed. The pressure exerted on the cake during formation is limited to the pumping force and filter closing system design. Filters are designed at pressures ranging from 50 to 225 psi (340 to 1600 kg/m^2). As the final filtration pressure increases, a corresponding increase in dry cake solids is obtained. Most municipal sludges can be dewatered to produce 40 to 50 percent cake solids with 225 psi (1600 kN/m^2) systems or 30 to 40 percent solids with 100 psi (690 kN/m^2) filters. Filtrate quality will vary from 10 mg/l suspended solids with precoat to 50–500 mg/l with unprecoated cloth depending on the media, type of solids, and type of conditioning. Conditioning chemicals are the same as used in vacuum filtration (lime, ferric chloride, or polymers). Materials such as ash have also been used. A pressure filter is shown in figure 11.23. Performance characteristics are shown in table 11.15.

Hydraulic presses have also been applied to further dewater filter cake from paper mill sludges for incineration. Board mill sludge has been dewatered to 40 percent solids from 30 percent solids at a pressure of 300 psi (2100 kg/m^2) and a pressing time of 5 min.

A belt filter press has recently been developed. Chemically conditioned sludge is fed through two filter belts and is squeezed by force to drive water through these belts as shown in figure 11.24. Variations of this device have been successfully used to dewater municipal and industrial sludge. Dewatering perfor-

SLUDGE	CONCENTRATION, percent	CAKE, percent	FILTRATION RATE lb/ft^2-hr (kg/m^2-hr)	CHEMICALS percent (by wt)	
				[a]	[b]
PRIMARY SEWAGE					
UNDIGESTED	6-10	66-69	6.9 (33.7)	1.0-2.0	6.0-9.0
DIGESTED	6-10	70-73	7.2 (35.2)	2.5-3.5	7.0-12.0
HIGH-RATE TRICKLING FILTER					
MIXED PRIMARY AND SECONDARY	7	68-75	7.1 (34.7)	1.5-2.5	7.0-11.0
DIGESTED MIXTURE	7	71	--	3.5	9.0
RAW PRIMARY OR RAW PRIMARY AND FILTER				[c]	
HOMOS	--	63-72	6-20 (29-98)	0.2-1.2	
DIGESTED PRIMARY	--	66-74	4-15 (19.5-73)	0.2-1.5	
DIGESTED PRIMARY AND ACTIVATED	--	68-76	4-8 (19.5-39)	0.5-2.0	

[a]Ferric chloride
[b]Lime
[c]Polyelectrolyte

TABLE 11.14
Vacuum Filtration Characteristics of Sludges

FIGURE 11.20
Capillary Suction Time Apparatus

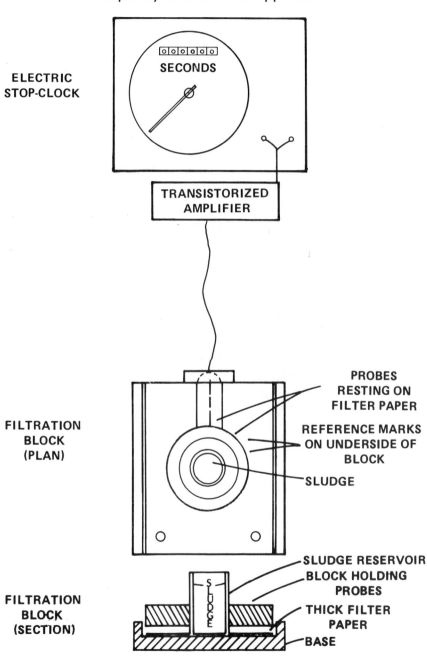

ELECTRIC
STOP-CLOCK

SECONDS

TRANSISTORIZED
AMPLIFIER

FILTRATION
BLOCK
(PLAN)

PROBES
RESTING ON
FILTER PAPER

REFERENCE MARKS
ON UNDERSIDE OF
BLOCK

SLUDGE

FILTRATION
BLOCK
(SECTION)

SLUDGE RESERVOIR
BLOCK HOLDING
PROBES
THICK FILTER
PAPER
BASE

FIGURE 11.21
Side View of a Filter Press

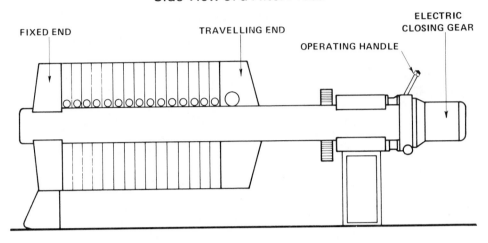

FIGURE 11.22
Cutaway View of a Filter Press

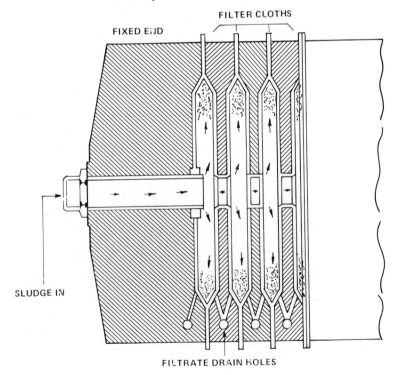

FIGURE 11.23
Pressure Filter

TABLE 11.15
Typical Filter Press Production Data[a]

Sludge Type	Suspended Solids (%)	Conditioning of Dry Solids (%)		Cake Solids (%)	Time Cycle (hr)
Raw primary	5–10	Ash	100	50	1.5
		FeCl₃ Lime	5 10	45	2.0
Raw primary with less than 50% AS[b]	3–6	Ash	150	50	2.0
		FeCl₃ Lime	5 10	45	2.5
Raw primary with more than 50% AS	1–4	Ash	200	50	2.0
		FeCl₃ Lime	6 12	45	2.5
Digested and Digested with less than 50% AS	6–10	Ash	100	50	1.5
		FeCl₃ Lime	5 10	45	2.0
Digested with more than 50% AS	2–6	Ash	200	50	1.5
		FeCl₃ Lime	7.5 15	45	2.5
AS	Up to 5	Ash	250	50	2.0
		FeCl₃ Lime	7.5 15	45	2.5

[a] From [8].
[b] AS = Activated sludge.

mance for a belt filter press on several sludges is shown in table 11.16. A comparison of the performance of solid bowl centrifuges, belt presses and pressure filters was made for 400 treatment plants by Reuter and reported by Bohnke [9]. This comparison is shown in figure 11.25.

Klein Co. developed a belt filter press (see figure 11.26) that employed not only the concept of cake shear with simultaneous application of pressure but also low pressure filtration and thickening by gravity drainage. An endless filter belt (A) runs over a drive and guide roller at each end (B and C) like a conveyor belt. The upper side of the filter belt is supported by several rollers (D). Above the filter bed a press belt (E) runs in the same direction and at the same speed. The drive roller for this belt (F) is coupled with the drive roller (B) of the filter belt. The press belt can be pressed on the filter belt by means of a pressure roller system

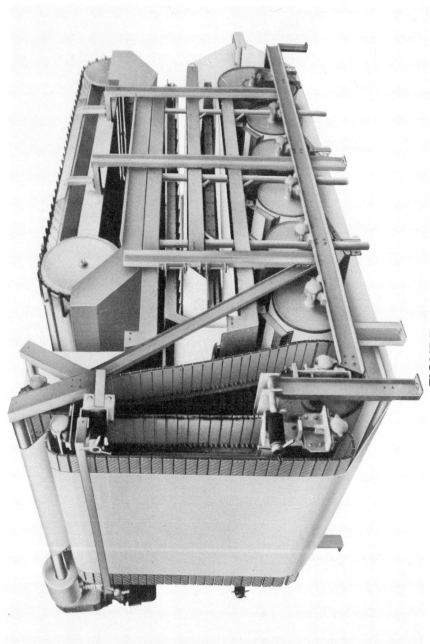

FIGURE 11.24
Belt Filter

(Courtesy of Komline-Sanderson Engineering Co.)

FEED SOLIDS, %	SECONDARY:PRIMARY RATIO	POLYMER DOSAGE[a]	PRESSURE psig[b]	CAKE SOLIDS, %	SOLIDS RECOVERY, %	CAPACITY[c]
9.5	100% PRIMARY	1.6	100	41	97-99	2,706
8.5	1:5	2.4	100	38	97-99	2,706
7.5	1:2	2.7	25-100	33-38	95-97	1,485
6.8	1:1	2.9	25	31	95	898
6.5	2:1	3.1	25	31	95	858
6.1	3:1	4.1	25	28	90-95	605
5.5	100% SECONDARY	5.5	25	25	95	546

TABLE 11.16

Belt Filter Performance for Municipal Sludge

[a]POUNDS PER TON DRY SOLIDS
[b]POUNDS PER SQUARE INCH GAUGE
[c]POUNDS DRY SOLIDS PER HOUR PER METER

FIGURE 11.25
Results in Relation to the Kind of Sludge with Spread

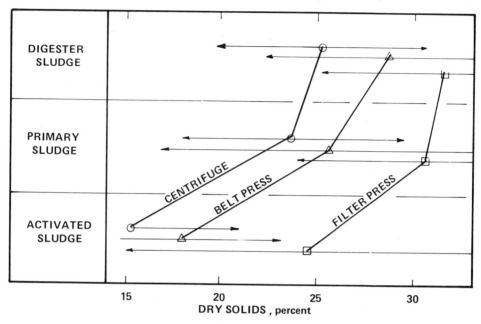

FIGURE 11.26
Schematic of Belt Filter Press

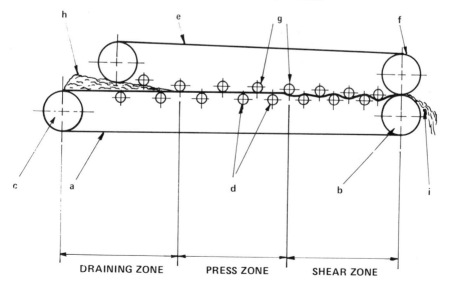

whose rollers (G) can be individually adjusted horizontally and vertically. The sludge to be dewatered (H) is fed on the upper face of the filter belt and is continuously dewatered between the filter and press belts. Note how the supporting rollers of the filter belt and the pressure rollers of the pressure belt are adjusted in such a way that the belts and the sludge between them describe an S-shaped curve. This configuration induces parallel displacement of the belts relative to each other due to the difference in radius producing shear in the cake. After dewatering in the shear zone, the sludge is removed by a scraper (I).

PROBLEM 11–4

A refinery has a sludge consisting of waste lime from a neutralization process and several oily-waste streams. The composite waste has a solids concentration of 52.6 g/l and an average flow of 29,600 gpd. The design coefficients and operating conditions for the vacuum filter are shown below. Find the required vacuum filter area based on a seven day per week operation.

Design Coefficients:

$$\mu = 1.0 \text{ centipoise}$$
$$\frac{l\text{-}s}{2} = 0.087$$
$$m = 0.548$$
$$n = 0.562$$
$$R_o = 3.5 \text{ sec}^2/g$$

Design Operating Conditions:

$$P = 9.8 \text{ psig}$$
Cake Solids = 34%
Cycle Time = 6 min

The initial solids concentration of 52.6 g/l is approximated to 5.26% and thus, the moisture content is 94.74%.
The solids deposited per unit volume filtrate is:

$$c = \left[\frac{C_i}{100 - C_i} - \frac{C_f}{100 - C_f} \right]^{-1}$$

$$= \left[\frac{94.74}{100 - 94.74} - \frac{66}{100 - 66} \right]^{-1}$$

$$= 0.0622 \text{ g/ml}$$

The filter loading is:

$$L = 35.7 \left(\frac{P^{1-s}}{\mu R_o} \right)^{1/2} \frac{cm}{t^n}$$

$$= 35.7 \; \frac{(9.8)^{0.087}}{(1 \cdot 3.5)^{1/2}} \; \frac{(0.0622)^{0.548}}{(6)^{0.562}} \; lb/ft^2\text{-hr}$$

$$= 1.86 \; lb/ft^2\text{-hr}$$

The weight of sludge solids is:

$$W = 52600 \cdot 0.0296 \cdot 8.34 \; lb/day$$

$$= 13{,}000 \; lb/day$$

The required vacuum filter area is:

$$A = \frac{W}{L}$$

$$= \frac{13000}{1.86 \times 24} \; ft^2$$

$$= 292 \; ft^2$$

SAND BED DRYING

For smaller sewage plants and some industrial waste treatment plants, the most common method of sludge dewatering is drying on open or covered sand beds. Drying of the sludge occurs by percolation and evaporation. The proportion of the water removed by percolation may vary from 20 to 55 percent, depending on the initial solids content of the sludge and on the characteristics of the solids. The design and use of drying beds are affected by climatic conditions (rainfall and evaporation). Sludge drying beds usually consist of 4 to 9 in. (10 to 23 cm) of sand over 8 to 18 in. (20 to 45 cm) of graded gravel or stone. The sand has an effective size of 0.3 to 1.2 mm and a uniformity coefficient less than 5.0. Gravel is graded from 1/8 to 1 in. (0.32 to 2.54 cm). The beds are provided with underdrains spaced from 9 to 20 ft apart (2.6 to 6.1 m). The underdrain piping may be vitrified clay laid with open joints having a minimum diameter of 4 in. (10 cm) and a minimum slope of about 1 percent. The filtrate is returned to the treatment plant.

Wet sludge is usually applied to the drying beds at depths of 8 to 12 in. (20 to 30 cm). Removal of the dried sludge in a "liftable state" varies with both individual judgment and final disposal means, but usually involves sludge of 30 to 50 percent solids. The dewatering characteristics of some sludges are shown in figure 11.27 and table 11.17.

FIGURE 11.27

Solids Drainage and Drying Characteristics on Sand Beds (Numbers Refer to Table 11.17)

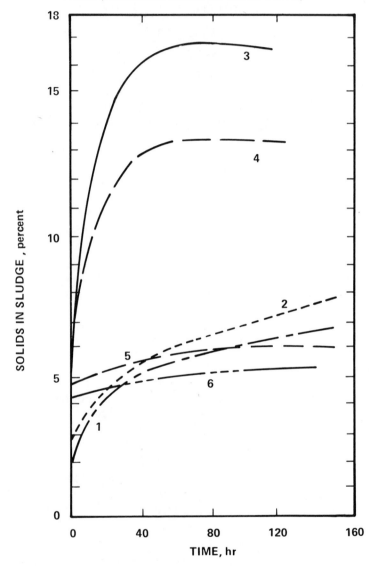

No.	Source or Type of Sludge	Solids Applied, %	Vol., %	Depth Applied, cm	Period to Liftable State, days	Solids Content Removed, %	Specific Resistance × 10^9, sec²/g (500 g/cm²)	Coefficient of Comp.
1.	Steam peeling of carrots	1.91	74.3	39.4	25	7.0	0.46	1.02
2.	Lime neutralization of pickling liquor	2.46	16.9	48.3	25	10.5	0.30	0.63
3.	Vegetable tanning	6.79	43.5	19.5	20	20.0	0.15	0.79
4.	Lime neutralization of regenerant liquors from ion-exchange demineralization of plating wastes	7.05	19.8	17.5	20	18.0	0.19	0.55
5.	Alum sludge	4.82	44.1	26.9	50	14.5	5.30	0.31
6.	Digested sewage sludge	4.28	56.8	30.5	74	25.0	13.50	0.60

TABLE 11.17

Waste-Sludge Drying Characteristics[a]

[a]From WPRB [10].

In many cases, the bed turnover can be substantially increased by the use of chemicals. Alum treatment can reduce the sludge drying time by 50 percent. The use of polymers can increase the rate of bed dewatering and also increase the depth of application. Bed yield has been reported to increase linearly with polymer dosage.

Drying beds have been designed on the empirical basis of square feet of bed area per capita (square meters per capita) or lb of dry solids/ft^2/yr (kg/m^2/yr). Values commonly employed in the United States are summarized in table 11.18.

A rational method has been developed by Swanwick [11] based on the observed dewatering characteristics of a variety of sewage and industrial waste sludges. In this procedure sludge after drainage (usually 18–24 hrs) is permitted to air dry to the desired consistency. The moisture difference (initial-final) is that which must be evaporated. Depending on the cumulative rainfall and evaporation for the geographical area in question the time required for various times of the year for evaporation of this moisture is confronted. The required bed area may then be determined.

TABLE 11.18
Sludge-Drying-Bed Design Parameters

Type of Digested sludge	Area, ft^2/capita (m^2/capita)	lb of dry solids/ft^2/year
Primary	1.0 (0.09)	30
Primary and standard trickling filter	1.6 (0.11)	25
Primary and activated sludge	3.0 (0.28)	20
Chemically precipitated sludge	2.0 (0.19)	22

COMPOSTING

Composting may be employed for dewatered organic sludge or mixtures of sludge and solid waste. Composting converts organics into humus material for soil conditioner and nutrient source. Composting is an aerobic thermophilic decomposition. The mixture must have a moisture content 50–60 percent and a carbon/nitrogen ratio that does not exceed 35. A digestion temperature of 45–70° C is generated by the thermophilic digestion that kills pathogenic organisms, which may be present in the sludge resulting in sterilization. In order to resume continuous digestion a recycle of digested material is necessary. Digestion time varies from one day to three weeks depending on the degree of treatment and type of feed material. A post digestion or maturing period follows. Odorous by-products may be destroyed in a soil filter. The pH should be maintained near pH 7 and aeration employed to maintain aerobic conditions.

Composting has been conducted in stacks (mixed for aeration) with pre- and posttreatment of the material, in rotating drums or in digestion elevators with vertical rotation shafts with several levels. In this case only one to two days retention is required with pretreatment and post digestion.

Many organic sludges can be incorporated into the soil without mechanical dewatering. Surface application can be accomplished by spreading from a truck or spraying. Sludge may also be injected into the soil 8 to 10 inches below the surface by a mobile unit. Injection offers the advantage of minimizing surface runoff and odor problems. An important consideration is the heavy metal content of the sludge. At a pH greater than 6.0 heavy metals will exchange for Ca^{++}, Mg^{++}, Na^+ and K^+. This natural ability to exchange heavy metals by the soil is called the cation exchange capacity (CEC) and is expressed in meq/100g dry soil. The amount of heavy metals from sludge incorporation that can be retained by the soil is up to the CEC, but will be influenced by such factors as pH, and aerobic or anaerobic conditions. The CEC of sandy soil may vary from 0–5 while clay soils will have a CEC between 15–20. The nutrient content of the sludge will support the growth of plants. The organic portion of the soil will also chelate heavy metals.

Prior to incorporation, sludges should receive a minimum degree of stabilization. Chow [13] has recommended aerobic digestion of 15 days to reduce the volatile content to less than 55 percent. The amount of heavy metals that can be applied to the land is expressed in metal equivalent (pounds per ton of dry sludge) [14].

$$\text{Metal Equivalent of Sludge} = \frac{Zn + 2\,Cu + 4\,Ni}{500}$$

Total Metal Equivalents that can be Applied (lbs/acre) = 65 · CEC
Cadmium is considered separately:
Cd = 10 lbs/acre (lifetime application) and 2 lbs/acre/yr maximum

This is based on the assumption that zinc, copper and nickel are the most toxic to plant growth. The equations also limit metal additions to 10 percent of the CEC. The U.S. Department of Agriculture has listed the maximum amount of heavy metals that can be applied to the land as shown in table 11.20.

TABLE 11.19
Suggested Maximum Heavy Metals Can Be Applied [15][16]

Metal (lbs/acre)	Soil Cationic Exchange Capacity (meq/100g)		
	0–5	5–15	15
Zn	225	450	900
Cu	110	225	450
Ni	45	90	180
Cd	4.5	9	18
Pb	450	900	1800

Each crop has a nutrient requirement (N, P, K, etc.). The annual quantity of sludge that can be incorporated depends on the available nitrogen content of the sludge and the nitrogen uptake of the selected crop. Excess application of sludge can result in oxidation of ammonia to nitrate, which can back into the groundwater. Since all the applied organic nitrogen is not available to the crops in the same year, there is a sequential removal of organic nitrogen. Normally, about 40 percent of the organic nitrogen applied in the first year is available for crop growth that year. Subsequently, 20, 10, 5, and 2.5 percent of the organic nitrogen is available for the 2nd, 3rd, 4th, and 5th years.

The Illinois EPA [17] recommends a minimum depth of earth cover to the annual water table of ten feet with a permeability rate of 2 to 20 inches/hr. A maximum land slope of 8 percent is recommended. A minimum soil pH of 6.5 should be maintained.

HEAT TREATMENT

Many sludges, including sewage sludge, are difficult to dewater and require high dosages of coagulating chemicals for this purpose. Two processes have been developed that heat the sludge for short periods under pressure. This coagulates the solids, breaks down the gel structure, and reduces the hydration and hydrophilic nature of the solids. This permits rapid dewatering without the use of chemical additives.

In the Porteus process sludge is passed through a heat exchanger into a reaction vessel, where steam is injected directly into the sludge. The retention time is 30 minutes at 350 to 390° F (188 to 199° C) and 180 to 210 psi (1240 to 1450 kN/m^2). After heat treatment, the sludge passes back through the heat exchanger into a thickener-decanter. For sewage sludge, a cake from a filter press of about 40 percent moisture is obtained. The decant and press liquor is high in BOD and requires return to the treatment process.

The low-pressure Zimpro system operates with pressures in the range of 150 to 300 psi (1030 to 2070 kN/m^2). Sludge and air are heated by an exchanger to 250 to 300° F (149 to 167° C) before entering the reactor. Reactor temperatures are maintained at 300 to 350° F (167 to 177° C) by the injection of steam. The primary difference between Zimpro and Porteus is the injection of air to the reactor in the Zimpro process. The exhaust gases are water scrubbed. The treated sludge is disposed of by vacuum filtration or on sand beds. The filtrate normally contains 2000 to 3000 mg/l of BOD and requires biological treatment. The Zimpro process is shown in figure 11.28. The principal advantages of heat treatment are that the sludge is sterilized, substantially deodorized, and dewaters readily on vacuum or pressure filters. Performance data is shown in table 11.19.

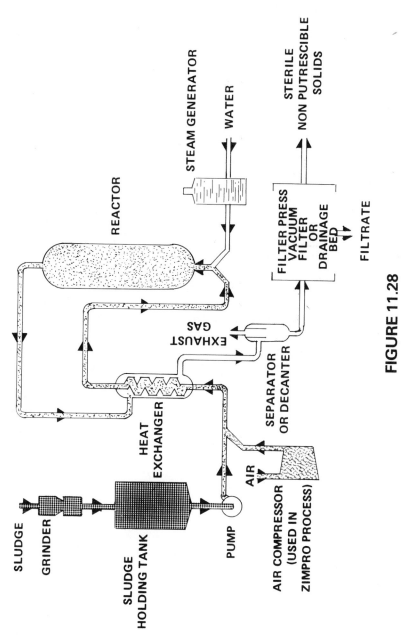

FIGURE 11.28

Sludge Heat-Treatment Process

TABLE 11.20
Pilot-Scale Test Results of Heat-Treated Sludge Dewatering

	Feed, %	Cake, %	Cycle, sec	Yield, lb/sq ft/hr
Vacuum filter				
Waste activated	12.5	36.7	135	7.2
Raw primary	6.0	43.8	67	39.5
Digested primary	10.5	38.5	100	21.7
Filter press				
Waste activated	6.84	53.3	120	8.7

	Feed, %	Cake, %	Feed Rate, lb/hr	Recovery, %
Solid bowl centrifuge				
Waste activated		38.8	66	91.7
Waste activated		36.0	102	96.7
Raw primary/				
waste activated		41.6	77	96.4
Raw primary/				
waste activated		51.3	181	89.4

PROBLEM 11–5

Design a land incorporation system for an excess activated sludge

  I. Sludge Characteristics

Quantity, Gallons/Day	6,560
Amount, Pounds/Day	3,500
NH_3, mg/l	235
Org-N, mg/l	865
SS, mg/l	63,000
PO_4, mg/l	30

  II. Metal Analysis of Sludge

Al	700 mg/kg (dry solid)
Cd	3.0
Ca	105,000
Cr	400
Cu	60
Fe	6000

Pb	30
Ni	150
Zn	120
K	150

III. Average CEC of the Soil = 16.8 meq/100g

Sludge Application

$$\text{Metal Equivalent} = \frac{Zn + 2Cu + 4Ni}{500}$$

$$= \frac{120 + 2 \times 60 + 4 \times 150}{500}$$

$$= 1.68 \text{ lbs/Ton Sludge}$$

$$\text{Total Sludge can be Applied} = 65 \times CEC \text{ lbs/A}$$

$$= 65 \times 16.8$$

$$= 1092 \text{ lbs/A}$$

$$\text{Sludge Application} = \frac{1092}{1.68} = \begin{array}{l} 650 \text{ Tons/A} \\ \text{(Life Time)} \end{array}$$

Maximum Allowable Sludge Application 650 Tons/Acre

Bermuda Grass   200 lbs/Acre for Nitrogen Loading

**Subsurface Incorporation   100% NH_3 Availability
40% Org-N Availability**

Agronomic Loading
Available N for the 1st Year Application

$$235 + 865 \times 0.4 = 581 \text{ mg/l}$$

$$= 0.00486 \text{ lbs/gal}$$

$$\text{Sludge Loading} = 200/0.00486 = 41,126 \text{ gal/A}$$

$$\text{Acres Required} = \frac{6,560 \times 365}{41,126}$$

$$= 58 \text{ acres}$$

Project Life of the Plot Based on CEC

$$\text{Sludge Concentration} = 63,000 \text{ mg/l} = 0.53 \text{ lbs/gal}$$

$$\text{Sludge Applied} = 0.53 \times 41,126 = 21,684 \text{ lbs/A}$$

$$= 10.8 \text{ Tons/A}$$

$$\text{Years of Application} = \frac{650}{10.8} = 60.2 \text{ years}$$

Based on Cd Content in the Sludge

$$3.0 \times \frac{31324}{2.2} \times \frac{1}{1000} \times \frac{1}{454} = 0.094 \text{ lbs/A}$$

$$\text{Years of Application} = 106 \text{ years}$$

Subsequent years of application should consider additional organic nitrogen conversion.

LAND DISPOSAL

Land disposal of wet sludges can be accomplished in a number of ways: lagooning or application of liquid sludge to land by truck or spray system or by pipeline to a remote agricultural or lagoon site.

Lagooning is commonly employed for the disposal of inorganic industrial waste sludges. Sewage and organic sludges usually receive aerobic and anaerobic digestion prior to lagooning to eliminate odors and insects. Lagoons may be operated as substitutes for drying beds in which the sludge is periodically removed and the lagoon refilled. In a permanent lagoon, supernatant liquor is removed, and, when filled with solids, the lagoon is abandoned and a new site selected. Sewage sludge stored in a lagoon can be dewatered from 95 percent moisture to 55 to 60 percent moisture in a two-to-three-year period.

In general, lagoons should be considered where large land areas are available and the sludge will not present a nuisance to the surrounding environment.

In several cases, biological sludges after aerobic or anaerobic digestion have been sprayed on local land sites from tank wagons or pumped through agricultural pipe. Multiple applications at low dosages form a thin sludge layer that is easily worked into the soil. Reported loadings range from 100 dry tons/acre (22.4 metric tons/1000 m^2) average conditions to 300 tons/acre (67.2 metric tons/1000 m^2) in areas of low rainfall.

Excess activated sludge has been disposed of in oxidation ponds in which algal activity maintains aerobic conditions in the overlaying liquid while the sludge undergoes anaerobic digestion. This procedure has been successfully employed for municipal activated sludge at Austin, Texas, and excess activated sludge from a petrochemical plant in Houston, Texas.

Pipeline transportation of wet sludge for land or lagoon disposal in remote areas is gaining increasing interest, particularly for large urban communities. Sewage sludge requires digestion prior to pumping. The relative costs of pipeline disposal and other methods for a city of 100,000 people is shown in figure 11.29.

FIGURE 11.29

Cost of Disposal of Approximately 50,000 gpd of Sludge
by Various Methods (City of 100,000 People)
(After Riddel and Cormack [12])

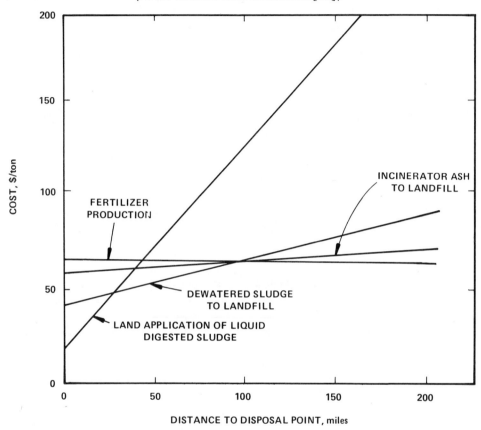

DISTANCE TO DISPOSAL POINT, miles

INCINERATION

After dewatering the sludge cake must be disposed of. This can be accomplished by hauling the cake to a land-disposal site or by incineration.

The variables to be considered in incineration are the moisture and volatile content of the sludge cake and the thermal value of the sludge. The moisture content is of primary significance because it dictates whether the combustion process will be self-supporting or whether supplementary fuel will be required. The thermal value of several sludges is shown in table 11.21.

TABLE 11.21
Heat Values of Sampled Sludges

Description of Tested Sludge	Com-bustibles, %	Ash, %	Sludge Heat Value, BTU/lb
Oil			17,500
Grease and scum	88.5	11.5	16,750
Raw wastewater	74.0	26.0	10,285
Fine screenings	86.4	13.6	7,820
Waste sulfite liquor solids			7,900
Primary wastewater sludge			8,990
Activated wastewater sludge			6,540
Semichemical pulp solids			5,812
Digested primary sludge	59.6	40.4	5,290
Grit	33.2	69.8	4,000

Incineration involves drying and combustion. Various types of incineration units are available to accomplish these reactions in single or combined units. In the incineration process, the sludge temperature is raised to 212° F (100° C), at which point moisture is evaporated from the sludge. The water vapor and air temperature is increased to the ignition point. Some excess air is required for complete combustion of the sludge. Self-sustaining combustion is often possible with dewatered waste sludges once the burning of auxiliary fuel raises incinerator temperature to the ignition point. The primary end products of combustion are carbon dioxide, sulfur dioxide, and ash.

Incineration can be accomplished in multiple-hearth furnaces in which the sludge passes vertically through a series of hearths. In the upper hearths, vaporization of moisture occurs and cooling of exhaust gases. In the intermediate hearths, the volatile gases and solids are burned. The total fixed carbon is burned in the lower hearths. Temperatures range from 1000° F (538° C) at the top hearth to 600° F (316° C) at the bottom. The exhaust gases pass through a scrubber to remove fly ash and other volatile products. A typical furnace is shown in figure 11.30. The effects of sludge moisture and volatile solids on gas consumption for municipal sludge is shown in figure 11.31.

In the fluidized bed, sludge particles are fed into a bed of sand fluidized by upward-moving air. A temperature of 1400 to 1500° F (760 to 815° C) is maintained in the bed, resulting in rapid drying and burning of the sludge. Ash is removed from the bed by the upward-flowing combustion gases. A flow sheet for this system is shown in figure 11.32.

FIGURE 11.30
Cross-Section of a Typical Multiple-Hearth Incinerator

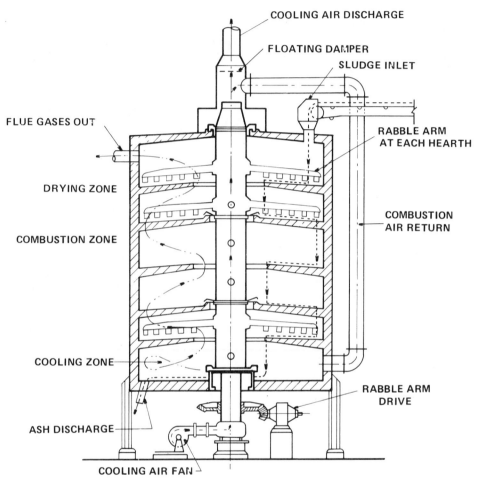

WET OXIDATION

Wet oxidation is a process by which the organic solids in a sludge are chemically oxidized in an aqueous phase by dissolved oxygen in a specifically designated reactor at elevated temperature and pressure.

The principal components of the process shown in figure 11.33 are a reactor, an air compressor, a heat exchanger, and a high-pressure sludge pump. The process involves pressurizing and preheating the sludge, injection of air and

FIGURE 11.31

The Effects of Sludge Moisture and Volatile Solids
Content on Gas Consumption

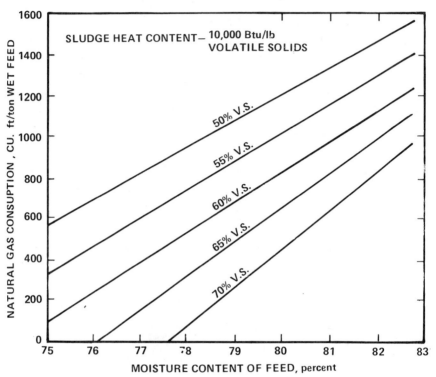

preheated sludge into the reactor, and oxidation in the reactor followed by gas-liquid and ash-liquid separation.

The primary design parameters are temperature, air supply, pressure, and feed solids concentration. The degree and rate of oxidation is significantly influenced by the reactor temperature. The pressure must be sufficient to condense the water vapor. Operation results have indicated COD reductions of 75 to 80 percent at temperatures of 525° F (247° C) and pressures of about 1750 psig (12100 kN/m²). The effluent liquor has a BOD between 5000 and 9000 mg/l. The residual solids can be dewatered by vacuum filtration or disposed of in lagoons or drying beds.

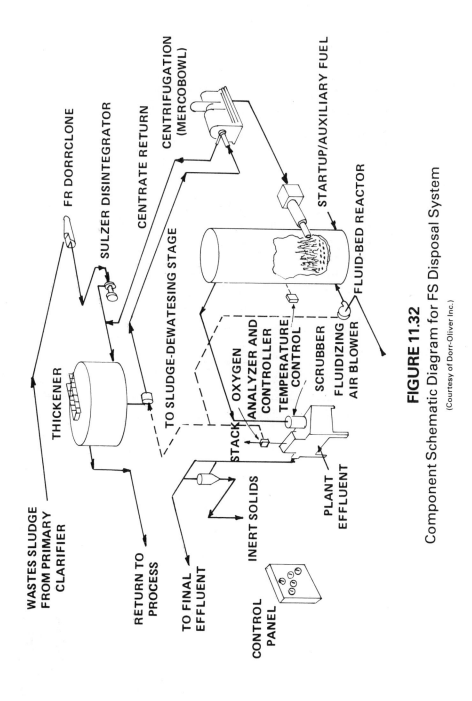

FIGURE 11.32

Component Schematic Diagram for FS Disposal System

(Courtesy of Dorr-Oliver Inc.)

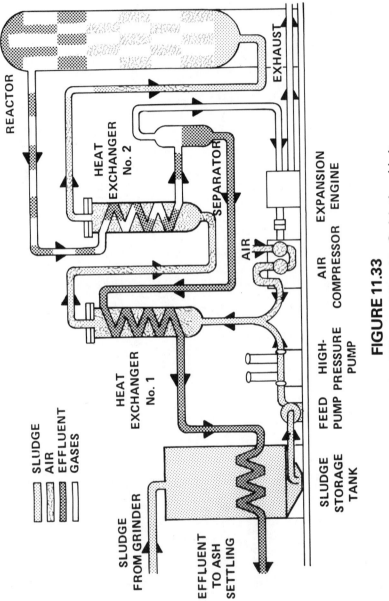

FIGURE 11.33

Flow Diagram of Zimpro Wet-Air-Oxidation Unit

(Courtesy of Zimpro Inc.)

REFERENCES

1. Krissel, J. F. and Westrick, J. J. 1973. Municipal waste treatment by physical-chemical methods. *Applications of New Concepts of Physical-Chemical Waste-Water Treatment* W. W. Eckenfelder, Jr., *et al.*, Eds. Oxford, England: Pergamon Press.
2. Burd, R. S. *A study of sludge handling and disposal.* 1968. Publ. WP-20-4, Washington, D.C.: U.S. Department of Interior, FWPCA.
3. Dick, R. I. 1972. Thickening. *Process design in water quality engineering.* E. L. Thackston and W. W. Eckenfelder, Jr., Eds. Austin, Texas: Jenkins Publishing.
4. Barnard, J. and Englande, A. J. 1972. Centrifugation. *Process Design in Water Quality Engineering.* E. L. Thackston and W. W. Eckenfelder, Jr., Eds. Austin, Texas: Jenkins Publishing.
5. Montanaro, W. J. 1964. Sludge dewatering by centrifuges. *Proc. 37th Annual Conf. Water Pollution Control Federation,* State College, Pa., Unpublished paper.
6. Coackley, P. 1956. Laboratory scale filtration experiments and their application to sewage sludge dewatering. *Biological Treatment of Sewage and Industrial Wastes,* vol. 2, J. McCabe and W. W. Eckenfelder, Jr., Eds., New York: Reinhold Publishing.
7. Schepman, B. A. and Cornell, C. F. 1956. Fundamental operating variables in sewage sludge filtration. *Sew. Ind. Wastes* 28: 1443.
8. Forster, H. W. Sludge dewatering by pressure filtration. *Water-1972.* AIChE Symp. Series, 129, (1973).
9. Bohnke, B. Treatment and removal of sewage sludge in the Federal Republic of Germany-Advanced Wastewater Treatment JRGWP Seminar, Tokyo, Japan (October 1976).
10. Water Pollution Research, Her Majesty's Stationary Office, London, England (1963).
11. Water Pollution Research, Her Majesty's Stationary Office, London, England (1966).
12. Riddel, M. D. and Cormack, J. W. 1966. Ultimate disposal of sludge in inland areas. *Proc. 39th Annual Meeting, Central States Water Pollution Control Assoc.,* Eau Claire, Wisconsin, Unpublished paper.
13. Chow, V. 1979. *Sludge Disposal on Land,* St. Paul, Minnesota: 3M Company.
14. Keeney, D. R., Lee, K. W. and Walsh, L. M. Guidelines for the Application of Wastewater Sludge to Agricultural Land in Wisconsin, Tech. Bulletin 88, Madison, Wisconsin: Wisconsin Department of Natural Resources.
15. Black, C. A. 1973. Methods of Soil Analyses, *American Society of Agronomy,* Madison, Wisconsin.
16. Recommendations for Application of Municipal Wastewater Sludges on Land, 1978. Roseville, Minnesota 55113, Minnesota Pollution Control Agency.
17. Illinois EPA Technical Policy WPC-3, 1977. Draft Copy 2nd Edition Design Criteria for Municipal Sludge Utilization on Agricultural Land.

Land Disposal
of Wastewaters

Disposal of wastewaters by irrigation on the land can be practiced in one of several ways depending on the topography of the land, the nature of the soil, and the characteristics of the waste:

1. Distribution of wastewater through spray nozzles over relatively flat terrain (spray irrigation—SI)
2. Distribution of wastewater over sloping land that runs off to a natural water course (overland runoff—OR)
3. Disposal through ridge and furrow irrigation channels (R & F)
4. Infiltration through porous soil layers (RI)

In the first method, screened wastewater is pumped through laterals and sprayed through sprinklers located at appropriate intervals as shown in figure 12.1. The wastewater percolates through the soil, and during this process the organics undergo biological degradation. The liquid is either stored in the soil layer, or discharged to the ground water. Most spray irrigation systems use a cover crop of grass or other vegetation to maintain porosity in the upper soil layers. There is a net waste loss by evapo-transpiration (evaporation to the atmosphere) and adsorption by the roots and leaves of plants, which may amount to as much as 10 percent of the waste flow.

In some cases wastewaters have been sprayed into woodland areas. Trees develop a high porosity soil cover and yield high transpiration rates. A small elm tree may take up as much as 3000 gpd (11.4 m³/day) under arid conditions.

The principle factors governing the capacity of a site to absorb wastewater are:

1. Character of the soil—Sandy soil has a high filtration rate while clay passes very little water. Spray irrigation systems are most applicable for silty

FIGURE 12.1
Spray Irrigation System

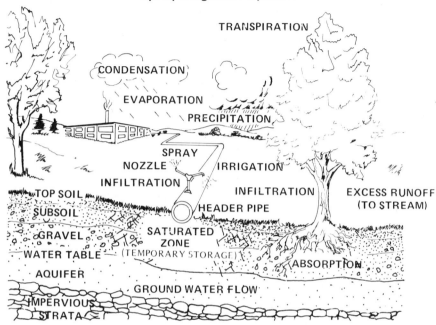

gravels; gravels, sands, and silt mixtures; and silty soils. In general this would consider a liquid application rate of 2 to 3.5 inches (5 to 9 cm) per day, corresponding to a percolation rate of 5 to 15 minutes for the liquid to fall one inch (2.5 cm) [1]. Stratification of the soil profile is important, since some soils exhibit clay lenses which present a barrier to flow.

2. Depth to ground water—The quantity of wastewater which can be sprayed on a given area is proportional to the depth of the soil through which it must travel to the ground water. Sufficient soil depth must be available to effect biological degradation of the organics and the removal of other contaminants. It has been recommended that 5 to 10 feet (1.5 to 3 m) of depth from the point of application to the ground water be available to avoid contamination.

3. Initial moisture content—The capacity of the soil to absorb water is proportional to the initial moisture content of the soil.

4. Terrain and ground cover—A cover crop increases the quantity of water that can be absorbed by a given area. A sloping site increases the run-off. The cover crop may consist of reed, canary, bermuda, or sudan grasses, or in some cases, wheat, corn, and oats. Reed and canary grass has been found to be an effective cover crop.

Application rates have varied from 0.2 inches (0.5 cm) per week to 6 inches (15 cm) per week with 2 inches (5 cm) per week being a common application rate. (One inch per acre equals 27,141 gallons.) Studies at Pennsylvania State University [2] with secondary effluent applied an application rate of 0.25 inches (0.64 cm) per hour over 8 hours, with a six-day rest period. At Muskeegan, Michigan, a total application rate, including both rainfall and wastewater of 4 inches (10 cm) per week, has been employed [1].

Overland Run-Off

Overland run-off may be employed on sloping sites with impermeable subsoils. A cover crop is essential to control erosion and to aid renovative responses of the soil.

Wastewater application usually employs spray nozzles. Based on present experience, a run-off lane of 175 to 300 feet (53 to 91 m) with a 2 to 6 percent slope is to be preferred. Application rates treating cannery wastewaters in Paris, Texas [3] varied from 2.5 inches (6.4 cm) per week during the summer to 1.25 inches (3.2 cm) during the winter. Application was at a rate of 0.125 inches (0.32 cm) per hour for 4 hours a day, five days a week. The Paris experience indicated that 61 percent of the effluent was found in the run-off, 18 percent lost through evaporation and 21 percent through soil percolation. The overland run-off site should contain 6 to 8 feet (0.9 to 2.4 m) of good top soil to aid biological decomposition of the wastewaters.

Rapid Infiltration

Rapid infiltration involves inundation of a site followed by rapid filtration through the soil layers. Recharge rates of 1 to 4 feet (0.3 to 1.2 m) per day have been practiced. A general rule might provide for 6 inches (15 cm) per day application for fourteen days followed by a fourteen-day rest period [1]. Rapid infiltration would be applicable for graded gravels and graded sands, with little or no fines. A percolation rate of 1 to 4 minutes for the liquid to fall 1 inch (2.5 cm), corresponding to 4 to 8 inches (10 to 20 cm) per day liquid application rate, would apply for rapid infiltration sites.

Other Considerations

In addition to soil conditions, there are several waste characteristics which require consideration in a land disposal system. Suspended solids should be removed from the wastewater either by screening or by sedimentation before it is sprayed. Solids tend to clog the spray nozzles and may mat the soil surface rendering it impermeable to further percolation. An excess acid or alkaline waste

Waste	Soil	Cover	Organic (lb/acre day)	Hydraulic (in./acre day)	Duration	Type	Removal efficiency (%)
Cannery	—	Grass	183 COD	0.88	—	Spray	81
Cannery	—	Grass	102 COD	0.49	—	Spray	81
Domestic	—	Forest	—	0.20	June–Dec	Spray	80
Domestic	Silt loam	Grass	35 COD	1.32	223 days	R&F	65
Domestic	Silt loam	Grass	46 COD	1.72	235 days	R&F	65
Cannery	Sandy silt	Grass	—	3.90	—	Spray	—
Cannery	Clay	Grass	218 BOD	0.88	Year-round	Spray	—
Cannery	Clay	Grass	74 COD	0.44	Year-round	Spray	98
Cannery	—	—	238 BOD	1.80	57 days	R&F	—
Cannery	—	—	2020 BOD[a]	1.25	35 days	R&F	—
Cannery	—	—	1360 BOD[a]	2.86	120 days	R&F	—
Milk	—	—	58 BOD	0.86	—	R&F	95
Meat	Sand	—	100 BOD	1.47	Year-round	R&F	91
Cannery	—	—	413 BOD	2.96	—	—	—
Cannery	—	—	864 BOD	3.35	—	—	—
Cannery	—	—	22 BOD	3.50	—	—	—
Cannery	—	—	40 BOD	0.37	—	—	—
Cannery	—	—	65 BOD	0.37	—	—	—
Cannery	—	—	807 BOD	3.61	—	—	—

[a]Exceed 30 ton/yr loading.
R&F: Ridge and furrow.
— Not given in literature source.

TABLE 12.1
Hydraulic and Organic Loading at
Selected Land Disposal Sites

would be harmful to the cover crop. High salinity impairs the growth of the cover crop and in clay soils causes sodium to replace calcium and magnesium by ion exchange. This causes soil dispersion and, as a result, drainage and aeration of the soil will be poor. A maximum salinity of 0.15 percent has been suggested to eliminate these problems [4].

Wastewaters that have been successfully disposed of by spray irrigation include cannery, pulp and paper, dairy, and tannery wastewaters. Spent sulfite liquor has been applied up to a rate of 1 lb of solids per day per square yard (0.54 kg/m²-day), resulting in 95 percent removal of BOD through a 10-foot (3 m) soil layer. Boxboard wastes have been sprayed on a blown silt loam with a gravel underlay and alfalfa cover crop at a short term rate of 0.7 inches (1.8 cm) per hour and a total daily rate of 0.21 to 0.56 inches (0.53 to 1.42 cm) per day. Kraft mill wastes have been found to require 1.5 acres per ton (5510 m²/metric ton) of production or more.

Typical data for spray irrigation of wastewaters is shown in table 12.1. Luley [5] showed that spraying a cannery wastewater over sloping ground reduced the BOD from 1095 mg/l to 80 mg/l with the final run-off being 37 percent of the initial wastewater flow, the balance being removed by soil absorption, evaporation, and transpiration.

The removal of pollutants from secondary effluents by land disposal under properly designed and operated conditions is shown in table 12.2.

TABLE 12.2

Pollutant Removals from Land Disposal of Wastewater[a]

	% Removal		
Parameter	**SI**	**OR**	**RI**
BOD	98+	98+	90–95
COD	95+	95+	90+
N	85+	85+	75–80
P	99+	85+	95+
Metals	95+	85+	95+
Suspended solids	99	95+	99
Pathogens	99	99	99

[a]From [1].

REFERENCES

1. Corps of Engineers. *Wastewater Management by Disposal on the Land.* 1972. Special Report No. 171. Hanover, N.H. Cold Regions Research and Engineering Laboratory.
2. Sopper, W. E. *Effects of trees and forests in neutralizing wastes.* 1971. Penn State University Reprint Series, No. 23.
3. Gilde, L. C. *et al.* 1971. A spray irrigation system for the treatment of cannery wastes. *J. Water Pollution Control Federation* 43: 2011.
4. Diagnosis and improvement of saline and alkali soils, U.S. Dept. of Agriculture Handbook 60 (1954).
5. Luley, H. G. 1963. Spray irrigation of vegetable and fruit processing wastes. *J. Water Pollution Control Federation* 35: 1252.

13

Deep Well Disposal

Deep-well disposal involves the injection of liquid wastes into a porous subsurface stratum that contains noncommercial brines [1]. The wastewaters are stored in sealed subsurface strata isolated from groundwater or mineral resources. Disposal wells may vary in depth from a few hundred feet (meters) to 15,000 ft. (4,570 m) with capacities ranging from less than 10 to more than 2,000 gpm (38 to 7550 l/min). Wastes disposed of in wells are usually highly concentrated toxic, acidic, or radioactive wastes or wastes high in inorganic content which are difficult or excessively expensive to treat by some other processes. The disposal system consists of the well and the pretreatment equipment necessary to prepare the waste for suitable disposal into the well.

A casing, generally of steel, is cemented in place to seal the disposal stratum from other strata penetrated during drilling of the well. An injection tube transports the waste to the disposal stratum as shown in figure 13.1.

Oil or fresh water is used to fill the annular space between the injection tube and the casing and extends to but is sealed from the injection stratum. Leaks in the injection tube or drainage to the casing can be detected by monitoring the pressure of the fluid.

The system includes a basin to level fluctuations in flow, pretreatment equipment, and high-pressure pumps. Pretreatment requirements are determined by the characteristics of the wastewater, compatibility of the wastewater and the formation water, and the characteristics of the receiving stratum. Pretreatment may include the removal of oils and floating material, suspended solids, biological growths, dissolved gases, precipitable ions, acidity, or alkalinity. A typical system is shown in figure 13.2.

The best disposal areas include sedimentary rock in the unfractured state, including sandstones, limestones, dolomites, and unconsolidated sands. Frac-

FIGURE 13.1

Schematic Diagram of a Waste-Injection Well Completed in Competent Sandstone [2]

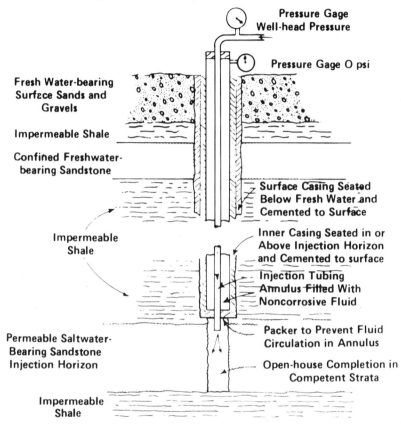

tured strata should be avoided because if a vertical fissure exists ground water contamination may result.

The well-head pressure is related to the difference between the bottom-hole pressure and the reservoir pressure. Cores of the injection location are needed to evaluate the porosity and permeability of the stratum and any possible reactions between the wastewater and the stratum.

Although wastewaters should be generally free of suspended solids, some vugular formations will accept suspended solids without problems or an increase in injection pressure. In some cases injection can be increased by well stimulation, which involves the injection of mineral acids to dissolve calcium carbonate and other acid-soluble particulates that tend to plug the stratum. Mechanical procedures involve scratching, swabbing, washing, and underreaming the well bore and shooting the uncased stratum with explosives or hydraulic fracturing.

FIGURE 13.2

Typical Subsurface Waste-Disposal System

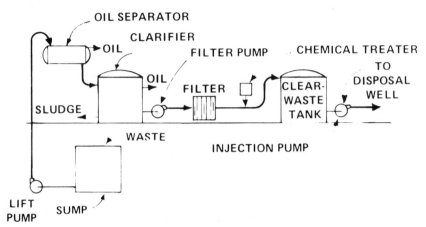

The cost of a well system is affected by the depth of the injection well, type of formation, geographic location, waste volume, required pretreatment, and injection pressure.

REFERENCES

1. Warner, D. L. *Deep Well Injection of Liquid Waste*. 1965. Environmental Health Series. Cincinnati, Ohio:1 U.S. Department of Health, Education, and Welfare.
2. Donaldson, E. C. 1964. Subsurface disposal of industrial wastes in the United States, Bureau of Mines Inform. Cir. 8212, U.S. Department of Interior, Washington, D.C.

Economics of Wastewater Treatment

Following the selection of alternative processes applicable for the treatment of a particular waste, the cost of each process constitutes the most significant criterion for selection of a final process design. Cost estimating techniques can carry the economic feasibility of alternative treatment methods to any desired degree of accuracy. The accuracy of treatment cost estimates can vary from very general estimates constructed from rule-of-thumb figures to detailed figures obtained from construction bids. Estimates with an intermediate degree of accuracy can be constructed from individual design parameters for specific treatment processes and unit cost curves developed from actual construction costs. Usually, estimates of this type are sufficiently accurate to permit the selection of a final treatment design based on economic considerations. The following information is required for the development of cost estimates:

1. wastewater characterization data
2. design parameters for applicable treatment processes
3. effluent standards to meet water quality criteria
4. unit cost information for applicable treatment processes

Design data required for individual treatment processes are discussed in previous chapters. This chapter deals with the application of these data to determine treatment cost estimates.

TREATMENT COST RELATIONSHIPS

Treatment cost relationships for municipal and industrial wastes are available from a wide variety of sources [1, 2, 3, 4]. Cost data are available for both the total plant (including allowances for the cost of laboratory facilities, control structures, and sludge handling facilities) and for individual process units. Total treatment costs for municipal facilities are usually not representative of the costs that will be incurred for the treatment of industrial wastes, since complete control structures, laboratory facilities, and pumping stations are often not required. While the size of a municipal plant (including activated sludge treatment) can usually be estimated using only the wastewater flow to be treated, the cost of activated sludge facilities for industrial applications must also include a consideration of the BOD concentration of the waste. Likewise, the cost of sludge handling facilities for industrial wastes can be estimated only after the properties of the particular sludge are known.

Unit Process Costs

A summary of treatment cost relationships including the basis on which costs are commonly estimated is given in table 14.1. Construction cost relationships based on wastewater flow for individual unit processes for municipal sewage treatment are shown in table 14.2.

TABLE 14.1
Bases for Unit Process Cost Relationships

Process	Cost Basis
Equalization	Volume
Oil Separation	Flow
Neutralization	Flow, Acidity, Suspended Solids
Primary Clarification	Surface Area
Aerated Lagoon	Aeration Volume
Activated Sludge	Aeration Volume
Aerators	Power Required
Secondary Clarification	Surface Area
Gravity Thickening	Surface Area
Aerobic Digestion	Aeration Volume
Vacuum Filtration	Surface Area
Chemical Coagulation	Flow, Chemical Requirements
Dissolved Air Flotation	Surface Area
Filtration	Flow or Surface Area
Sludge Disposal	Weight or Volume of Sludge

TABLE 14.2
Unit Process Construction Costs for Municipal Sewage Treatment[a]

Liquid Stream	Equation[b]
Preliminary Treatment	$C = 5.79 \times 10^4 Q^{1.17}$
Influent Pumping	$C = 1.47 \times 10^5 Q^{1.03}$
Flow Equalization	$C = 1.09 \times 10^5 Q^{0.49}$
Primary Sedimentation	$C = 6.94 \times 10^4 Q^{1.04}$
Activated Sludge	$C = 2.27 \times 10^5 Q^{0.87}$
Rotating Biological Contactor	$C = 3.19 \times 10^5 Q^{0.92}$
Filtration	$C = 1.85 \times 10^5 Q^{0.84}$
Clarification	$C = 1.09 \times 10^5 Q^{1.01}$
Chemical Addition	$C = 2.36 \times 10^4 Q^{1.68}$
Stabilization Pond	$C = 9.05 \times 10^5 Q^{1.27}$
Aerated Lagoon	$C = 3.35 \times 10^5 Q^{1.13}$
Chlorination	$C = 5.27 \times 10^4 Q^{0.97}$
Land Treatment of Effluent	$C = 3.67 \times 10^5 Q^{1.02}$
Outfall Pumping	$C = 3.32 \times 10^4 Q^{1.26}$
Effluent Outfall	$C = 7.39 \times 10^4 Q^{1.37}$
Outfall Diffuser	$C = 3.24 \times 10^4 Q^{0.91}$

Solids Stream	
Gravity Thickening	$C = 3.28 \times 10^4 Q^{1.10}$
Flotation Thickening	$C = 2.99 \times 10^4 Q^{1.14}$
Other Sludge Handling	$C = 4.26 \times 10^4 Q^{1.36}$
Aerobic Digestion	$C = 1.47 \times 10^5 Q^{1.14}$
Anaerobic Digestion	$C = 1.12 \times 10^5 Q^{1.12}$
Heat Treatment	$C = 1.51 \times 10^5 Q^{0.81}$
Mechanical Dewatering	$C = 3.44 \times 10^4 Q^{1.61}$
Air Drying	$C = 9.89 \times 10^4 Q^{1.35}$
Incineration	$C = 8.77 \times 10^4 Q^{1.33}$

Others	
Laboratory/Maintenance Building	$C = 1.65 \times 10^5 Q^{1.02}$

[a] From [5].
[b] Costs in EPA, second-quarter 1977 dollars, Q in MGD.

Because of the prevalence of the activated sludge process, additional comments should be made regarding cost estimates for this process and associated sludge handling procedures. Estimated operation and maintenance costs for activated sludge should include allowances for nutrient addition, where required, and for power costs. In cases where nutrient addition is necessary, allowances of $0.15/lb for nitrogen as aqua ammonia and $0.42/lb for phosphorus as 85 percent phosphoric acid (based on 1974 price levels) should be included. Power costs may be calculated from the horsepower requirements for aeration and mixing using unit costs applicable to a particular location.

Sometimes it may be necessary to construct additional cost relationships for specific needs. Manufacturers of water pollution control equipment are usually willing to furnish prices for specified equipment.

In most instances allowances must be made for piping required to convey wastes from one unit process to another. While pumping of the main waste stream is included in cost relationships for such processes as filtration, dissolved air flotation, and carbon adsorption that require a pressurized waste stream, allowances for pumping are usually not included in other cost relationships. If extraordinary pumping is required, appropriate allowances should be included in cost estimates.

Costing Procedure for Total Treatment Costs

A general procedure for the development of treatment cost estimates is shown below. Modifications of this procedure may be applied to cases requiring special treatment and to allow for different degrees of accuracy.

1. Collection of wastewater characterization data—Characterization data for the wastewater to be treated must be assembled for the design of individual treatment processes. The completeness of characterization data available significantly affects the accuracy of the treatment cost estimate. If characterization data for a particular industry are unavailable, it may be possible to select typical values for similar industries from available literature. If changes in the characteristics or quantity of wastewater are anticipated during the useful life of the treatment system, these factors should be considered in the final design.

2. Selection of treatment processes—Design parameters for treatment processes which have been found to be applicable to the treatment of the particular waste are chosen. If alternative treatment sequences are being compared, it should be determined that each sequence is capable of reducing the constituents present in the raw waste stream to required effluent values. Factors affecting the selection of alternative treatment processes include the volume of waste to be treated, the nature of consituents present in the waste stream, the reliability of various treatment processes, the possibilities afforded by each process for the recovery of saleable by-products, and the flexibility of candidate processes in handling possible changes of waste characteristics.

3. Selection of plant design size—The design size for each process must be determined by considering the characteristics of each individual waste stream.

If increases in the plant waste flow are anticipated, the final design should include the capacity to handle the greater flow at a future date. Generally, hydraulic structures are designed to handle the waste flow not exceeded 90 percent of the time, basin volumes are sized on the fiftieth-percentile flow, and oxygen requirements for biological systems are sized for the ninetieth-percentile demand. The required accuracy of the cost estimate determines the extent to which these factors are considered in developing process designs.

4.   Selection of unit process cost models—Treatment cost relationships for the processes being considered should be obtained from reliable sources. Unit cost information is available from current treatment cost literature, from literature describing industrial waste treatment practices and costs for specific industries, and from manufacturers of industrial wastewater treatment equipment.

5.   Construction of treatment cost estimates—Final design factors used as a basis for estimating treatment costs are combined with unit cost information to determine the treatment cost for each unit process. Construction cost estimates should include the cost of each unit process plus allowances added to cover contingencies, engineering, administration of the construction contract, land costs, and other miscellaneous costs. Operation and maintenance costs should cover all expenses required for the operation of unit treatment processes plus allowances for the transport and disposal of sludges and brines resulting from treatment operations.

Capital Costs

Factors that must be included to determine the total construction cost of a treatment facility include:

1. Unit construction costs
2. Additional costs for pumping, piping, and electrical work required
3. Land costs
4. Contingencies, engineering, administration of the contract, and miscellaneous costs

To cover contingencies, engineering, administration of the contract, and miscellaneous costs, 35 percent may be added to the combined cost of unit treatment processes. However, this factor suffers from generality and should be evaluated in individual cases. Additional costs that may need to be added to total construction estimates include instrumentation not usually included in treatment facilities, allowances for control facilities and laboratories, and special costs for landscaping. Instrumentation costs are usually small compared to total construction costs, and should be included when a high degree of instrumentation is desired. For instance, the cost for complete instrumentation of ion exchange facilities usually adds approximately 10 percent to the total construction cost.

Costs for the construction of offices, maintenance shops, laboratories, and the cost of landscaping will vary significantly depending on facilities already available at particular locations and the emphasis to be placed on the general appearance of the treatment facility.

Operation and Maintenance Costs

Factors that contribute to operational and maintenance costs for treatment facilities include:

1. Labor costs, including provisions for operational, general maintenance, and administrative personnel
2. Chemical costs
3. Materials, including items required for general maintenance and substances which must be replaced periodically, such as activated carbon and filter media
4. Transportation of sludges and other materials for ultimate disposal
5. Power costs

In some cases all of these items are included in one relationship reflecting the total cost for operation and maintenance. However, in other cases, chemical and power costs must be calculated separately. Treatment costs may vary significantly depending on the price of these items in different locations. In addition, the quantity of chemicals and power required is affected by specific wastewater characteristics. For physical-chemical treatment processes, it may be desirable to construct several cost estimates to determine the sensitivity of total treatment costs to the variation of chemical costs.

Cost Indexes

Frequently it is necessary to update costs to current values using construction and labor cost indexes. The most widely used indexes of construction are the Engineering News-Record (ENR) Construction Cost Index and the Environmental Protection Agency—Treatment Plant Indexes (LCAT and SCCT). The ENR Construction Cost Index is based on the cost of 200 hours of common labor calculated using an average of the prevailing rate in 20 cities, 25 cwt of standard structural steel shapes at the mill price, 22.56 cwt of Portland cement at the 20 cities' average price, and 1,088 board feet of 2×4 lumber at the 20 cities' average price. Ten-year ENR Construction Cost Indexes are tabulated in table 14.3. The first EPA-Treatment Plant Index is the Large City Advanced Treatment (LCAT) Index which is based on a model 50-MGD treatment plant consisting of bar screens, grit chamber, primary clarification, conventional activated sludge, lime clarification, recalcination, multi-media gravity filtration, chlorination, gravity thickening, vacuum filtration, and multiple hearth incineration. The LCAT

Year	Jan.	Feb.	Mar.	Apr.	May	June	July	Aug.	Sept.	Oct.	Nov.	Dec.	Annual Average
1968	1107	1114	1117	1124	1142	1154	1158	1171	1186	1190	1191	1201	1155
1969	1216	1229	1238	1249	1258	1270	1283	1292	1285	1299	1305	1305	1269
1970	1309	1311	1314	1329	1351	1375	1414	1418	1421	1434	1445	1445	1386
1971	1465	1467	1496	1513	1551	1589	1618	1629	1654	1657	1665	1672	1581
1972	1686	1691	1887	1707	1735	1761	1772	1777	1786	1794	1808	1816	1753
1973	1838	1850	1859	1874	1880	1896	1901	1902	1929	1933	1935	1939	1895
1974	1940	1940	1940	1961	1961	1994	2040	2076	2089	2100	2094	2098	2019
1975	2103	2128	2128	2135	2164	2201	2243	2270	2271	2289	2287	2293	2209
1976	2300	2310	2317	2327	2328	2410	2414	2445	2465	2478	2486	2490	2400
1977	2494	2505	2513	2514	2515	2541	2579	2611	2644	2675	2659	2669	2577

[a]Index for 1913 = 100.

TABLE 14.3

Engineering News-Record Construction Cost Index[a]

Indexes for 25 large cities were developed using a Kansas City Index of 100 in the third quarter of 1973 as a base. The second EPA-Treatment Plant Index is the Small City Conventional Treatment (SCCT) Index which uses a 5.0-MGD activated sludge facility as a model. It includes the same processes as the LCAT model excluding those designed to produce greater than secondary treatment, namely, lime clarification, recalcination, multi-media gravity filtration, and multiple hearth incineration. The base city for the SCCT Indexes maintained for 25 small cities nationwide is St. Joseph, Missouri with an index of 100 in the third quarter 1973. EPA indexes are updated quarterly and published in the Engineering News-Record periodically. Historical values of these two indexes appear in table 14.4.

The ENR Skilled Labor Index is usually used to update operation and maintenance costs. The value of this index from 1968 to the present time for U.S. average is presented in table 14.5.

Presentation of Costs

Treatment costs can be presented in several ways depending on the desired use. In many cases it is most useful to express information on the basis of the cost required to treat a given volume of water or the total cost required for a one-year period. In order to present costs in this manner, it is necessary to amortize capital expenditures using appropriate capital recovery factors. Thus, capital costs are expressed as a uniform series of annual payments required to repay construction costs over the useful life of the project.

Operation and maintenance costs are usually estimated on a $/day or $/yr basis. Total treatment costs expressed as $/yr may easily be converted to $/1,000 gal or $/MG gal for a given waste flow. In converting the costs from an annual to unit flow basis, it should be realized that while some industries operate on a five-

TABLE 14.4
EPA Treatment Plant Index of 25 Cities' Average

Year	SCCT Index[a]				LCAT Index[b]			
	1	2	3	4	1	2	3	4
1973	–	–	93	93	–	–	102	102
1974	95	104	110	112	106	114	122	123
1975	109	108	109	110	120	118	118	120
1976	113	115	118	119	123	126	129	130
1977	121	123	126	128	132	134	139	140

[a] Third-quarter 1973, St. Joseph, Missouri = 100.
[b] Third-quarter 1973, Kansas City = 100.

Year	Jan.	Feb.	Mar.	Apr.	May	June	July	Aug.	Sept.	Oct.	Nov.	Dec.	Annual Average
1968	104	104	104	104	106	108	108	110	111	111	111	112	108
1969	112	113	113	114	114	118	120	121	122	123	123	124	118
1970	125	125	125	126	129	131	136	137	138	138	139	139	132
1971	142	142	143	145	149	151	152	153	154	154	154	155	150
1972	156	158	158	159	160	162	163	164	165	166	168	169	162
1973	169	170	170	170	171	172	173	175	176	176	177	178	173
1974	178	179	179	179	179	182	186	188	188	190	191	191	184
1975	192	193	193	194	195	200	202	204	204	206	206	207	199
1976	207	207	207	208	209	215	218	218	219	219	220	221	214
1977	221	221	221	222	222	226	230	230	232	233	233	234	227

[a]Index for 1967 = 100.

TABLE 14.5

Engineering News-Record Skilled Labor Index[a]

or six-day week, many treatment processes may be operated on a seven-day week basis. This is especially true of biological treatment processes.

An alternative method for the presentation of costs that is useful when comparing several treatment alternatives is present worth. The present worth of a system equals the sum of the initial construction cost plus the present worth of annual operation and maintenance and replacement costs as required during the course of the project. Operation and maintenance and replacement costs must be reduced to equivalent expenditures at the beginning of the project. Provisions for escalation of costs during the lifetime of a project are usually not made when computing costs on this basis. Therefore, the implicit assumption is made that treatment costs rise in proportion to price levels in general.

A graphical representation for various levels of treatment and the cost of achieving these effluent qualities is shown in figures 14.1 and 14.2. Similar representation can be formulated for any process combination, flow, and constituent concentrations. This type of cost presentation is particularly useful in establishing practicable levels of treatment.

Estimation of Total Plant Costs

In some cases it may be desirable to estimate the cost of a treatment plant based on the cost of an identical facility having a different capacity. In these cases it has been found that the cost of different sized plants varies with the ratio of the capacities of the two plants raised to the 0.6 power. This relationship may be mathematically represented as follows:

$$(\$2)_{cap} = (\$1)_{cap} \left(\frac{C_2}{C_1} \right)^{0.6}$$

where $(\$1)_{cap}, (\$2)_{cap}$ = capital cost of plants 1 and 2, respectively
       C_1, C_2 = capacity of plants 1 and 2, respectively

Although this relationship was developed specifically for construction costs in chemical process industries, analysis of capital cost relationships for wastewater treatment processes indicates that costs generally vary with the plant capacity ratio raised to the 0.5 to 0.7 power. However, this is a very approximate guideline and should be used with care.

A similar relationship has been derived for operation and maintenance costs. Because there is much less economy of scale for operational costs, these costs generally vary with the 0.85 power of the plant capacity ratio. The relationship used for the estimation of operation and maintenance costs for treatment facilities of different sizes is:

$$(\$2)_{OM} = (\$1)_{OM} \left(\frac{C_2}{C_1} \right)^{0.85}$$

where $(\$1)_{OM}, (\$2)_{OM}$ = operation and maintenance costs for plants 1 and 2, respectively

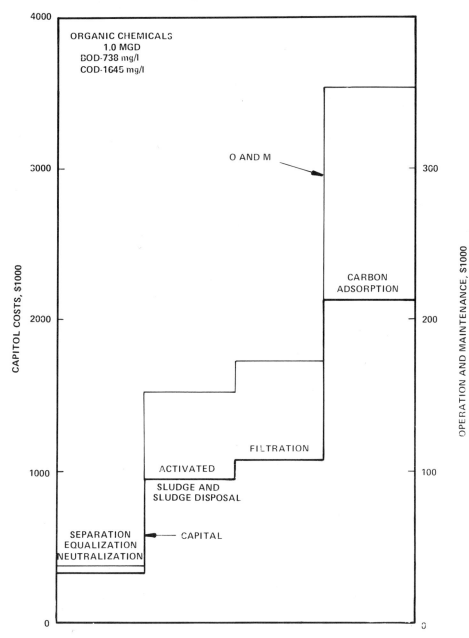

FIGURE 14.1

Capital and Operating Costs for Treatment of an
Organic Chemicals Wastewater Facility
(1975 Dollars)

FIGURE 14.2
Relationship Between Effluent Quality and Capital and Operating Cost-Municipal Sewage

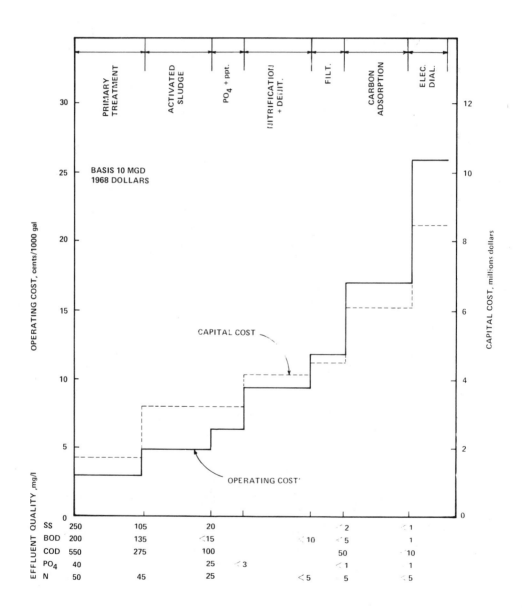

COST EFFECTIVENESS
AND WATER QUALITY

There is an increasing awareness of cost effectiveness in wastewater treatment plant design and operation. A water quality limiting discharge usually involves effluent requirements more stringent than those imposed by effluent guidelines and in many cases requires wastewater treatment beyond conventional secondary treatment processes. This discussion relates to parameters pertinent to the oxygen balance in the receiving water and the cost effectiveness of wastewater treatment relative to this impact.

Water Quality

Several factors should be considered in the water quality limiting case. A mixed wastewater may contain many organics of varying biodegradability. The overall rate will be the sum of the individual rates. For example, a recent study involving treatment in the activated sludge process using a mixture of glucose, phenol, and sulfonilic acid showed individual removal rates of 0.072 mg/mg/day, 0.049 mg/mg/day, and 0.015 mg/mg/day, respectively with an overall removal rate of 0.130 mg/mg/day. The removal rate coefficient K can be computed from equation (9–18) for individual and mixed substrates.

Removal of the more readily degradable organics through a treatment facility, that is, the glucose and the phenol in the previous example, causes a reduction in the deoxygenation rate coefficient (K_1) of the residual organics discharged to the receiving stream. For example, a raw wastewater might exhibit a BOD rate coefficient of 0.25/day, while after treatment and removal of the readily degradable organics, the resulting BOD rate coefficient might be 0.08/day. Hypothetically, discharging the same pounds of BOD causes less of an oxygen deficit if the deoxygenation rate is low as opposed to a high K_1 rate, as shown in figure 14.3. This illustrates the fallacy of only considering the pounds of BOD discharged without regard to the resulting deoxygenate rates in the receiving stream.

Therefore, relative to the impact on the receiving stream, both the quantity of BOD as well as the biodegradation rate of the residual BOD should be considered. The effects of biodegradability on activated sludge plant costs are shown in figure 14.4. It becomes apparent that for cost effective design, higher concentrations of effluent soluble BOD can be discharged for wastewaters of low biodegradability without as significant an effect on the oxygen balance in the receiving stream.

FIGURE 14.3
Effect of Deoxygenation Rate on Oxygen Sag Curve

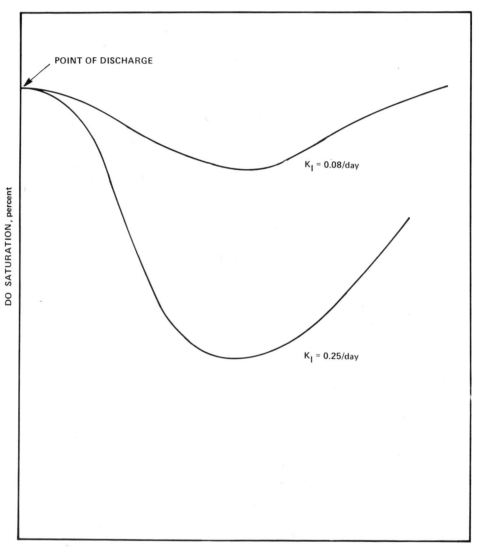

FIGURE 14.4
Effect of Biodegradability on Activated Sludge Costs

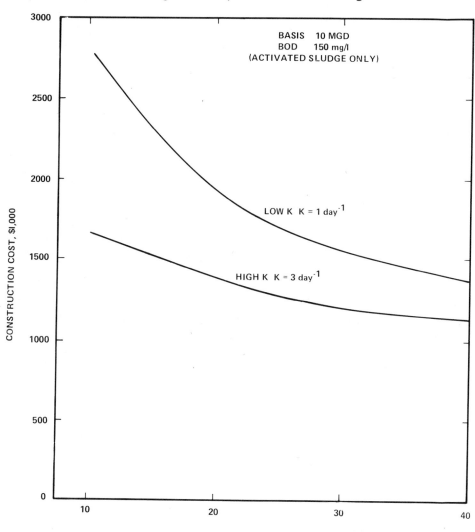

EFFLUENT SOLUBLE BOD, mg/l

There is an increasing emphasis today on nitrification in a wastewater treatment facility, particularly as it relates to the oxygen balance in the receiving water. Removal of carbonaceous organics may move the nitrification oxidation upstream closer to the wastewater discharge resulting in greater depletions of oxygen. This phenomena is accentuated when the wastewater treatment plant is nitrifying and thereby discharging increased numbers of nitrifying organisms to the stream. Temperature has a major effect on the nitrification process, both in the treatment plant and the receiving stream. A cost effective wastewater treatment plant might be designed to produce nitrification during the summer months when oxygen depletion into the receiving stream would be greatest. Nitrification would not be significant during the winter months when the nitrification rate in the stream and the resulting numbers of nitrifiers from the plant would be minimal. The effect of temperature on nitrification design and resulting costs is shown in table 14.6.

Significant capital and operating costs for construction and power are required to achieve nitrification under cold conditions. Another factor that must be considered in the oxygen balance in the receiving stream is the fact that many substances that depress the reaeration coefficient, such as surface active agents present in raw wastewaters, are removed in wastewater treatment. The net effect is to increase the reaeration coefficient in the receiving water with increasing degrees of wastewater treatment. This has the net effect of permitting higher organic loads without further oxygen depletion. Because of the nature of the biodegradation process, both in the wastewater treatment plant and the receiving stream, cost effective design and operation should lead to a two-tiered standard in those parts of the country where cold weather temperatures affect the biological oxidation process, both in the treatment facility and in the receiving water.

TABLE 14.6
Effect of Temperature on Nitrification Design and Costs

Raw Waste Load	10 mg/l 150 mg/l BOD 20 mg/l NH$_3$–N	
Effluent Quality	15 mg/l BOD 2 mg/l NH$_3$–N	
Temperature, °C	10	25
Sludge Age, days	14	5
Aeration Basin Volume, MG	2.0	1.0
Aeration, hp	200	170
Construction Cost	$696,000	$481,000

Effluent Guidelines

The second case considers those plants subject to effluent guideline limitations. It should be recognized that most industrial plants are or will be coming into compliance with Best Practicable Control Technology (BPT) regulations, effective July 1, 1977. This would imply that most industrial plants discharging organic wastewaters will have installed biological wastewater treatment, and that any consideration of additional reduction in pollutional loads should consider the existence of a biological wastewater treatment facility at that time. The original effluent limitations relating to Best Available Technology Economically Achievable (BATEA) generally considered some in-plant reductions in wastewater volume and principally add-on, end-of-pipe treatment units such as filtration and carbon adsorption. Most of the cases in industry that have been evaluated would indicate that other approaches to effluent quality improvement may be considerably more cost effective than indiscriminant add-on treatment facilities. These approaches are:

1. In-plant changes to eliminate or reduce pollutional loads
2. Installation of treatment systems for process modification at specific discharge sources to eliminate, reduce, or modify the wastewater characteristics to render them more compatible with existing wastewater treatment facilities
3. Add-on tertiary treatment units to the existing wastewater treatment facility

A detailed study was conducted for the Effluent Standards and Water Quality Information Advisory Committee (EPA) [6] in order to define the cost effectiveness of pollution reduction by in-plant changes with existing treatment facilities as compared to additional end-of-pipe wastewater treatment as defined by the BATEA criteria. This study considered an activated sludge plant in place at the time that improved effluent quality was to be considered. The alternatives considered are coagulation, filtration, carbon adsorption, and in-plant changes to reduce wastewater flow and strength. The in-plant changes include equipment revision and additions, unit shut downs, scrubber replacement, segregation, collection and incineration, raw material substitutions, reprocessing, and miscellaneous small projects. Table 14.7 summarizes the results of this study. Cost relationships for COD removal for the various options are shown in figure 14.5.

It is apparent that little benefit in effluent quality is gained by adding carbon adsorption to the activated sludge plant effluent over in-plant changes with biological and chemical treatment. As can be seen, additional end-of-pipe treatment, including coagulation, filtration, and carbon adsorption, resulted in an increased removal of 23,930 pounds of COD per day at an annual additional cost of $4.1 million dollars per year. In contrast, in-plant changes with minimal additional treatment, that is, chemical coagulation and filtration, resulted in a COD reduction of 19,260 pounds per day at a cost of $1.3 million dollars per year. It is readily

Graph Code	Pollution Reduction Scheme	Influent		Effluent				Annual Costs[a]		
		Flow (MGD)	Sol. BOD$_5$ (1000 lb/day)	Sol. BOD$_5$	Tot. BOD$_5$	COD	TSS	Capital	Operating	Total
				← (1000 lb/day) →			→	← Millions of Dollars →		
BPT	Activated Sludge	11.1	55.7	3.50	6.30	41.78	8.51	—	—	—
A	Activated Sludge & Coagulation	11.1	55.7	3.50	5.00	36.33	4.62	0.21	0.22	0.43
F	All in-plant changes & Activated Sludge & Coagulation	8.3	37.1	2.35	3.50	25.43	3.46	0.71	0.22	0.93
G	All in-plant changes & Activated Sludge & Coagulation & Filtration	8.3	37.1	2.35	2.81	22.52	1.38	0.93	0.40	1.33
H	Activated Sludge & Coagulation & Filtration & Carbon Adsorption	11.1	55.7	2.27	2.57	17.85	0.92	1.20	2.90	4.10

[a]Costs presented are those above the cost of the installed BPT facility. All costs are 1975 dollars.

TABLE 14.7
Effluent Quality Cost Effective Alternatives

FIGURE 14.5

Economic Alternatives for COD Reduction

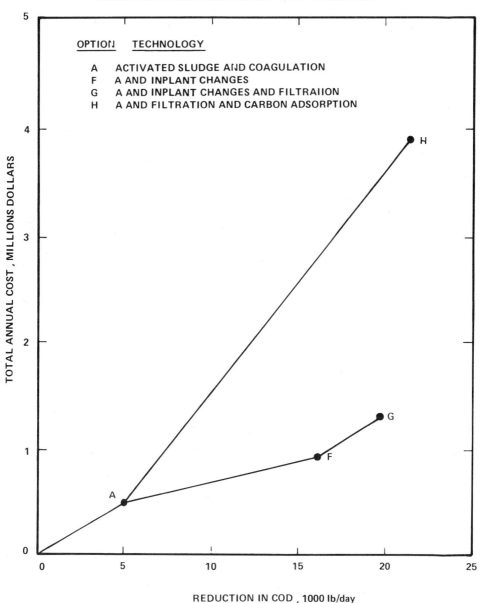

OPTION	TECHNOLOGY
A	ACTIVATED SLUDGE AND COAGULATION
F	A AND INPLANT CHANGES
G	A AND INPLANT CHANGES AND FILTRAIION
H	A AND FILTRATION AND CARBON ADSORPTION

REDUCTION IN COD , 1000 lb/day

apparent from figure 14.5 that a cost effective analysis would mitigate against the application of carbon adsorption for the minimal improvement in effluent quality achieved.

In most cases, effluent treatment facilities have been designed to treat total wastewater discharges to levels consistent with effluent guideline limitations established by EPA for specific industrial categories and subcategories. In many cases, removal of particular constituents that are inhibitory to the biological treatment process or possess a very low degradation rate, can, by treatment of these constituents at their source, result in a marked improvement in performance and an increase in capacity of existing biological wastewater treatment facilities. Table 14.8 illustrates a case in which one wastewater stream markedly reduced the overall biodegradation rate in the biological wastewater treatment facility.

Removal of this constituent by separate carbon adsorption treatment rendered the total wastewater stream considerably more degradable. This would substantially reduce the effluent pollutant levels from the biological treatment facility or permit higher organic loadings through the facility with resultant reductions in effluent discharges. An evaluation of both the biodegradability and the effect of wastewater constituents on biodegradation on specific sources within the industrial facility should in many cases lead to marked improvement in both wastewater treatment plant operation and effluent variability. This is particularly true where new products are to be introduced into the plant which may affect the overall biodegradation characteristics.

Tischler and Ford [7] have presented a comprehensive analysis of cost effectiveness in the petroleum refining industry. Table 14.9 shows present costs for biological treatment at petroleum refineries. Using a model refinery with a crude throughput of 42,500 BPSD with a wastewater flow of 1 MGD the effluent qualities after varying degrees of treatment are shown in table 14.10. The incremental costs of wastewater treatment are shown in table 14.11 and the incremental cost effectiveness of wastewater treatment is shown in table 14.12.

TABLE 14.8
Reaction Rate Coefficient With and Without Carbon Pretreatment of a Pesticide Wastewater

Temp °C	Non-Carbon Treated (K, day^{-1})	Carbon Treated (K, day^{-1})
28	2.25	23.1
8	0.81	6.5

System Type	Flow (MGD)	Organic Removal		Amortized Capital Cost[a]			Annual Operation and Maintenance Costs		
		BOD₅ (lb/day)	COD (lb/day)	($/1000 gal)	($/lb BOD)	($/lb COD)	($/1000 gal)	($/lb BOD)	($/lb COD)
Rotating Biological Surface	0.9	840	1,200	0.23	0.24	0.16	0.06	0.120	0.050
Rotating Biological Surface	4.3	5,300	—	0.19	—	—	—	—	—
Activated Sludge	4.6	9,100	26,200	0.12	0.06	0.02	0.10	0.050	0.020
Activated Sludge	3.9	41,500	83,000	0.19	0.02	0.01	0.07	0.006	0.003
Activated Sludge	1.9	8,600	13,000	0.24	0.05	0.03	0.12	0.030	0.020
Activated Sludge	2.3	3,100	—	0.19	0.14	—	—	—	—

[a]Adjusted to 1977 and 10 percent interest for 15 years.

TABLE 14.9

Case Histories of Biological Treatment at Petroleum Refineries

TABLE 14.10

Model Refinery Wastewater Characteristics at Varying Levels of Treatment

Parameter	Annual Average			
	After Air Flotation	After Biological Treatment	After Tertiary Filtration	After Tertiary Carbon Adsorption
Flow, MGD	1	1	1	1
BOD_5, mg/l	150	25	15	5
COD, mg/l	390	125	97	40
TSS, mg/l	30	40	12	6

TABLE 14.11

Incremental Costs of Wastewater Treatment for Model Refinery[a]

Treatment Type	Capital Cost ($1,000)	Annual Operation and Maintenance Cost ($1,000)	Annual Energy Cost ($1,000)
Activated Sludge	461	32	9
Tertiary Granular-media Filtration	220	6	2
Tertiary Activated Carbon	740	92	52

[a]Basis: 1977 Gulf Coast Costs.

TABLE 14.12

Incremental Cost Effectiveness of Wastewater Treatment
for Model Refinery[a]

Cost Effectiveness Parameter	Treatment Method		
	Activated Sludge	Tertiary Filtration	Tertiary Carbon Adsorption Columns
Flow, $/1000 gal			
Capital	0.16	0.08	0.27
Annual O&M	0.09	0.02	0.25
BOD_5, $/lb removed			
Capital	0.16	0.95	3.20
Annual O&M	0.08	0.20	3.04
COD, $/lb removed			
Capital	0.07	0.34	0.56
Annual O&M	0.04	0.07	0.53
TSS, $/lb removed			
Capital	–	0.34	5.31
Annual O&M	–	0.07	5.04

[a]Basis: 1977 Gulf Coast, Capital amortized at 10 percent for 15 years.

REFERENCES

1. Smith, R. 1968. Cost of conventional and advanced treatment of wastewater. *J. Water Pollution Control Federation* 40: 1546.
2. Barnard, J. L. and Eckenfelder, W. W. Jr., Treatment-cost relationships for industrial waste treatment. 1971. Technical Report No. 23, EWRE. Nashville, Tenn.: Vanderbilt University.
3. Burrows, D. and Ziegler, F. G. Analysis of wastewater treatment costs for selected industries. Prepared for CONSAD Research Co., Pittsburgh, Penn., by AWARE, Inc., Nashville, Tenn. (May 1974).
4. Edwards, A. W. *et al.* Industrial water cost studies. Prepared for the Office of Research and Development, U.S. EPA, Washington, D.C., by the Center for Industrial Water Quality Management, Vanderbilt University, Nashville, Tenn. (July 1975).
5. "Construction costs for municipal wastewater treatment plants: 1973–1977," Technical Report MCD-37, U.S. EPA, Washington, D.C. (January 1978).
6. An approach for establishing best available technology, Effluent Standards and Water Quality Information Advisory Committee Report to U.S. EPA (1975).
7. Tischler, L. F. and Ford, D. L. "Treatment cost and energy effectiveness as a function of effluent quality," *Proc. 6th Annual Industrial Pollution Conf. and Exposition,* WWEMA, St. Louis, Missouri (April 1978).

Selected Readings

GENERAL

Advances in Water Pollution Research, Vol. I, II, and III. 1962. Proceedings of the International Conference held in London, England, September 1962. Oxford: Pergamon Press, 1963.

Advances in Water Pollution Research, Vol. I, II, and III. 1964. Proceedings of the Second International Conference held in Tokyo, Japan, August 1964. Oxford: Pergamon Press, 1966.

Advances in Water Pollution Research, Vol. I, II, and III. 1966. Proceedings of the Third International Conference held in Munich, West Germany, September 1966. Washington, D.C.: Water Pollution Control Federation, 1967.

Advances in Water Pollution Research, 1969. Proceedings of the Fourth International Conference held in Prague, Czechoslovakia, April 1969. Oxford: Pergamon Press, 1969.

Advances in Water Pollution Research, Vol. I and II. 1970. Proceedings of the Fifth International Conference held in San Francisco and Hawaii, July-August 1970. Oxford: Pergamon Press 1971.

Advances in Water Pollution Research, Vol. I and II. 1972. Proceedings of the Sixth International Conference held in Jerusalem, Israel, June 1972. Oxford: Pergamon Press, 1973.

Babbitt, H. E., and Baumann, E. R. *Sewerage and Sewage Treatment.* 8th ed. New York: Wiley, 1958.

Clark, J. W., Viessman, W. Jr., and Hammer, M. *Water Supply and Pollution Control.* 3rd ed. Scranton, Penn: International Textbook, 1977.

Fair, G. M., Geyer, J. C., and Okun, D. A., *Water and Wastewater Engineering, Vol. 1. Water Supply and Wastewater Removal, Vol. 2. Water Purification and Wastewater Treatment and Disposal.* New York: Wiley, 1966.

Gehm, H. W., and Bregman, J. I. Eds. *Handbook of Water Resources and Pollution Control.* New York: Van Nostrand Reinhold 1976.

Glossary of Water and Wastewater Engineering. American Public Health Association, American Society of Civil Engineers, American Water Works Association, and Water Pollution Control Federation, 1969.

Imhoff, K., Mulleer, W. J., and Thistlethwayte, D. K. B. *Disposal of Sewage and Other Waterborne Wastes.* 2nd ed. Ann Arbor, Mich.: Ann Arbor Science, 1971.

McGauhey, P. H., *Engineering Management of Water Quality.* New York: McGraw-Hill, 1968.

McKinney, R. E., *Microbiology for Sanitary Engineers.* New York: McGraw-Hill, 1962.

Metcalf & Eddy, Inc., *Wastewater Engineering: Collection, Treatment, Disposal.* New York: McGraw-Hill, 1972.

Mitchell, R., Ed., *Water Pollution Microbiology.* New York: Wiley, 1972.

Proceedings of the Seventh International Association on Water Pollution Research Conference, Paris, France, September 1974, Progress in Water Technology, Vol. 7, Nos. 2–6, Oxford, England: Pergamon Press, 1976.

Proceedings of the Eighth International Association on Water Pollution Research Conference, Sydney, Australia, October 1976, Progress in Water Technology, Vol. 9, Nos. 1–4, Oxford, England: Pergamon Press, 1977.

Ramalho, R. S. *Introduction to Wastewater Treatment Processes.* New York: Academic Press, 1977.

Rich, L. G. *Environmental Systems Engineering.* New York: McGraw-Hill, 1973.

Rich, L. G. *Unit Operations of Sanitary Engineering.* New York: Wiley, 1961.

Rich, L. G. *Unit Processes of Sanitary Engineering.* New York: Wiley, 1963.

Sawyer, C. N., and McCarty, P. L. *Chemistry for Environmental Engineering.* 3rd ed. New York: McGraw-Hill, 1978.

Schroeder, E. D., *Water and Wastewater Engineering Treatment.* New York: McGraw-Hill, 1977.

American Public Health Association. *Standard Methods for the Examination of Water and Wastewater.* 14th ed. New York, 1976.

Steel, E. W., *Water Supply and Sewerage.* 4th ed. New York: McGraw-Hill, 1960.

Tchobanoglous, G., Smith, R., and Crites, R. *Wastewater Management.* Detroit Gale Research Co., 1976.

Tebbutt, T. H. Y., *Principles of Water Quality Control.* 2nd ed. Oxford: Pergamon Press, 1976.

Thomann, R. V., *Systems Analysis and Water Quality Control.* New York: McGraw-Hill, 1972.

Water-1968 AIChE Symposium Series, Vol. 64, No. 90. New York: American Institute of Chemical Engineers, 1968.

Water-1969 AIChE Symposium Series, Vol. 65, No. 97. New York: American Institute of Chemical Engineers, 1969.

Water-1970 AIChE Symposium Series, Vol. 67, No. 107. New York: American Institute of Chemical Engineers, 1971.

Water-1971 AIChE Symposium Series, Vol. 68, No. 124. New York: American Institute of Chemical Engineers, 1972.

Water-1972 AIChE Symposium Series, Vol. 69, No. 129. New York: American Institute of Chemical Engineers, 1973.

Water-1973 AIChE Symposium Series, Vol. 70, No. 136. New York: American Institute of Chemical Engineers, 1974.

Water-1974: New York: I. Industrial Wastewater Treatment, AIChE Symposium Series, Vol. 70, No. 144, American Institute of Chemical Engineers, 1974.

Water-1974: II. Municipal Wastewater Treatment, AIChE Symposium Series, Vol. 71, No. 145. New York: American Institute of Chemical Engineers, 1975.

Water-1975 AIChE Symposium Series, Vol. 72, No. 151. New York: American Institute of Chemical Engineers, 1976.

Water-1976: I. Physical, Chemical Wastewater Treatment, AIChE Symposium Series, Vol. 73, No. 166. New York: American Institute of Chemical Engineers, 1977.

Water-1976: II. Biological Wastewater Treatment, AIChE Symposium Series, Vol. 73, No. 167. New York: American Institute of Chemical Engineers, 1977.

GENERAL CONCEPTS OF WATER QUALITY MANAGEMENT

Coulston, F., and Mrak, E. *Water Quality: Proceedings of an International Forum.* New York: Academic Press, 1977.

Jenkins, S. H., Ed. *Water Quality: Management and Pollution Control Problems.* Vol. 3, Progress in Water Technology, Oxford:1 Pergamon Press, 1973.

Kneese, A. V., and Bower, B. T. *Managing Water Quality:1 Economics, Technology, Institution.* Baltimore: The Johns Hopkins University Press, 1968.

Kneese, A. V., and Smith, S. C. Eds. *Water Research*. Baltimore: The Johns Hopkins University Press, 1966.

Loehr, R. C. *Agricultural Waste Management: Problems, Processes and Approaches*. New York: Academic Press, 1974.

Loehr, R. C. *Pollution Control for Agriculture*. New York: Academic Press, 1977.

McGauhey, P. H. *Engineering Management of Water Quality*. New York: McGraw-Hill, 1968.

McKee, J. E., and Wolf, H. W. Eds. *Water Quality Criteria* 2nd ed. Pasadena: California Institute of Technology, 1963.

Proceedings of the National Symposium on Quality Standards for Natural Waters. Ann Arbor: University of Michigan, 1966.

Proceedings of the Third National Conference on Complete Water Reuse, Cincinnati, Ohio, June 1976, American Institute of Chemical Engineers, New York, 1976.

Thomann, R. V. *Systems Analysis and Water Quality Management*. New York: McGraw-Hill, 1972.

Water Quality Criteria 1972, Report of the Committee on Water Quality Criteria to the Environmental Protection Agency, Washington, D.C., 1973.

Whipple, W., Jr. *Planning of Water Quality Systems*. Lexington, Mass.: D.C. Heath, 1977.

SEWAGE AND INDUSTRIAL WASTE CHARACTERIZATION

Biology of Water Pollution. U.S. Department of Interior, Federal Water Pollution Control Administration, Washington, D.C., 1967.

Camp, T. R., and Meserve, R. L. *Water and Its Impurities*. 2nd ed. Stroudsburg, Penn.: Dowden, Hutchinson, and Ross, 1974.

Department of Scientific and Industrial Research. *Water Pollution Research* (annual report). London: Her Majesty's Stationery Office.

Faust, S. D., and Hunter, J. V. Eds. *Principles and Applications of Water Chemistry*. New York: Wiley, 1967.

Klein, L., *River Pollution: 1. Chemical Analysis*. Washington, D.C.: Butterworth Inc., 1959.

Mancy, K. H., Ed. *Instrumental Analysis for Water Pollution Control*. Ann Arbor, Mich.: Ann Arbor Science, 1971.

Mancy, K. H., and Weber, W. J. Jr. *Analysis of Industrial Wastewater*. New York: Wiley, 1972.

Manual of Methods for Chemical Analysis of Water and Wastes. Technology Transfer, U.S. Environmental Protection Agency, 1974.

Metcalf & Eddy, Inc. *Wastewater Engineering: Collection, Treatment, Disposal.* New York: McGraw-Hill, 1972.

Ramalho, R. S. *Introduction to Wastewater Treatment Processes.* New York: Academic Press, 1977.

Sawyer, C. N., and McCarty, P. L. *Chemistry for Environmental Engineering.* 3rd ed. New York: McGraw-Hill, 1978.

Standard Methods for the Examination of Water and Wastewater. 1976. 14th ed. New York: American Public Health Association.

Stumm, W., and Morgan, J. J. *Aquatic Chemistry.* New York: Wiley, 1970.

ANALYSIS OF POLLUTION EFFECTS
IN NATURAL WATERS

Advances in Water Pollution Research. Vol. I and III. 1962. Proceedings of the International Conference held in London, England, September 1962. Oxford: Pergamon Press, 1963.

Advances in Water Pollution Research, Vol. I and III. 1964. Proceedings of the Second International Conference held in Tokyo, Japan, August 1964. Oxford: Pergamon Press, 1966.

Advances in Water Pollution Research, Vol. I and III. 1966. Proceedings of the Third International Conference held in Munich, West Germany, September 1966. Washington, D.C.: Water Pollution Control Federation, 1967.

Advances in Water Pollution Research. 1969. Proceedings of the Fourth International Conference held in Prague, Czechoslovakia, April 1969. Oxford: Pergamon Press, 1969.

Advances in Water Pollution Research, Vol. I and II. 1970. Proceedings of the Fifth International Conference held in San Francisco and Hawaii, July-August 1970. Oxford: Pergamon Press, 1971.

Advances in Water Pollution Research, Vol. I. 1972. Proceedings of the Sixth International Conference held in Jerusalem, Israel, June 1972. Oxford: Pergamon Press, 1973.

Department of Scientific and Industrial Research. *Water Pollution Research* (annual report). London: Her Majesty's Stationery Office.

Gloyna, E. F., and Eckenfelder, W. W., Jr., Eds. *Advances in Water Quality Improvement.* Austin: University of Texas Press, 1968.

Heukelekian, H., and Dondero, N. C. Eds. *Principles and Applications in Aquatic Microbiology.* New York: Wiley, 1964.

Klein, L., *River Pollution: Causes and Effects*. Washington, D.C.: Butterworth, 1962.

Klein, L., *River Pollution: Control*. Washington, D.C.: Butterworth, 1966.

Nemerow, N. L. *Scientific Stream Pollution Analysis*. New York: McGraw-Hill, 1974.

Oxygen Relationships in Streams, R. A. Taft Center Technical Report, W58-2, U.S. Department of Health, Education, and Welfare, Washington, D.C., 1958.

Proceedings of the Seventh International Association on Water Pollution Research Conference, Paris, France, September 1974, Progress in Water Technology, Vol. 7, Nos. 2–6, Oxford, England: Pergamon Press, 1976.

Proceedings of the Eighth International Association on Water Pollution Research Conference, Sydney, Australia, October 1976, Progress in Water Technology, Vol. 9, Nos. 1–4, Oxford, England: Pergamon Press, 1977.

Thomann, R. V. *Systems Analysis and Water Quality Control*. New York: McGraw-Hill, 1972.

Velz, C. J. *Applied Stream Sanitation*. New York: Wiley, 1970.

CHARACTERISTICS OF MUNICIPAL WASTEWATERS

Babbitt, H. E., and Baumann, E. R., *Sewerage and Sewage Treatment*. 8th ed. New York: Wiley, 1958.

Bolton, R. L., and Klein, L. *Sewage Treatment: Basic Principles and Trends*. Ann Arbor, Mich.: Ann Arbor Science, 1971.

Department of Scientific and Industrial Research. *Water Pollution Research* (annual report). Her Majesty's Stationery Office, London, England.

Imhoff, K., Mulleer, W. J., and Thistlethwayte, D. K. B. *Disposal of Sewage and Other Waterborne Wastes*. 2nd ed. Ann Arbor, Mich.: Ann Arbor Science, 1971.

Steel, E. W. *Water Supply and Sewerage*. 4th ed. New York: McGraw-Hill, 1960.

INDUSTRIAL WASTEWATERS

Azad, H. S., Ed. *Industrial Wastewater Management Handbook*. New York: McGraw-Hill, 1976.

Besselievre, E. B., and Schwartz, M. *Treatment of Industrial Wastes*. 2nd ed., New York: McGraw-Hill, 1976.

Callely, A. G., Forster, C. F., and Stafford, D. A. Eds. *Treatment of Industrial Effluents*. Somerset, N.J.: Halsted Press, 1976.

Department of Scientific and Industrial Research. *Water Pollution Research* (annual report). Her Majesty's Stationery Office, London, England.

Goransson, B., Ed. *Industrial Wastewater and Wastes–2*. Oxford: Pergamon Press, 1977.

Gurnham, C. F. *Principles of Industrial Waste Treatment*. New York: Wiley, 1955.

Gurnham, C. F., ed., *Industrial Wastewater Control*. Vol. II: Chemical Technology. New York: Academic Press, 1965.

Handbook for Monitoring Industrial Wastewater, Technology Transfer. U.S. Environmental Protection Agency, August 1973.

Imhoff, K., Mulleer, W. J., and Thistlethwayte, D. K. B. *Disposal of Sewage and Other Waterborne Wastes*. 2nd ed. Ann Arbor, Mich.: Ann Arbor Science, 1971.

Isaac, P. C. G. Ed. *Waste Treatment*. Oxford: Pergamon Press, 1960.

Krotchak, D. *Management and Engineering Guide to Economic Pollution Control: A General Approach to Industrial Waste Problems with Case Histories*. Warren, Mich.: Clinton Industries, 1972.

Mancy, K. H., and Weber, W. J. Jr. *Analysis of Industrial Wastewater*. New York: Wiley, 1972.

Nemerow, N. L. *Industrial Water Pollution, Origins, Characteristics, and Treatment*. Reading, Mass.: Addison-Wesley, 1978.

Proceedings, Annual Ontario Industrial Waste Conference, Ontario Ministry of the Environment, Canada.

Proceedings, Industrial Effluents, Progress in Water Technology, Vol. 8, No.2/3, Pergamon Press, Oxford, England, 1977.

Proceedings of Annual Industrial Waste Conference, Purdue University, West Lafayette, Ind.

Reuse of Water in Industry. London: Butterworth, 1963.

Ross, R. D., Ed. *Industrial Waste Disposal*. New York: Van Nostrand Reinhold, 1968.

Water Reuse. AIChE Symposium Series Vol. 63, No. 78. New York: American Institute of Chemical Engineers, 1967.

Water Technology in the Pulp and Paper Industry. TAPPI Monograph Series, No. 18. New York: New Technical Association of the Pulp and Paper Industry, 1957.

WASTEWATER TREATMENT PROCESSES

Azad, H. S., Ed. *Industrial Wastewater Management Handbook*. New York: McGraw-Hill, 1976.

Babbitt, H. E., and Baumann, E. R. *Sewerage and Sewage Treatment*. 8th ed. New York: Wiley, 1958.

Bolton, R. L., and Klein, L. *Sewage Treatment: Basic Principles and Trends*. Ann Arbor, Mich.: Ann Arbor Science, 1971.

Callely, A. G., Forster, C. F., and Stafford, D. A. Eds. *Treatment of Industrial Effluents*. Somerset, N.J.: Halsted Press, 1976.

Camp, T. R., and Meserve, R. L. *Water and Its Impurities*. 2nd ed. Stroudsburg, Penn.: Dowden, Hutchinson, and Ross, 1974.

Clark, J. W., Viessman, W. Jr., and Hammer, M. *Water Supply and Pollution Control*. 3rd ed. Scranton, Penn.: International Textbook, 1977.

Eckenfelder, W. W., Jr., and Ford, D. L. *Water Pollution Control: Experimental Procedures for Process Design*. Austin, Texas: Jenkins, 1970.

Fair, G. M., Geyer, J. C., and Okun, D. A. *Water and Wastewater Engineering*. Vol. 2, Water Purification and Wastewater Treatment and Disposal. New York: Wiley, 1966.

Gehm, H. W., and Bregman, J. I. Eds. *Handbook of Water Resources and Pollution Control*. New York: Van Nostrand Reinhold, 1976.

Hammer, M. J. *Water and Waste-Water Technology*. New York: Wiley, 1975.

Isaac, P. C. G., Ed. *Waste Treatment*. Oxford: Pergamon Press, 1960.

Metcalf & Eddy, Inc. *Wastewater Engineering: Collection, Treatment, Disposal*. New York: McGraw-Hill, 1972.

Parker, H. W. *Wastewater Systems Engineering*. Englewood Cliffs, N.J.: Prentice-Hall, 1975.

Patterson, J. W. *Wastewater Treatment Technology*. Ann Arbor, Mich.: Ann Arbor Science, 1975.

PRETREATMENT AND PRIMARY TREATMENT

Adams, C. E., Jr., and Eckenfelder, W. W. Jr. *Process Design Techniques for Industrial Waste Treatment*. Nashville, Tenn.: Enviro Press, 1974.

Advances in Water Pollution Research. 1962. Vol. II. Proceedings of the International Conference held in London, England, September 1962. Oxford: Pergamon Press, 1963.

Advances in Water Pollution Research. 1964. Vol. II. Proceedings of the Second International Conference held in Tokyo, Japan, August 1964. Oxford: Pergamon Press, 1966.

Advances in Water Pollution Research. 1966. Vol. II. Proceedings of the Third International Conference held in Munich, West Germany, September 1966. Washington, D.C.: Water Pollution Control Federation, 1967.

Advances in Water Pollution Research. 1969. Proceedings of the Fourth International Conference held in Prague, Czechoslovakia, April 1969. Oxford: Pergamon Press, 1969.

Advances in Water Pollution Research. 1970. Vol. I and II. Proceedings of the Fifth International Conference held in San Francisco and Hawaii, July-August 1970. Oxford: Pergamon Press, 1971.

Advances in Water Pollution Research. 1972. Vol. I and II. Proceedings of the Sixth International Conference held in Jerusalem, Israel, June 1972. Oxford: Pergamon Press, 1973.

Babbitt, H. E., and Baumann, E. R. *Sewerage and Sewage Treatment.* 8th ed. New York: Wiley, 1958.

Clark, J. W., Viessman, W. Jr., and Hammer, M. *Water Supply and Pollution Control.* 3rd ed. Scranton, Penn.: International Textbook, 1977.

Department of Scientific and Industrial Research. *Water Pollution Research* (annual report). Her Majesty's Stationery Office, London.

Fair, G. M., Geyer, J. C., and Okun, D. A. *Water and Wastewater Engineering.* Vol. 2. Water Purification and Wastewater Treatment and Disposal. New York: Wiley, 1966.

Isaac, P. C. G., Ed. *Waste Treatment.* Oxford: Pergamon Press, 1960.

Metcalf & Eddy, Inc. *Wastewater Engineering: Collection, Treatment, Disposal.* New York: McGraw-Hill, 1972.

Proceedings, Annual Ontario Industrial Waste Conference, Ontario Ministry of the Environment, Canada.

Advances in Water Pollution Research. 1964. Vol. II. Proceedings of the Second International Conference held in Tokyo, Japan, August 1964. Oxford: Pergamon Press, 1966.

Advances in Water Pollution Research. 1966. Vol. II. Proceedings of the Third International Conference held in Munich, West Germany, September 1966. Washington, D.C.: Water Pollution Control Federation, 1967.

Advances in Water Pollution Research. 1969. Proceedings of the Fourth International Conference held in Prague, Czechoslovakia, April 1969. Oxford: Pergamon Press, 1969.

Advances in Water Pollution Research. 1970. Vol. I and II. Proceedings of the Fifth International Conference held in San Francisco and Hawaii, July-August 1970. Oxford: Pergamon Press, 1971.

Advances in Water Pollution Research. 1972. Vol. I and II. Proceedings of the Sixth International Conference held in Jerusalem, Israel, June 1972. Oxford: Pergamon Press, 1973.

Antonie, R. L. *Fixed ·Biological Surfaces-Wastewater Treatment.* West Palm Beach, Fla.: CRC Press, 1976.

Busch, A. W. *Aerobic Biological Treatment of Waste Waters: Principles and Practices.* Houston, Texas: Oligodynamics Press, 1971.

Canale, R. P., Ed. *Biological Waste Treatment.* New York: Wiley, 1971.

Eckenfelder, W. W., Jr., and Ford, D. L. *Water Pollution Control: Experimental Procedures for Process Design.* Austin, Texas: Jenkins, 1970.

Eckenfelder, W. W., Jr., and McCabe, B. J. Eds. *Advances in Biological Waste Treatment.* Oxford: Pergamon Press, 1963.

Fair, G. M., Geyer, J. C., and Okun, D. A. *Water and Wastewater Engineering.* Vol. 2. Water Purification and Wastewater Treatment and Disposal. New York: Wiley, 1966.

Gloyna, E. F., and Eckenfelder, W. W. Jr. Eds. *Advances in Water Quality Improvement.* Austin: University of Texas Press, 1968.

Gloyna, E. F., Malina, J. F. Jr., and Davis, E. M. Eds. *Ponds as a Wastewater Treatment Alternative.* Center for Research in Water Resources. Austin: University of Texas, 1976.

Gould, R. F., Ed. *Anaerobic Biological Treatment Processes.* Washington, D.C.: American Chemical Society, 1971.

Isaac, P. C. G., Ed. *Waste Treatment.* Oxford: Pergamon Press, 1960.

McCabe, B. J., and Eckenfelder, W. W. Jr., Eds. *Biological Treatment of Sewage and Industrial Wastes.* Vol. I. Aerobic Oxidation. New York: Reinhold, 1956.

McCabe, B. J., and Eckenfelder, W. W. Jr., Eds. *Biological Treatment of Sewage and Industrial Wastes.* Vol. II. Anaerobic Digestion and Solids-Liquid Separation. New York: Reinhold, 1958.

McWhirter, J. R., Ed. *The Use of High-Purity Oxygen in the Activated Sludge Process.* Vol. I and II. West Palm Beach, Fla.: CRC Press, 1978.

Metcalf & Eddy, Inc. *Wastewater Engineering: Collection, Treatment, Disposal.* New York: McGraw-Hill, 1972.

Proceedings, A Consolidated Approach to Activated Sludge Process Design, Progress in Water Technology, Vol. 7, No. 1, Pergamon Press, Oxford, England, 1975.

Proceedings in the Seventh International Association on Water Pollution Research Conference, Paris, France, September 1974, Progress in Water Technology, Vol. 7, Nos. 2–6, Pergamon Press, Oxford, England, 1976.

Proceedings of the Eighth International Association on Water Pollution Research Conference, Sydney, Australia, October 1976, Progress in Water Technology, Vol. 9, Nos. 1–4, Pergamon Press, Oxford, England, 1977.

Proceedings of Annual Industrial Waste Conference, Purdue University, West Lafayette, Ind.

Proceedings of the Seventh International Association on Water Pollution Research Conference, Paris, France, September 1974, Progress in Water Technology, Vol. 7, Nos. 2–6, Pergamon Press, Oxford, England, 1976.

Proceedings of the Eighth International Association on Water Pollution Research Conference, Sydney, Australia, October 1976, Progress in Water Technology, Vol. 9, Nos. 1–4, Pergamon Press, Oxford, England, 1977.

Ramalho, R. S. *Introduction to Wastewater Treatment Processes.* New York: Academic Press, 1977.

Rich, L. G. *Unit Processes of Sanitary Engineering.* New York: Wiley, 1963.

Thackston, E. L., and Eckenfelder, W. W. Jr., Eds. *Process Design in Water Quality Engineering: New Concepts and Developments.* Austin, Texas: Jenkins, 1972.

Wastewater Treatment Plant Design, American Society of Civil Engineers and Water Pollution Control Federation, 1977.

OXYGEN TRANSFER AND AERATION

Aeration in Wastewater Treatment, Water Pollution Control Federation, Washington, D.C., 1971.

Department of Scientific and Industrial Research. *Water Pollution Research* (annual report). Her Majesty's Stationery Office, London, England.

Eckenfelder, W. W., Jr., and Ford, D. L. *Water Pollution Control: Experimental Procedures for Process Design.* Austin, Texas: Jenkins, 1970.

Gibbon, D. L., Ed. *Aeration of Activated Sludge in Sewage Treatment.* Oxford: Pergamon Press, 1974.

Gloyna, E. F., and Eckenfelder, W. W. Jr. *Advances in Water Quality Improvement.* Austin: University of Texas Press, 1968.

Metcalf & Eddy, Inc. *Wastewater Engineering: Collection, Treatment, Disposal.* New York: McGraw-Hill, 1972.

Ramalho, R. S. *Introduction to Wastewater Treatment Processes.* New York: Academic Press, 1977.

Schroeder, E. D. *Water and Wastewater Engineering Treatment.* New York: McGraw-Hill, 1977.

Speece, R. E., and Malina, J. F. Jr., Eds. *Application of Commercial Oxygen to Water and Wastewater Systems.* Austin: University of Texas Press, 1973.

Thackston, E. L., and Eckenfelder, W. W. Jr., Eds. *Process Design in Water Quality Engineering: New Concepts and Developments.* Austin, Texas: Jenkins, 1972.

BIOLOGICAL WASTE TREATMENT

Adams, C. E., Jr., and Eckenfelder, W. W. Jr. *Process Design Techniques for Industrial Waste Treatment.* Nashville, Tenn.: Enviro Press, 1974.

Advances in Water Pollution Research. Vol. II. Proceedings of the International Conference held in London, England, September 1962. Oxford: Pergamon Press, 1963.

Process Design Manual for Nitrogen Control, Technology Transfer, U.S. Environmental Protection Agency, October 1975.

Ramalho, R. S. *Introduction to Wastewater Treatment Processes.* New York: Academic Press, 1977.

Rich, L. G. *Unit Processes of Sanitary Engineering.* New York: Wiley, 1963.

Schroeder, E. D. *Water and Wastewater Engineering Treatment.* New York: McGraw-Hill, 1977.

Thackston, E. L., and Eckenfelder, W. W. Jr., Eds. *Process Design in Water Quality Engineering: New Concepts and Developments.* Austin, Texas: Jenkins, 1972.

The Activated Sludge Process in Sewage Treatment—Theory and Application. Ann Arbor: University of Michigan Press, 1966.

Tourbier, J., and Pierson, R. W. Jr., Eds. *Biological Control of Water Pollution.* Philadelphia: University of Pennsylvania Press, 1976.

Wastewater Treatment Plant Design. American Society of Civil Engineers and Water Pollution Control Federation, 1977.

Water-1968. AIChE Symposium Series, Vol. 64, No. 90. New York: American Institute of Chemical Engineers, 1968.

Water-1969. AIChE Symposium Series, Vol. 65, No. 97. New York: American Institute of Chemical Engineers, 1969.

Water-1970. AIChE Symposium Series, Vol. 67, No. 107. New York: American Institute of Chemical Engineers, 1971.

Water-1971. AIChE Symposium Series, Vol. 68, No. 124. New York: American Institute of Chemical Engineers, 1972.

Water-1972. AIChE Symposium Series, Vol. 69, No. 129. New York: American Institute of Chemical Engineers, 1973.

Water-1973. AIChE Symposium Series, Vol. 70, No. 136. New York: American Institute of Chemical Engineers, 1974.

Water-1974. I. Industrial Wastewater Treatment. AIChE Symposium Series, Vol. 70, No. 144. New York: American Institute of Chemical Engineers, 1974.

Water-1974. II. Municipal Wastewater Treatment. AIChE Symposium Series, Vol. 71, No. 145. New York: American Institute of Chemical Engineers, 1975.

Water-1975. AIChE Symposium Series, Vol. 72, No. 151. New York: American Institute of Chemical Engineers, 1976.

Water-1976. II. Biological Wastewater Treatment. AIChE Symposium Series, Vol. 73, No. 167. New York: American Institute of Chemical Engineers, 1977.

PHYSICAL CHEMICAL TREATMENT

Adams, C. E., Jr., and Eckenfelder, W. W. Jr. *Process Design Techniques for Industrial Waste Treatment*. Nashville, Tenn.: Enviro Press, 1974.

Advanced Waste Treatment Research, Water Pollution Control Research Series AWTR 1–18, U.S. Department of Interior, Federal Water Pollution Control Administration, Washington, D.C., 1964–1967.

Advances in Water Pollution Research. 1972. Vol. II. Proceedings of the International Conference held in London, England, September 1962. Oxford: Pergamon Press, 1963.

Advances in Water Pollution Research. 1964. Vol. II. Proceedings of the Second International Conference held in Tokyo, Japan, August 1964. Oxford: Pergamon Press, 1966.

Advances in Water Pollution Research. 1966. Vol. II. Proceedings of the Third International Conference held in Munich, West Germany, September 1966. Washington, D.C.: Water Pollution Control Federation, 1967.

Advances in Water Pollution Research. 1969. Proceedings of the Fourth International Conference held in Prague, Czechoslovakia, April 1969. Oxford: Pergamon Press, 1969.

Advances in Water Pollution Research. 1970. Vol. I and II. Proceedings of the Fifth International Conference held in San Francisco and Hawaii, July-August 1970. Oxford: Pergamon Press, 1971.

Advances in Water Pollution Research. 1972. Vol. I and II. Proceedings of the Sixth International Conference held in Jerusalem, Israel, June 1972. Oxford: Pergamon Press, 1973.

Culp, R. L., Wesner, G. M., and Culp, G. L. *Handbook of Advanced Wastewater Treatment*. 2nd ed. New York: Van Nostrand Reinhold, 1978.

Eckenfelder, W. W., Jr., and Cecil, L. K. Eds. *Applications of New Concepts of Physical-Chemical Wastewater Treatment, Progress in Water Technology.* vol. 1. Oxford: Pergamon Press, 1973.

Evans, F. L., III, Ed. *Ozone in Water and Wastewater Treatment.* Ann Arbor, Mich.: Ann Arbor Science, 1972.

Fair, G. M., Geyer, J. C., and Okun, D. A. *Water and Wastewater Engineering.* Vol. 2, Water Purification and Wastewater Treatment Disposal. New York: Wiley, 1966.

First International Symposium on Ozone for Water and Wastewater Treatment held in Washington, D.C., December 1973, International Ozone Institute, Waterbury, Conn., 1975.

Gloyna, E. F., and Eckenfelder, W. W. Jr., Eds. *Water Quality Improvement by Physical and Chemical Processes.* Austin: University of Texas Press, 1970.

Humenick, M. J. *Water and Wastewater Treatment: Calculations for Chemical and Physical Processes.* New York: Marcel Dekker, 1977.

Mattson, J. S., and Mark, H. B. Jr. *Activated Carbon, Surface Chemistry and Adsorption from Solution.* New York: Marcel Dekker, 1971.

Metcalf & Eddy, Inc. *Wastewater Engineering: Collection, Treatment, Disposal.* New York: McGraw-Hill, 1972.

Nemerow, N. L. *Industrial Water Pollution, Origins, Characteristics, and Treatment.* Reading, Mass.: Addison-Wesley, 1978.

Proceedings, Advanced Treatment and Reclamation of Wastewater, Progress in Water Technology, Vol. 10, Nos. 1/2, Pergamon Press, Oxford, England, 1978.

Proceedings of the Seventh International Association on Water Pollution Research Conference, Paris, France, September 1974, Progress in Water Technology, Vol. 7, Nos. 2–6, Oxford, England: Pergamon Press, 1976.

Proceedings of the Eighth International Association on Water Pollution Research Conference, Sydney, Australia, October 1976, Progress in Water Technology, Vol. 9, Nos. 1–4, Oxford, England: Pergamon Press, 1977.

Process Design Manual for Carbon Adsorption, Technology Transfer, U.S. Environmental Protection Agency, October 1973.

Process Design Manual for Phosphorus Removal, Technology Transfer, U.S. Environmental Protection Agency, April 1976.

Process Design Manual for Nitrogen Control, Technology Transfer, U.S Environmental Protection Agency, October 1975.

Ramalho, R. S. *Introduction to Wastewater Treatment Processes.* New York: Academic Press, 1977.

Rice, R. G., and Browning, M. E. Eds. *First International Symposium on Ozone for Water and Wastewater Treatment.* Waterbury, Conn.: International Ozone Institute, 1975.

Rich, L. G. *Unit Processes for Sanitary Engineering*. New York: Wiley, 1963.

Schroeder, E. D. *Water and Wastewater Engineering Treatment*. New York: McGraw-Hill, 1977.

Second International Symposium on Ozone Technology held in Montreal, Canada, May 1975, International Ozone Institute, Syracuse, N.Y., 1976.

Thackston, E. L., and Eckenfelder, W. W. Jr., Eds. *Process Design in Water Quality Engineering: New Concepts and Developments*. Austin, Texas: Jenkins, 1972.

Water-1968. AIChE Symposium Series, Vol. 64, No. 90. New York: American Institute of Chemical Engineers, 1968.

Water-1969. AIChE Symposium Series, Vol. 65, No. 97. New York: American Institute of Chemical Engineers, 1969.

Water-1970. AIChE Symposium Series, Vol. 67, No. 107. New York: American Institute of Chemical Engineers, 1971.

Water-1971. AIChE Symposium Series, Vol. 68, No. 124. New York: American Institute of Chemical Engineers, 1972.

Water-1972. AIChE Symposium Series, Vol. 69, No. 129. New York: American Institute of Chemical Engineers, 1973.

Water-1973. AIChE Symposium Series, Vol. 70, No. 136. New York: American Institute of Chemical Engineers, 1974.

Water-1974. I. Industrial Wastewater Treatment. AIChE Symposium Series, Vol. 70, No. 144. New York: American Institute of Chemical Engineers, 1974.

Water-1974. II. Municipal Wastewater Treatment, AIChE Symposium Series, Vol. 71, No. 145. New York: American Institute of Chemical Engineers, 1975.

Water-1975. AIChE Symposium Series, Vol. 72, No. 151. New York: American Institute of Chemical Engineers, 1976.

Water-1976. I. Physical, Chemical Wastewater Treatment, AIChE Symposium Series, Vol. 73, No. 166. New York: American Institute of Chemical Engineers, 1977.

Weber, W. J., Jr. *Physicochemical Processes for Water Quality Control*. New York: Wiley, 1972.

Zweibel, I., and Sweed, N. H., Eds. *Adsorption and Ion Exchange*. AIChE Symposium Series, Vol. 71, No. 152. New York: American Institute of Chemical Engineers, 1975.

SLUDGE HANDLING AND DISPOSAL

Adams, C. E., Jr., and Eckenfelder, W. W. Jr. *Process Design Techniques for Industrial Waste Treatment*. Nashville, Tenn.: Enviro Press, 1974.

Advances in Water Pollution Research. 1962. Vol. II. Proceedings of the International Conference held in London, England, September 1962. Oxford: Pergamon Press, 1963.

Advances in Water Pollution Research. 1964. Vol. II. Proceedings of the Second International Conference held in Tokyo, Japan, August 1964. Oxford: Pergamon Press, 1966.

Advances in Water Pollution Research. 1966. Vol. II. Proceedings of the Third International Conference held in Munich, West Germany, September 1966. Washington, D.C.: Water Pollution Control Federation, 1967.

Advances in Water Pollution Research. Proceedings of the Fourth International Conference held in Prague, Czechoslovakia, April 1969. Oxford: Pergamon Press, 1969.

Advances in Water Pollution Research. 1970. Vol. I and II. Proceedings of the Fifth International Conference held in San Francisco and Hawaii, July-August 1970. Oxford: Pergamon Press, 1971.

Advances in Water Pollution Research. 1972. Vol. I and II. Proceedings of the Sixth International Conference held in Jerusalem, Israel, June 1972. Oxford: Pergamon Press, 1973.

Babbitt, H. E., and Baumann, E. R. *Sewerage and Sewage Treatment.* 8th ed. New York: Wiley, 1958.

Burd, R. S. *A Study of Sludge Handling and Disposal.* Publication WP-20-4. Washington, D.C.: U.S. Department of Interior, Federal Water Pollution Control Administration, 1968.

Fair, G. M., Geyer, J. C., and Okun, D. A. *Water and Wastewater Engineering.* Vol. 2, Water Purification and Wastewater Treatment and Disposal. New York: Wiley, 1966.

James, R.W. *Sewage Sludge Treatment.* Park Ridge, N.J.: Noyes Data, 1972.

McCabe, B. J., and Eckenfelder, W. W. Jr., Eds. *Biological Treatment of Sewage and Industrial Wastes.* Vol. II. Anaerobic Digestion and Solids-Liquid Separation. New York: Reinhold, 1958.

Metcalf & Eddy, Inc. *Wastewater Engineering: Collection, Treatment, Disposal.* New York: McGraw-Hill, 1972.

Proceedings of the National Conference on Municipal Sludge Management, Information Transfer, Washington, D.C., 1974.

Proceedings of the Seventh International Association on Water Pollution Research Conference, Paris, France, September 1974, Progress in Water Technology, Vol. 7, Nos. 2–6, Oxford, England: Pergamon Press, 1976.

Proceedings of the Eighth International Association on Water Pollution Research Conference, Sydney, Australia, October 1976, Progress in Water Technology, Vol. 9, Nos. 1–4, Oxford, England: Pergamon Press, 1977.

Process Design Manual for Sludge Treatment and Disposal, Technology Transfer, U.S. Environmental Protection Agency, October 1974.

Ramalho, R. S. *Introduction to Wastewater Treatment Processes*. New York: Academic Press, 1977.

Thackston, E. L., and Eckenfelder, W. W. Jr., Eds. *Process Design in Water Quality Engineering: New Concepts and Developments*. Austin, Texas: Jenkins, 1972.

Vesilind, P. A. *Treatment and Disposal of Wastewater Sludge*. Ann Arbor, Mich.: Ann Arbor Science, 1974.

Wastewater Treatment Plant Design, American Society of Civil Engineers and Water Pollution Control Federation, 1977.

LAND DISPOSAL OF WASTEWATERS

Land Application of Waste Materials, Soil Conservation Society of America, Ankeny, Iowa, 1976.

Loehr, R. C., Ed. *Land as a Waste Management Alternative*. Ann Arbor, Mich.: Ann Arbor Science, 1977.

Loehr, R. C. *Pollution Control for Agriculture*. New York: Academic Press, 1977.

Proceedings, International Conference on Land for Waste Management, National Research Council of Canada, 1974.

Process Design Manual for Land Treatment of Municipal Wastewater, Technology Transfer, U.S. Environmental Protection Agency, October 1977.

Sanks, R. L., and Asano, T. *Land Treatment and Disposal of Municipal and Industrial Wastewater*. Ann Arbor, Mich.: Ann Arbor Science, 1976.

Sopper, W. E., and Kardos, L. T. Eds. *Recycling Treated Municipal Wastewater and Sludge through Forest and Cropland*. University Park: Pennsylvania State University Press, 1973.

Wilson, C. W., and Beckett, F. E. Eds. *Municipal Sewage Effluent for Irrigation*. Ruston, La.: Louisiana Tech Alumni Foundation, 1968.

DEEP WELL DISPOSAL

Braunstein, J., Ed. *Underground Waste Management and Artificial Recharge*. Vol. 1 and 2. Tulsa, Okla.: American Association of Petroleum Geologists, 1973.

Walker, W. R., and Cox, W. E. *Deep Well Injection of Industrial Wastes: Government Controls and Legal Constraints*. Blacksburg, Va.: Virginia Water Resources Research Center, 1976.

ECONOMICS OF WASTEWATER TREATMENT

Analysis of Operation and Maintenance Costs for Municipal Wastewater Treatment Systems, Technical Report MCD-39, U.S. EPA, Washington, D.C., 1978.

Construction Costs for Municipal Wastewater Conveyance Systems: 1973-1977, Technical Report MCD-38, U.S. EPA, Washington, D.C., 1978.

Construction Costs for Municipal Wastewater Treatment Plants: 1973-1977, Technical Report MCD-37, U.S. EPA, Washington, D.C., 1978.

Davis, R. K. *The Range of Choice in Water Management.* Baltimore: The Johns Hopkins University Press, 1968.

The Cost of Clean Water. Vol. III, Industrial Waste Profiles: No. 1, Blast Furnaces and Steel Mills; No. 2, Motor Vehicles and Parts; No. 3, Paper Mills; No. 4, Textile Mill Products; No. 5, Petroleum Refining; No. 6, Canned and Frozen Fruits and Vegetables; No. 7, Leather Tanning and Finishing; No. 8, Meat Products; No. 9, Dairies; No. 10, Plastic Materials and Resins; U.S. Department of Interior, Federal Water Pollution Control Administration, Washington, D.C., 1968.

Index

707